Acoustics and Noise Control

Acoustics and Noise Control

Second edition

B J SMITH BSc, PhD, FIOA, FCIOB, FCIBSE, MInstP, FIMgt, FBIM, CEng
formerly Director of Studies and Head of School of Building, Ulster Polytechnic

R J PETERS BSc, BA, MSc, DIC, CEng, PhD, FIOA, MInstP, ARCS
Reader in the Built Environment, North-East Surrey College of Technology (NESCOT)

STEPHANIE OWEN LLB (Hons), Solicitor
Visiting Lecturer, North-East Surrey College of Technology (NESCOT)

Harlow, England • London • New York • Boston • San Francisco • Toronto
Sydney • Tokyo • Singapore • Hong Kong • Seoul • Taipei • New Delhi
Cape Town • Madrid • Mexico City • Amsterdam • Munich • Paris • Milan

Pearson Education Limited
Edinburgh Gate
Harlow
Essex CM20 2JE
England

and Associated Companies throughout the world

Visit us on the World Wide Web at:
www.pearsoned.co.uk

First published 1982
Second edition published 1996

British Library Cataloguing in Publication Data
A CIP record for this book is available from the British Library

ISBN-10: 0-582-08804-6
ISBN-13: 978-0-582-08804-7

Set by 4 in 9/11 Compugraphic Times and Melior
Printed in Malaysia, GPS

13 12
09 08 07 06

Contents

Preface

The science of acoustics and noise control has developed enormously since the publication of the Wilson Report on Noise in 1966. One of the problems identified in that report was the general lack of knowledge about noise and its control. My first book, *Environmental Physics: Acoustics*, published in 1971, was intended primarily for students of architecture and building on higher diploma and first degree courses. In the event, many studying for higher degrees found the book of use as well. Developments during the following decade meant a need for an update which resulted in *Acoustics & Noise Control* written with co-authors Bob Peters and Stephanie Owen.

Developments in acoustics and noise control have continued in the nineteen eighties and early nineteen nineties. Hence in this new volume most of the original work in *Acoustics & Noise Control* has been retained, but has been updated in line with these developments. The chapter on the law relating to noise has largely been rewritten and a new chapter on Instrumentation has been added.

The changes should make the book of continued use to students starting a study of acoustics and noise control, as well as those working on a higher level. The questions at the end of each chapter should be of particular help to students. Those studying for the Institute of Acoustics Diploma in Acoustics & Noise Control should find it a textbook which covers much of the course. The chapter on law continues to be one of the few concise references available on the subject and should be of use to barristers and solicitors who are concerned with noise cases. Written again by a solicitor with expert knowledge of the subject, it will be of interest to lay and professional people alike.

It has again been a privilege to work on the book with my co-authors, Bob Peters and Stephanie Owen, whose expertise and enthusiasm are enormous.

B.J. Smith
December 1995

Acknowledgements

We are grateful to the following for permission to reproduce copyright material: Bruel & Kjaer Ltd, Harrow, for figures 7.17, 7.18, 7.23, 9.9, 9.10, 9.11 and 9.25; British Standards Institute for tables 2.6, 7.6 and 7.7, and figures 7.25, 7.27, 7.28, 7.29, 7.30 and 7.31. Extracts from British Standards are reproduced with the permission of BSI. Complete copies can be obtained by post from BSI Customer Services, 389 Chiswick High Road, London, W4 4AL.

We are also grateful to the Institute of Acoustics for permission to reproduce questions from past exam papers.

Terminology and notation

It is current practice in the literature on the subject to represent all decibel quantities by L, with appropriate subscripts such as L_p, L_W, L_{eq}, $L_{EP,d}$, L_{AE}. The term SPL, for sound pressure level is still, however, in wide use. In this edition of the book, SPL is retained for convenience and clarity, but L_W is used for sound power level.

Again in line with common usage, pascals (Pa) have been used instead of N/m^2. A list of the main symbols has been provided.

Symbols

A list of the main symbols used in the book is given below. Most symbols are in common usage in most literature on the subject. The list is not exhaustive and other symbols are defined in the text.

a acceleration
A A frequency weighting
A acoustic absorption, acceleration amplitude
c velocity of sound
C C frequency weighting
d depth or thickness
D D frequency weighting
D level difference
f frequency
F fast time weighting
g gram
g acceleration due to gravity
Hz hertz
I impulse time weighting
I sound intensity
k kilo (e.g. kg, km, kHz)
k wavenumber, $k = 2\pi/\lambda$
l length
L level in decibels
m metres
N newtons
p sound pressure
P atmospheric pressure
Pa pascals
Q dynamic magnification factor
R sound reduction index/transmission loss
r distance
S slow frequency weighting
S area (m^2)
t time
T reverberation time
u acoustic particle velocity
v vibration velocity
V vibration velocity amplitude
x distance, displacement
X displacement amplitude
α acoustic absorption coefficient
ω angular frequency, $\omega = 2\pi f$
λ wavelength, $\lambda = 2\pi/k$

Where the same symbol has more than one meaning, dual representation has been retained to comply with common usage. The meaning will be clear from the context and is made clear in the text.

1 The measurement of sound

Sound is an aural sensation caused by pressure variations in the air which are always produced by some source of vibration. They may be from a solid object or from turbulence in a liquid or gas. These pressure fluctuations may take place very slowly, such as those caused by atmospheric changes, or very rapidly and be in the ultrasonic frequency range. The velocity of sound is independent of the rate at which these pressure changes take place and depends solely on the properties of the air in which the sound wave is travelling.

1.1 Frequency

This is the number of vibrations or pressure fluctuations per second. The unit is the hertz (Hz).

1.2 Wavelength

This is the distance travelled by the sound during the period of one complete vibration.

1.3 Velocity of sound in air

Velocity = frequency × wavelength

$$c = f\lambda$$

where c = velocity of sound in m/s

f = frequency in Hz

λ = wavelength in metres (m)

It is sufficiently accurate for the purpose of building acoustics to consider the velocity of sound to be a constant at 330 m/s. The wavelength of a sound of 20 Hz frequency

$$= \frac{330}{20} \text{ m}$$

$$= 16{\cdot}5 \text{ m}$$

The wavelength of a sound of 20 kHz frequency

$$= \frac{330}{20\,000}$$

$$= 0{\cdot}0165 \text{ m}$$

$$= 16{\cdot}5 \text{ mm}$$

These are the extremes of wavelength for audible sounds.

1.4 Propagation of sound waves

Air cannot sustain a shear force so that the only type of wave possible is longitudinal, where the vibrations are in the direction of the motion. This is illustrated in Fig. 1.1. These pressure fluctuations are of a vibrational nature causing the neighbouring air pressure to change but no movement of the air takes place. Air pressure, which can be assumed to be steady, has these pressure fluctuations superimposed upon it.

Reflection of sound takes place when there is a change of medium. The larger the change the greater the amount of reflection and the smaller the transmission. The laws of reflection for sound are similar to those for light:

1. The angle of incidence is equal to the angle of reflection.
2. The incident wave, the reflected wave and the normal all lie in the same plane.

There is a limitation on the first of these. The reflecting surface must have dimensions of at least the same order of size as the wavelength of the sound. If the reflecting object is much smaller than the wavelength, then diffraction will take place.

1.5 Simple harmonic motion

A pure sound consists of regular vibrations such that the displacement of the vibrating object from its original position is given by:

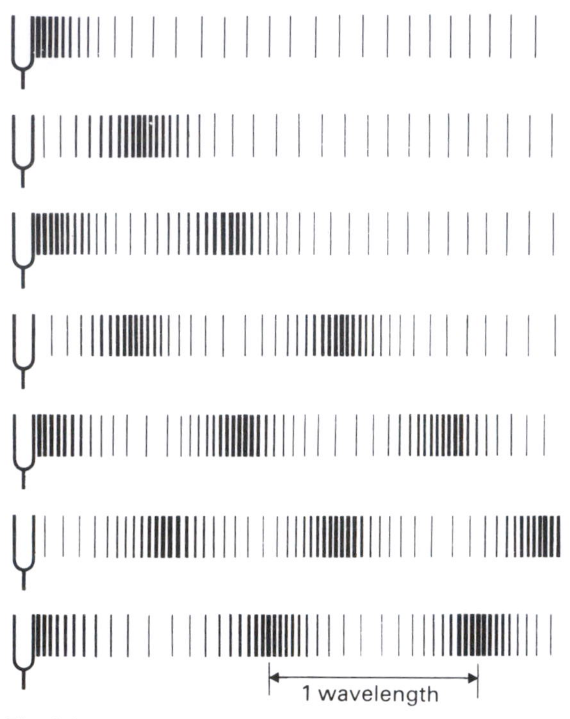

Fig. 1.1 Propagation of a sound wave

displacement, $x = X \sin 2\pi f \,.\, t$

where f = frequency in Hz

t = time in seconds from original position

X = maximum displacement or amplitude

The pressure fluctuations in the air are due to molecules of air vibrating back and forth about their original position but passing on some of their energy of movement. If a particular molecule has a displacement at time t of

$$x = X \sin 2\pi f \,.\, t$$

then it is moving at a velocity of vibration given by

$$\frac{dx}{dt} = 2\pi f X \cos 2\pi f t \qquad (t = 0 \text{ when } s = 0)$$

and is being accelerated at a rate

$$\frac{d^2x}{dt^2} = -4\pi^2 f^2 X \sin 2\pi f t$$

It can be seen from Fig. 1.2 that the average displacement and pressure fluctuation is zero due to equal positive and negative changes. To overcome this problem it is convenient to make measurements of the root mean square pressure change (RMS value). For pure tones the RMS value is equal to 0·707 times the peak value or amplitude of the wave. The most commonly used measurable aspects of sound are

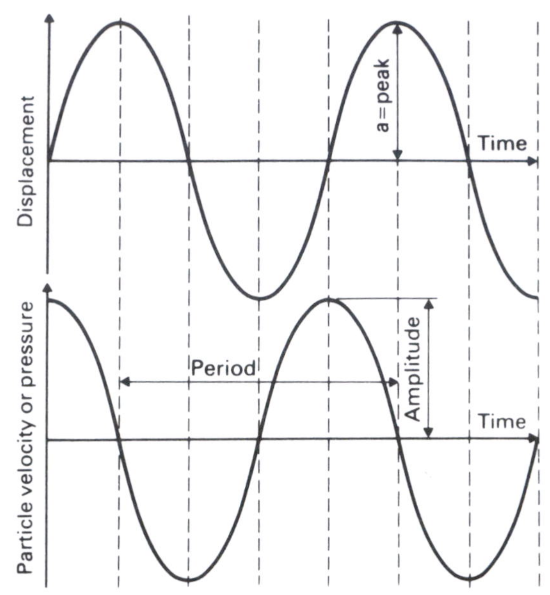

Fig. 1.2 Displacement and pressure variations

particle displacement, particle velocity, particle acceleration and sound pressure. As the ear is a pressure sensitive mechanism, it is most convenient to use pressure as the measure of sound magnitude. The sound intensity is the sound power per unit area in a sound wave and is related directly to the square of the sound pressure.

1.6 Decibels

Magnitudes of sound pressure affecting the ear vary from 2×10^{-5} Pa at the threshold, up to 200 Pa in the region of instantaneous damage. This may be compared with normal atmospheric pressure of 10^5 Pa. Because of this inconveniently large order of values involved and also because the ear response is not directly proportional to pressure, a different scale is used. A psychologist, Weber, suggested that the change of subjective response (R) is proportional to the fractional change of stimulus (S) and this has been shown to be largely true:

$$\delta R \propto \frac{\delta S}{S}$$

By integration it is to be expected that the actual response will be proportional to the logarithm of the stimulus (Fechner)

$$R = k \log S$$

The measure of sound pressure level (SPL) used in practice, decibels, uses a logarithmic scale:

$$\text{SPL} = 20 \log_{10} \left(\frac{p_1}{p_0}\right)$$

It will be noticed that this is not an absolute scale but a simple comparative scale relating two different pressures. For convenience p_0 is taken as the pressure at the average threshold of hearing at 1000 Hz frequency (e.g. 2×10^{-5} Pa).

1.7 Sound intensity

Intensity, $I \propto p^2$

$$\therefore \quad \frac{I_1}{I_0} = \left(\frac{p_1}{p_0}\right)^2$$

$$\text{but,} \quad \text{dB} = 10 \log_{10} \left(\frac{p_1}{p_0}\right)^2$$

$$= 10 \log_{10} \frac{I_1}{I_0}$$

where I_0 = threshold intensity, measured in W/m^2.

For a plane wave, intensity is equal to the square of the pressure divided by the product of density times velocity of sound for the material. This product, density of air × speed of sound in air, is known as the characteristic acoustic impedance of air.

$$I = \frac{p^2}{\rho c}$$

where ρ = density

c = velocity of sound

ρc = 410 rayls in air at normal temperatures and pressures

$$\therefore \quad I = \frac{p^2}{410} \text{ W/m}^2$$

$$\text{Threshold intensity} = \frac{(2 \times 10^{-5})^2}{410} \text{ W/m}^2$$

$$= \frac{400 \times 10^{-12}}{410}$$

$$\simeq 10^{-12} \text{ W/m}^2$$

Example 1.1

The RMS pressure of a sound is 200 Pa. What is the sound pressure level (SPL)? (Reference pressure 2×10^{-5} Pa)

$$\text{SPL} = 20 \log_{10} \frac{200}{2 \times 10^{-5}}$$

$$= 20 \log_{10} 10^7$$

$$= 20 \times 7$$

$$= 140 \text{ dB}$$

Note: 200 Pa, or 140 dB, is the peak action level of the Noise at Work Regulations 1989.

Example 1.2

What is the intensity of a sound whose RMS pressure is 200 Pa?

$$I = \frac{p^2}{\rho c}$$

$$= \frac{(200)^2}{410}$$

$$= \frac{40\,000}{410}$$

$$= 97{\cdot}8 \text{ W/m}^2$$

Example 1.3

What is the sound pressure level in decibels of a sound whose intensity is $0{\cdot}01$ W/m^2?

$$\text{SPL} = 10 \log_{10} \frac{0{\cdot}01}{10^{-12}}$$

$$= 10 \log_{10} 10^{10}$$

$$= 10 \times 10$$

$$= 100 \text{ dB}$$

Example 1.4

What is the increase in sound pressure level (in dB) if the intensity is doubled?

$$\text{Increase in SPL} = 10 \log_{10} \frac{2I}{I_0} - 10 \log_{10} \frac{I}{I_0}$$

$$= 10 \log_{10} 2$$

$$= 10 \times 0{\cdot}3010$$

$$= 3 \text{ dB}$$

Example 1.5

What is the increase in sound pressure level (in dB) if the pressure is doubled?

$$\text{Increase in SPL} = 20 \log_{10} \frac{2p}{p_0} - 20 \log_{10} \frac{p}{p_0}$$

$$= 20 \log_{10} 2$$

$$= 20 \times 0{\cdot}3010$$

$$= 6 \text{ dB}$$

Subjectively an increase of 3 dB is just noticeable, whereas an increase of 10 dB, a tenfold increase in intensity, is judged by most people as a doubling of loudness. A 1 dB increase is just detectable under the most favourable laboratory conditions.

1.8 Sound pressure level and sound intensity level

The sound intensity level of the sound in Example 1.2 is given as follows:

$$\text{Sound intensity level} = 10 \log_{10} \frac{I}{I_0}$$

$$= 10 \log_{10} \left(\frac{97 \cdot 8}{1 \times 10^{-12}} \right)$$

$$= 10 \log_{10} 10^{14}$$

$$= 10 \times 14$$

$$= 140 \text{ dB}$$

This is numerically the same value as the sound pressure level shown in Example 1.1. The reason is that the reference quantities, p_0 and I_0, are related by the plane wave relationship:

$$I = \frac{p^2}{\rho c}$$

Thus, for a plane wave (but not necessarily for other types of wave), the magnitude of the sound intensity level is the same as the sound pressure level. However, sound intensity level is a vector quantity, having direction as well as magnitude. A sound level meter does not give information about the direction of the flow of sound energy as does a sound intensity meter. The relationship between sound intensity and sound pressure levels is considered further in Chapters 8 and 9.

1.9 Addition and subtraction of decibels

The previous two examples illustrate that decibel values cannot be added arithmetically, due to the fact that they involve logarithmic scales. Intensities can be added arithmetically but the squares of individual pressures must be added.

$$\therefore \quad I = I_1 + I_2$$

$$\text{or} \quad p = \sqrt{(p_1^2 + p_2^2)}$$

Example 1.6

If three identical sounds are added, what is the increase in level in decibels?

$$\text{Increase in SPL} = 10 \log_{10} \frac{3I_1}{I_0} - 10 \log_{10} \frac{I_1}{I_0}$$

$$= 10 \log_{10} 3$$

$$= 10 \times 0 \cdot 4771$$

$$= 4 \cdot 8 \text{ dB}$$

$$\simeq 5 \text{ dB}$$

When adding decibel values it is necessary to use a logarithmic scale and it is convenient for this purpose to use a chart such as Fig. 1.3.

Example 1.7

Two cars are producing individual sound pressure levels of 77 dB and 80 dB measured at the pavement. What is the resultant sound pressure level when they pass each other?

$$\text{Difference} = 80 - 77$$

$$= 3 \text{ dB}$$

From Fig. 1.3, amount to be added to the higher level = 1·75 dB

Resultant SPL = 81·75 dB = 82 dB (to nearest decibel)

Example 1.8

In a certain factory space the noise level with all machines running is 101 dB. One machine alone produces a level of 99 dB. What would be the level in the factory with all machines running except this one?

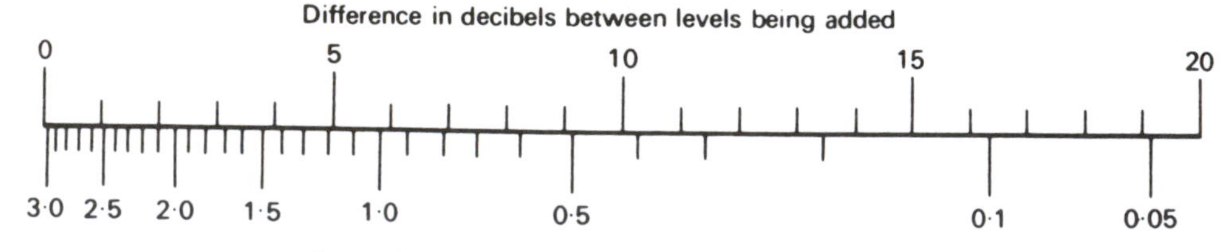

Fig. 1.3 Scale for combining sound pressure levels

Difference in sound pressure level = 101 − 99

= 2 dB

Amount to be subtracted from the higher level (from Fig. 1.3) = 2·4 dB

Resultant sound pressure level = 96·6 dB

= 97 dB (to nearest decibel)

1.10 Averaging decibels

The average of a number of decibels may be found from the following equation:

$$L_{AV} = 10 \log_{10} \left(\frac{10^{L_1/10} + 10^{L_2/10} + \ldots + 10^{L_n/10}}{n} \right)$$

$$= 10 \log_{10} (10^{L_1/10} + 10^{L_2/10} + \ldots + 10^{L_n/10}) - 10 \log_{10} n$$

$$= \text{sum of decibels} - 10 \log_{10} n$$

where n = number of different sounds
L_{AV} = average of the decibels
L_1 = first SPL in dB
L_2 = second SPL in dB

Example 1.9

Calculate the average of the following sound pressure levels: 80 dB, 82 dB, 84 dB, 86 dB and 88 dB.

The levels may be combined in pairs using Fig. 1.3.

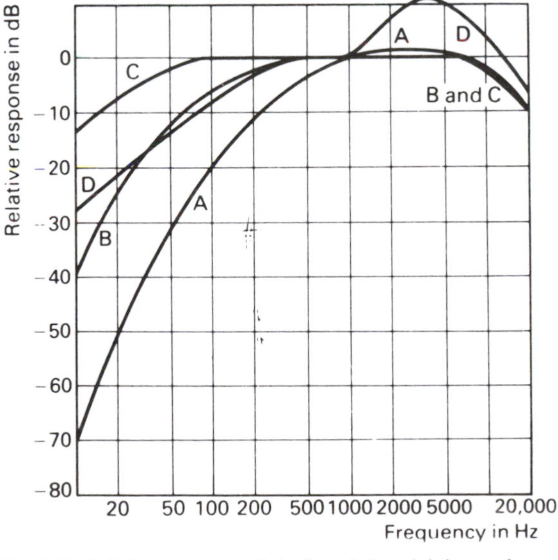

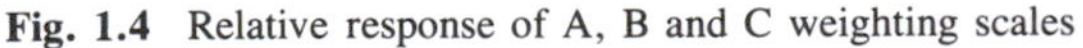

Fig. 1.4 Relative response of A, B and C weighting scales

SPL in dB	Add dB	SPL in dB	Add dB	SPL in dB	Add dB	SPL in dB
80						
	2	84	0	84		
82						
					0·8dB	91·8
84						
	2	88				
86			3	91		
88	0	88				

Hence, the average, $L_{AV} = 91 \cdot 8 \text{ dB} - 10 \log_{10} 5$

= 84·9 dB

= 85 dB (to nearest decibel)

Alternatively, the sum and average may be found using the formula and a calculator. Hence,

$$L_{AV} = 10 \log_{10} \left(\frac{10^{L_1/10} + 10^{L_2/10} + \ldots + 10^{L_5/10}}{5} \right)$$

$$= 10 \log_{10} \left(\frac{10^8 + 10^{8 \cdot 2} + 10^{8 \cdot 4} + 10^{8 \cdot 6} + 10^{8 \cdot 8}}{5} \right)$$

= 84·9 dB

= 85 dB (to nearest decibel)

1.11 Sound level meters and weighting scales

The sound level meter used for the measurement of RMS sound pressure levels consists of a microphone, amplifier and a meter. The microphone converts the sound pressure waves into electrical voltage fluctuations which are amplified and operate the meter. Unfortunately, no meter could indicate accurately over such a large range as may be needed from 30 dB to 120 dB or more. To overcome this the amplification is altered as required in steps of 10 dB and the meter only has to read the difference between the amplifier setting and the sound pressure level. Most meters will have connections to which filters can be added to select particular frequencies of the sound. An output is common to allow the sound to be recorded on tape or the levels plotted on a chart.

Besides a linear reading of sound pressure level, most meters have A and B scales where the response varies with frequency, as shown in Fig. 1.4 and Table 1.1. It can be seen that a 30 Hz note of sound pressure level 70 dB would indicate 70 − 40 = 30 dB(A), or 70 − 17 = 53 dB(B) or 70 − 3 = 67 dB(C). The C scale is taken to be linear for most practical purposes, but in fact is so only for frequencies from 200 to 1250 Hz. The attenuation either side is small. The B scale is intended to give responses on the sound level meter corresponding to the 70 dB equal-loudness contour for pure tones (see Chapter 2). The A

Table 1.1 'A', 'B', 'C' and 'D' weightings

Frequency (Hz)	Curve A (dB)	Curve B (dB)	Curve C (dB)	Curve D (dB)
10	−70·4	−38·2	−14·3	−26·5
12·5	−63·4	−33·2	−11·2	−24·5
16	−56·7	−28·5	−8·5	−22·5
20	−50·5	−24·2	−6·2	−20·5
25	−44·7	−20·4	−4·4	−18·5
31·5	−39·4	−17·1	−3·0	−16·5
40	−34·6	−14·2	−2·0	−14·5
50	−30·2	−11·6	−1·3	−12·5
63	−26·2	−9·3	−0·8	−11
80	−22·5	−7·4	−0·5	−9
100	−19·1	−5·6	−0·3	−7·5
125	−16·1	−4·2	−0·2	−6·0
160	−13·4	−3·0	−0·1	−4·5
200	−10·9	−2·0	0	−3·0
250	−8·6	−1·3	0	−2·0
315	−6·6	−0·8	0	−1·0
400	−4·8	−0·5	0	−0·5
500	−3·2	−0·3	0	0
630	−1·9	−0·1	0	0
800	−0·8	0	0	0
1 000	0	0	0	0
1 250	0·6	0	0	2·0
1 600	1·0	0	−0·1	5·5
2 000	1·2	−0·1	−0·2	8·0
2 500	1·3	−0·2	−0·3	10
3 150	1·2	−0·4	−0·5	11
4 000	1·0	−0·7	−0·8	11
5 000	0·5	−1·2	−1·3	10
6 300	−0·1	−1·9	−2·0	8·5
8 000	−1·1	−2·9	−3·0	6·0
10 000	−2·5	−4·3	−4·4	3·0
12 500	−4·3	−6·1	−6·2	0
16 000	−6·6	−8·4	−8·5	−4·0
20 000	−9·3	−11·1	−11·2	−7·5

response corresponds approximately to the 40 dB equal-loudness contour.

It has been shown that the readings on the A scale, dB(A), correspond most closely to the response of the ear. For many practical purposes, when simple direct readings are needed this is the best scale to use. It is shown in the next chapter that the response of the ear is dependent on frequency, and readings of sound pressure on a linear scale can be most misleading in subjective acoustics. It was for this reason that the weighting scales were originally devised.

The 'D' weighting is a specialized characteristic, being the proposed standard for aircraft noise measurements. It is approximately the inverse of the 40 noy contour (see p. 29). The reason for this is that the 40 noy contour is fairly representative of the bandwidth and level of aircraft flyover noise. Also, aircraft noise spectra biased by this weighting have given good agreement with perceived noise level calculations except for a constant 10 dB difference between absolute values.

The 'A', 'B', 'C' and 'D' weightings are shown numerically in Table 1.1.

1.12 Calibration of sound level meters

This is done by means of a source of known noise level, such as a pistonphone (Fig. 1.5).

Calibration by the pistonphone is simple in that there is no difficulty in positioning the source and meter. The pistonphone fits over the microphone and produces a note of 250 Hz at 124 dB. Exterior noise is unimportant, but the calibrator can only be used on microphones to which it will fit. A small motor operates a cam, producing the sound by sinusoidal movement of two small pistons. This method is also very convenient as a standard source of sound when making calibrated tape recordings for later analysis in the laboratory. Several other acoustic calibrators are also used.

1.13 Analysis of sound

Sound does not normally consist of single-frequency notes, but a highly complex combination of tones. Often, it is necessary to know at least what bands of frequencies are present. Mostly it is sufficient to know the magnitude of the sounds contained within octave bands: 75–150 Hz,

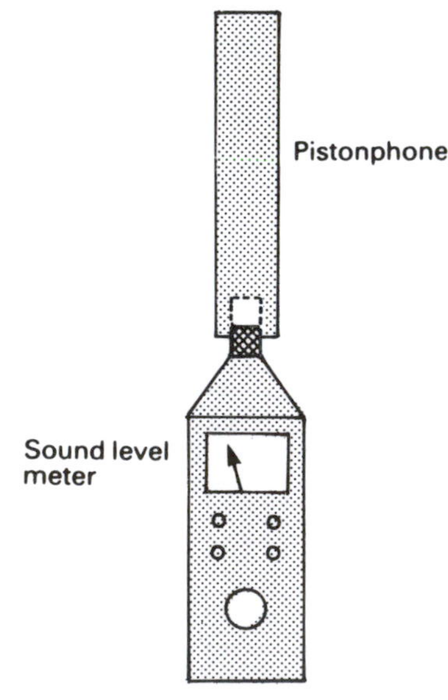

Fig. 1.5 Calibration of a sound level meter using a pistonphone

150–300 Hz, 300–600 Hz, 600–1200 Hz, 1200–2400 Hz, 2400–4800 Hz, 4800–9600 Hz. It can be seen that one octave band consists of all sounds from any frequency to twice that same frequency. Similarly for third-octave bandwidths. In each case it is convenient to refer only to the centre frequency within each band; e.g. 63, 125, 250, 500, 1000, 2000, 4000, 8000 Hz for octaves and 63, 80, 100, 125, 160, 200, 250, 315, 400, 500, 630, 800, etc., for third octaves.

1.14 Filters

Using suitable electrical filters, it is possible to separate the appropriate band of frequencies from the remainder, thus measuring the magnitude of that one group. In practice, filters working to one octave or third-octave bandwidths are most common for building acoustics. Even third-octave analysis will involve three times as much work and time as octave analysis. Narrower bandwidth filters down to one or two cycles width are available, but they are of very limited application because of the amount of time involved in their use. The centre frequencies of octave, half-octave and third-octave bandwidths used in standard filters are given in Table 1.2.

Table 1.2 Preferred frequencies for acoustic measurements (Hz)

Octave 16, 31·5, 63, 125, 250, 500, 1000, 2000, 4000, 8000
Half octave 16, 22·4, 31·5, 45, 63, 90, 125, 180, 250, 355, 500, 710, 1000, 1400, 2000, 2800, 4000, 5600, 8000
Third octave 16, 20, 25, 31·5, 40, 50, 63, 80, 100, 125, 160, 200, 250, 315, 400, 500, 630, 800, 1000, 1250, 1600, 2000, 2500, 3150, 4000, 5000, 6300, 8000

These are the geometric centre frequencies of filter pass bands.

Example 1.10

A certain noise was analysed into octave bands. The sound pressure levels in each were measured as shown below. What was the total level?

Centre frequency (Hz)	125	250	500	1000	2000	4000
SPL (dB)	80	82·5	77·5	70	65	60

Method 1
Total Intensity $I = I_1 + I_2 + I_3 + \ldots + I_6$

But SPL $= 10 \log_{10} \frac{I}{I_0}$ dB

so $\frac{I}{I_0} = \text{antilog}_{10} \frac{\text{SPL}}{10}$

$= 10^{\text{SPL}/10}$

Frequency (Hz)	SPL (dB)	$\frac{I}{I_0} \times 10^8$
125	80	1·0000
250	82·5	1·7780
500	77·5	0·5623
1000	70	0·1000
2000	65	0·0316
4000	60	0·0100
		3·4819

Total sound pressure level $= 10 \log_{10} 3\cdot4819 \times 10^8$

$= 10 \times 8\cdot5419$

$= 85\cdot5$ dB

or, by calculator

$\text{SPL} = 10 \log_{10} [10^{8\cdot0} + 10^{8\cdot25} + \ldots + 10^{6\cdot0}]$

Method 2
The levels may be combined in pairs using Fig. 1.3.

<table>
<tr><td>Frequency (Hz)</td><td>125</td><td>250</td><td>500</td><td>1000</td><td>2000</td><td>4000</td></tr>
<tr><td>SPL (dB)</td><td>80</td><td>82·5</td><td>77·5</td><td>70</td><td>65</td><td>60</td></tr>
<tr><td>Difference (dB)</td><td colspan="2">2·5</td><td colspan="2">7·5</td><td colspan="2">5·0</td></tr>
<tr><td>Add dB</td><td colspan="2">1·95</td><td colspan="2">0·75</td><td colspan="2">1·2</td></tr>
<tr><td>Result</td><td colspan="2">84·5</td><td colspan="2">78·25</td><td colspan="2">67·2</td></tr>
<tr><td>Add dB</td><td colspan="4">0·95</td><td colspan="2"></td></tr>
<tr><td>Result</td><td colspan="4">85·5</td><td colspan="2">67·2</td></tr>
<tr><td>Result</td><td colspan="6">85·5</td></tr>
</table>

Example 1.11

Calculate the sound pressure level in dB(A) of a noise with the following analysis:

Centre frequency (Hz)	31·5	63	125	250	500	1000	2000	4000
SPL (dB)	60	60	65	70	65	65	45	40

From Fig. 1.4 the dB(A) values in each octave are found, then added as shown in the previous example.

Frequency (Hz)	Level dB(A)	Sum dB(A)	Sum dB(A)	Sum dB(A)
31·5	21			
		34·2		
63	34			
			61·3	
125	49			
		61·3		
250	61			
				67·9
500	62			
		66·8		
1000	65			
			66·8	
2000	46			
		47·2		
4000	41			

Total level in dB(A) = 67·9
= 68 (to nearest dB(A))

1.15 Time-varying noise

The level of many noises varies with time, for example, traffic sounds. It is not easy to find a measure which can accurately quantify with a single number what is heard. To overcome the problem, statistical measurements are made in dB(A) because that corresponds approximately to the response of the ear.

L_{10} = sound level in dB(A) which is exceeded for 10% of the time
L_{50} = sound level in dB(A) which is exceeded for 50% of the time
L_{90} = sound level in dB(A) which is exceeded for 90% of the time

It will be realized that L_{90} is a measure of the background noise, whereas L_{10} is a measure of the peaks. Experiment 2 in Appendix 3 shows how they may be measured using a simple sound level meter. In practice, this may be more conveniently done using a sound level analyser, which will sample the noise level several times per second and perform the calculation to produce the required index.

1.16 Equivalent continuous noise level, L_{eq}

The equivalent continuous noise level, L_{eq}, is the sound pressure level of a steady sound that has, over a given period, the same energy as the fluctuating sound in question. It is an average and is measured in dB(A).

In theory, it can be calculated using the formula:

$$L_{eq} = 10 \log_{10} (t_1 \times 10^{L_1/10} + t_2 \times 10^{L_2/10} + t_3 \times 10^{L_3/10} + \ldots + t_n \times 10^{L_n/10})/T$$

where t_1 = time at L_1 dB(A)
t_2 = time at L_2 dB(A)
t_3 = time at L_3 dB(A)
T = total time over which the L_{eq} is required.

With rapidly fluctuating sounds it is not practical to calculate L_{eq}. Integrating meters are available which can make the measurements and calculate the L_{eq} for the given time.

Example 1.12

Calculate the L_{eq} over an eight-hour day for a worker exposed to the following noise levels and duration.

dB(A)	Time (hour)
94	3
89	2
98	0·5
83	2·5

$$L_{eq} = 10 \log_{10} (t_1 \times 10^{L_1/10} + t_2 \times 10^{L_2/10} + t_3 \times 10^{L_3/10} + t_4 \times 10^{L_4/10})/T$$

$$= 10 \log_{10} (3 \times 10^{9\cdot4} + 2 \times 10^{8\cdot9} + 0\cdot5 \times 10^{9\cdot8} + 2\cdot5 \times 10^{8\cdot3})/8$$

$$= 92\cdot034$$

$$= 92 \text{ dB(A)}$$

1.17 Personal daily noise exposure level

The L_{eq} over 8 hours, used for assessing noise exposure in the workplace is given the special name personal daily noise exposure level, $L_{EP,d}$, in the Noise at Work Regulations 1989. The value in Example 1.12 exceeds the second action level of 90 dB(A) in the regulations and would therefore impose particular duties on employer and employee. The first action level is 85 dB(A).

An alternative method to calculation is using a nomogram as shown in Fig. 1.6 from the guidance notes to the Noise at Work Regulations. A fractional dose is obtained from

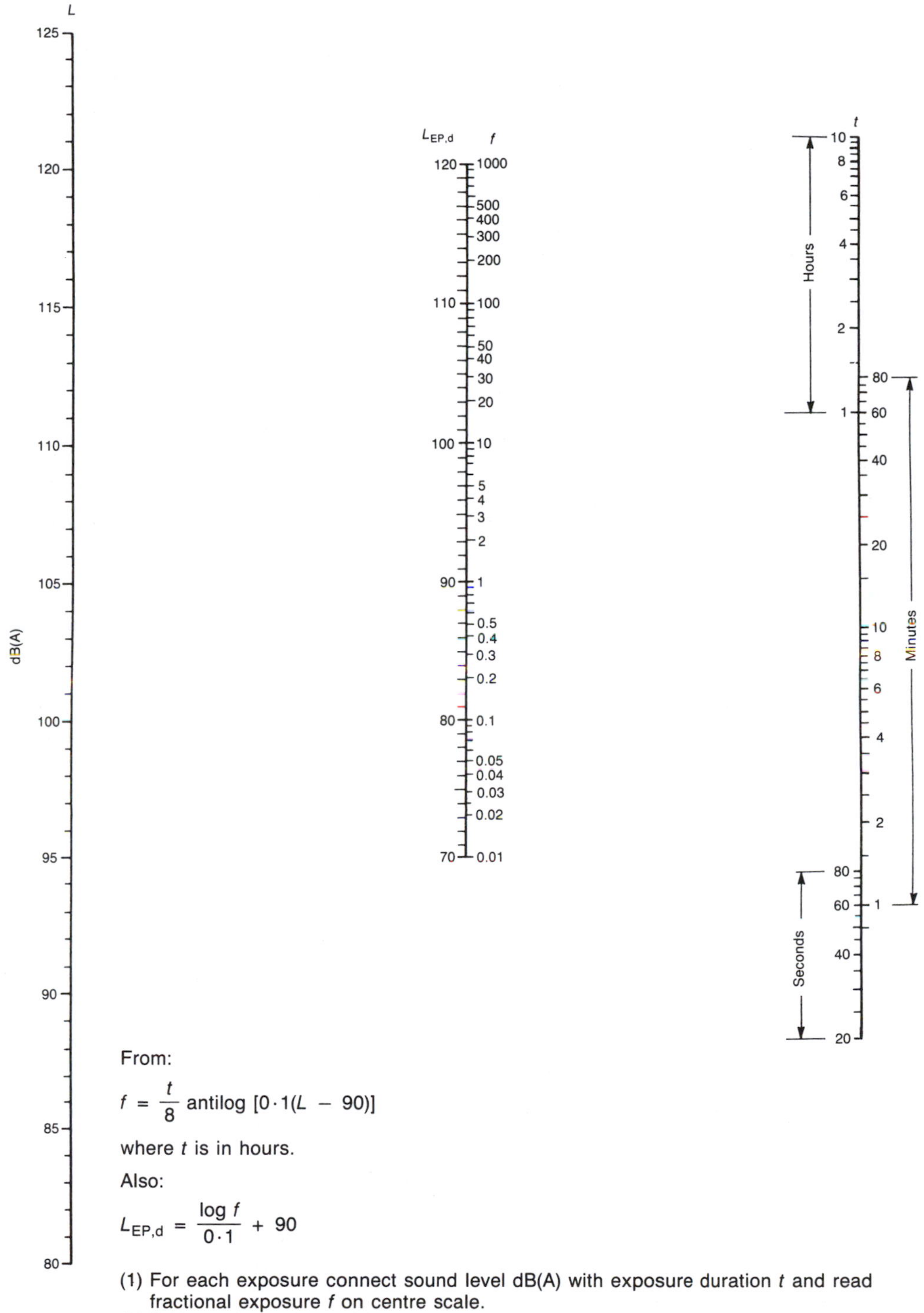

From:

$$f = \frac{t}{8} \text{ antilog } [0{\cdot}1(L - 90)]$$

where t is in hours.

Also:

$$L_{EP,d} = \frac{\log f}{0{\cdot}1} + 90$$

(1) For each exposure connect sound level dB(A) with exposure duration t and read fractional exposure f on centre scale.
(2) Add together values of f received during one day to obtain total value of f.
(3) Read $L_{EP,d}$ opposite total value of f.

Fig. 1.6 Nomogram for calculation of $L_{EP,d}$

the nomogram for each component of the exposure. The fractional doses are added together and converted, using the nomogram, to the $L_{EP,d}$ value.

Example 1.13

Use the nomogram (Fig. 1.6) to find L_{eq} for Example 1.12.

Place a transparent ruler between 94 dB(A) on the left-hand line to the 3 h time on the right-hand line. Read off the t value from the middle line. Repeat the procedure for other values.

dB(A)	Time (hour)	t value
94	3	1
89	2	0·2
98	0·5	0·4
83	2·5	0·07
	Total	1·67

Read off the $L_{EP,d}$ value adjacent to 1·67 on the middle line. Hence the personal dose level or L_{eq} = 92 dB(A).

Example 1.14

Noise from a building site is caused by five items of plant. The periods of operation of each item of plant during the working day and the noise level each produces at a noise sensitive property at the boundary of the site are shown below. Calculate the equivalent continuous noise level over a twelve-hour working day.

Compressor 83 dB(A) operating for 5 h
Excavator 85 dB(A) operating for 2 h
Dumper truck 76 dB(A) operating for 6 h
Pump 75 dB(A) operating for 7 h
Pile-driver 88 dB(A) operating for 1·5 h

$$L_{eq} = 10 \log_{10} (t_1 \times 10^{L_1/10} + t_2 \times 10^{L_2/10} + \ldots + t_n \times 10^{L_n/10}/T$$

$$= 10 \log_{10} (5 \times 10^{8\cdot3} + 2 \times 10^{8\cdot5} + 6 \times 10^{7\cdot6} + 7 \times 10^{7\cdot5} + 1\cdot5 \times 10^{8\cdot8})/12$$

$$= 84\cdot032273$$

$$= 84\cdot0 \text{ dB(A) (to the nearest } 0\cdot5 \text{ dB(A))}$$

Note that in Example 1.12 the averaging period, T, is equal to the sum of the individual component periods (i.e. $T = t_1 + t_2 + \ldots + t_n$), but as shown by Example 1.14, this need not be the case.

Sometimes it is required to average the same amount of noise energy over a different time period.

Example 1.15

Calculate the effect of spreading the building-site operations of Example 1.14 over an eighteen-hour period, instead of a twelve-hour period.

One could repeat the calculation of Example 1.14, using the increased averaging time of 18 h, but this is not necessary. It can be shown quite easily that the difference between two equivalent levels averaged over T_1 and T_2 hours is given by:

$$L_{Aeq\,T_1} - L_{Aeq\,T_2} = 10 \log_{10} \frac{T_2}{T_1}$$

$$= 10 \log_{10} \frac{18}{12}$$

$$= 1\cdot7609126$$

$$= 2 \text{ dB (to the nearest } 0\cdot5 \text{ dB(A))}$$

so that the effect of extending the time period to 18 h in this case is to reduce the equivalent continuous noise level by 2 dB to 82·0 dB(A).

Example 1.16

What is the maximum time for which an employee may spend in a particular workshop where the noise level is 106 dB(A) without using hearing protection if his or her noise dose is not to exceed an L_{eq} of 90 dB(A) over the period of the eight-hour working shift?

(a) Assume that for the rest of the shift the employee works in a quiet environment, in practice this means less than 80 dB(A).
(b) Assume that for the rest of the shift the employee is subjected to a constant level of 85 dB(A).

(a) A rough estimate of the allowed time may be obtained using the fact that the energy content of the noise exposure, and thus the L_{eq} value, will remain the same if the intensity of the sound is doubled, but for only half the duration, i.e. if the level is increased by 3 dB for half the time. Thus, 90 dB(A) for 8 h is equivalent to:

93 dB(A) for 4 h
96 dB(A) for 2 h
99 dB(A) for 1 h
102 dB(A) for 30 min
105 dB(A) for 15 min
108 dB(A) for 7·5 min

and so on.

Therefore, in this case the exposure to 106 dB(A) may be allowed for somewhere between 7·5 and 15 min. It may be calculated more precisely as follows:

$$L_{Aeq\,T_1} - L_{Aeq\,T_2} = 10\log_{10}\frac{T_2}{T_1}$$

where T_1 is the required time (in hours)

$T_2 = 8$ h

$L_{Aeq\,T_1} = 106$ dB(A)

$L_{Aeq\,T_2} = 90$ dB(A)

Therefore: $$106 - 90 = 10\log_{10}\frac{8}{T_1}$$

$$16 = 10\log_{10}\frac{8}{T_1}$$

From which: $$\frac{8}{T_1} = 10^{1\cdot6} = 39\cdot8$$

$$T_1 = \frac{8}{39\cdot8} = 0\cdot201\text{ h}$$

$$= 12\cdot1\text{ min}$$

(b) Let the required time in this case be t hours, so that the employee is exposed to 106 dB(A) for t hours and 85 dB(A) for $(8 - t)$ hours.

The eight-hour L_{eq} is given by:

$$L_{eq} = 10\log_{10}\left(\frac{t \times 10^{10\cdot6} + (8-t)\times 10^{8\cdot5}}{8}\right) = 90$$

from which (by dividing by 10 and taking antilogs):

$$t \times 10^{10\cdot6} + (8-t)\times 10^{8\cdot5} = 8 \times 10^{9}$$

This equation may be solved to give a value for t of 0·1385 h, or 8·3 min.

A simpler method involves making the assumption that the exposure of 106 dB(A) for t hours is in addition to an exposure to 85 dB(A) for the full 8 h. If the 85 dB(A) is subtracted from 90 dB(A) (by decibel subtraction using Fig. 1.3), it gives a value of 88·3 dB(A), which is the L_{eq} value to be produced by the workshop exposure alone, the 106 dB(A). The calculation then proceeds as in part (a) but using the value of 88·3 dB(A) instead of 90 dB(A). This gives an allowed time of 0·1374 h or 8·2 min, which agrees with the more exact calculation to within a tenth of a minute.

Example 1.17

The noise level at a site on which it is proposed to build a housing estate arises mainly from trains on a nearby railway line. There are three types of train using the line, fast express trains, slower suburban trains and freight trains. It is proposed to predict the equivalent continuous noise level at the site over a 24 h period from sample noise measurements of each of the three noise events. The results of these measurements are:

for fast trains $L_{eq} = 85$ dB(A) over a period of 12 s
for slow trains $L_{eq} = 78$ dB(A) over a period of 18 s
for freight trains $L_{eq} = 76$ dB(A) over a period of 24 s

During the 24 h period there are 120 fast trains, 200 slow trains and 80 freight trains. Calculate the equivalent continuous noise level over a 24 h period.

There are a number of alternative ways of doing this calculation.

Method 1
The 24 h L_{eq} is separately calculated for each type of train, then these individual values are combined. The total duration for each type of event is obtained by multiplying the single event duration by the number of events.

For the fast trains the total duration is 1440 s (12 s per train × 120 trains) and:

$$L_{eq} = 10\log_{10}\left(\frac{120 \times 12 \times 10^{8\cdot5}}{24 \times 60 \times 60}\right) = 67\cdot2\text{ dB(A)}$$

For the slow trains:

$$L_{eq} = 10\log_{10}\left(\frac{200 \times 18 \times 10^{7\cdot8}}{24 \times 60 \times 60}\right) = 64\cdot2\text{ dB(A)}$$

For the freight trains:

$$L_{eq} = 10\log_{10}\left(\frac{80 \times 24 \times 10^{7\cdot6}}{24 \times 60 \times 60}\right) = 59\cdot4\text{ dB(A)}$$

The total L_{eq} is obtained by combining these three levels:

$$L_{total} = 10\log_{10}(10^{6\cdot72} + 10^{6\cdot42} + 10^{5\cdot94})$$

$$= 69\cdot5\text{ dB(A)}$$

Note that the 24 h period has been converted to seconds because the individual event durations were in seconds.

Method 2
Method 2 simply involves combining the two stages of Method 1 and calculating the total L_{eq} for the three types of event, all in one go.

$$L_{eq} = 10\log_{10}\left(\frac{t_1 \times 10^{L_1/10} + t_2 \times 10^{L_2/10} + t_3 \times 10^{L_3/10}}{T}\right)$$

$$= 10\log_{10}(120 \times 12 \times 10^{8\cdot5} + 200 \times 18 \times 10^{7\cdot8} + 80 \times 24 \times 10^{7\cdot6})/24 \times 60 \times 60$$

$$= 69\cdot5\text{ dB(A) (as before)}$$

1.18 Noise exposure from single discrete events

In many situations, as in the last example, the total noise exposure over a period of time is made up from a number of different individual events such as the passing of an aircraft overhead or a train nearby, or a short burst of machinery noise, and L_{eq} measurements of the noise from different events will be made over different durations. It would be convenient for comparison of different types of event if the noise from all events could be averaged over the same duration. For convenience, a time of 1 s is chosen.

The **sound exposure level**, L_{AE}, of a single discrete noise event is the level which if maintained constant for a period of 1 s would contain as much A-weighted sound energy as is contained in the actual noise event. The name **single event noise exposure level** and the symbol L_{AX} are also in use; they mean exactly the same. The idea of L_{AE} is illustrated in Fig. 3.8.

The relationship between the L_{eq} value produced by an event over a period of time and the L_{AE} value for the event is given by:

$$L_{AE} = L_{Aeq,T} + 10 \log_{10} T$$

where T must be in seconds.

Some sound level meters have the facility to measure and indicate L_{AE} values directly but, if not, the value may be calculated from the above formula.

In the case of Example 1.17, the L_{AE} values for the three types of train are:

for the fast trains $L_{AE_1} = 85 + 10 \log_{10} 12 = 95{\cdot}8$ dB(A)

for the slow trains $L_{AE_2} = 78 + 10 \log_{10} 18 = 90{\cdot}6$ dB(A)

for the freight trains $L_{AE_3} = 76 + 10 \log_{10} 24 = 89{\cdot}8$ dB(A)

It is possible to calculate the L_{eq} over a given period of time from the number of events occurring in that period, and their L_{AE} values. This gives a third method for doing Example 1.17.

Method 3

$$L_{Aeq,T} = 10 \log_{10} (n_1 \times 10^{L_{AE_1}/10} + n_2 \times 10^{L_{AE_2}/10} + n_3 \times 10^{L_{AE_3}/10})/T$$

$$= 10 \log_{10} (120 \times 10^{9{\cdot}58} + 200 \times 10^{9{\cdot}06} + 80 \times 10^{8{\cdot}98})/24 \times 60 \times 60$$

$$= 69{\cdot}5 \text{ dB(A) (as before)}$$

1.19 Sound power

Total sound power in watts is equal to the intensity in watts/m^2 multiplied by the area in m^2. A large orchestra may produce 10 W, a jet plane up to 100 kW, whereas the loudest voice would only reach about 1 mW.

It is convenient to put the sound power level on a logarithmic scale relative to a standard power level. The reference level is 10^{-12} W.

$$\text{Sound power level, } L_W = 10 \log_{10} \frac{W}{W_0}$$

$$= 10 \log_{10} \frac{W}{10^{-12}}$$

$$= 10 \log_{10} W - 10 \log_{10} 10^{-12}$$

$$= 10 \log_{10} W + 120$$

Example 1.18

Determine the sound power level of 0·001 watts.

$$L_W = 10 \log_{10} 0{\cdot}001 + 120$$

$$= 10 \log_{10} 10^{-3} + 120$$

$$= -30 + 120$$

$$= 90 \text{ dB re } 10^{-12}\ W$$

It will have been noticed that both sound power level and sound pressure level are given in dB. Whenever giving sound power levels it is essential to state the reference level.

1.20 Sound power and sound intensity

The sound intensity from a point source of sound radiating uniformly into free space can be found from the power output and the distance from the source, r.

$$\text{Intensity, } I = \frac{\text{Sound power } W \text{ (watts)}}{4\pi r^2}$$

If the sound is produced at ground level, assuming that the ground is perfectly reflecting, then the energy is only radiated into a hemisphere instead of a complete sphere. In this case the formula for intensity becomes

$$I = \frac{W}{2\pi r^2}$$

Example 1.19

Calculate the intensity and SPL of a sound at a distance of 10 m from a uniformly radiating source of 1 watt power.

$$I = \frac{W}{4\pi r^2}$$

$$= \frac{1{\cdot}0}{4\pi (10)^2}$$

$$= 7{\cdot}95 \times 10^{-4} \text{ W/m}^2$$

$$\text{SPL} = 10 \log_{10} \frac{7 \cdot 95 \times 10^{-4}}{10^{-12}}$$

$$= 10 \log_{10} 7 \cdot 95 \times 10^{8}$$

$$= 10 \times 8 \cdot 9004$$

$$= 89 \text{ dB}$$

1.21 Sound power level and sound pressure level

It is often convenient to express the total sound output from a source in terms of sound power level. In order to obtain this, measurements of the sound pressure level at given distances must be made.

If the situation of a sound source of power output W is radiating uniformly in space, then:

Sound power level $L_W = 10 \log_{10} W + 120$

Sound pressure level, SPL

$$= 10 \log_{10} \left(W \middle/ \frac{4\pi r^2}{10^{-12}} \right)$$

$$= 10 \log_{10} W - 10 \log_{10} 4\pi r^2 + 120$$

$$= (10 \log_{10} W + 120) - 10 \log_{10} 4\pi r^2$$

$$= L_W - 20 \log_{10} r - 10 \log_{10} 4\pi$$

$$\therefore \quad \text{SPL} = L_W - 20 \log_{10} r - 11$$

If the more common situation applied where the sound is radiated over non-absorbent ground, then the relation would become

$$\text{SPL} = L_W - 20 \log_{10} r - 8$$

Example 1.20

A compressor with an A-weighted sound power level of 140 dB is radiating uniformly over a flat non-absorbent surface. Calculate the sound level at a distance of: (a) 10 m; (b) 45 m.

(a) At a distance of 10 m,

$$\text{sound level} = 104 - 20 \log_{10} r - 8$$

$$= 104 - 20 \log_{10} 10 - 8$$

$$= 104 - 20 - 8$$

$$= 76 \text{ dB(A)}$$

(b) At 45 m,

$$\text{sound level} = 104 - 20 \log_{10} 45 - 8$$

$$= 104 - 20 \times 1 \cdot 6532 - 8$$

$$= 104 - 33 - 8$$

$$= 104 - 41$$

$$= 63 \text{ dB(A)}$$

In this case it will be noticed that an A-weighted value for sound power level is given.

Example 1.21

Find the power output of a sound which radiates uniformly into unobstructed space if the pressure at a distance of 5 m is 1 Pa. (Assume $\rho c = 400$ rayls.)

$$\text{Intensity, } I = \frac{P^2}{\rho c}$$

$$\therefore \quad \frac{P^2}{\rho c} = \frac{W}{4\pi r^2}$$

$$\therefore \quad W = \frac{4\pi r^2 P^2}{\rho c}$$

$$= \frac{4\pi 25 \cdot 1^2}{400}$$

$$= 0 \cdot 786 \text{ watts}$$

Questions

(1) Calculate the sound pressure level in dB of a sound whose RMS pressure is 7·2 Pa.

(2) Determine the sound pressure level in dB of a sound whose intensity is 0·007 W/m^2 (re 10^{-12} W/m^2).

(3) The noise level from a factory with ten identical machines measured near some residential property was found to be 54 dB. The maximum permitted is 50 dB at night. How many machines could be used during the night?

(4) Find the total sound pressure level in dB for a sound with the following analysis. Calculate also the total intensity in W/m^2.

Centre frequency (Hz)	Level (dB)
125	55
250	63
500	71
1000	68
2000	59

(5) A motor car was found to produce the following noise. Calculate the total noise level in dB (linear) and dB(A).

Octave band centre frequency (Hz)	Level (dB)
63	95
125	84
250	80
500	68
1000	65
2000	61
4000	60
8000	60

(6) Calculate the intensity of a sound whose RMS pressure is 0·0045 Pa if ρc = 410 rayls for air.

(7) Find the RMS pressure of a sound whose intensity is 1 W/m^2.

(8) Determine the sound power level range (re 10^{-12} W) of the human voice which is from about 10 to 50 microwatts.

(9) Calculate the intensity and SPL of a sound at a distance of 20 m from a uniformly radiating source of 1·26 W power.

(10) Two sounds of 4 W and 10 W power are produced at ground level at a distance of 10 m and 20 m respectively from a listener. If the ground is level, unobstructed and non-absorbing, what will be the SPL of the sound heard by the listener?

(11) What power output is needed to produce an SPL of 60 dB at a distance of 100 m if both source and hearer are at ground level? Assume that the ground is level, unobstructed and non-absorbing.

(12) Assuming that it radiates sound uniformly, find the sound power from a motor car whose SPL at a distance of 7·5 m is 87 dB.

(13) Calculate the sound pressure level at a distance of 80 m from a generator producing a sound power level of 112 dB. Assume that there are no obstructions and that the ground is flat.

(14) (a) Give the reasons why weighted frequency response is incorporated in the construction of a sound level meter and describe the characteristics of the 'A', 'B' and 'C' weightings.

(b) Explain why it is necessary to consider background noise levels when selecting plant sound levels.

(c) The combined noise level in an office with a unit air-conditioner installed is taken as 65 dB. The noise level in the office with the conditioner off is taken as 55 dB. Calculate the level of the conditioner only in a quiet room. *(IOB)*

(15) Give an account of a simple sound level meter, indicating clearly the main functions and characteristics of the various parts of the instrument.

Explain how the temporal analysis of fluctuating

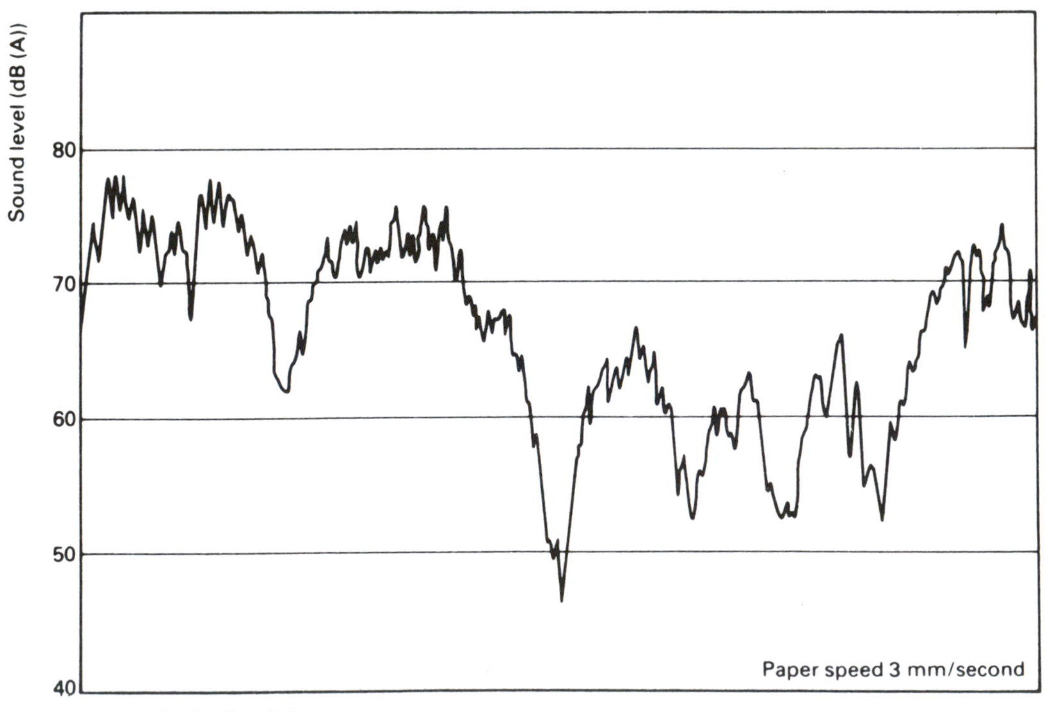

Fig. 1.7 Temporal noise level variations

noise levels can be carried out using only a portable sound level meter.

Using the data provided in Fig. 1.7, estimate the 'noise climate' ($L_{10} - L_{90}$) corresponding to the sound level fluctuations indicated. (*IOA*)

(16) The noise exposure pattern of a worker over an eight-hour shift is as follows:

85 dB(A) for 3·5 h
88 dB(A) for 3 h
92 dB(A) for 1·0 h
97 dB(A) for 0·5 h

Calculate the L_{eq} over 8 h.

(17) Calculate the equivalent continuous noise level over a twelve-hour working period at a dwelling close to the boundary of a building site if the component noise levels, measured at the house, are as follows:

83 dB(A) from a crane which operates for 4 h a day
79 dB(A) from a concrete mixer which operates for 7 h a day
81 dB(A) from a vibratory poker which operates for 15 min every hour

(18) The twelve-hour L_{eq} at the boundary of a building site is 75 dB(A). This includes the effect of a compressor which produces a steady level of 80 dB(A) and operates for 2·5 h. Calculate the twelve-hour L_{eq} if the compressor were not operating.

(19) A certain factory operation requires a short 30 s burst of noise every so often. The noise level produced at nearby houses by these bursts is 80 dB(A). How many bursts per hour may be permitted if the noise level is to be kept below an hourly L_{eq} of 65 dB(A)?

(20) The noise level inside a test bed is 97 dB(A). For how long each day may a technician work inside the test bed without hearing protection if the technician's noise exposure is to be less than an eight-hour L_{eq} of 90 dB(A)?

(a) Assume that for the rest of the day the technician works in a relatively quiet environment (less than 80 dB(A)).

(b) Assume that for the rest of the day the technician is subjected to a steady level of 86 dB(A).

(21) The sound exposure level L_{AE} for a single helicopter take-off is 103 dB(A) at the specified monitoring point. How many take-offs may be permitted during an eighteen-hour period each day if the L_{eq} over this period due to the take-offs is to be limited to 75 dB(A)?

(22) Three types of train run along a railway track which passes close to a plot of ground. Estimate the L_{eq} over one hour at the plot, given the following data. The L_{AE} values were measured at the plot of ground in question.

Train	L_{AE} (dB(A))	Number of trains per hour
Type 1	96	5
Type 2	99	3
Type 3	90	7

2 Aural environment

In the first chapter the physical properties of sounds and their measurements were discussed. In this chapter the subjective effects of noise and their cause are examined.

2.1 The ear

The ear is a transducer converting sound pressure waves into signals which are sent to the brain (see Figs 2.1 and 2.2).

Sound first reaches the outer and visible part of the ear known as the concha. A concave shape of a certain size will act as a focusing device only for wavelengths up to the same order of size, and so the concha will tend to scatter the longer wavelengths while reflecting shorter ones into the meatus. The meatus is the tube connecting the outer ear to the eardrum and, because of its size, it resonates to a frequency of about 3 kHz. The eardrum separates the outer from the inner ear. Major atmospheric pressure changes can be equalized on either side of the eardrum through the Eustachian tube by the act of swallowing. The problem at this stage is a high impedance mismatch due to the outer and middle ear being filled with air and the inner ear filled with liquid. The small bones which connect the eardrum with the oval window effectively make the middle ear equivalent to a 'step-up transformer' of about twenty times. This just compensates for a theoretical loss

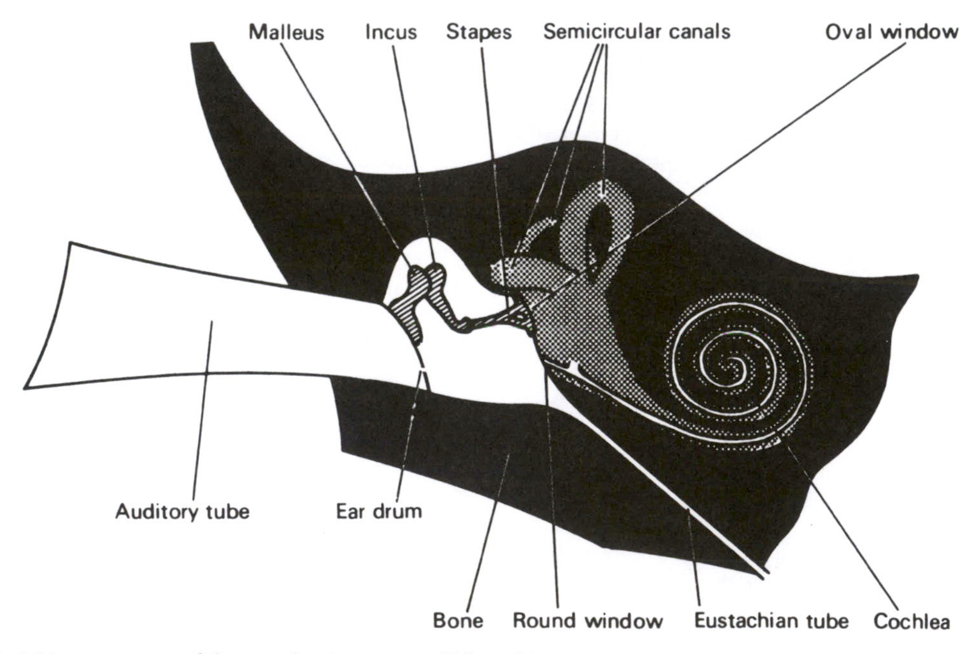

Fig. 2.1 Main components of the ear, showing outer, middle and inner parts

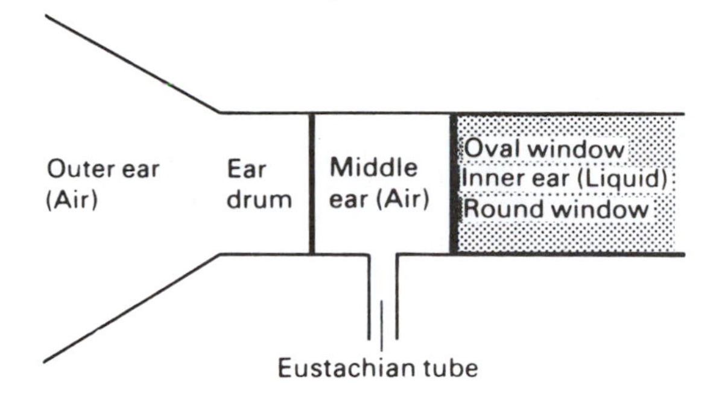

Fig. 2.2 Three main sections of the ear shown schematically

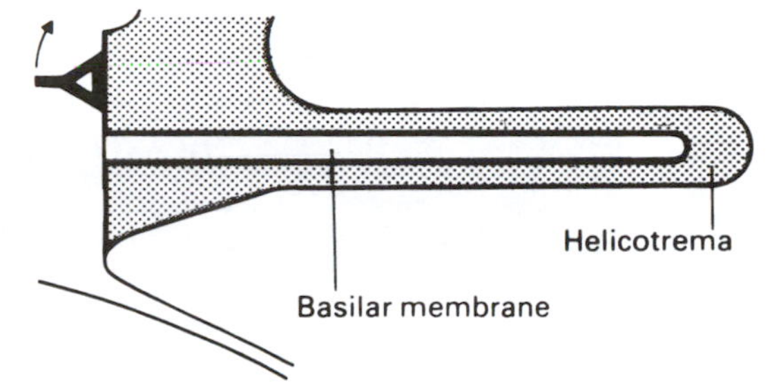

Fig. 2.3 Diagram of the cochlea unwound to show its main components

of approximately 30 dB between air and fluids of the inner ear.

The lever system consists of three bones when in theory only one is needed (as is the case with birds). It does give three important advantages:

1. Minimum bone conduction.
2. More linear response for different frequencies.
3. A protective overload device possible as the ossicles change their mode of operation above 140 dB sound pressure level.

The middle ear also possesses another protective device consisting of two small muscles which adjust the eardrum and stapes for levels of sound above 90 dB which last more than 10 ms.

The inner ear is a system of liquid-filled canals protected both mechanically and acoustically by being located inside the temporal bone of the skull. The cochlea is a hollow coil of bone filled with liquid, with a total length of about 40 mm (see Fig. 2.3). This is divided along its length by the basilar membrane with a small gap at its far end known as the helicotrema. Acoustic energy is converted into impulses transmitted to the brain at the basilar membrane on which about 24 000 nerve endings terminate.

Intense sounds can damage or even destroy any of the moving parts of the ear. In the more common case of hearing damage, because of prolonged exposure to high levels of noise, it is the hair cells that are damaged.

2.2 Audible range

This depends upon the age and physical condition of a person, but can be from 20 Hz to 20 kHz. The sensitivity varies considerably over this frequency range, especially near the threshold of hearing where there is a variation of the order of 70 dB, as shown in Fig. 2.4.

2.3 Pitch

Frequency is an objective measure of the number of vibrations per second, whereas the term pitch is subjective and, although dependent mainly on frequency, is also affected by intensity.

2.4 Loudness

The loudness of a sound is a subjective effect which is a function of the ear and brain as well as amplitude and frequency of the vibration. In practice, it is usual to consider people with normal hearing and correlate only amplitude and frequency with loudness. Pure tones of different frequencies are compared with that of 1000 Hz by adjusting the amplitude to obtain equal-loudness contours. The loudness level is given in phons. These are numerically equal to the sound pressure level in dB at 1 kHz, as shown in Fig. 2.5.

The phon scale is quite arbitrary and has no physical or physiological basis, but is convenient because of its relation to decibels (at 1 kHz). Unfortunately, it is not conveniently additive for different sounds, and another scale, that of sones, which overcomes this difficulty is used (see Fig. 2.6). The values of sones may be added arithmetically to obtain loudness levels, e.g. 50 sones is twice as loud as 25 sones.

It is obviously very easy to find the loudness in phons or sones of pure tones from Fig. 2.5. When the noise is other than pure, it is commonly assessed in terms of Stevens phons. An octave analysis (or third octave) of the noise needs to be done in the eight bands 20–75, 75–150, 150–300, 300–600, 600–1200, 1200–2400, 2400–4800 and 4800–10 000 Hz. Each level in dB is converted to sones by means of Fig. 2.7. Then the total loudness in sones, S_t is given by:

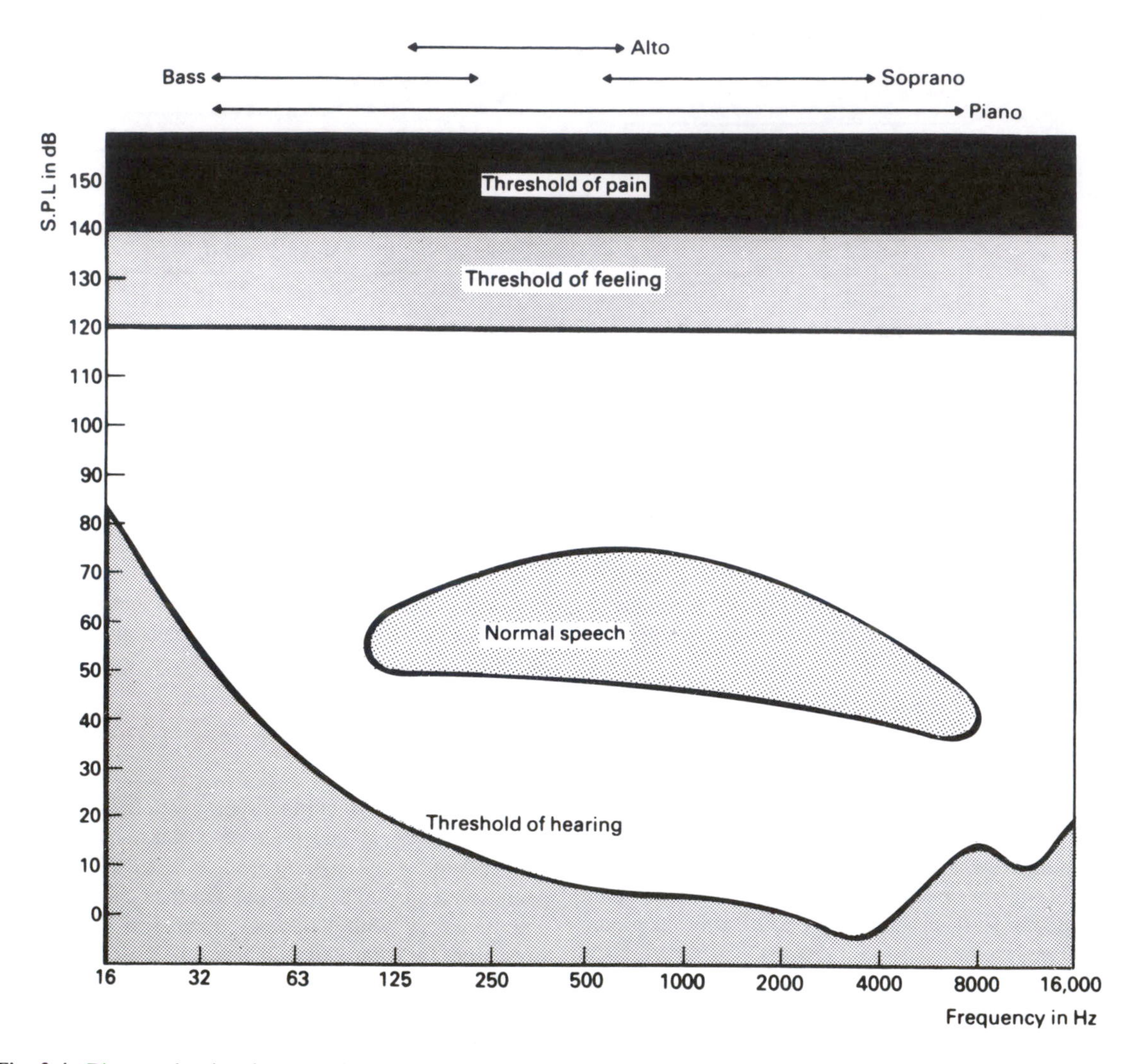

Fig. 2.4 Diagram showing the approximate threshold of hearing for young people in the age range 18–25. The thresholds of feeling and pain occur at about 120 and 140 dB respectively; the range of levels and frequencies of normal speech are shown along with frequency limits for the piano and singers

$$S_t = S_m + F(\Sigma S - S_m)$$

where S_m = loudness of loudest band

and F = 0·15 for third-octave bandwidth

= 0·2 for half-octave bandwidth

= 0·3 for octave bandwidth

and $\Sigma S = S_1 + S_2 + S_3 + S_4 + S_5 + S_6 + S_7 + S_8$

(S_1, S_2, etc. = loudness in appropriate bands)

This total level in sones, S_t, may be then converted into phons by means of Fig. 2.6.

Example 2.1

A certain motor car was found to produce the following noise levels:

Octave band centre frequency (Hz)	*dB level*
63	95
125	84
250	80
500	68
1000	65
2000	61
4000	60
8000	60

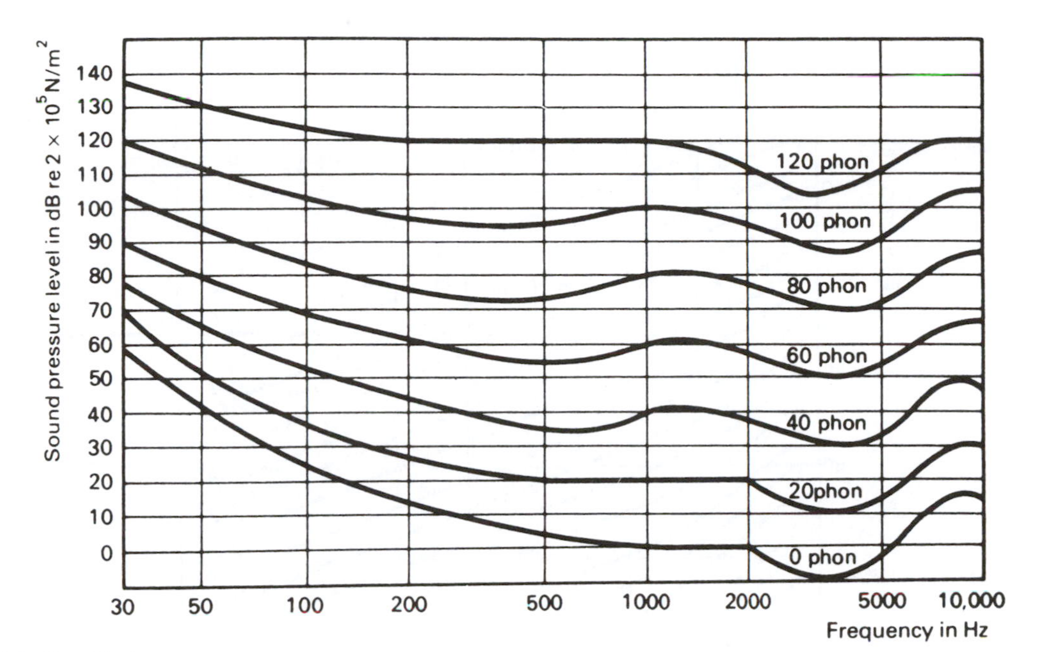

Fig. 2.5 Equal-loudness contours

Calculate the level in phons.

Using Fig. 2.7, convert the dB readings into sones.

Octave band centre frequency (Hz)	*dB level*	*Sones*
63	95	20
125	84	12
250	80	11
500	68	7
1000	65	6
2000	61	6·5
4000	60	7
8000	60	8
	$\Sigma S =$	77·5

S_m (the loudest) = 20 sones

$$S_t = 20 + 0\cdot3(77\cdot5 - 20)$$
$$= 20 + 0\cdot3(57\cdot5)$$
$$= 37\cdot25 \text{ sones}$$

From Fig. 2.6 the level in phons = 92

2.5 Octave-band analysis

In the last section the need was seen for an octave analysis in order to calculate loudness. (It could have been done using a third-octave analysis, but this is long and tedious and seldom necessary.)

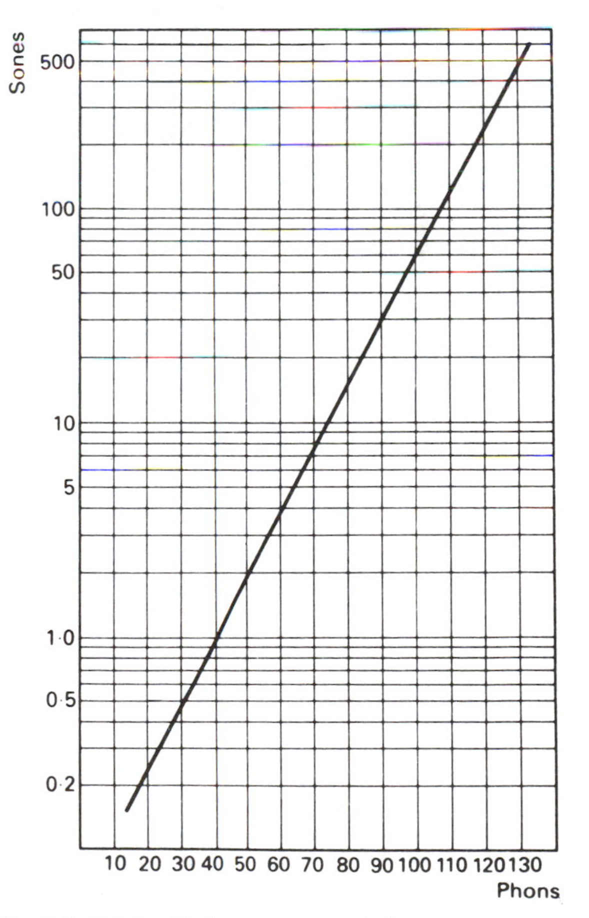

Fig. 2.6 Relationship between sones and phons

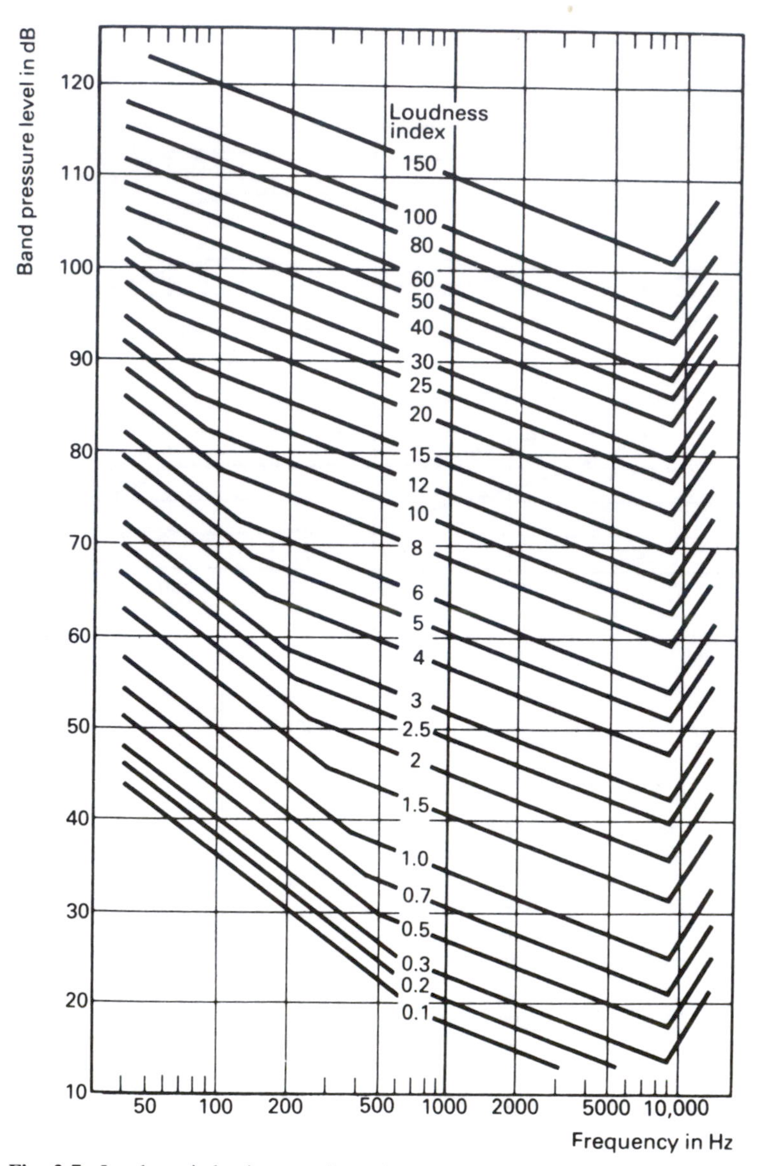

Fig. 2.7 Loudness index in sones from the SPL of the frequency band dB

There are two ways to measure the levels in each octave band. The first is by the use of a sound level meter set on to the linear scale using a set of octave band filters. For noises which are reasonably continuous, this is simple. Where the noise is more conveniently recorded and analysed in the laboratory, a calibrated recording must be made (see Fig. 2.8). The output of the sound level meter is connected to a portable tape recorder. A noise of known level must be recorded on the tape. This is normally done by means of a pistonphone or falling-ball calibrator. This signal, attenuated by a known amount, using the sound level meter is fed to the tape recorder. The actual noise is recorded with a suitable attenuation on the sound level meter. The whole is later played back through an octave filter set (or audio-frequency analyser which operates on the same principle) and the levels recorded on paper (see Fig. 2.9). A very short piece of tape can be played repeatedly for each octave band.

2.6 Masking

This is the effect of one noise on another. Speech, for instance, can become masked by road traffic or aircraft noise. The masking noise raises the threshold of audibility

Table 2.1 Loudness of common noises

Noise	Sones	Phons
Large jet plane 80 m overhead	700	134
Heavy road traffic at kerbside	79	103
Light road traffic at kerbside	16	80
Students' refectory	20	84
Normal speech (male) at 1 m	11	75
Inside noisy motor car	40	94
Machine shop	97	106

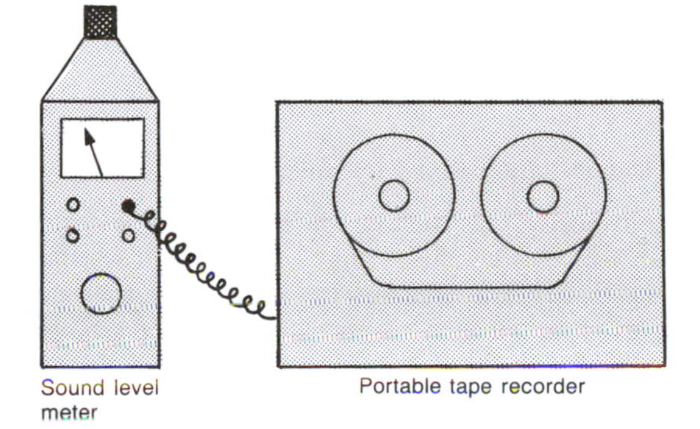

Fig. 2.8 Calibrated recording of sound on tape using a sound level meter to give a known attenuation. The calibrated signal is put on the tape using a pistonphone as sound source

for other sounds. A masking noise is most effective with a sound of similar frequency, but a low frequency noise is more effective in masking one of high frequency than the reverse.

Masking can obviously be inconvenient where very good hearing conditions are needed, but it can also be useful when speech privacy is required. For instance, it may be utilized by playing soft music in the waiting room of a doctor's surgery to prevent those waiting hearing details of examinations. A better solution would be good sound insulation. It thus follows that the amount of sound insulation needed depends upon the ambient noise level, which can provide a useful masking effect. Open-plan offices have been designed with reasonable audible privacy on the basis of the admission of a certain amount of exterior traffic noise. Although the calculation of the amount by which a masking noise raises the threshold of audibility is complicated, it should be noted that even relatively low levels of background noise have a considerable masking effect.

2.7 Audiometry

Audiometry is the term used to describe the measurement of hearing sensitivity. The instrument used for this purpose is an audiometer, which produces pure tones of various frequencies at different known pressure levels. The subject is asked to say, for each ear, which level at each frequency he or she can just detect. With no hearing defect the audiometer results follow the bottom curve of Fig. 2.4. If the subject has some hearing defect, the audiometer levels

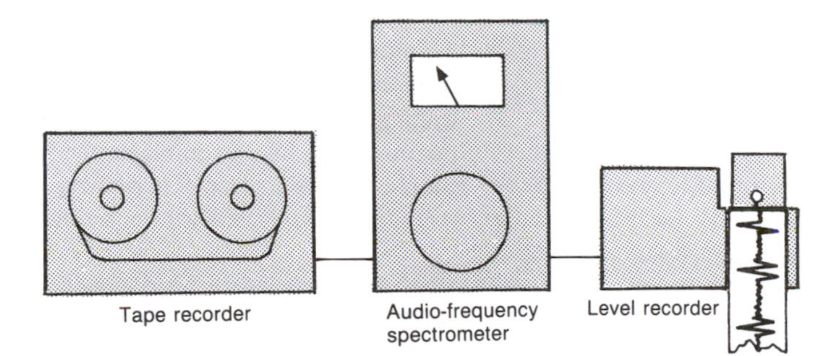

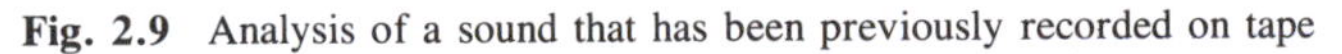

Fig. 2.9 Analysis of a sound that has been previously recorded on tape

Table 2.2 Typical octave band analysis levels in dB

	Centre frequency of octave band (Hz)							
Noise	63	125	250	500	1000	2000	4000	8000
Light traffic at kerbside	81	81	75	70	72	71	63	60
Heavy traffic at kerbside	96	93	90	88	89	84	78	76
Machine shop	68	72	90	87	86	88	90	84
Students' refectory	68	70	75	75	68	64	56	49

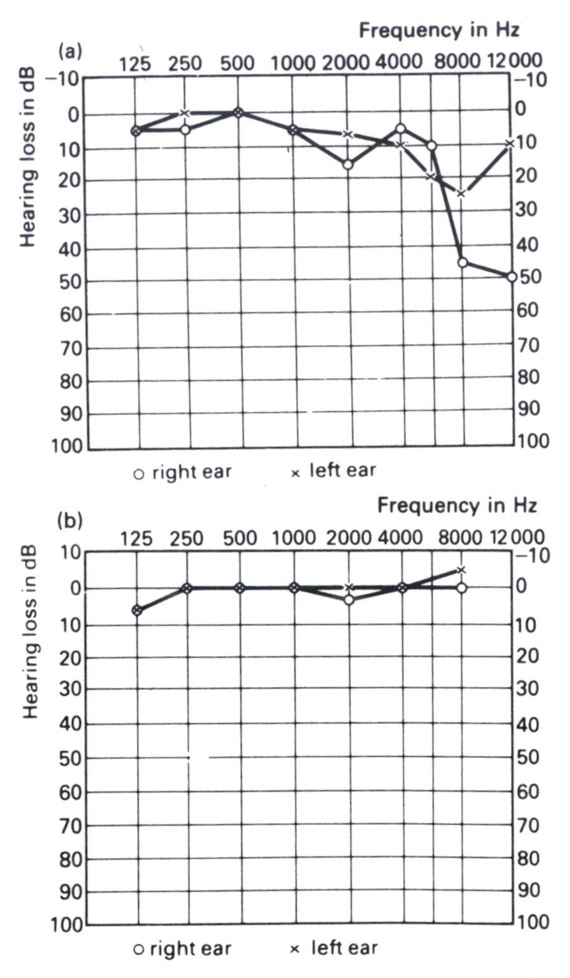

Fig. 2.10 Audiometric test result showing (a) some hearing loss at high frequencies and (b) normal hearing

would need to be raised at some or all frequencies above the normal threshold. The amount the level needs to be increased gives the hearing loss. Examples of audiometric tests for left and right ears are given in Fig. 2.10.

It is important that audiometric testing is only carried out in a suitably quiet background to avoid masking effects. This is particularly important for industrial audiometry where the subject may have pre-employment tests at a young age when there should not be any hearing loss. For these to be valid there must be no interference due to masking sound. Suggested maximum levels are given in Table 2.3.

2.8 Hearing defects

These are usually divided into two general types: first, those which are not related to noise exposure; and second, those which are directly attributable to damage caused by noise.

Table 2.3 Maximum allowable sound pressure levels for industrial audiometry

Octave band Centre frequency (Hz)	Maximum band SOL without noises excluding headset (dB)	Maximum band SPL with noise-barrier headset (e.g. Amplivox Audio cups) (dB)
31·5	76	(76)*
63	61	(61)*
125	46	55
250	31	44
500	7	31
1000	1	31
2000	4	43
4000	6	50
8000	9	44

*No data available on headset attenuation.

1 Defects not caused by noise

Presbycusis This is a loss which is normally associated with age. Whether this is simply due to age or whether it is due to the effects of normal levels of environmental noise seems to be a matter of speculation at the present time. Audiometric tests on one primitive tribe show that their hearing at the age of 70 is comparable with that of Americans at 30. In any event it becomes impossible to separate completely noise-induced hearing loss from presbycusis in most cases. Typical presbycusis losses are shown in Fig. 2.11. These should apply to both men and women in the absence of noise-induced hearing loss.

Tinnitus This defect, which is experienced by most people from time to time, is usually in the form of a high-pitched ringing in the ears. Where people suffer frequently from tinnitus, they may even blame the noise on some other source, such as a piece of machinery. Tinnitus is often but not always associated with noise-induced hearing loss.

Deafness There are three main types of deafness: conductive deafness, nerve deafness and cortical deafness.

Conductive deafness is due to defects in those parts of the ear (external canal, ear drum and ossicles) which conduct the sound waves in the air to the inner ear. Examples are a thickening of the eardrum, stiffening of the joints of the ossicles or a blocking of the external canal by wax. These affect all frequencies evenly, but the loss is limited to between 50 and 55 dB due to conduction through the head.

Otosclerosis makes the stapes immobile. It can be helped by fenestration, where a new window is introduced into the lateral semicircular canal.

A perforated eardrum can be caused by disease or by

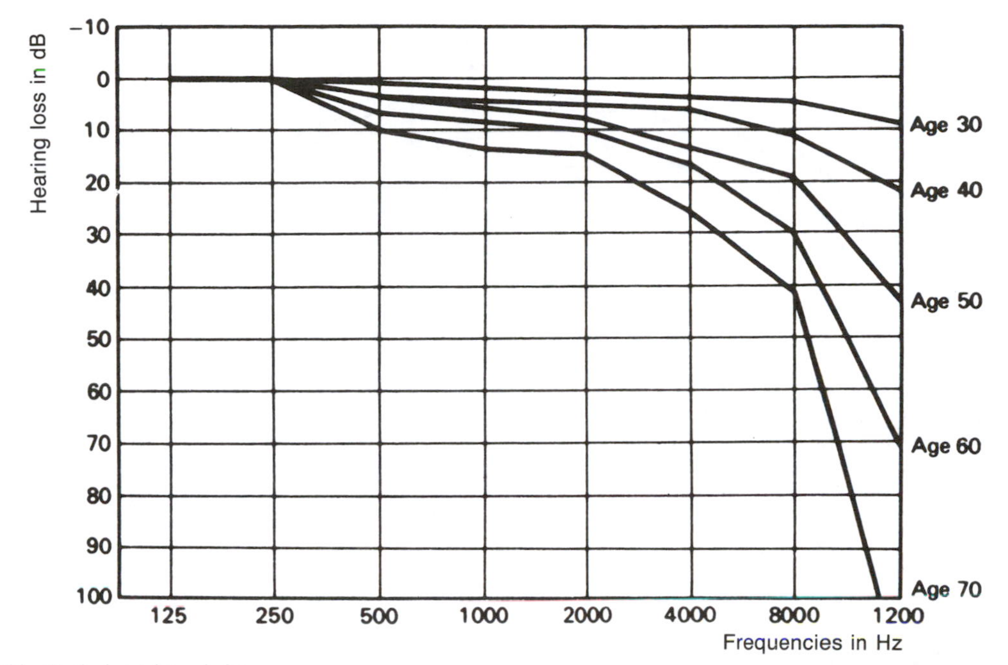

Fig. 2.11 Typical presbycusis loss

an explosion, but it can heal, and recently artificial eardrums have been used.

Conductive hearing loss causes a loss despite amplitude. A 20 dB loss would mean that a sound of 20 dB is needed for threshold, 40 dB to hear a level of 20 dB, 60 dB to hear 40 dB, etc. (at a particular frequency). People afflicted may not hear normal speech but may easily hear loud speech in a noisy factory.

Nerve deafness is either due to loss of sensitivity in the sensory cells in the inner ear or to a defect in the auditory canal. There is no medical remedy and the hearing loss is usually different for different frequencies.

Cortical deafness chiefly affects old people, and is due to a defect in the brain centres.

2 Deafness caused by noise

It is not known exactly how loud a sound will cause immediate permanent deafness, but it is of the order of 150 dB. The harmful effect of long duration noise is even more difficult to assess. The Wilson Report suggested maximum levels as shown in Table 2.4 for more than 5 hours working per day in order to prevent any noticeable hearing loss. The Noise at Work Regulations 1989 prescribe action levels for noise exposure at work.

As it can be seen, noise-induced hearing loss depends on frequency. Narrowband noise is far more serious than broadband noise. A rough guide of 80 dB for the limit of 8 hours/day at 5 days/week for 40 years seems to be a reasonable level for normal noise.

Temporary threshold shift (TTS) When a person of normal hearing is exposed to intense noise for a few hours, he or she suffers a temporary loss of hearing sensitivity called temporary threshold shift. The threshold of his or her hearing has been raised. After a sufficiently long rest from noise he or she usually recovers. If the noise is of a pure tone of low level, the threshold shift is greatest at the same frequency as the noise. However, for pure tones of high intensity the shift of threshold is greatest at about a half octave above (at about 1·4 kHz for a noise of 1 kHz). The most susceptible frequency range, however, is from 3 to 6 kHz. With broadband noise, deterioration first occurs in this region, with 4 kHz being often worst effected.

Table 2.4 Maximum levels of noise

Frequency (Hz)	37·5–150	150–300	300–600	600–1200	1200–2400	2400–4800
Value (dB)	100	90	85	85	80	80

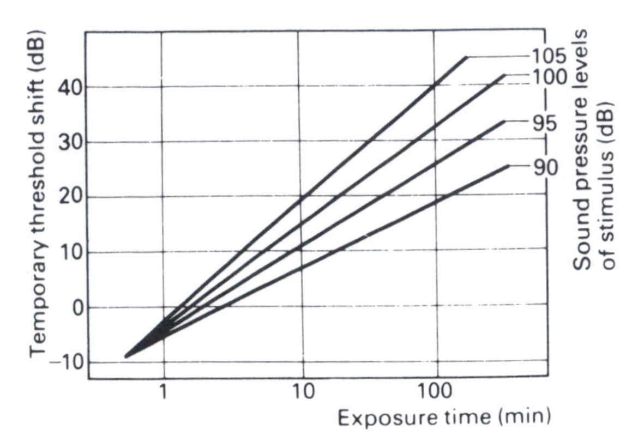

Fig. 2.12 Temporary threshold shift at 4000 Hz 2 min after the end of exposure to an octave band of noise of 1200–2400 Hz at the sound pressure levels and durations indicated

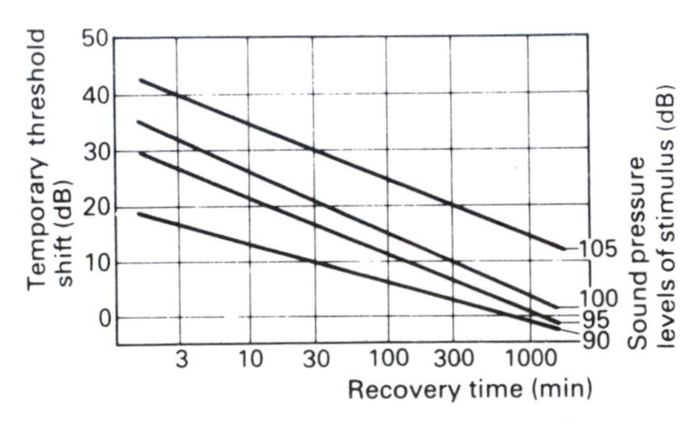

Fig. 2.13 Recovery from temporary threshold shift at 4000 Hz, at various intervals after the end of exposure to an octave band of steady noise of 1200–2400 Hz, at the sound pressure levels indicated and for the same exposure duration

The amount of TTS is related to the time of exposure. Figure 2.12 indicates the TTS which may be expected at 4 kHz after exposure to sound in the octave band 1·2 to 2·4 kHz. Recovery is indicated in Fig. 2.13.

Permanent threshold shift In the case of someone exposed to intense occupational noise during the working day for a matter of years, the stage is often reached where the temporary threshold shift has not completely recovered overnight before the next exposure. It appears that 'persistent threshold shift' is followed eventually by 'permanent threshold shift'. That permanent damage is being done to hearing may often be indicated by signs of dullness of hearing, often together with tinnitus after exposure of noise at work.

Noise-induced hearing loss is characterized by a 4 kHz drop in sensitivity, as can be seen in Fig. 2.14. Continued exposure leads to a widening of the dip as can be seen in Figs 2.15, 2.16 and 2.17.

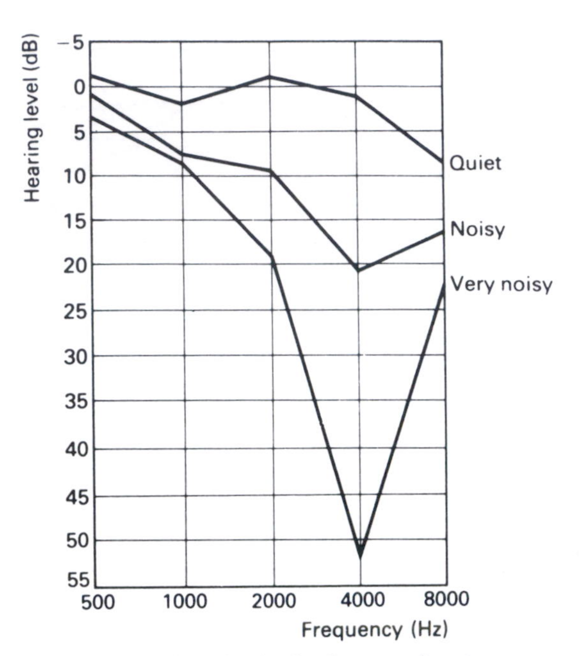

Fig. 2.14 Typical hearing levels of groups of workers exposed to different noise conditions for 10 years

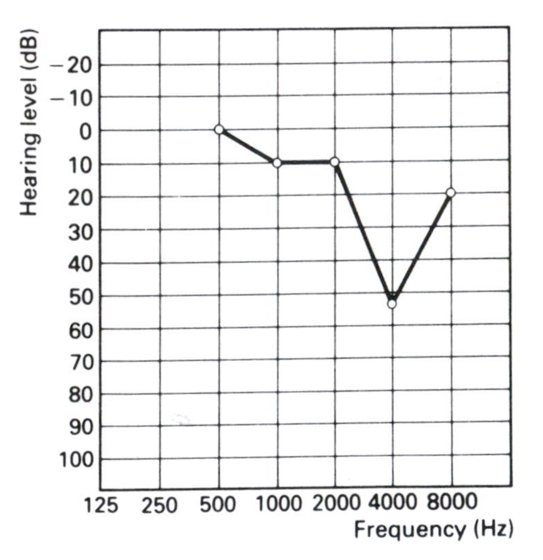

Fig. 2.15 Audiogram after working 20 years in weaving industry

The actual hearing level of a person will be worse than might have been expected because it is the sum of the permanent threshold shift and presbycusis loss. Hearing level becomes a function of three variables: noise level, exposure time and age.

It is possible to assess the risk of hearing loss when exposed to a certain level of broad-band noise for a known time. First calculate the noise exposure from:

1. $E = L + 10 \log_{10} T$

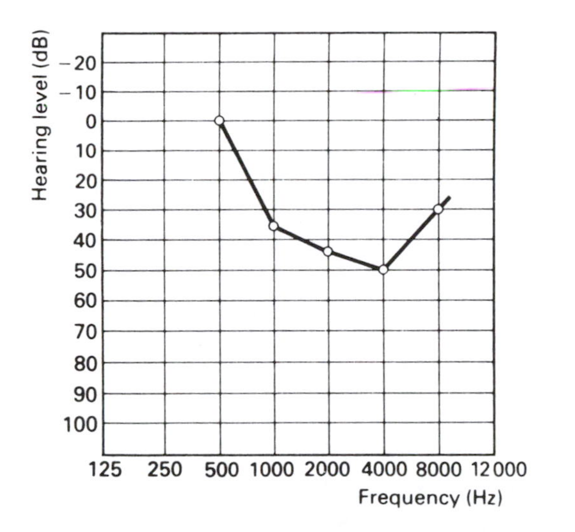

Fig. 2.16 Audiogram after working 35 years in weaving industry

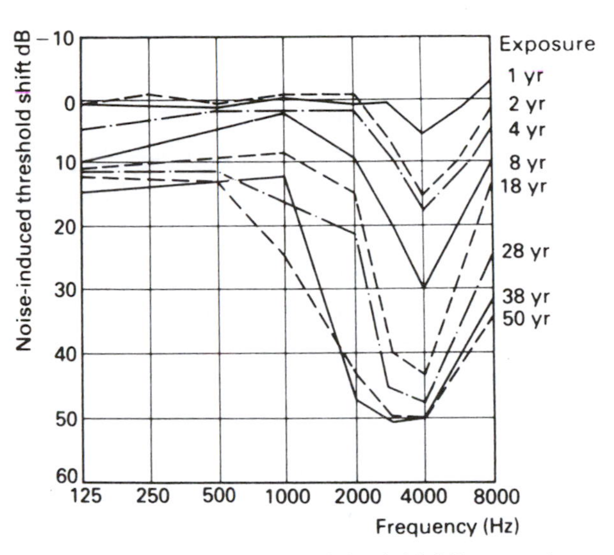

Fig. 2.17 Typical noise-induced threshold shift progressing with years of exposure

where E = noise exposure

L = sound level to which exposed in dB(A)

T = number of years exposure (5 days per week, 50 weeks/year).

2. The mean loss at a given frequency may then be determined from Fig. 2.18.
3. The percentage of the population that may be expected to have a certain hearing level at any given frequency may then be found from Fig. 2.19.

   The median noise-induced hearing loss at a given frequency is found from the appropriate position on the median line in Fig. 2.19. By following parallel to the curves, the centile of interest may be studied.
4. The actual hearing level may then be determined by adding the presbycusis loss. The average value for this may be calculated from:

$$\text{Presbycusis loss} = \frac{K}{1000}(N - 20)^2$$

where N = age in years (>20)

K = constant dependent upon audiometric frequency (see Table 2.5).

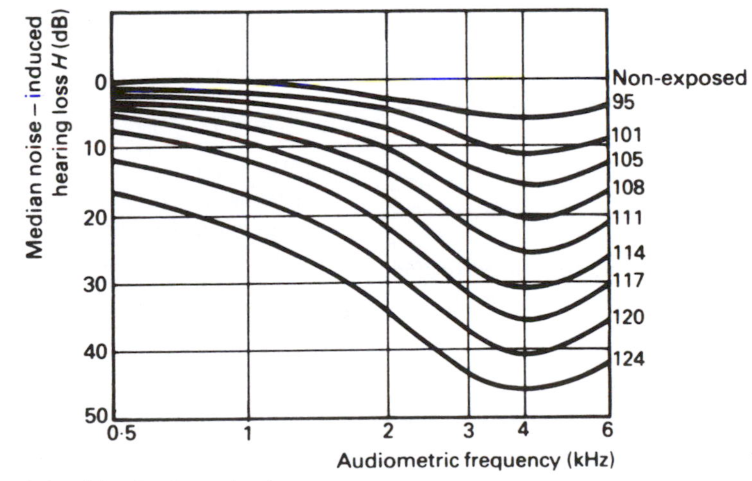

Fig. 2.18 Median noise-induced hearing loss related to exposure

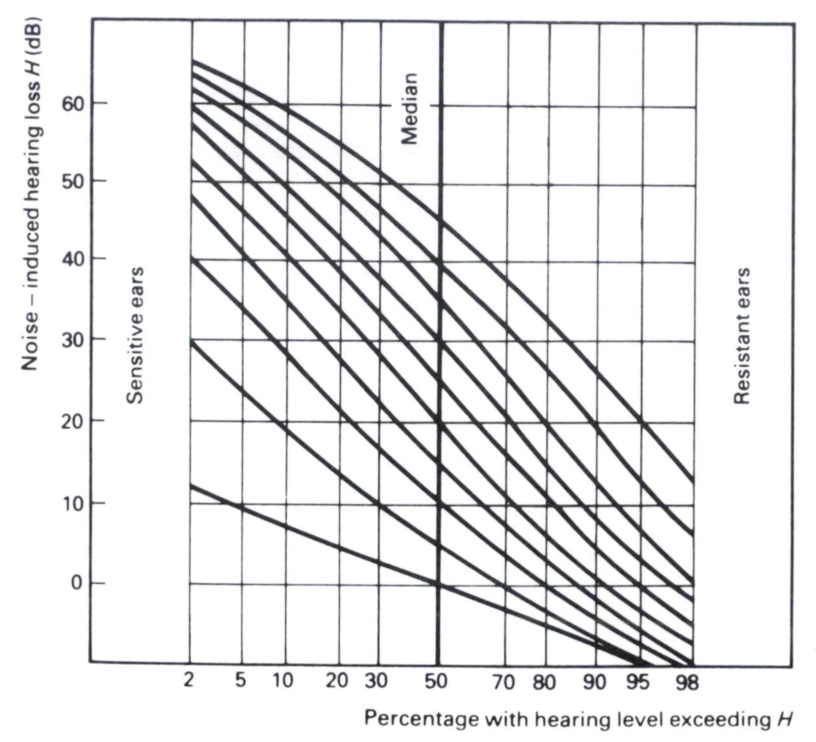

Fig. 2.19 Diagram for calculating risk of hearing loss

Table 2.5 Coefficients for calculation of presbycusis loss

Audiometric frequency (kHz)	*K*
0·5	4
1	4·3
2	6
3	8
4	12
6	14

Example 2.2

(a) Determine the average hearing loss at 4 kHz for a person of 46 having been exposed to a sound level of 96 dB(A) for 25 years.

(b) What percentage of people might be expected to suffer from a total loss of 50 dB at this frequency?

(a) (1) $E = L + 10 \log_{10} T$

$= 96 + 10 \log_{10} 25$

$= 96 + 10 \times 1{\cdot}3979$

$= 96 + 14$

$= 110$

(2) From Fig. 2.18, the median loss at 4 kHz would be 23 dB.

(4) Presbycusis loss $= \dfrac{12}{1000}(46 - 20)^2$

$= \dfrac{12}{1000} \times 26^2$

$= 8$ dB

$\therefore$ Mean hearing loss at 4 kHz $= 23 + 8$

$= 31$ dB

(b) If the total loss is 50 dB, then the noise-induced loss would be 42 dB. Thus from Fig. 2.19, start at 23 dB on median vertical line, follow parallel to curves to 42. Hence, about 12 per cent or one in eight of the workers could be expected to suffer a total hearing loss of 50 dB at 4 kHz.

Noise susceptibility Most people have average noise susceptibility, but some appear to have 'tough' ears which suffer less from permanent threshold shift. Similarly there are those with 'tender' ears. In order to protect the latter, pre-employment tests followed by routine retests at frequent intervals are needed.

Hearing conservation and surveys A hearing conservation programme is in three parts:

1. Analysis of noise exposure.
2. Control of noise exposure. (This may include reduction of noise at source, reduction of transmission of noise, ear protection — by earplugs, earmuffs or change of job.)
3. Routine periodic audiometry.

2.9 Measurement of noise exposure

It is usually considered that provided a person is not exposed to more than 90 dB(A) over his or her working week, the risk of hearing damage is very small indeed (less than 1 in 20 of the population). The difficulty in most practical situations is the considerable variation in sound level during any working week. To overcome this, it is necessary to determine the *equivalent continuous sound level* (L_{eq}), which is defined as the sound level of a steady sound that has, over the given period, the same energy as the fluctuating sound in question. For the purpose of hearing conversation, this period is taken as 40 h in one week.

The equivalent continuous sound level in dB(A) may be determined either by using an integrating meter or a sound level meter, using a slow response and timing the duration of the different levels.

An integrating meter is basically a sound level meter with an 'A' weighting together with the appropriate calculation facilities built in. A common variant is the dosimeter which calculates continuously the percentage of the maximum permissible dose which a worker has received. It normally includes a small microphone fixed as near an ear as practical with the meter placed in a breast pocket (see also Chapter 9).

If measurements are made using an ordinary sound level meter, then the results need to show the length of time during the 40 h week which the level at the ear was in classes centred at 80, 85, 90, 95, 100, 105, 110, 115, 120 dB(A). Hence the partial index may be calculated from:

$$E_i = \frac{\Delta t_i}{40} 10^{0\cdot 1(L_i - 70)}$$

where E_i is the partial noise exposure index

L_i is the sound level in dB(A)

Δ_t is the duration in hours/week

Thence $L_{eq} = 70 + 10 \log_{10} \Sigma E_i$

Example 2.3

Calculate the 40 h L_{eq} of a sound of 120 dB(A) of 10 min duration.

$$E_i = \frac{\Delta t_i}{40} 10^{0\cdot 1(L_i - 70)}$$

$$= \frac{1}{6} \cdot \frac{1}{40} \cdot 10^{0\cdot 1 \times 50}$$

$$= 417$$

$$L_{eq} = 70 + 10 \log_{10} 417$$

$$= 70 + 26\cdot 2$$

$$= 96 \text{ dB(A)}$$

Example 2.4

Calculate the 40 h L_{eq} for noises of 85 dB(A) lasting 30 h, 95 dB(A) lasting 9 h, and 100 dB(A) lasting 1 h total during a 40 h week.

$$E_i = \frac{\Delta t_i}{40} 10^{0\cdot 1(L_i - 70)}$$

$$E_1 = \frac{30}{40} 10^{0\cdot 1 \times 15}$$

$$= \frac{3}{4} \times 31\cdot 6$$

$$= 23\cdot 7$$

$$E_2 = \frac{9}{40} \times 10^{0\cdot 1 \times 25}$$

$$= 71\cdot 2$$

$$E_3 = \frac{1}{40} \times 10^{0\cdot 1 \times 30}$$

$$= 25$$

$$\therefore \quad \Sigma E = 23\cdot 7 + 71\cdot 2 + 25$$

$$= 119\cdot 9$$

$$L_{eq} = 70 + 10 \log_{10} 119\cdot 9$$

$$= 90\cdot 79$$

$$= 91 \text{ dB(A)}$$

Note: Figure 1.6 gives a nomogram to save the use of formulae, but with the use of calculators it is easier to calculate L_{eq}, as in Example 1.13.

2.10 Other physiological effects of noise

Infrasound Infrasound consists of pressure waves with a frequency below 20 Hz and therefore below the normal threshold of hearing. Recently, it has been established that infrasonic frequencies can occur in buildings with long ventilation ducts. A frequency of 7 Hz is particularly unpleasant and can cause a throbbing in the head due, it is suggested by Gavreau, to the fact that this coincides with the medium frequency of alpha waves of the brain.

Frequencies below 1 Hz are produced by wind effects on buildings.

Vibration Human reaction to vibration depends upon both amplitude and frequency. It appears that the effect of vibration on the people within a building will be far more serious than the effect on the building. Tests that have been carried out show that similar painful human reactions have been produced by vibration of about 0·075 mm amplitude at 20 Hz and about 0·015 mm at 50 Hz.

Very large amplitudes of mechanical vibration, as with exceptionally high noise levels, may affect other sensory receptors such as touch.

It would appear that in general the subjective effect of vibration in buildings will set the limit for continuous vibration rather than structural considerations (see also Chapter 7).

2.11 Psycho-acoustics

Apart from the physiological effects of noise there is the psychological noise problem. This is even more difficult to define, because of the tremendous variability between individuals and environment. Music may sound enjoyable at a party but be annoying when someone is trying to work or sleep, and perhaps be indirectly harmful to health when ill. In a factory the noise a machine produces may be helpful to its operator but have a dangerous and annoying masking effect on someone else. We might happily work in levels of 110 dB(A) in a machine shop, although our hearing may ultimately be damaged, but be disturbed by a tap dripping.

Various attempts have been made to correlate objective measures of a noise with its subjective effect on some sort of 'annoyance' or similar scale. The results obtained are usually only applicable for that particular type of noise in that particular environment. One of the following four types of scale is used: nominal, ordinal, interval or ratio. A nominal scale simply finds out that noise A is different from noise B, and therefore such a scale is of limited use. An ordinal scale determines the relative magnitudes. Interval scales are much more useful in that they show the size of graduations on the scale, but do not determine where the scale begins. A ratio scale finds the intervals and also the zero.

In one subjective test involving aircraft noise for the Committee on the Problem of Noise, people were asked to rate certain noises on a scale — not noticeable, noticeable, intrusive, annoying, very annoying or unbearable. This is an ordinal scale and not an interval scale because the sizes of the gaps between each of the named points on the scale are not known and are very unlikely to be equal. In this test it was, however, assumed to be an equal-interval scale, and later this was justified. It is very difficult and often impossible for subjects to rate noise on a ratio scale outside a laboratory, and to obtain even an interval scale may be difficult.

Another method of finding the effect of noise on people can be by determining the number of complaints that have been made about the particular noise. It is doubtful whether this relates to the average population who probably seldom complain, but it may be a useful guide to show the amount of annoyance.

Indirect methods may also be useful. One such method used to find the effect of motor traffic noise on people was to inquire whether householders slept at the front or rear of their house. The fact that many had moved from the main bedroom at the front to a smaller room at the rear indicated that the noise of traffic was having an annoying effect.

In practice, it may be necessary to use a combination of these methods. The present standards of sound insulation for dwellings are the results of social surveys based on them.

2.12 Perceived noise level (PNdB)

The perceived noise level of a particular noise is the sound pressure level of a band of noise from 910 to 1090 Hz that sounds as 'noisy' as the sound under comparison. The noisiness is given in noys.

Calculation of PNdB

1. Find the maximum sound pressure level in each octave band centred at 63, 125, 250, 500, 1000, 2000, 4000 and 8000 Hz.
2. From Fig. 2.20, use the frequency and sound pressure level in dB to find the noisiness from the contours, e.g., 80 dB for an octave centred at 1000 Hz gives a noy value of 17.
3. Find the total noisiness N from

$$N = N_{max} + 0{\cdot}3(\Sigma N - N_{max})$$

where N_{max} = highest noy value and

ΣN = sum of the noy values in all octave bands.

It will be seen that the formula is exactly equivalent to that used in the calculation of loudness, except for the use of equal-noisiness contours in the place of equal-loudness contours.

4. Noys may be converted to PNdB using

$$N = 2^{(x-40)/10}$$

$$\text{hence, } x = 40 + \frac{100}{3}\log_{10} N$$

where x = perceived noise level in PNdB.

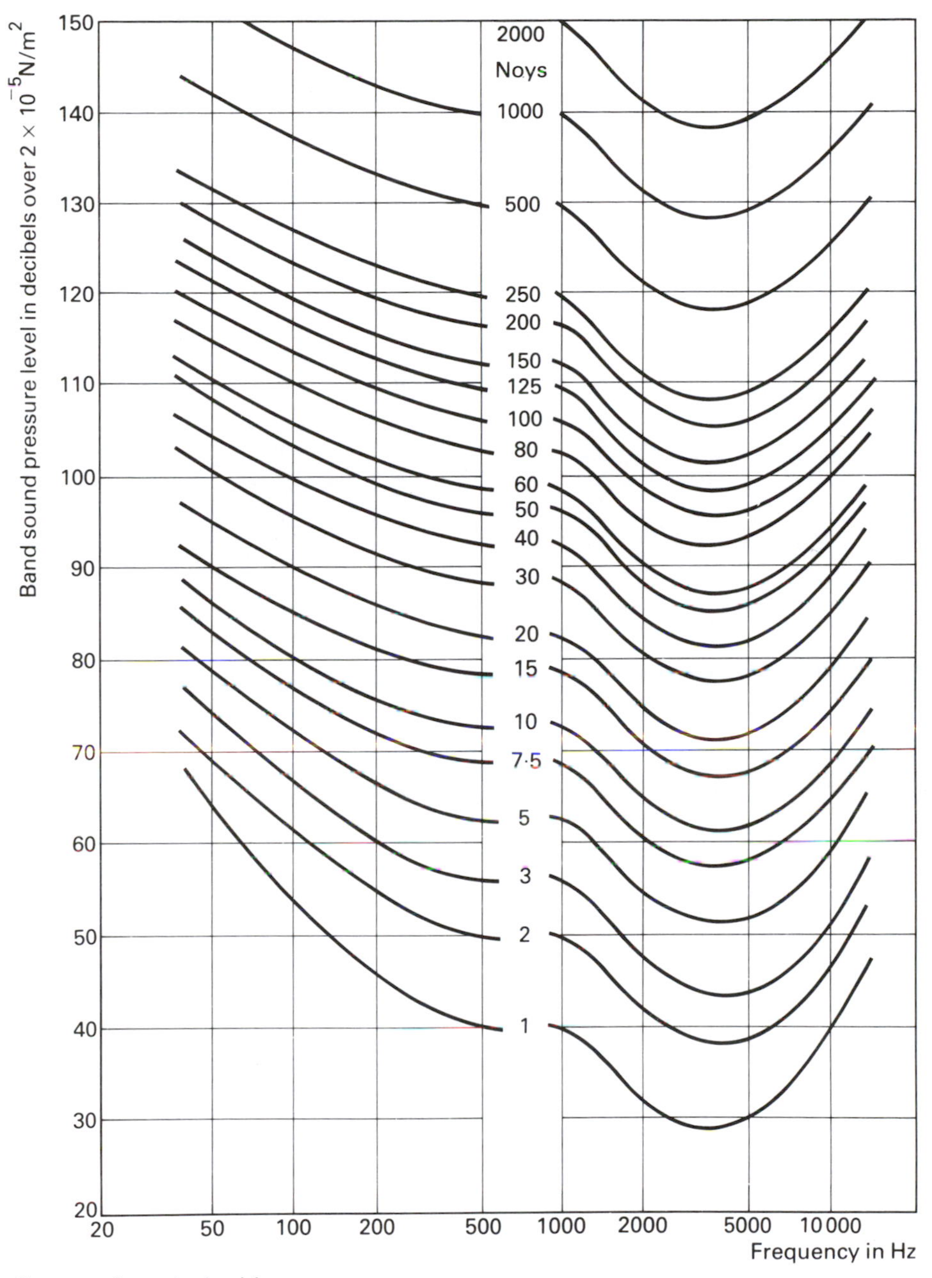

Fig. 2.20 Contours of perceived noisiness

Alternatively, Table 2.6 may be used to find the value of x. Note the similarity between sones for loudness and noys for noisiness, and between phons for loudness and PNdB for perceived noise level.

Example 2.5

Calculate the perceived noise level in PNdB of the motor car whose analysis is given in Example 2.1.

Table 2.6 Perceived noise level as a function of total perceived Noisiness from ISO 3891 (1978)

N				*N*			
Lower	*Mid*	*Upper*	*PNdB*	*Lower*	*Mid*	*Upper*	*PNdB*
1·0	1·0	1·0	40	15·5	16·0	16·6	80
1·1	1·1	1·1	41	16·7	17·1	17·7	81
1·1	1·1	1·2	42	17·8	18·4	19·0	82
1·2	1·2	1·3	43	19·1	19·7	20·4	83
1·3	1·3	1·4	44	20·5	21·1	21·8	84
1·4	1·4	1·5	45	21·9	22·6	23·4	85
1·5	1·5	1·6	46	23·5	24·2	25·1	86
1·6	1·6	1·7	47	25·2	26·0	26·9	87
1·7	1·7	1·8	48	27·0	27·8	28·8	88
1·9	1·9	1·9	49	28·9	29·8	30·9	89
2·0	2·0	2·1	50	31·0	32·0	33·1	90
2·1	2·1	2·2	51	33·2	34·3	35·5	91
2·3	2·3	2·4	52	35·6	36·8	38·1	92
2·5	2·5	2·5	53	38·2	39·4	40·8	93
2·6	2·6	2·7	54	40·9	42·2	43·7	94
2·8	2·8	2·9	55	43·8	45·2	46·8	95
3·0	3·0	3·1	56	46·9	48·5	50·2	96
3·2	3·2	3·4	57	50·3	52·0	53·8	97
3·5	3·5	3·6	58	53·9	55·7	57·7	98
3·7	3·7	3·9	59	57·8	59·7	61·8	99
4·0	4·0	4·1	60	61·9	64·0	66·3	100
4·2	4·3	4·4	61	66·4	68·6	71·0	101
4·5	4·6	4·7	62	71·1	73·5	76·1	102
4·8	4·9	5·1	63	76·2	78·8	81·6	103
5·2	5·3	5·5	64	81·7	84·4	87·4	104
5·6	5·6	5·8	65	87·5	90·5	93·7	105
5·9	6·1	6·3	66	93·8	97·0	100·4	106
6·4	6·5	6·7	67	100·5	104·0	107·6	107
6·8	7·0	7·2	68	107·7	111·4	115·3	108
7·3	7·5	7·7	69	115·4	119·4	123·6	109
7·8	8·0	8·3	70	123·7	128·0	132·5	110
8·4	8·6	8·9	71	132·6	137·2	142·0	111
9·0	9·2	9·5	72	142·1	147·0	152·2	112
9·6	9·8	10·2	73	152·3	157·6	163·1	113
10·3	10·6	10·9	74	163·2	168·9	174·8	114
11·0	11·3	11·7	75	174·9	181·0	187·4	115
11·8	12·1	12·5	76	187·5	194·0	200·8	116
12·6	13·0	13·5	77	200·9	207·9	215·3	117
13·6	13·9	14·4	78	215·4	222·8	230·7	118
14·5	14·9	15·4	79	230·8	238·8	247·3	119

Octave band centre frequency (Hz)	*dB level*	*Noys*
63	95	21·0
125	84	13·8
250	80	14·0
500	68	7·5
1000	65	6·0
2000	61	6·1
4000	60	9·3
8000	60	16·5
	$\Sigma N =$	94·2

$$N_{max} = 21{\cdot}0$$

$$\therefore \quad N = 21{\cdot}0 + 0{\cdot}3(94{\cdot}2 - 21)$$

$$= 21{\cdot}0 + 0{\cdot}3(73{\cdot}2)$$

$$= 21{\cdot}0 + 21{\cdot}96$$

$$= 42{\cdot}96 \text{ noys}$$

From Table 2.6, 42·96 noys = 94 PNdB.

2.13 Noise and number index (NNI)

The noise and number index, now largely historical, was originally derived to show the effect of aircraft noise on people. It was a composite measure taking into account both average peak noise level in PNdB of aircraft as well as the number of aircraft involved.

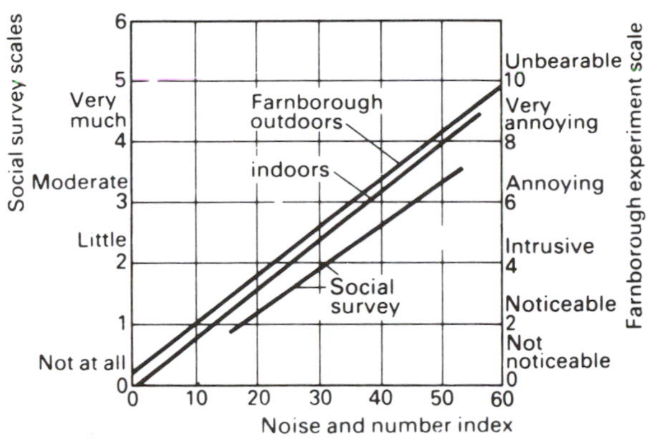

Fig. 2.21 Relations between annoyance rating and noise and number index obtained from social survey and Farnborough experiments

$$\text{NNI} = \text{average peak noise level} + 15 \log_{10} N - 80$$

where N is the number of aircraft heard and average peak noise level is a logarithmic average; hence:

$$\text{average peak noise level} = 10 \log_{10} \frac{1}{N} \sum_{1}^{N} 10^{L/10}$$

and L is in PNdB.

The Committee on the Problem of Noise found, in the survey around London Heathrow Airport, that there was very good correlation between annoyance scores and the noise and number index. Thus it is possible to predict the amount of annoyance to be expected from peak noise levels and numbers of aircraft (see Fig. 2.21). Similarly, it is possible to predict the noise and number index at various places at future times (see Figs 2.22, 2.23, 2.24 and 2.25).

It would appear that there is some disagreement at the

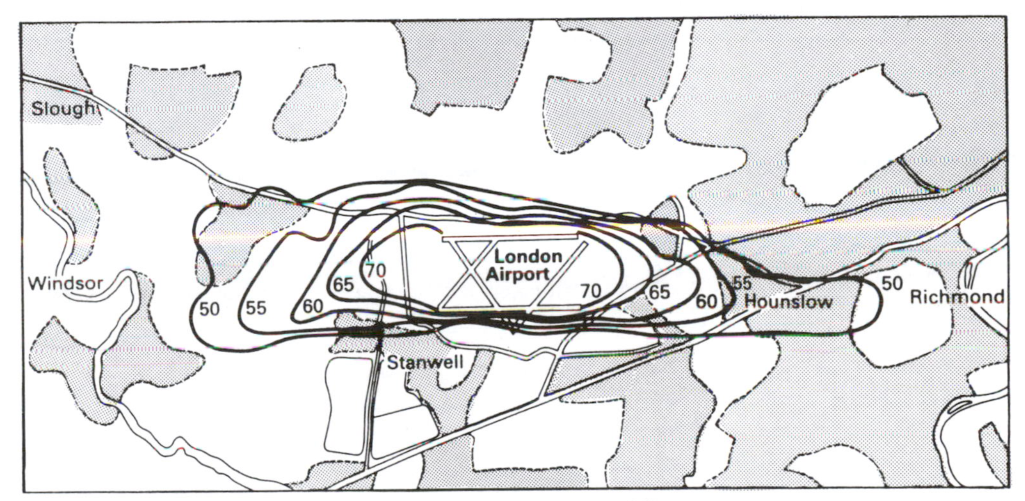

Fig. 2.22 London Airport: approximate noise and number index contours 1961

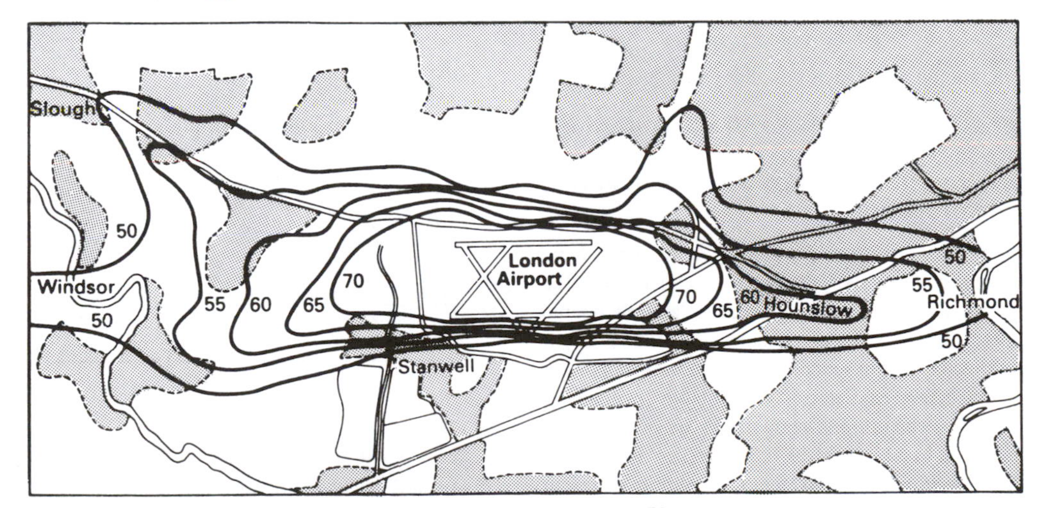

Fig. 2.23 London Airport: approximate noise and number index contours 1970

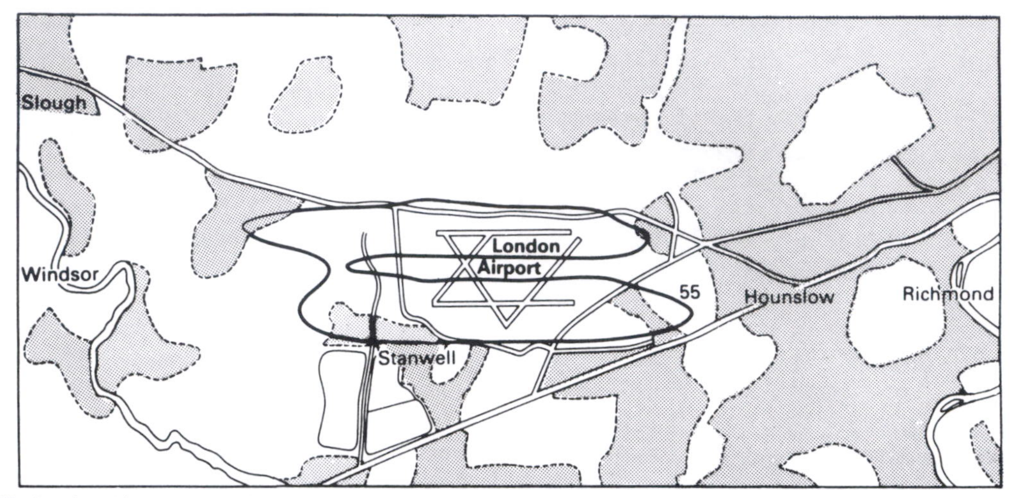

Fig. 2.24 London Airport: approximate position of 55 NNI contour in 1977

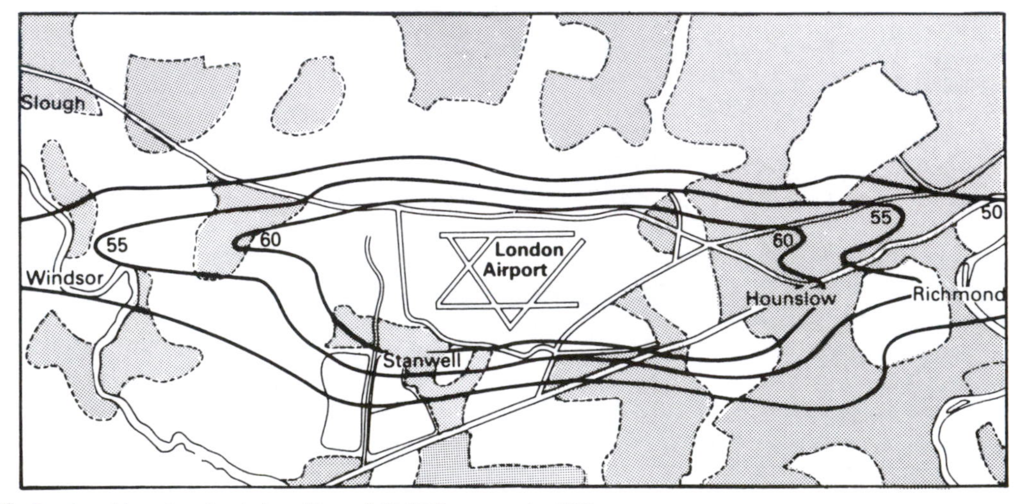

Fig. 2.25 London Airport: estimated position of 55 NNI contour in 1990

present time on the accuracy of prediction of the noise and number index.

2.14 Relation of noise levels to particular environments

So far the effects of noise on people in general terms have been considered. However, it must be realized that the effects are not only due to the type of noise and the people listening, but also to their environment. For instance, it would be expected that in schools one of the most important factors is the effect that noises might have in masking speech. Conversely, in open-plan offices or certain parts of hospitals this masking effect on speech is desirable.

The various criteria to be taken into account in different environments are discussed in the later chapters.

Questions

(1) Calculate the loudness in phons of noise which has the following analysis:

Octave band centre frequency (Hz)	63	125	250	500	1000	2000	4000	8000
dB Level	73	70	69	71	70	65	71	56

(2) If the noise in Question (1) is heard on the other side of a partition having the following sound reduction values, calculate the expected loudness in the receiving room.

Octave band centre frequency (Hz)	63	125	250	500	1000	2000	4000	8000
Insulation (dB)	30	34	36	41	51	58	62	66

(3) Calculate the perceived noise level in PNdB of the noise in Question (1).

(4) Calculate the perceived noise level in PNdB of the noise in the receiving room in Question (2).

(5) An octave-band analysis of sound in a machine shop was made and the following results obtained:

Octave band centre frequency (Hz)	63	125	250	500	1000	2000	4000	8000
SPL (in dB)	68	72	90	87	86	88	90	84

Calculate the loudness in phons.

(6) Find the perceived noise level in PNdB of the analysis in Question (5).

(7) Explain the difference between temporary threshold shift and permanent threshold shift. Discuss methods by which noise-induced hearing loss may be avoided in people working in noisy industrial situations.

(8) Describe how a calibrated tape recording of road traffic noise may be made, with a view to later analysis in the laboratory.

(9) State the reasons for the use of a self-recording audiometer as given in the Health and Safety Executive Discussion Document 'Audiometry in Industry'. Describe a test for conductive-hearing loss.

Sketch a self-recorded audiogram of a typical case of early noise-induced hearing loss and give the probable cause in terms of the damage to the structure of the ear.

Explain the meaning of the term 'equal energy principle' when determining the maximum exposure times for various noise levels. Evaluate the permitted daily exposure-time for a person subjected to a steady SPL of 109 dB(A).

(10) (a) Sketch audiograms to show clearly the conditions of:
(i) presbycusis
(ii) noise-induced hearing loss

(b) Outline the damaging processes in the ear that cause these two conditions. How do the two conditions relate to one another?

(c) Discuss the practical and social implications on individuals of severe loss of hearing due to these conditions.

(*IOA*)

(11) (a) Describe the components of the ear and describe their rôle in transforming sound in the external air to impulses in the auditory nerve.

(b) Name *two* pathological conditions, other than noise-induced, associated with conductive hearing loss, and *one* associated with sensory hearing loss. In each case state the effect of the condition on the auditory mechanism.

(c) Briefly describe a method which may be used to distinguish between conductive and sensory hearing loss, and state the rationale of this method.

(d) Briefly describe the structural changes which occur in the inner ear as noise-induced hearing loss develops as a result of prolonged exposure to noise.

(*IOA*)

(12) (a) Discuss the structure of the ear with respect to the following dysfunctions:
(i) acoustic trauma
(ii) conductive hearing loss
(iii) tinnitus

(b) Explain the principle behind the provision of ear defenders.

(c) Where protection is required for persons working underwater, what additional factors have to be considered?

(13) (a) Describe the physical processes involved in the transformation of pressure waves in the inner ear into nerve impulses.

(b) Discuss the physiological characteristics associated with noise-induced hearing loss and indicate the frequency range at which it is most likely to be detected.

(c) Define conductive and sensory hearing loss.

(d) Discuss the advantages and disadvantages of the self-recording audiometer.

(*IOA*)

(14) (a) List and briefly discuss the effects of sound of both low and high intensity on people.

(b) Distinguish between conductive and sensorineural hearing loss.

(c) In what ways may noise cause (i) conductive, and (ii) sensorineural hearing loss?

(d) Briefly describe the following pathological conditions and in each case comment on their effect on the hearing mechanism: otosclerosis, Ménière's disease, wax plug, chronic catarrh.

(*IOA*)

(15) Explain what is meant by an equal loudness contour and sketch the approximate shapes of three such contours for low, medium and high loudness levels respectively. Explain the relationship between these and the standard 'A' frequency weighting.

An octave analysis of the noise in a workshop yielded the following results:

Centre frequency (Hz)	63	125	250	500	1000	2000	4000	8000
Band level (dB)	85	87	76	73	72	70	60	58

Calculate (a) the loudness in Stevens sones, (b) the corresponding loudness in phons, and (c) the sound level in dB(A). ('A' weightings and loudness indices provided)

(IOA)

(16) Figure 2.26 shows the structure of the organ of Corti of the cochlea of the ear.

(a) Name the parts labelled A to E.

(b) Briefly describe how these structures are involved in the conversion of sound waves into nerve impulses.

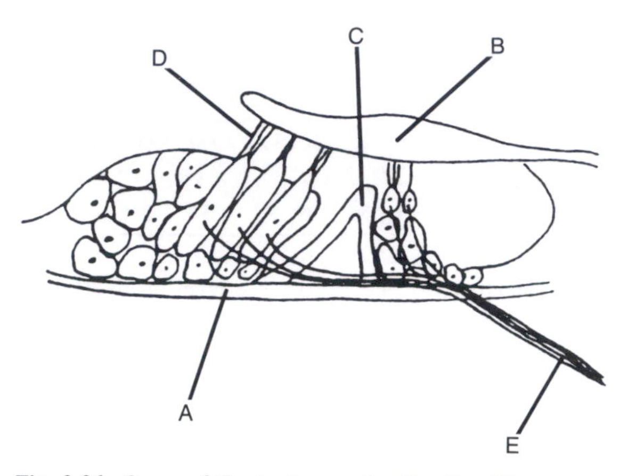

Fig. 2.26 Organ of Corti: diagram for Question 17

(c) Distinguish between conductive and sensory hearing loss.

(d) Name *two* pathological conditions, other than noise-induced, associated with sensory hearing loss.

(e) Briefly describe the structural changes which occur in the inner ear as noise-induced hearing loss develops due to prolonged exposure to noise.

(IOA)

3 Room acoustics

This chapter is concerned with the behaviour of sound within an enclosed space with a view to obtaining the optimum acoustic effect on the occupants. The various effects which rooms may have on the subjective properties of the sound are therefore studied.

3.1 Requirements

1. An adequate amount of sound must reach all parts of the room. Most attention in this respect needs to be given to those seats furthest from the source.
2. An even distribution of sound should be achieved throughout the room, irrespective of distance from the source.
3. Other noise which might tend to mask the required sound must be reduced to an acceptable level in all parts of the room.
4. The rate of decay of sound within the room (reverberation time) should be the optimum for the required use of the room. This is to ensure clarity for speech or 'fullness' for music.
5. Acoustical defects to be avoided include:
   (a) Long delayed echoes
   (b) Flutter echoes
   (c) Sound shadows
   (d) Distortion
   (e) Sound concentrations.

3.2 Behaviour of sound

Reflection In the first chapter it was shown that sound can be reflected in a similar way to light, the angle of incidence being equal to the angle of reflection. However, it must be remembered that for this to be true the reflecting object must be at least the same size as the wavelength concerned. It can often be very useful to carry out a limited geometrical analysis. This can prevent the problem of long delayed reflections and focusing effects. It is impractical to take a geometrical analysis beyond the first or second reflections, but it can prevent gross errors in design.

These errors are often caused by the focusing effects of concave shapes which may produce places with very loud sounds or dead spots, as shown in Fig. 3.1. It is generally unwise to have concave surfaces in a hall unless the focus is well outside. Convex surfaces can be useful in providing a diffusing surface in order to reflect the sound evenly in the hall.

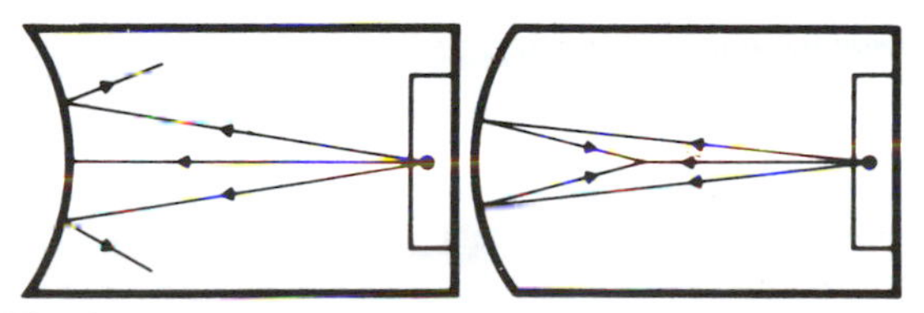

Fig. 3.1 Diffusing effect of convex surfaces and focusing effect of concave surfaces

Long delayed reflections In large halls care must be taken to make certain that no strong reflections of sound are received by the audience after about 50 ms, otherwise confusion is likely between the direct and the reflected sound for speech. Average speech is at the rate of about 15 to 20 syllables per second or roughly one syllable every 70 ms to 50 ms, respectively. This corresponds to a delay of about 17 m. A member of the audience sitting at 8·5 m from a good reflecting rear wall of a hall will find it difficult to understand speech. In larger halls other surfaces become important, such as the side walls and ceiling, as shown in Fig. 3.2. These strong reflections can be prevented by covering the surfaces concerned with absorbent material or by making them into diffusing surfaces by means of a convex shape. Quick reflections from the corners can be a problem but are easily overcome by the use of an acoustic plaster or some other absorbent material.

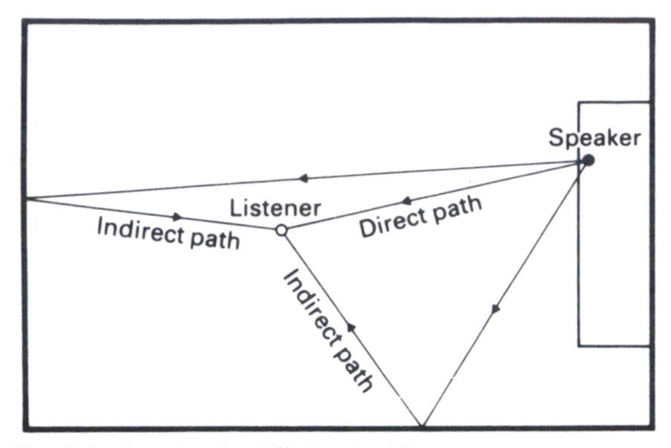

Fig. 3.2 Sound paths differing by 17 m or more can cause confusion of speech

The simple solution appears to be to cover as much of the surfaces as possible with an absorbent material. Too much, however, would lead to a very unpleasant effect and also make speech without amplification too quiet to hear near the back of the hall. The aim should be to use the minimum amount of absorbent material so that the hall can have the minimum volume for a given number of people.

Flutter echoes These consist of a rapid succession of noticeable echoes which can be detected after short bursts of sound such as a hand clap (see Fig. 3.3). They can be avoided by making certain that the sound source is not between parallel reflecting surfaces. Sound-absorbent material on one of the offending walls would also cure this defect, but the distribution of sound would become less uniform.

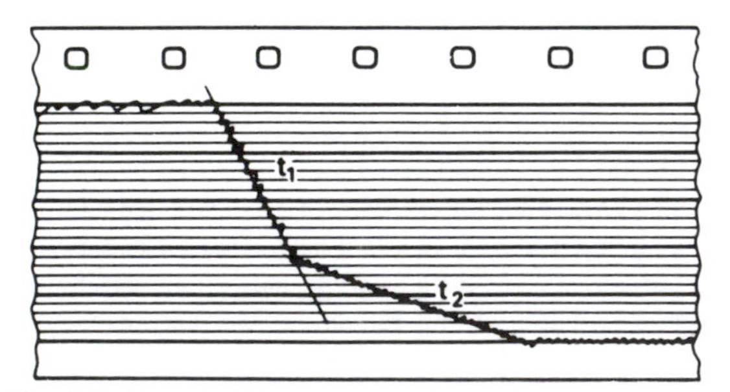

Fig. 3.3 Recorded decay curve showing effects of a flutter echo

Flutter echoes are particularly difficult to control in small rooms such as music practice rooms where much absorbent material is undesirable. The answer in this case is to avoid parallel walls and parallel ceilings and floors, as shown in Fig. 3.4.

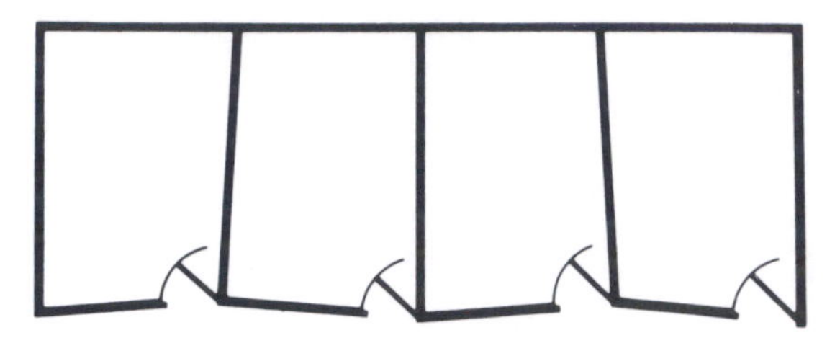
Fig. 3.4 Flutter echoes can be avoided by making the walls a few millimetres out of parallel

3.3 Ripple-tank studies

The effect of awkward shapes can be studied using a ripple tank. This consists of a thin transparent tray of water in which the shape sits. The source of 'sound' is a vibrator causing surface ripples. Effects of focusing, diffusion or interference can be investigated in this simple manner. A stroboscope shows this more clearly. It must be appreciated that this method only shows the effect for two dimensions and does not simulate absorption.

3.4 Room modes

In any particular room the behaviour of a sound is unique and highly complex because of the absorption and reflection properties of all the shapes within it. The fact that sound energy is not evenly distributed can be demonstrated by having a single-frequency note played and listening to it while walking around the room. Positions of higher and lower levels will be noticed. Of course, these positions are not the same for different frequencies. These variations in level should not be too great at audience listening positions.

3.5 Acoustic reflectors

Without amplification systems there will be a limited amount of sound power available. For unaided speech this is about 10–50 μW. In a larger hall the available sound power must be distributed evenly. There should be little problem near the front but further back it may be necessary to have sound directed by means of a specially shaped reflector. In many cases the ceiling can be utilized provided that long delayed reflections are prevented (see Fig. 3.5).

3.6 Shape of hall

Three basic plans are in common use for large halls. These are rectangular, fan and horseshoe shaped. In a hall which seats under 1000 people the shape is not so critical. As the size increases there becomes a preference for the fan shape, so that the audience is seated slightly closer to the sound source (see Fig. 3.6). Care must be taken that the rear of

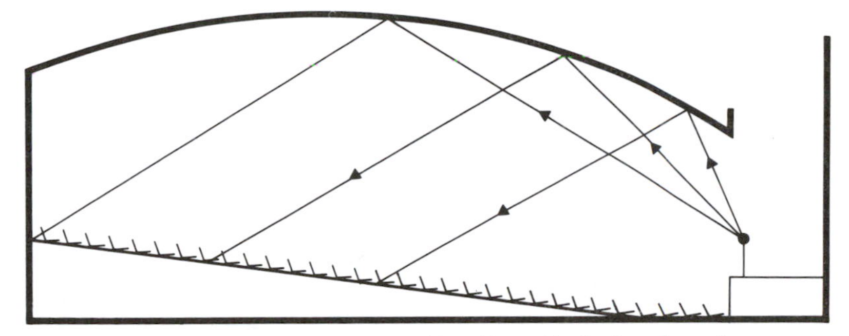

Fig. 3.5 Correctly shaped ceiling or ceiling reflector can provide uniform distribution of sound energy

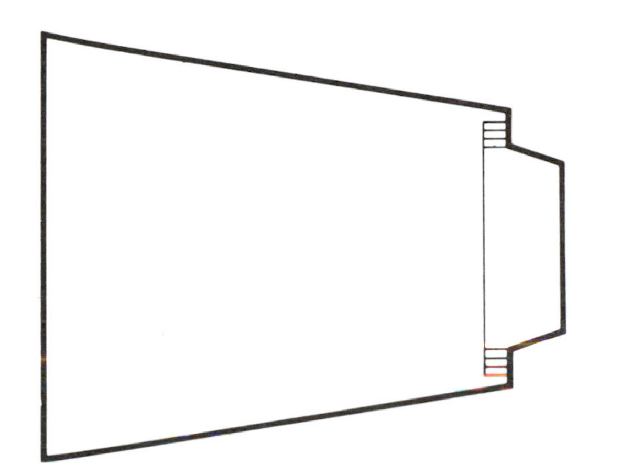

Fig. 3.6 Fan-shaped hall

such a hall is not concave. Problems can arise from reflections from side walls which may have to be broken up with either large diffusing surfaces or by the use of absorbent material. An example of a fan-shaped hall is the F. R. Mann Concert Hall, Tel Aviv.

Opera-houses are often built in a horseshoe shape. The concave surfaces are broken up by tiers of boxes around the walls. The audience provides the absorption of sound. This type of finish is excellent for opera where clarity of sound is more important than fullness of tone, but would not be considered so good for orchestral music. Covent Garden Opera House is an example of a horseshoe shape.

The traditional rectangular shape has many advantages in construction and as long as reflectors are used over the sound source the difficulty of obtaining sufficient loudness near the back can be overcome. The Royal Festival Hall, London, is an example of a rectangular hall (see Fig. 3.7).

Fig. 3.7 Rectangular-shaped hall

Other shapes have been used or suggested. Conventional shapes are probably more popular because the difficulties are less of an unknown quantity. The Royal Albert Hall is one example of an oval construction, but has acoustic problems. The Philharmonic Concert Hall, Berlin, is of irregular shape. The Philharmonic Hall, New York, a combination of the rectangular and fan shape. Traditional dimensions have the ratio 2:3:5 for height:width:length.

Although traditionally rectangular in shape, churches have been built which are round. The problem of the focusing of sound by reflection has been overcome by careful use of absorbents.

3.7 Seating arrangements

Rows of people, particularly at grazing incidence to the sound, represent a most efficient absorber. It is essential in all but the smallest halls (above about 200 people) to rake the seating if at all possible. As a simple rule, adequate vision should ensure an adequate sound path. This will mean that the line of sight needs to be raised by 80 to 100 mm for each successive row (see Fig. 3.8).

3.8 Volume

In order that the optimum listening conditions are obtained, it is essential that a hall has the correct order of volume for its use. These are given in Table 3.1.

It will be seen that the volume per person is dependent upon the purpose for which the building is to be used. Music played in a hall with too small a volume is likely to lack fullness, whereas speech in a hall with a very large volume

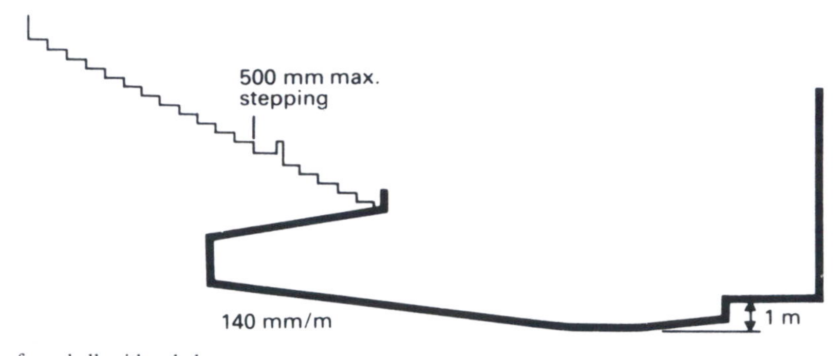

Fig. 3.8 Suitable rake for a hall with a balcony

Table 3.1 Optimum volume/person (m^3) for various types of hall

	Minimum	Optimum	Maximum
Concert halls	6·5	7·1	9·9
Italian-type opera houses	4·0	4·2–5·1	5·7
Churches	5·7	7·1–9·9	11·9
Cinemas	—	3·1	4·2
Rooms for speech	—	2·8	4·9

for its seating capacity can be expected to lack clarity. Tables 3.2 and 3.3 give a list of the vital acoustic statistics for a few of the well-known concert halls and opera-houses. Nearly all of the best concert halls and opera-houses in the world fit into the general pattern of volumes recommended in Table 3.1.

Volumes of churches vary enormously, with many famous cathedrals having a capacity up to four times that recommended. St Paul's Cathedral, for example, has a volume of roughly 150 000 m^3. Many other English cathedrals have a volume of about 30 000 m^3 or more.

3.9 Reverberation time

This is the time it takes for a sound to decay by 60 dB. One of the few concert-hall criteria which can be defined precisely and measured with reasonable accuracy is the reverberation time. It is certainly one of the most important criteria. Sound does not die away the instant it is produced but will continue to be heard for some time because of reflections from walls, ceilings, floors and other surfaces. It will mix with later direct sound and is known as rever-

Table 3.2 Acoustical data for some well known concert halls

Name	Volume (m^3)	Audience capacity	Volume per aud. seat (m^3)	Mid frequency R.T. in seconds (full hall)
St. Andrew's Hall, Glasgow (built 1877)	16 100	2 133	7·6	1·9
Carnegie Hall, New York (1891)	24 250	2 760	8·8	1·7
Symphony Hall, Boston (1900)	18 740	2 631	7·1	1·8
Tanglewood Music Shed Lennox, Mass. (1938)	42 450	6 000	7·1	2·05
Royal Festival Hall (1951)	22 000	3 000	7·3	1·47
Liederhalle, Grosser Saal, Stuttgart (1956)	16 000	2 000	8·0	1·62
F. R. Mann Concert Hall, Tel Aviv (1957)	21 200	2 715	7·8	1·55
Beethovenhalle, Bonn (1959)	15 700	1 407	11·2	1·7
Philharmonic Hall, New York (1962)	24 430	2 644	9·3	2·0
Philharmonic Hall, Berlin (1963)	26 030	2 200	11·8	2·0

Table 3.3 Acoustical data for some well known opera houses

Name	Volume (m^3)	Audience capacity	Volume per aud. seat (m^3)	Mid frequency R.T. in seconds (full hall)
Teatro alla Scala, Milan (1778)	11 245	2 289	4·91	1·2
Academy of Music, Philadelphia (1857)	15 090	2 836	5·32	1·35
Royal Opera House (1858)	12 240	2 180	5·6	1·1
Theatre National de L'Opera, Paris (1875)	9 960	2 131	4·67	1·1
Metropolitan Opera House, New York (1883)	19 520	3 639	5·36	1·2

berant sound. People expect some reverberant sound which can assist understanding or help convey at atmosphere to an audience, from the haunted house with a long reverberation time to the padded cell. One's sense of well-being or otherwise is undoubtedly affected by the length of the reverberation time.

Ideal reverberation times have been suggested by various workers using empirical methods. One such due to Stephens and Bate is:

$$T = r(0{\cdot}012\sqrt[3]{V} + 0{\cdot}1070)$$

where T = reverberation time in seconds

V = volume of the hall in m^3

r = 4 for speech

= 5 for orchestra

= 6 for choir

An increase of about 40 per cent is advisable at the lower frequencies. Another method uses a set of graphs, as shown in Fig. 3.9. A few simple calculations show that these give similar results in most cases.

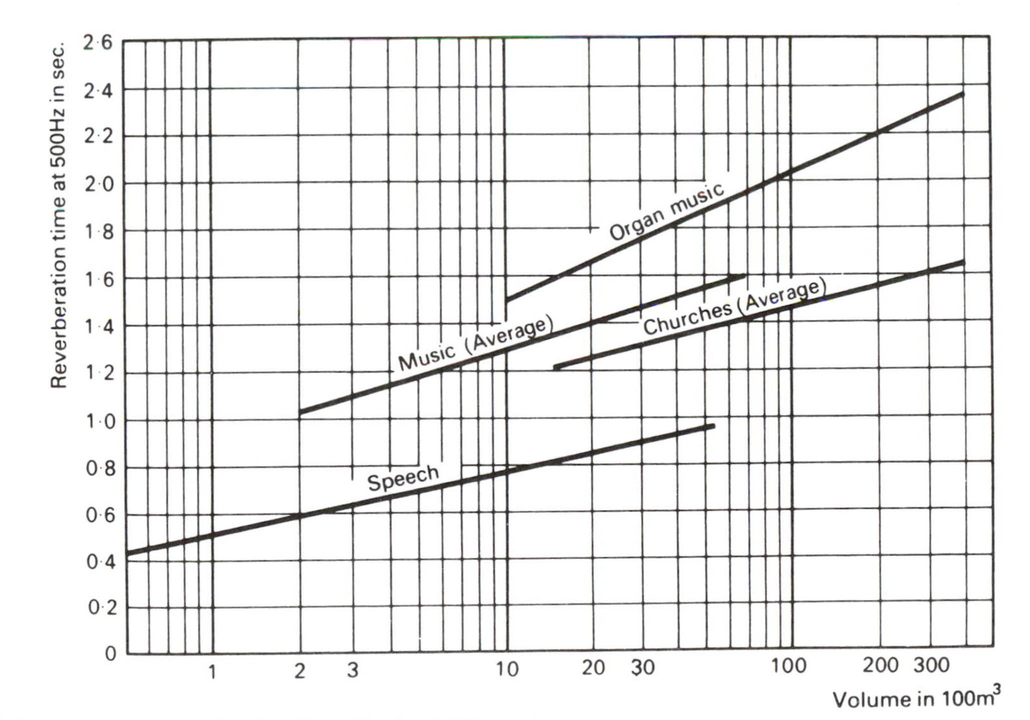

Fig. 3.9 Optimum reverberation time for auditoria of different sizes

Example 3.1

Suggest the optimum volume and reverberation time for a concert hall to be used mainly for orchestral music and to hold 450 people.

From Table 3.1 a volume of 7·5 m^3 per person would be reasonable. This gives a total volume of $450 \times 7 \cdot 5$ m^3 = 3375 m^3. Using the Stephens and Bate formula:

$$T = r(0 \cdot 012\sqrt[3]{V} + 0 \cdot 1070)$$

where $r = 5$ for orchestral music

optimum reverberation time

$$\begin{aligned} T &= 5(0 \cdot 012 \times \sqrt[3]{3375} + 0 \cdot 1070) \\ &= 5(0 \cdot 012 \times 15 + 0 \cdot 1070) \\ &= 5(0 \cdot 2840) \\ &= 1 \cdot 4 \text{ s} \end{aligned}$$

Sabine's formula The actual reverberation time is calculated using the Sabine formula:

$$T = \frac{0 \cdot 16 V}{A}$$

where T = reverberation time in seconds

V = volume of hall in m^3

A = absorption units in m^2

The number of absorption units is calculated using tables giving absorption coefficients for each of the materials used in the hall. The absorption coefficient for a material is the fraction of incident sound which is not reflected. Examples of these tables are given in Tables 3.4 and 3.5. The number of absorption units will normally be calculated for three or four selected frequencies: 125 Hz, 500 Hz, 2000 Hz and, if necessary 4000 Hz.

Example 3.2

Calculate the reverberation time at 125 Hz, 500 Hz and 2000 Hz for a hall of volume 2500 m^3 (to hold 250 people) having the following surface finishes:

Plaster on brickwork	265 m^2
3 mm glass window	43 m^2
Stage, boards on joist	70 m^2
25 mm wood-wool slabs	60 m^2
Plate glass screen	96 m^2
Ceiling plaster	310 m^2
Wood block floor	300 m^2

Assume the shading of the floor by the audience effectively reduces its absorption by 40 per cent at 125 Hz and 500 Hz, and by 60 per cent at 2000 Hz.

Total absorption at 125 Hz = 157·8 m^2
Total absorption at 500 Hz = 191·9 m^2
Total absorption at 2000 Hz = 224·6 m^2

The actual reverberation times by Sabine's formula, $T = (0 \cdot 16V)/A$, are

$$T = \frac{0 \cdot 16 \times 2500}{157 \cdot 8} = 2 \cdot 5 \text{ s for 125 Hz}$$

and $$T = \frac{0 \cdot 16 \times 2500}{191 \cdot 9} = 2 \cdot 1 \text{ s for 500 Hz}$$

and $$T = \frac{0 \cdot 16 \times 2500}{224 \cdot 6} = 1 \cdot 8 \text{ s for 2000 Hz}$$

Absorbent		Absorption					
		125 Hz		500 Hz		2000 Hz	
Item	Area (m^2)	Absorption coefficient	Absorption	Absorption coefficient	Absorption	Absorption coefficient	Absorption
Plaster	265	0·02	5·3	0·02	5·3	0·04	10·6
3 mm glass	43	0·3	12·9	0·1	4·3	0·05	2·2
Stage boards on joists	70	0·15	10·5	0·1	7·0	0·1	7·0
25 mm wood-wool slabs	60	0·1	6·0	0·4	24·0	0·6	36·0
Plate glass	96	0·1	9·6	0·04	3·8	0·02	1·9
Ceiling plaster	310	0·2	62·0	0·1	31·0	0·04	12·4
Wood block floor minus shading	300	0·05–40%	9·0	0·05–40%	9·0	0·1–60%	12·0
Audience	250	0·17/person	42·5	0·43/person	107·5	0·47/person	117·5
Air	2500 m^3	—	—	—	—	0·01	25·0
Total absorption			157·8		191·9		224·6

Table 3.4 Absorption coefficients of common building materials

Material and method of fixing	Absorption coefficients: Low frequency 125 Hz	Medium frequency 500 Hz	High frequencies 2000 Hz	High frequencies 4000 Hz
Boarded roof; underside of pitched slated or tile roof	0·15	0·1	0·1	0·1
Boarding ('match') about 20 mm thick over air space on solid wall	0·3	0·1	0·1	0·1
Brickwork — plain or painted	0·02	0·02	0·04	0·05
Clinker ('breeze') concrete unplastered	0·2	0·6	0·5	0·4
Carpet (medium) on solid concrete floor	0·1	0·3	0·5	0·6
Carpet (medium) on joist or board and batten floor	0·2	0·3	0·5	0·6
Concrete, constructional or tooled stone or granolithic finish	0·01	0·02	0·02	0·02
Cork slabs, wood blocks, linoleum or rubber flooring on solid floor (or wall)	0·05	0·05	0·1	0·1
Curtains (medium fabrics) hung straight and close to wall	0·05	0·25	0·3	0·4
Curtains (medium fabrics) hung in folds or spaced away from wall	0·1	0·4	0·5	0·6
Felt, hair, 25 mm thick, covered by perforated membrane (vis. muslin) on solid backing	0·1	0·7	0·8	0·8
Fibreboard (normal soft) 13 mm thick mounted on solid backing	0·05	0·15	0·3	0·3
Ditto, painted	0·05	0·1	0·15	0·15
Fibreboard (normal soft) 13 mm thick mounted over air space on solid backing or on joists or studs	0·3	0·3	0·3	0·3
Ditto, painted	0·3	0·15	0·1	0·1
Floor tiles (hard) or 'composition' flooring	0·03	0·03	0·05	0·05
Glass; windows glazed with up to 3 mm glass	0·2	0·1	0·05	0·02
Glass; 7 mm plate or thicker in large sheets	0·1	0·04	0·02	0·02
Glass used as a wall finish (viz. 'Vitrolite') or glazed tiles or polished marble or polished stone fixed to wall	0·01	0·01	0·01	0·01
Glass wool or mineral wool 25 mm thick on solid backing	0·2	0·7	0·9	0·8
Glass wool or mineral wool 50 mm thick on solid backing	0·3	0·8	0·75	0·9
Glass wool or mineral wool 25 mm thick mounted over air space on solid backing	0·4	0·8	0·9	0·8
Plaster, lime or gypsum on solid backing	0·02	0·02	0·04	0·04
Plaster, lime or gypsum on lath, over air space on solid backing, or on joists or studs including decorative fibrous and plaster board	0·3	0·1	0·04	0·04
Plaster, lime gypsum or fibrous, normal suspended ceiling with large air space above	0·2	0·1	0·04	0·04
Plywood mounted solidly	0·05	0·05	0·05	0·05
Plywood panels mounted over air space on solid backing, or mounted on studs, without porous material in air space	0·3	0·15	0·1	0·05
Plywood panels mounted over air space on solid backing, or mounted on studs with porous material in air space	0·4	0·15	0·1	0·05
Water — as in swimming baths	0·01	0·01	0·01	0·01
Wood boards on joists or battens	0·15	0·1	0·1	0·1
Wood-wool slabs 25 mm thick (unplastered) solidly mounted	0·1	0·4	0·6	0·6
Wood-wool slabs 80 mm thick (unplastered) solidly mounted	0·2	0·8	0·8	0·8
Wood-wool slabs 25 mm thick (unplastered) mounted over air space on solid backing	0·15	0·6	0·6	0·7

Table 3.5 Absorption of special items

	Absorption unit (m^2): Low frequency 125 Hz	Medium frequency 500 Hz	High frequencies 2000 Hz	High frequencies 4000 Hz
Air (per m^3)	—	—	0·007	0·020
Audience seated in fully upholstered seats (per person)	0·19	0·47	0·51	0·47
Audience seated in wooden or padded seats (per person)	0·16	0·4	0·43	0·4
Seats (unoccupied) fully upholstered (per seat)	0·12	0·28	0·31	0·37
Seats (unoccupied) wooden or padded or metal and canvas (per seat)	0·07	0·15	0·18	0·19
Theatre proscenium opening with average stage set (per m^2)	0·2	0·3	0·4	0·5

Eyring's formula Sabine's formula has the great advantage of simplicity and accuracy as long as the average absorption of all the surfaces within a room is less than about 0·2.

Sabine assumed that the sound within a room decayed continuously, whereas Eyring considered intermittent decay at reflections.

Let I_0 be the original intensity of sound at time $t = 0$ and the average absorption coefficient of the reflecting surface be $\bar{\alpha}$.

Therefore intensity after 1 reflection $= (1 - \bar{\alpha})I_0$

intensity after 2 reflections $= (1 - \bar{\alpha})^2 I_0$

intensity after n reflections $= (1 - \bar{\alpha})^n I_0$

But reverberation time is the time taken for a 60 dB decay or a decay to 10^{-6} of the original intensity

$$\therefore \quad (1 - \bar{\alpha})^n I_0 = 10^{-6} I_0$$

$$\therefore \quad (1 - \bar{\alpha})^n = 10^{-6}$$

$$\therefore \quad n \log_e(1 - \bar{\alpha}) = -6 \log_e 10$$

$$\text{or} \quad n = \frac{-6 \log_e 10}{\log_e(1 - \bar{\alpha})}$$

It can be shown that the mean free path in a rectangular room of volume V and surface area $S = 4V/S$ m. Hence, the average time between reflections

$$= \frac{4V}{S} \cdot \frac{1}{c}$$

$$= \frac{4V}{cS}$$

where c = velocity of sound in air

$$\therefore \quad n \text{ reflections take a time } \frac{4Vn}{cS} \text{ seconds}$$

$\therefore$ the reverberation time,

$$T = \frac{4V(-6 \log_e 10)}{cS \log_e (1 - \bar{\alpha})}$$

$$= \frac{-24V \log_e 10}{cS \log_e (1 - \bar{\alpha})} \qquad \left[c = 330 \text{ m/s} \quad \frac{24 \log_e 10}{330} = 0 \cdot 16 \right]$$

$$= \frac{0 \cdot 16V}{-S \log_e (1 - \bar{\alpha})}$$

It should be noticed that the expansion of $\log_e(1 - \bar{\alpha})$ is $-\bar{\alpha} - \bar{\alpha}^2/2 - \bar{\alpha}^3/3 - \bar{\alpha}^4/4$ etc. (Table 3.6), and for small values of $\bar{\alpha}$ (say 0·2) all terms after the first may be neglected. The reverberation time then becomes:

$$T = \frac{0 \cdot 16V}{(-s)(-\bar{\alpha})}$$

$$= \frac{0 \cdot 16V}{A}$$

A further correction may need to be added for higher frequencies to allow for air absorption.

$$\text{Then,} \quad T = \frac{0 \cdot 16V}{-S \log_e(1 - \bar{\alpha}) + xV}$$

where x is the sound absorption/unit volume of air (Table 3.7). If the value of $\bar{\alpha}$ is less than about 0·2 but frequencies above 1000 Hz are being considered, then a modified form of Sabine's formula is convenient:

Table 3.6 Values of $\log_e (1 - \bar{\alpha})$ corresponding to values of $\bar{\alpha}$

$\bar{\alpha}$	$-\log_e(1-\bar{\alpha})$	$\bar{\alpha}$	$-\log_e(1-\bar{\alpha})$	$\bar{\alpha}$	$-\log_e(1-\bar{\alpha})$
0·01	0·0100	0·21	0·2355	0·41	0·5270
0·02	0·0202	0·22	0·2482	0·42	0·5441
0·03	0·0304	0·23	0·2611	0·43	0·5615
0·04	0·0408	0·24	0·2741	0·44	0·5792
0·05	0·0513	0·25	0·2874	0·45	0·5972
0·06	0·0618	0·26	0·3008	0·46	0·6155
0·07	0·0725	0·27	0·3144	0·47	0·6342
0·08	0·0833	0·28	0·3281	0·48	0·6532
0·09	0·0942	0·29	0·3421	0·49	0·6726
0·10	0·1052	0·30	0·3565	0·50	0·6924
0·11	0·1164	0·31	0·3706	0·51	0·7125
0·12	0·1277	0·32	0·3852	0·52	0·7331
0·13	0·1391	0·33	0·4000	0·53	0·7542
0·14	0·1506	0·34	0·4151	0·54	0·7757
0·15	0·1623	0·35	0·4303	0·55	0·7976
0·16	0·1742	0·36	0·4458	0·56	0·8201
0·17	0·1861	0·37	0·4615	0·57	0·8430
0·18	0·1982	0·38	0·4775	0·58	0·8665
0·19	0·2105	0·39	0·4937	0·59	0·8906
0·20	0·2229	0·40	0·5103	0·60	0·9153

Table 3.7 x per m^3 at a temperature of 20°C

Frequency (Hz)	Relative humidity (%) 30	40	50	60	70	80
	$\times 10^{-3}$	$\times 10^{-3}$	$\times 10^{-3}$	$\times 10^{-3}$	$\times 10^{-3}$	$\times 10^{-3}$
1000	3·28	3·28	3·28	3·28	3·28	3·28
2000	11·48	8·2	8·2	6·56	6·56	6·56
4000	39·36	29·52	22·96	19·68	16·4	16·4

$$T = \frac{0{\cdot}16V}{A + xV}$$

Example 3.3

Calculate the amount of absorption contributed at 2000 Hz by the 30 000 m³ of air in Coventry Cathedral when the relative humidity is 60 per cent. If its reverberation time empty at 2000 Hz is 4 s, find the number of square metres of absorbent in the structure.

$$\text{Absorption} = 6{\cdot}56 \times 10^{-3} \times 30\,000 \text{ m}^2$$

$$= 6{\cdot}56 \times 30$$

$$= 197 \text{ m}^2$$

Using Sabine's formula:

$$T = \frac{0{\cdot}16V}{A + xV}$$

$$4 = \frac{0{\cdot}16 \times 30\,000}{A + 197}$$

$$A + 197 = \frac{0{\cdot}16 \times 30\,000}{4}$$

$$A = 4 \times 300 - 197$$

$$= 1200 - 197$$

$$= 1003 \text{ m}^2$$

(The limits of accuracy suggest an answer of about 1000 m^2.)

3.10 Measurement of reverberation time

The chief problem is the accurate measurement of very small times which may be as little as 0·5 s. Basically, the method is simple. A sound is produced of suitable amplitude and then the rate of decay of its different frequencies is measured, as shown in Fig. 3.3.

Ideally, the source of sound will be octave or third-octave band random noise from a white-noise generator, through an amplifier to a pair of loudspeakers, as shown in Fig. 3.10. This may be inconvenient and an alternative is to use a pistol as shown in Fig. 3.11.

The measurement can be done by the method shown in Fig. 3.10 or Fig. 3.11. This consists of a microphone connected to a frequency analyser connected in turn to a level recorder. The level recorder will need to have a logarithmic potentiometer in the circuit to convert the pressure measurement into dB. Modern microprocessor-based equipment is able to provide graphs of the decay curve on a video screen and automatically calculate values of reverberation time.

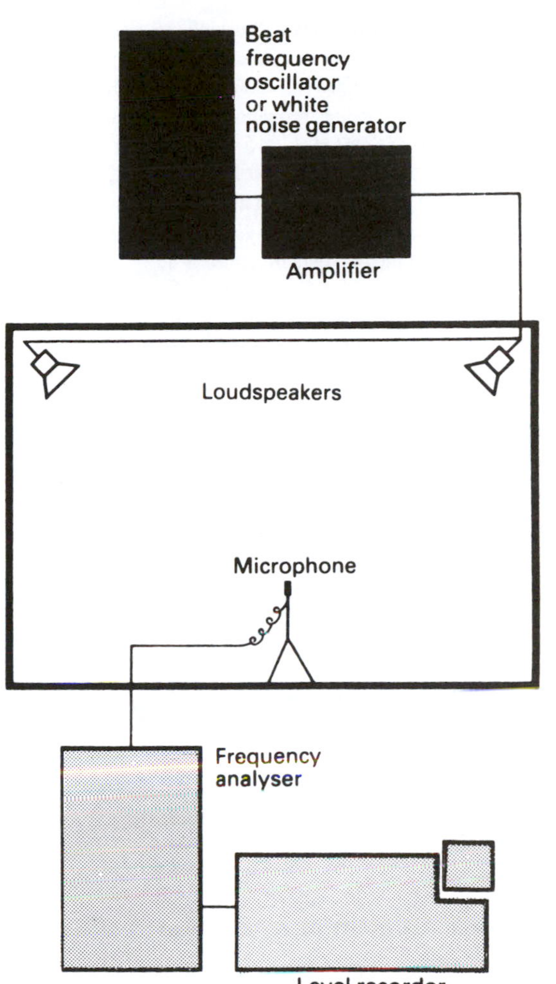

Fig. 3.10 Measurement of reverberation time using a beat-frequency oscillator or white-noise generator as sound source

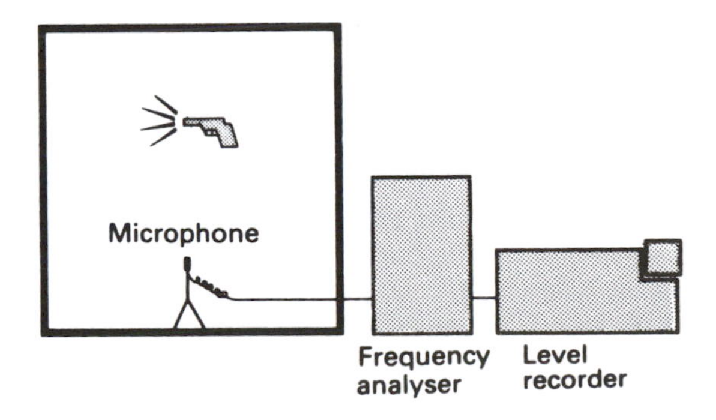

Fig. 3.11 Measurement of reverberation using a pistol as sound source

In each case measurements will usually be made in octave bandwidths, whose centre frequencies are 125 Hz, 250 Hz, 500 Hz, 1000 Hz, 2000 Hz, 4000 Hz or, in third-octave bandwidths, whose centre frequencies are 100 Hz, 125 Hz, 160 Hz, 200 Hz, 250 Hz, 315 Hz, 400 Hz, 500 Hz, 630 Hz, 800 Hz, 1000 Hz, 1250 Hz, 1600 Hz, 2000 Hz, 2500 Hz, 3150 Hz, 4000 Hz. This is done by switching the filters in the frequency analyser. A typical result for a certain room can be seen in Table 3.8.

Table 3.8 Typical reverberation times for a room

Third-octave bandwidth centre frequency (Hz)	Reverberation time (s)
100	1·55
125	1·60
160	1·45
200	1·30
250	1·20
315	1·05
400	1·05
500	1·00
630	1·10
800	1·00
1000	0·90
1250	1·05
1600	1·05
2000	1·05
2500	1·00
3150	0·95
4000	0·95

3.11 Measurement of absorption coefficient

The Sabine formula shows that the reverberation time in a room is proportional to its volume and inversely proportional to absorption. It is essential that the absorption of the surface finishes in a hall are known at the design stage so that the expected reverberation time may be calculated. For this a value for the material known as the absorption coefficient is needed. This is usually defined as the fraction of non-reflected sound energy to the incident sound energy. There are two main methods of measurement — the reverberation chamber or the impedance tube.

1 Reverberation chamber method

This is the better method because it allows for all angles of incidence, but has the disadvantage of requiring a room of about 200 m^3 so that measurements down to 100 Hz may be made. (The lowest frequency should not be lower than about $125\sqrt[3]{(180/V)}$ Hz to ensure a diffuse sound field, where V = the volume of the room.) The room itself will have walls and ceiling all slightly out of parallel. There will also be some diffusing objects in the form of curved sheets of plywood, or perspex a few millimetres thick. A long reverberation time is essential for accurate measurements.

The method is very simple. A measurement of reverberation time is made first without, and then with, the absorbent material in the chamber.

$$\therefore \quad T_1 = \frac{0\cdot 16V}{A}$$

$$\text{and} \quad T_2 = \frac{0\cdot 16A}{A + \delta A}$$

where V = volume of reverberation chamber

T_1 = reverberation time of chamber without absorbent material

T_2 = reverberation time of chamber with absorbent material

A = absorption of chamber without absorbent material

δA = extra absorption due to the material

$$\therefore \quad A = \frac{0\cdot 16V}{T_1} \qquad (1)$$

$$A + \delta A = \frac{0\cdot 16V}{T_2} \qquad (2)$$

$$\therefore \quad \delta A = 0\cdot 16V\left(\frac{1}{T_2} - \frac{1}{T_1}\right)$$

In practice, some slight correction needs to be made for the behaviour of sound in the chamber which can make a difference of nearly 5 per cent. These corrections are given in BS 3638:1987 Method for the measurement of sound absorption coefficients (ISO) in a reverberation room.

$$\text{Where } \delta A = \left(55\cdot 3\frac{V}{c}\right)\left(\frac{1}{T_2} - \frac{1}{T_1}\right)$$

and c = velocity of sound in air

$$\text{the absorption coefficient, } \alpha = \frac{\delta A}{S}$$

where S = surface area under measurement, which should be a single area between 10 and 12 m^2.

2 Impedance tube method

This method only measures the absorption coefficient at normal incidence. It is a useful indication of the sort of absorbent properties which a material may have. Its main use is in theoretical work, research work or in quality control for the production of acoustic absorbent materials.

Pure tones produced by an oscillator are used to excite the loudspeaker, as shown in Fig. 3.12, producing standing waves in the tube. Partial reflection will take place at the absorbent surface resulting in a standing wave pattern, as shown in Fig. 3.13.

If the displacement at any time of the incident wave is represented by

$$d_1 = a \sin(wt - kx) \qquad \left[k = \frac{2\pi}{\lambda},\ \omega = 2\pi f\right]$$

and the displacement of the reflected wave by

$$d_2 = fa \sin(wt + kx)$$

where a = initial maximum amplitude

fa = maximum amplitude of the reflected wave

then the resulting displacement at any point is given by

$$d = d_1 + d_2$$

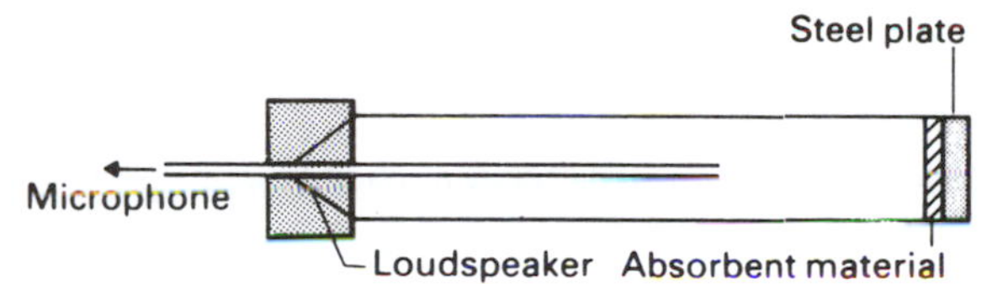

Fig. 3.12 Impedance or standing wave tube for absorption coefficient

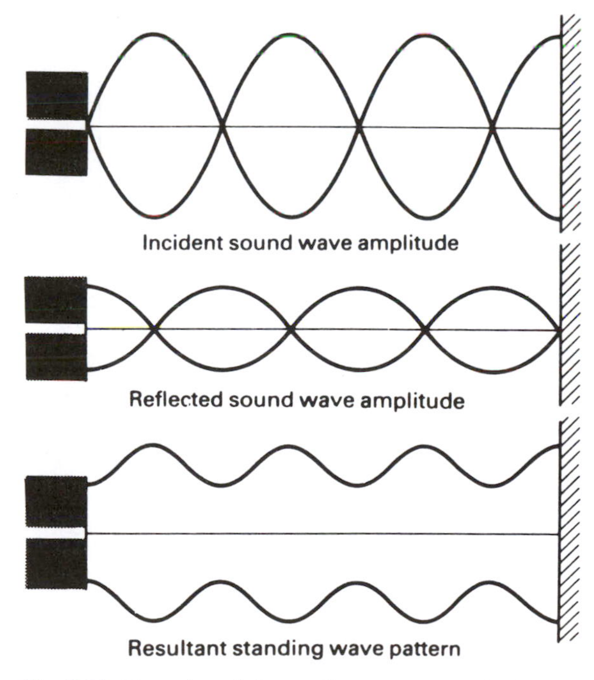

Fig. 3.13 Formation of the standing wave pattern in the impedance tube

$$= a \sin(wt - kx) + fa \sin(wt + kx)$$

$$= a(1 + f) \sin wt \cos kx + a(1 - f) \cos wt \sin kx$$

It can be seen that the maximum and minimum values will be $a(1 + f)$ and $a(1 - f)$ respectively, and $\lambda/4$ apart, the first being at 0, $\lambda/2$, λ, $3\lambda/2$, 2λ, etc., the second at $\lambda/4$, $3\lambda/4$, $5\lambda/4$, $7\lambda/4$, etc.

If the maximum and minimum amplitudes are A_1 and A_2,

$$\text{then } \frac{A_1}{A_2} = \frac{a(1 + f)}{a(1 - f)}$$

$$\text{or} \quad f = \frac{A_1 - A_2}{A_1 + A_2}$$

But the energy can be shown to be proportional to the square of the amplitude

$$\therefore \quad r = f^2 = \left(\frac{A_1 - A_2}{A_1 + A_2}\right)^2$$

where r is the fraction of reflected energy.

The absorption coefficient, $\alpha = 1 - r$

$$\therefore \quad \alpha = 1 - \left(\frac{A_1 - A_2}{A_1 + A_2}\right)^2$$

$$= \frac{(A_1 + A_2)^2 - (A_1 - A_2)^2}{(A_1 + A_2)^2}$$

$$= \frac{2A_1 2A_2}{(A_1 + A_2)^2}$$

$$= \frac{4A_1 A_2}{(A_1 + A_2)^2}$$

If the ratio of maximum:minimum, (A_1/A_2), is measured, the formula is more conveniently written

$$\frac{4\left(\frac{A_1}{A_2}\right)}{\left(\frac{A_1}{A_2} + 1\right)^2} = \frac{4}{\left(\frac{A_1}{A_2} + \frac{A_2}{A_1} + 2\right)}$$

It will normally be found that results from the standing wave tube, though reproducible, are less than those from the reverberation chamber. The size of the tube is also important. The maximum diameter of the sample should not be greater than about half of the wavelength under investigation. Thus, for measurements in a tube of 100 mm diameter, the upper limiting frequency is about 1600 Hz and for 6500 Hz the maximum diameter should be about 25 mm. It is possible to compare tube measurements of absorption coefficient with reverberation chamber measurements, but the values in the latter can vary depending upon the distribution of the absorbers in the room.

3.12 Types of absorbers

Absorbers may be divided into three main types:

1. Porous materials
2. Membrane absorbers
3. Helmholtz resonators

The porous materials usually have some absorption at all frequencies, membrane absorbers have far more critical absorption characteristics around the resonant frequency of the panel, whilst Helmholtz resonator absorbers have even more critical absorption characteristics.

1 Porous absorbers

These consist of such materials as fibreboard, mineral wools, insulation blankets, etc., and all have one important thing in common — their network of interlocking pores. They act by converting sound energy into heat. Materials with closed cells such as the foamed plastics are far less effective as absorbers. Typical characteristics are shown in Fig. 3.14. Sound absorption is far more efficient at high than low frequencies. It may be slightly improved by increased thickness or mounting with an airspace behind. Porous absorbers are available in three types: prefabricated

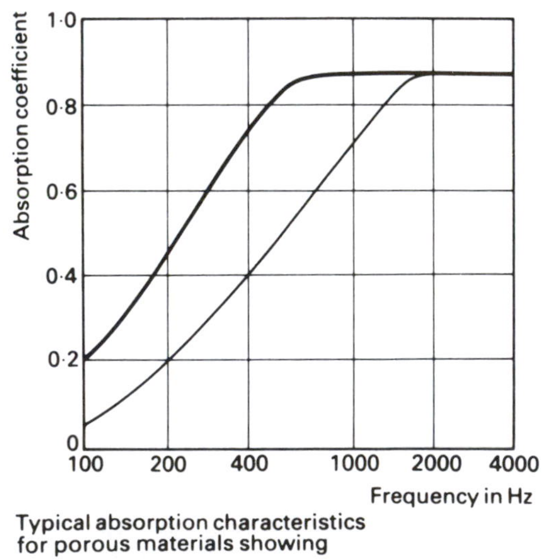

Fig. 3.14 Typical absorption characteristics for porous materials showing (1) thin material; (2) thick material with its increase in absorption at lower frequencies

(tiles), plasters and spray on materials, and acoustic blankets (glasswool).

The method of fixing can make a considerable difference to the efficiency of these materials, particularly at low frequencies. In general, it can be said that mounting the absorbent away from the wall surface results in a marked increase in low frequency absorption.

Where lack of space on walls or ceilings prevents the addition of absorbents, they may be used in the form of space absorbers. These can be made from perforated sheets of steel, aluminium or hardboard in various shapes, such as cubes, prisms, spheres or cones, and filled with glass-wool or other suitable material. It is possible to make them with the underside reflecting while the top is absorbent. This can be very helpful in preventing long delayed sound from a dome in a hall reaching the listeners and at the same time providing more reflection of sound to certain parts of the audience.

2 Membrane or panel absorbers

These are useful because they can have good absorption characteristics in the low frequency range. The absorption is highly dependent upon frequency and is normally in the range 50 to 500 Hz (see Fig. 3.15).

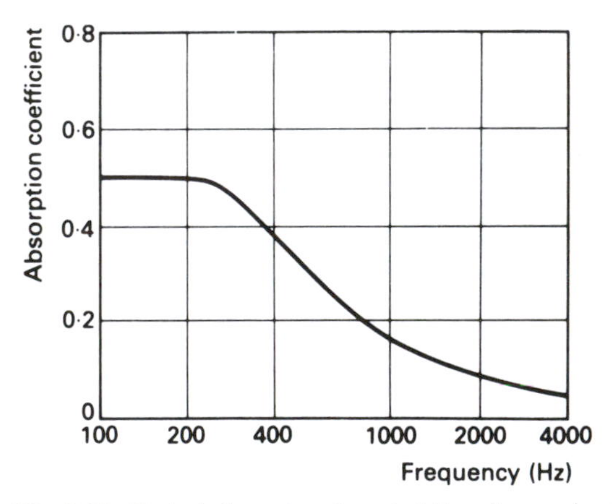

Fig. 3.15 Typical absorption characteristics of a membrane absorber showing increased efficiency at lower frequencies

The approximate resonant frequency, f, can be calculated from

$$f = \frac{60}{\sqrt{(md)}}$$

where m = mass of the panel in kg/m^2

and d = depth of the air space in m

In practice, this is only an approximation as the method of fixing and stiffness of individual panels can have a large effect. Panel absorbers can often be present fortuitously in the form of suspended ceilings or even closed double windows.

Example 3.4

A double-glazed window with internal glass of mass 7 kg/m^2 has an air gap of 200 mm and is lined with an acoustic absorbent. Find the approximate expected resonant frequency.

$$f = \frac{60}{\sqrt{(7 \times 0{\cdot}2)}}$$

$$= \frac{60}{\sqrt{1{\cdot}4}}$$

$$= \frac{60}{1{\cdot}183}$$

$$= 50{\cdot}72 \text{ Hz} \quad \text{(say 51 Hz)}$$

3 Helmholtz or cavity resonators

These are containers with a small open neck and they work by resonance of the air within the cavity. Porous material is introduced into the neck to increase the efficiency of absorption. It can be shown that for a narrow-necked resonator of the type shown in Fig. 3.16, the resonant frequency, f, is approximately

$$f = \frac{cr}{2\pi}\sqrt{\left[\frac{2\pi}{(2l + \pi r)V}\right]}$$

where c = velocity of sound in air

r = radius of neck

l = length of neck

V = volume of cavity

If there is no neck this reduces to:

$$f = \frac{c}{2\pi}\sqrt{\left[\frac{2r}{V}\right]}$$

Efficient absorption is only possible over a very narrow band, as shown in Fig. 3.17, and it is necessary to have many resonators tuned to slightly different frequencies if effective control of reverberation time is to be obtained. Cavity resonators are useful in controlling long reverberation times at isolated frequencies and are used in a number of concert halls.

3.13 Assisted resonance

In theory, as has been shown, it is possible to design a hall with the ideal amount of absorption. In some cases, variations in the amount of absorption with slight differences in construction can affect the final result. The optimum reverberation time is also dependent on the function of the hall (e.g. speech or music) so that some means of adjustment is useful.

One such system involves a time delay arrangement between a microphone and loudspeaker. The sound is played back at a suitably reduced amplitude after the necessary delay. The effect that must be conveyed subjectively is that of a steady decay of sound and not that of a delayed echo. Echo densities of 1000 per second or more may occur naturally in a hall. Subjectively, these are inter-

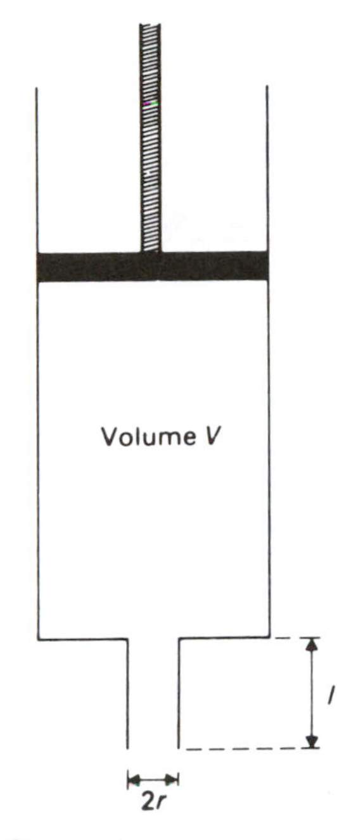

Fig. 3.16 Helmholtz resonator

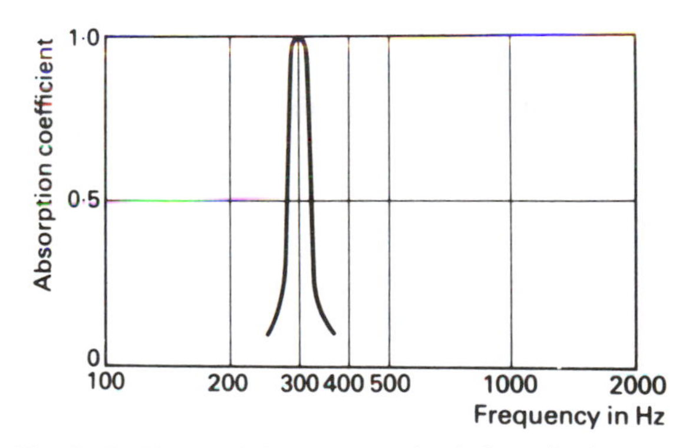

Fig. 3.17 Characteristic very narrowband absorption by Helmholtz or cavity resonator

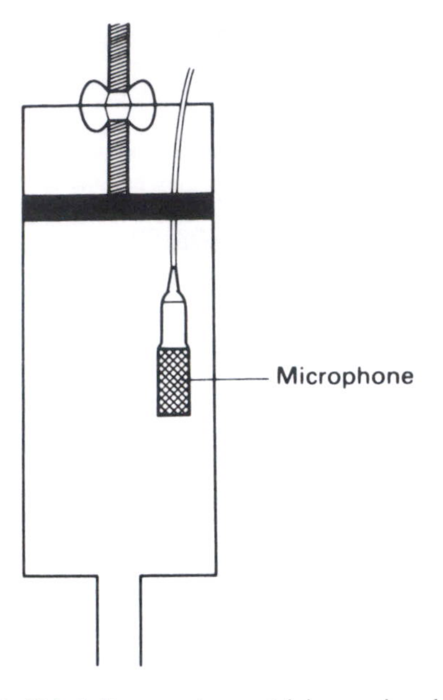

Fig. 3.18 Helmholtz resonator containing a microphone; the resonator can be adjusted to the correct frequency

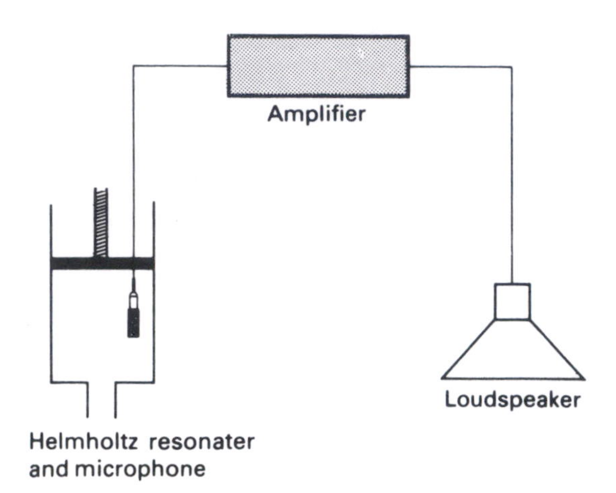

Fig. 3.19 Assisted resonance

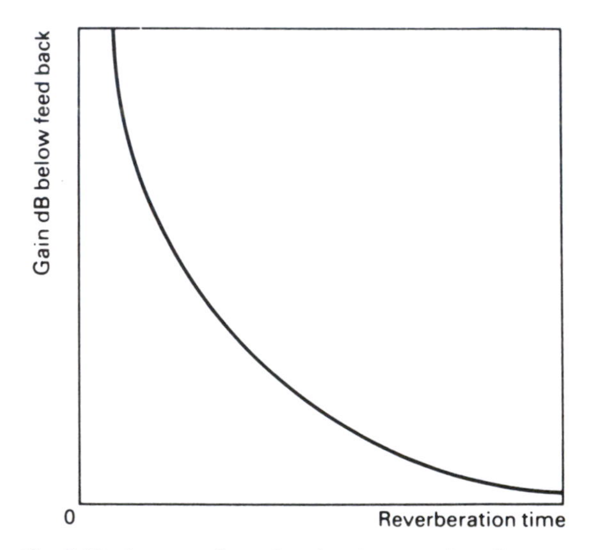

Fig. 3.20 Increase of reverberation time at a given frequency for change in amplifier gain with assisted resonance

preted as a continuous decay. Research done on artificial reverberation suggests that about 1000 echoes per second are needed for flutter-free reverberation and that the ear cannot distinguish any difference with a higher number. Echo densities of this magnitude are difficult to achieve economically. In addition, it should be remembered that having an isolated loudspeaker from which reverberation is controlled would give the wrong spatial distribution. Thus, it is necessary to have many loudspeakers each fed with different reverberated signals. This is known as ambiophonic reverberation. High quality equipment is needed throughout the whole chain of apparatus.

Another system which is in use in The Royal Festival Hall appears to be very successful. This method uses a separate circuit for each very narrow band of sound. The circuit consists of a microphone within a Helmholtz resonator (Fig. 3.18) connected to an amplifier (and filter) which feeds a loudspeaker (Fig. 3.19). The microphone and loudspeaker are set up at similar maximum mode positions for their given frequencies and control is achieved by the gain of the amplifier. To avoid altering the natural acoustics of the hall other than the reverberation time, it is important that the microphone and loudspeaker are exactly in phase or adjustment made to allow for this at the amplifier. In theory, it should be possible to increase the natural reverberation time to any value up to infinity (the feedback position). Control of the system becomes difficult if the gain is near to the feedback position, as shown in Fig. 3.20.

The Helmholtz resonator is acting as a narrowband filter, with the result that high quality microphones and loudspeakers are not required as a linear response is not needed. For a hall with a reverberation time of approximately 1·5 s it appears that one channel is needed for about each 3 Hz at the lower frequencies, decreasing to about one channel to 10 Hz at the higher frequencies.

Questions

(1) St Paul's Cathedral has a volume of about 150 000 m^3. Its reverberation time at 500 Hz is 11·7 s when empty, and 6·3 s when full. How many

people would you expect to be present when it is full? (Assume that each person contributes 0·4 m^2 of absorption units.)

(2) A college lecture theatre is to be built to hold 200 people and will be used mainly for speech.
   (a) Determine a suitable volume and reverberation time.
   (b) How many absorption units would be needed in the construction to achieve optimum conditions when the hall is about two-thirds full? (Assume that each person contributes 0·4 m^2 of absorption units at 500 Hz.)

(3) A certain concert hall has a volume of 5·7 m^3 per person and holds 1800 people. The reverberation time is 1·6 s at mid frequencies. Assuming that Sabine's formula is accurate and that to achieve fullness of sound an orchestra requires one instrument for each 20 m^2 of absorption units, find the optimum size of orchestra.

(4) The absorption coefficient of a certain material was measured in a reverberation chamber of volume 1300 m^3 and the following average reverberation times were obtained:

Frequency (Hz)	R.T. in s (empty)	R.T. in s with 30 m^2 absorber
125	16·8	10·2
250	20·1	10·4
500	18·5	9·4
1000	14·5	8·0
2000	9·1	6·1

The average absorption coefficient of all the surfaces is less than 0·2. Find the absorption coefficients of the material at the frequencies given.

(5) Calculate the optimum and actual reverberation times at 500 Hz for a hall of volume 2500 m^3 so as to be satisfactory for choral music using the following data. Allow 40 per cent for shading of the floor.

Data:

Item	Area (or no.) (m^2)	Absorption coefficient at 500 Hz
Wood block floor	160	0·05
Stage	80	0·30
Unoccupied seats	50	0·15/seat
Audience	200	0·4/person
Ceiling plaster	160	0·1
Canvas scenery	96	0·3
Perforated board	120	0·35
Glass	40	0·10
Plaster on brickwork	200	0·02

How many additional absorption units are required to make the actual reverberation time equal to the optimum?

(6) (a) Calculate the optimum reverberation time for speech in a hall of volume 4000 m^3.
   (b) Calculate the actual reverberation time at 500 Hz, in a hall with the following surface finishes and seating conditions:

Item	Absorption coefficient at 500 Hz
750 m^2 brick walls	0·02
540 m^2 plaster on solid backing	0·02
65 m^2 glass windows	0·10
70 m^2 curtain	0·40
130 m^2 acoustic board	0·70
300 m^2 wood block floor (allow 40 per cent for shading)	0·05

In addition there are 500 occupied seats each contributing 0·4 m^2 units of absorption. The volume of the hall is 4000 m^3.

(7) The reverberation time was measured in a lecture room of volume 150 m^3 and was found to be:

Octave band centre frequency (Hz)	R.T. in s
125	1·0
250	1·1
500	0·95
1000	1·0
2000	0·9
4000	0·8

Calculate the amount of extra absorption needed for each octave band.

(8) A room 16 m long, 10 m wide and 5 m high which was previously used as a laboratory is to be converted to use as a lecture room for 200 people. The original wall and floor surfaces are hard plaster and concrete whose average absorption coefficient is 0·05. Acoustical tiles of absorption coefficient 0·75 are available for wall or ceiling finishes. What is the desirable reverberation time for the new use of the room? Calculate the area of the tile to be applied to achieve this. Absorption of seated audience (per person) is 0·4 m^2 units.

(9) (a) A hall 60 m long, 25 m wide and 8 m high has seating for 1200, and generally hard surfaces whose average absorption coefficient is 0·05. Calculate the reverberation time with a two-thirds capacity audience for the frequency at which this data applies. The audience has an absorption of

0·4 m^2 units per person, but effectively reduces floor absorption by 40 per cent. The empty seats have an absorption of 0·28 m^2 units.

(b) What reduction in noise level would occur if the ceiling was then covered with acoustic tiles whose absorption coefficient is 0·6?

(10) A hall is to be built to hold about 600 people. Assuming that the normal use will be for speech, suggest the main points to be considered in order to obtain a suitable acoustic environment (without the use of amplification).

(11) A certain concert hall has a volume of 9 m^3 per person when full. It was designed to hold up to 1100 audience, 50 choir and 35 orchestral players. If the ratio of volume/absorption is 12·5/1, how does the actual reverberation time full compare with the optimum suggested by the Stephens and Bate formula?

What would the reverberation time during a solo performance be in the absence of choir or orchestra if each person contributes 0·4 m^2 units of absorption?

(12) A lecture room 16 m long, 12·5 m wide and 5 m high has a reverberation time of 0·7 s. Calculate the average absorption coefficient of the surfaces using the Eyring formula.

(13) (a) Describe the basic requirements for acceptable auditorium acoustics for speech.

(b) A theatre of dimensions 30 m long, 35 m wide and 10 m high has a seating capacity of 1000 persons on wooden chairs. The area of floor is 90 per cent of the total and consists of timber boards on joists. The remaining 10 per cent is stage. The walls and ceiling are concrete. The entire ceiling, rear wall and side walls are to be treated with the same material. What values of absorption coefficient are required for this material to achieve a reverberation time of 1 s, at each frequency, in the theatre with a capacity audience?

Material	Absorption coefficients 125 Hz	500 Hz	2 kHz
Timber boards on joists	0·15	0·1	0·1
Concrete	0·02	0·02	0·05
Stage	0·4	Nil	Nil
	Absorption units (m^2)		
Audience/person	0·15	0·4	0·45

(*HND*)

(14) A church is to be built at the bottom of a hill near a roundabout very close to the intersection of several busy main roads. It is to hold 250 people and can be expected to be full on most occasions. Suggest the main factors to achieve a good acoustical design. It may be assumed that a compromise will be made between the requirements for speech and those for choral music. (Each person is equivalent to 0·4 m^2 of absorption.)

(*HND*)

(15) It is common practice to use Sabine's formula to assess the reverberation time in small to medium size halls up to 1000 m^3.

(a) Explain the limitations of Sabine's formula and why it must be used with great care.

(b) Suggest any variations to the formula which could give a result more in keeping with the practical situation.

Sabine's formula: $RT = 0\cdot16\frac{V}{\alpha S}$

where RT = reverberation time in seconds
V = volume (m^3)
α = the arithmetically averaged absorption coefficient
S = surface area (m^2)

(*IOB*)

(16) (a) Describe the sound field produced by a source in a room. Explain how modification of the absorption properties of the room will affect this field.

(b) A room of 4000 m^3 has a reverberation time of 2 seconds. How much sound-absorbing material would need to be added to reduce the reverberant noise level by 6 dB? Of what importance is the noise spectrum and the absorption frequency characteristic?

(*IOB*)

(17) How can a scale model of a room be made to reproduce many of the acoustical characteristics of the full-sized original?

What precautions should be taken in the construction of the model and the execution of any test?

Give the advantages and disadvantages of different scale factors.

Discuss any advantages in the use of a medium other than air in the model.

(*IOA*)

(18) Reverberation time is defined as the time for a signal to decay to one-millionth of its original value. It is considered to be the most important single factor in acoustic design. Indicate to what extent you consider this statement to be true for:

(a) a large lecture theatre;
(b) a concert hall.

What other factors might also be taken into account in each case?

(*IOA*)

Table 3.9 Reverberation times (RT) in rooms (s)

Frequency (Hz)	125	500	2000	8000
Empty room RT	8·0	4·0	2·4	1·2
Carpeted room RT	6·4	1·8	0·7	0·65
Absorption coefficient (ISO values)	0·05	0·35	0·7	0·45

(19) Show with a block schematic diagram the equipment needed for the measurement of reverberation time and the principal requirements of each item. A simple rectangular room of dimensions 6 m × 5 m × 4 m high is carpeted. The reverberation times measured before and after carpeting are as given in Table 3.9. Compare your calculated values for the absorption coefficients with those given in the table for the same type of carpet measured in a reverberation chamber by the ISO method. Comment on the results.

(*IOA*)

(20) Discuss in detail one method for measuring the sound absorption coefficient of a material.

An empty enclosure with all its surfaces, including the floor, of similar material and construction has a reverberation time of 4·8 seconds. When 20 per cent of the original surfaces are covered by sound-absorbing material mounted on the walls and ceiling, a reverberation time of 1·2 seconds is obtained.

If the dimensions of the enclosure are 8 m by 6 m by 4 m high, calculate the absorption coefficients for:

(a) the original surfaces; and

(b) the sound-absorbing material.

(*IOA*)

4 Structure-borne sound

Dilational waves in fluids (gases and liquids) within the audible frequency range can be detected by the human ear. This is obviously not the case with solids, although they may be detected by a sense of touch. Vibrations in solids do not therefore, in the strict sense, come within the category of sound. However, because of much mathematical similarity and also because these vibrations usually cause excitation of fluids in contact with the solids, it is convenient to refer to them as sound. In talking of structure-borne or solid-borne sound it is the effect of the solid coupling mechanism between source and hearer which is the chief consideration.

In this mechanism there are three important parts:

1. the coupling between the source of vibration and the solid structure;
2. the type of vibrational waves produced in the solid;
3. the coupling with the fluid at the hearing end.

A full mathematical treatment of the different types of wave propagation is not given in this chapter and readers interested in this are referred to the appropriate works on the subject.

4.1 Wave propagation in solids

There are three main types of waves formed in solids: longitudinal, transverse and bending. In general, the longitudinal waves are modified at all but high frequencies due to the limited thickness dimension of the solid. They do not cause much direct radiation of sound into the air, but are important because of their ability to excite other parts of the structure into bending vibrations. It is these bending waves which are the most important because of their ability to radiate sound.

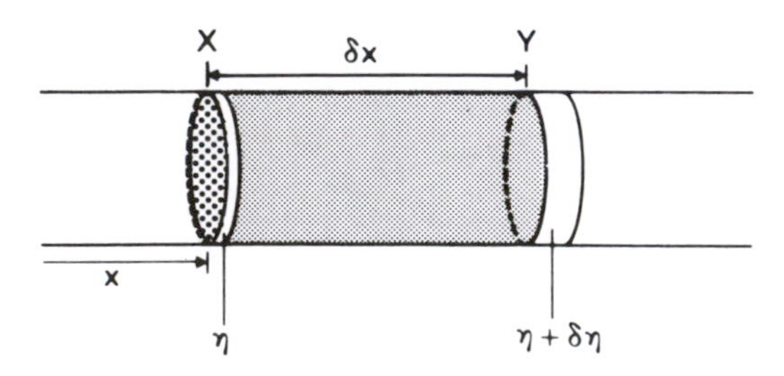

Fig. 4.1 Transmission of longitudinal vibrations in a bar

1 Longitudinal waves

Velocity of sound in a bar Consider a thin uniform rod of cross-sectional area A subject to longitudinal vibration (see Fig. 4.1). At some instant in time point X is subject to a horizontal displacement η from its original position x from the end of the rod. In similar manner the displacement of point Y is $\eta + \delta\eta$, the original distance between X and Y being δx.

$$\therefore \quad \text{the strain at X at this instant} = \frac{\partial \eta}{\partial x}$$

$$\text{and the rate of change of strain} = \frac{\partial^n \eta}{\partial x^2}$$

$$\therefore \quad \text{the strain at Y} = \frac{\partial \eta}{\partial x} + \frac{\partial^2 \eta}{\partial x^2}\,\delta x$$

$$\therefore \quad \text{the stress at X} = E\frac{\partial \eta}{\partial x}$$

$$\text{and the stress at Y} = E\frac{\partial \eta}{\partial x} + E\frac{\partial^2 \eta}{\partial x^2}\,\delta x$$

$$\therefore \quad \text{the change of stress} = E\frac{\partial^2 \eta}{\partial x^2}\,\delta x$$

$$\therefore \quad \text{the force between the two points} = AE\frac{\partial^2 \eta}{\partial x^2}\,\delta x$$

By Newton's law, force = mass × acceleration

$$= \rho A \delta x \frac{\partial^2 \eta}{\partial t^2}$$

where ρ = density

t = time

A = cross-sectional area

E = Young's modulus of elasticity

$$\therefore \quad \rho A \delta x \frac{\partial^2 \eta}{\partial t^2} = AE \frac{\partial^2 \eta}{\partial x^2} \delta x$$

$$\frac{\partial^2 \eta}{\partial t^2} = \frac{E}{\rho} \frac{\partial^2 \eta}{\partial x^2}$$

This is the wave equation for longitudinal waves in a bar. Using this equation it can be shown that the velocity, c, of waves in the bar is given by:

$$c = \sqrt{\left(\frac{E}{\rho}\right)}$$

In practice this relationship can be used to find Young's modulus of elasticity for materials such as concrete and glass. Table 4.1 gives values for the velocity of sound in some common materials.

Table 4.1 Velocity of sound in some common materials

Material	Velocity in bar (m/s)
Glass	5300
Steel	5200
Aluminium	5100
Cast iron	3400
Reinforced concrete	3700
Bricks with mortar	2350
Wood — with grain	3600–4900
across grain	1000–2500
Rubber, hard	1400
soft	50
Sand	97–210

Velocity of sound in plates The derivation is exactly similar to that in the preceding section, but taking into account the fact that the material will be restrained in another direction, and will thus be in a state of biaxial stress. Under these conditions E must be replaced by E', where:

$$E' = \frac{E}{(1 - \mu^2)}$$

where μ = Poisson's ratio

$$\therefore \quad V = \sqrt{\left[\frac{E}{\rho(1 - \mu^2)}\right]}$$

Infinite media In this case the material will be in triaxial stress and the relationship becomes:

$$V = \sqrt{\left[\frac{E(1 - \mu)}{(1 + \mu)(1 - 2\mu)\rho}\right]}$$

It will be clear to the reader that this last case is not normally important in building structures. In the case of a brick wall of thickness 250 mm with a velocity in the region of 2500 m/s these conditions would only be approached at frequencies above 10 kHz.

The velocity of sound in solids will be of the order of ten times higher than in air, and Poisson's ratio will be of the order of 0·3 for most building materials. The latter makes little difference to the velocity.

2 Transverse waves

A solid is able to resist both change of volume and change in shape. As a result of the latter it can transmit tangential (shear) stresses. It can be shown that the velocity of propagation is:

$$V_T = \sqrt{\left[\frac{E}{2(1 + \mu)\rho}\right]}$$

This gives a velocity of propagation very much lower than that for longitudinal waves. When $\mu = 0{\cdot}3$ the velocity of transverse waves will be about half that of longitudinal waves.

3 Bending waves

These are given a distinctive name because although transverse in appearance they are quite different from the waves that have just been considered (see Fig. 4.2). They are also by far the most important by virtue of their large deflections and hence their efficient radiating properties. There is a very small driving impedance. (Acoustic impedance is the ratio of sound pressure at the surface to the strength of the source.)

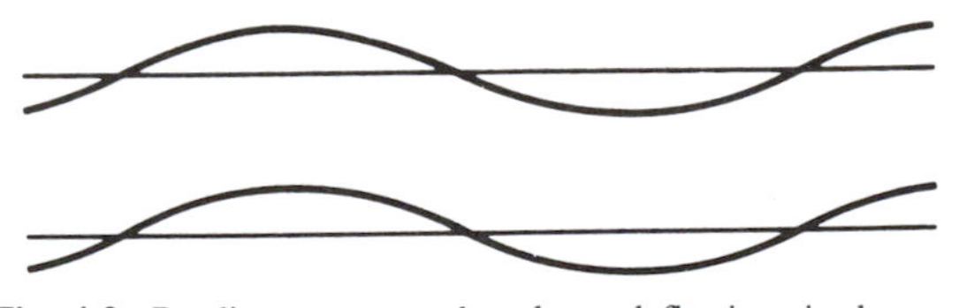

Fig. 4.2 Bending waves produce large deflections in the structure and cause a considerable radiation of sound energy

The velocity of bending waves C_B are frequency dependent and given by:

$$C_B = \left(\frac{EI}{A\rho}\right)^{1/4} \omega^{1/2}$$

where E = Young's modulus of elasticity

I = second moment of area

A = cross-sectional area of plate, beam or shell

ρ = density

ω = frequency in radians/s

Alternatively, this phase velocity of bending waves may be related to the longitudinal bar velocity, C_L, by:

$$C_B = (C_L\omega)^{1/2}\left(\frac{I}{A}\right)^{1/4}$$

In the case of a flat plate this can be reduced to:

$$C_B = (1\cdot 8hfC_L)^{1/2}$$

where h = thickness of the plate.

4.2 Attenuation of solid-borne waves

Metals such as aluminium and steel have an attenuation less than air but greater than water. This means that for structural members there is very little inherent attenuation. Fortunately some is provided by joints and changes in dimension. Losses in building materials are often fairly large, particularly in certain granular and plastic materials.

Limited available data suggests that the attenuation by building materials such as concrete or brick is only of the order of 10^{-2} to 10^{-1} dB/m. In practice it would seem possible to obtain much larger losses from 1 to 3 dB/m due to other factors. These include friction at joints and changes of cross-section. Attenuation of bending waves is about half that of longitudinal waves.

4.3 Changes of cross-section

The reduction factor is frequency independent for both longitudinal and bending waves. Figure 4.3 shows the reduction due to change in cross-section. It is fairly clear that for the orders of change that are practical in building construction the reduction is negligible.

4.4 Corners

Reduction achieved by corners is dependent upon the fixing conditions. In the case of two identical beams joined by a swivel link at right angles, the reduction R, in dB, can be shown to be:

$$R = 10\log_{10}\left[\frac{2\dfrac{C_B}{C_L}}{1+\dfrac{C_B}{C_L}+\dfrac{1}{2}\left(\dfrac{C_B}{C_L}\right)^2}\right]$$

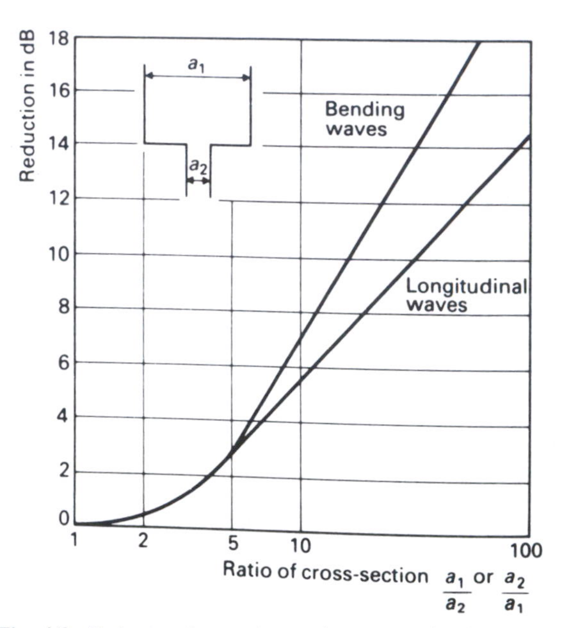

Fig. 4.3 Reduction due to changes in cross-sectional area

$$= 10\log_{10}\left[\frac{2\left(1\cdot 8\dfrac{hf}{C_L}\right)^{1/2}}{1+\left(1\cdot 8\dfrac{hf}{C_L}\right)^{1/2}+0\cdot 9\dfrac{hf}{C_L}}\right]$$

This reduction is reversible in that it applies to a longitudinal wave producing a bending wave or vice versa.

If the two bars were rigidly joined then the mathematical analysis is far more complicated because there are six boundary conditions to be considered. When the two members are identical a 3 dB reduction results. Similarly, at a T-junction a 6·5 dB reduction between the vertical and the arms of the T is found. In the case of X-junctions a 9 dB reduction is obtained.

It is fairly clear that the practical reduction achieved is quite small, and where compressional waves produce bending waves even of lower amplitude an increased level of room noise may result. A small reduction may be all that is needed. Thoughtful positioning of domestic central heating pumps and convection-preventing valves in relation to bends can prevent an annoying amount of noise being radiated into living rooms.

4.5 Change of medium

The reflection coefficient within a medium for normal incidence:

$$r = \frac{\text{reflected sound energy}}{\text{incident sound energy}}$$

Table 4.2 Coupling of the radiating surface to air

	Air 18°C	Water	Concrete	Iron
Density (kg/m^3)	1·2	1000	2300	7800
Stiffness for longitudinal waves (kg/ms)	$1{\cdot}4 \times 10^5$	$2{\cdot}1 \times 10^9$	$2{\cdot}3 \times 10^{10}$	$2{\cdot}2 \times 10^{11}$
Longitudinal velocity (m/s)	342	1460	3160	5900
Impedance (kg/m^{2}s)	410	$1{\cdot}46 \times 10^6$	$7{\cdot}3 \times 10^6$	$4{\cdot}6 \times 10^7$

$$= \frac{(R_2 - R_1)^2}{(R_2 + R_1)^2}$$

where R_1 and R_2 are the characteristic impedances of the two materials and may be calculated by the product ρC (density × velocity). Hence

$$r = \left(\frac{\rho_2 C_2 - \rho_1 C_1}{\rho_2 C_2 + \rho_1 C_1}\right)^2 \quad \text{(see Table 4.2)}$$

It does not matter whether the sound is travelling from a more dense to a less dense medium or vice versa. The applications are limited and the formula only applies under conditions where the two media may be considered infinite. The reason for this is that with thin sections, such as wall partitions, the reflected waves will have an interfering effect. If this is taken into account and the complex series summed then it can be shown that:

$$T = \frac{1}{1 + (\omega m/2R_1)^2}$$

where $T = 1 - r$, and is known as the transmission coefficient. It is normal to consider the transmission coefficient rather than the reflection coefficient.

ω = angular velocity in radians/s

m = mass of the partition/unit area

$R_1 = \rho c$ = characteristic impedance of air

In practice this can be reduced to:

$$T = \frac{1}{(\omega m/2R_1)^2}$$

as the second term is large compared with 1.

The actual reduction is most conveniently put on a logarithmic scale in dB and becomes:

$$\text{Reduction in dB, } R = 10 \log_{10} \frac{1}{T}$$

$$= 10 \log_{10} \left(\frac{\omega m}{2R_1}\right)^2$$

$$= 20 \log_{10} \left(\frac{\omega m}{2\rho c}\right)$$

This shows that in theory doubling the frequency or the mass should have the same effect, about 6 dB better insulation. Unfortunately, the theory does not agree with measurements made on many samples. In practice sound is incident at all angles and not just at right angles.

For other angles of incidence to the normal, the transmission coefficient,

$$T_\theta = \frac{1}{1 + (\omega m \cos \theta/2\rho c)^2}$$

By summing the transmission for θ from 0 to 90°, it can be shown that the reduction of sound in dB for all angles of incidence,

$$R = R_0 - 10 \log_{10} 0{\cdot}23 R_0$$

where R_0 = reduction of sound at normal incidence

$$= 10 \log_{10} \left(\frac{\omega m}{2\rho c}\right)^2$$

The result gives an increase of about 5 dB per doubling of frequency or mass per unit area. This is much nearer the measured value. In fact, integration over the full range 0 to 90° is not quite correct as little sound would be present at grazing incidence.

4.6 Statistical energy analysis (SEA)

It is much more difficult to predict structure-borne noise levels than airborne sound transmission (see Chapter 5). However, in recent years the technique of statistical energy analysis has been increasingly developed for this purpose, a subject beyond the scope of this text.

4.7 Resonance of partitions

In the previous section it was assumed that a panel was completely controlled by its mass. This is certainly true for most practical purposes over much of its frequency range, which for many constructions is within limits of audibility. Below these frequencies a panel becomes largely stiffness-controlled and acts in the manner of a spring. This means that there will be resonant frequencies which depend on stiffness and fixing conditions as well as mass.

Fig. 4.4 Patterns are formed in the sand at the resonant frequencies of a plate set in vibration using a vibrator controlled by an audio-frequency generator

These resonant conditions can be demonstrated by means of a rectangular metal plate clamped horizontally through a centre hole to a vibrator (Goodman or Advance). The frequency may be adjusted by a simple audio-frequency generator. Sand placed on the plate will form patterns at certain frequencies, the lowest resonant condition producing the simplest. If the edges are fixed the resonant frequencies are changed and a different pattern is formed (see Fig. 4.4).

If a rectangular panel is supported at its edges but not clamped, its resonant frequencies may be calculated from:

$$f = 0{\cdot}45 C_L h \left[\left(\frac{N_x}{l_x}\right)^2 + \left(\frac{N_y}{l_y}\right)^2\right] \text{ Hz}$$

where C_L = longitudinal velocity in m/s

h = thickness in m

l_x = length in m

l_y = height in m

N_x, N_y = integers 1, 2, 3, etc., the lowest, 1 and 1, giving the lowest resonant frequency

Example 4.1

Calculate the lowest resonant frequency for a brick partition 120 mm thick, 4 m by 2 m in area with a longitudinal wave velocity of 2350 m/s. (Assume it is supported at its edges.)

$$f = 0{\cdot}45 \times 2350 \times 0{\cdot}12[(\tfrac{1}{4})^2 + (\tfrac{1}{2})^2]$$

$$= 0{\cdot}45 \times 2350 \times 0{\cdot}12 \times \tfrac{5}{16}$$

$$= 40 \text{ Hz}$$

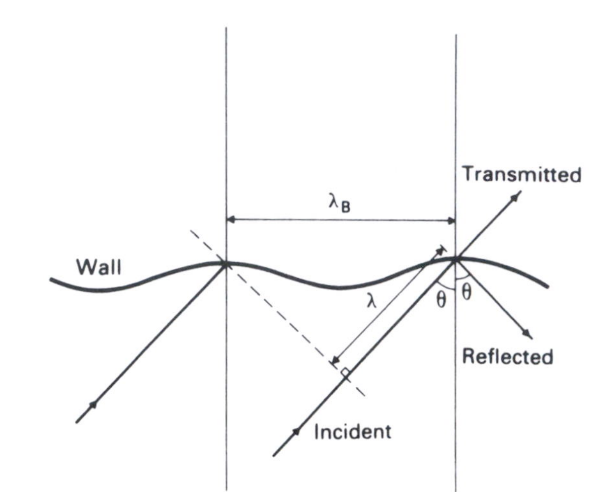

Fig. 4.5 Coincidence frequency. Incident sound wave reaches the wall at an angle and the projection of the natural wavelength of vibration of the wall is equal to the sound wavelength

4.8 Coincidence frequency

Under certain conditions the projected wavelength of incident sound is exactly equal to the length of resonant bending waves in the partition, as shown in Fig. 4.5.

$$\lambda_B \sin\theta = \lambda$$

$$\therefore \quad \lambda_B = \frac{\lambda}{\sin\theta}$$

where λ_B = bending wavelength

λ = wavelength of sound in air

θ = angle of incidence

but $C_B = (1{\cdot}8hfC_L)^{1/2}$

$C_B = f\lambda_B$

$c = f\lambda$

$$\therefore \quad C_B = \frac{\lambda_B c}{\lambda}$$

$$= \frac{c}{\sin\theta}$$

$$\therefore \quad \frac{c}{\sin\theta} = (1{\cdot}8hfC_L)^{1/2}$$

$$\therefore \quad f = \frac{c^2}{1{\cdot}8hC_L \sin^2\theta}$$

where c = velocity of sound in air

C_B = bending wave velocity

h = thickness of panel

C_L = longitudinal wave velocity

This frequency f at which the effect is a maximum is known as the coincidence frequency. Because f is inversely proportional to C_L, and C_L is related to stiffness, the greater the stiffness the lower the coincidence frequency.

4.9 Critical frequency

This is the lowest frequency at which coincidence occurs.

Thus $\sin\theta = 1$

$\therefore \quad \theta = 90°$

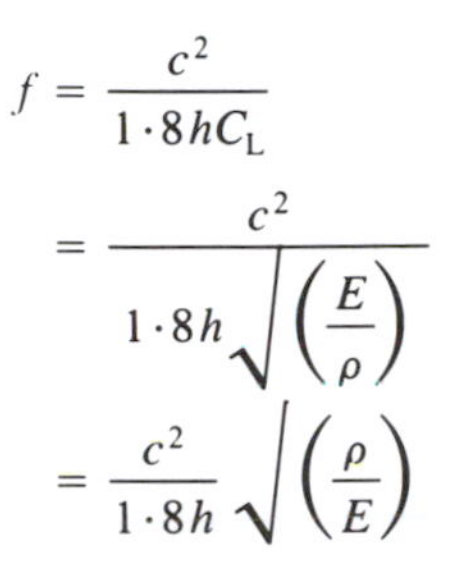

$$f = \frac{c^2}{1\cdot 8hC_L}$$

$$= \frac{c^2}{1\cdot 8h\sqrt{\left(\frac{E}{\rho}\right)}}$$

$$= \frac{c^2}{1\cdot 8h}\sqrt{\left(\frac{\rho}{E}\right)}$$

Example 4.2

What is the expected critical frequency for a 120 mm thick brick wall? Assume a longitudinal wave velocity in brick of 2350 m/s and that the velocity of sound in air is 330 m/s.

$$f = \frac{330^2}{1\cdot 8 \times 0\cdot 120 \times 2350}$$

$$= 214\cdot 5 \text{ Hz}$$

Critical frequency is also considered in Chapters 5 and 8.

4.10 Coupling of source to the structure

This is exactly similar to the mass law part of the preceding section except that the other media will not always be air. It is essential to obtain a high-impedance mismatch between the source and the structure. In the case of machinery a resilient mount will normally be needed.

It can be seen from Fig. 4.6 that the natural frequency of the resilient mount must be much lower than the driving frequency of the machine. The resonant frequency of the mount can be found from its static deflection:

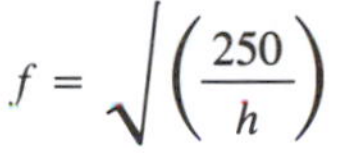

$$f = \sqrt{\left(\frac{250}{h}\right)}$$

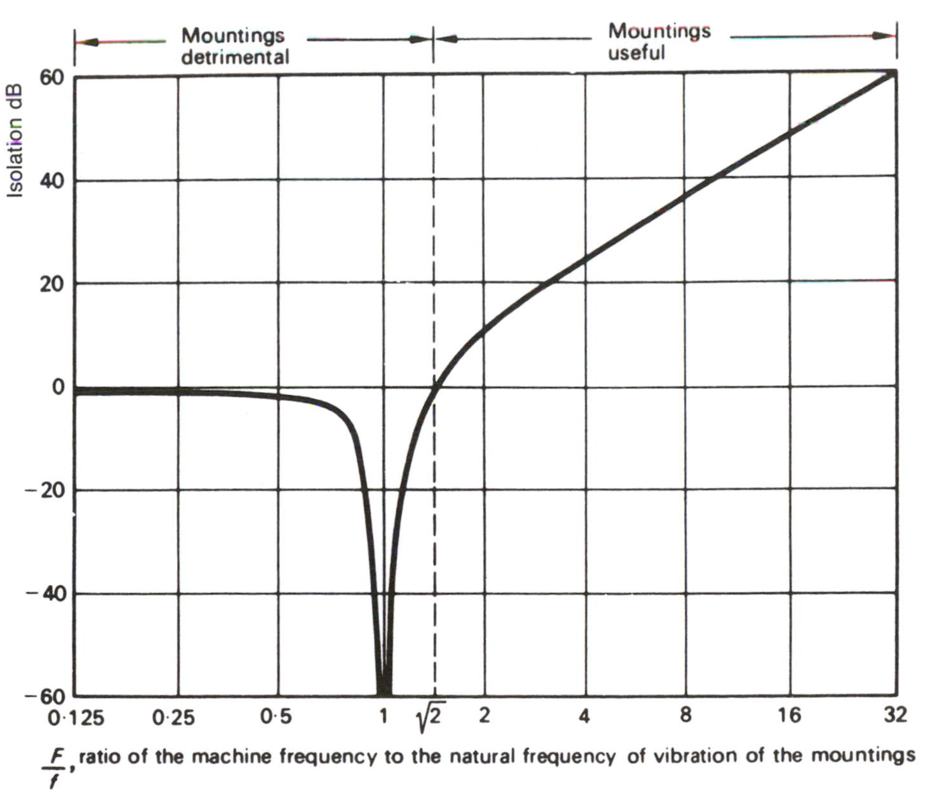

Fig. 4.6 Energy reduction in decibels by the use of resilient machine mountings

where h = static deflection in mm under the machine.

Deflections may be much greater in the case of non-elastic materials such as rubber ($\times 2$) and cork ($\times 4$).

Example 4.3

A certain machine with a slightly out-of-balance motor rotating at 1800/min is fixed on a perfectly elastic mount with a static compression of 2·50 mm. Calculate the reduction in dB by use of the mounting.

$$\text{Forcing frequency, } F = \frac{1800}{60} = 30 \text{ Hz}$$

Resonant frequency of mount,

$$f = \sqrt{\left(\frac{250}{2 \cdot 50}\right)} = \sqrt{100} = 10 \text{ Hz}$$

$$\text{Ratio } \frac{F}{f} = \frac{30}{10} = 3$$

∴ Isolation (from Fig. 4.6) = 17·5 dB

4.11 Impact sound insulation

The most common cause of structure-borne sound is that of footsteps, particularly in multi-storey dwellings. Clearly what is needed is a suitably large impedance between the source of impact and the radiating surfaces. This may be achieved by mass, by the use of resilient materials or a combination of both. Lightweight constructions are likely to cause problems because they need to be very stiff for structural reasons, with a consequential loss in sound insulation.

4.12 Measurement of impact sound insulation

With impact sound it is the level of noise produced in the receiving room which is of interest and not the level of the impact sound itself. The standard method is to produce a constant level by impact with the source-room floor. The level in the receiving room is measured and this is compared with levels known to be satisfactory (see Fig. 4.7).

The microphone and measuring apparatus are first calibrated at a known frequency by a pistonphone. This known level is marked on to paper using a level recorder. With the standard footsteps machine switched on, the levels in the lower room are measured in appropriate frequency

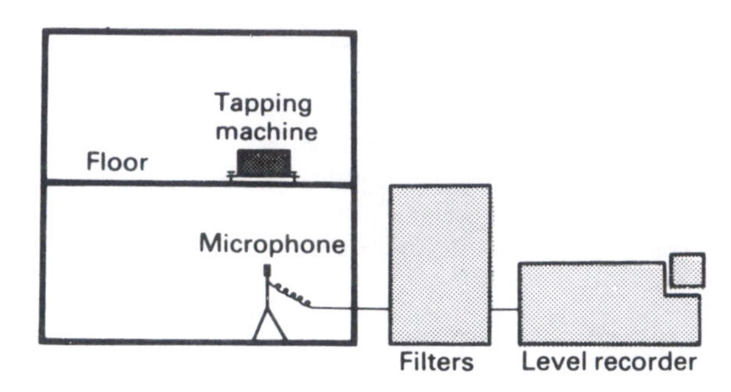

Fig. 4.7 Measurement of impact sound insulation

intervals. These may be in octave or third-octave bandwidths from 100 to 3150 Hz.

Usually six measurements will be made in each frequency band up to 500 Hz and three in each band above. The average should be calculated from

$$L_R = 10 \log_{10} \frac{p_1^2 + p_2^2 + \ldots + p_n^2}{np_0^2}$$

$$= 10 \log_{10} \left(\frac{10^{L_1/10} + 10^{L_2/10} + \ldots + 10^{L_n/10}}{n} \right)$$

where p_0 = reference sound pressure level
$= 2 \times 10^{-5}$ Pa

This is long and tedious, and provided that the maximum differences are less than 6 dB, it is sufficient to take the decibel average.

The results obtained in this way would be affected by the absorption of sound in the receiving room and not merely by the effectiveness of the floor construction. For comparison purposes the results are standardized to a fixed amount of room absorption. The room absorption is compared to 10 m² of absorption. This is the amount that would be present in the living room of an average dwelling.

$$\therefore \quad L_c = L_R - 10 \log_{10} \frac{10}{A}$$

where L_c = corrected level
L_R = measured level
A = area of absorption in the receiving room in m²

In Britain, the correction to allow for the reverberation time of the receiving room is slightly different for domestic dwellings. It has been found that most dwelling rooms have a reverberation time very close to 0·5 s. For simplicity, insulation standards between domestic dwellings are corrected to this 0·5 s reverberation time.

$$\text{Hence,}\quad L_c = L_R - 10\log_{10}\left(\frac{T}{0\cdot 5}\right)$$

where T = actual reverberation time of the receiving room in s.

On the assumption that Sabine's formula is correct, the two corrections are identical:

$$\frac{T}{0\cdot 5} = \frac{0\cdot 16V/A}{0\cdot 16V/10}$$

$$\frac{T}{0\cdot 5} = \frac{10}{A}$$

Departures from Sabine's formula explained in the previous chapter will lead to minor discrepancies in the correction term.

4.13 Impact sound insulation standards

A floor between two flats or maisonettes clearly has to minimize the impact sound and also to provide adequate reduction of airborne sound. The Building Regulations Part E on sound in the approved document indicate two methods of showing compliance. Firstly, by copying a standard form of floor construction or using a form of floor construction that is tested and achieves an acceptable maximum impact level, known as the weighted standardized sound pressure level. This figure is obtained by comparing the corrected measured levels, $L_c = L_R - 10\log_{10}(T/0\cdot 5)$, at each third-octave with the ISO reference levels (shown in Table 4.4 and Figs 4.8 and 4.9). This comparison may be carried out either by calculation or using the graph. The corrected measured levels in the receiving room are plotted then the reference graph (Fig. 4.8) is adjusted up or down in 1 dB steps until, on average, it is 2 dB below the plotted graph. The decibel figure at 500 Hz identifies the weighted standardized impact sound pressure level. The maximum values allowed by the Building Regulations are shown in Table 4.3.

Example 4.4

The mean corrected levels of measurements in eight receiving rooms found from the impact results are shown in the table below. Calculate whether this floor construction complies with the requirements of the Building Regulations.

This exceeds the maximum of a 2 dB average unfavourable deviation (total 32 dB) and therefore does not comply with the maximum weighted standardized sound pressure level of 62 dB required by the Building Regulations. Comparison with the '63 dB curve' gives a total adverse deviation of 31 dB, ie less than 32 dB, and so the weighted

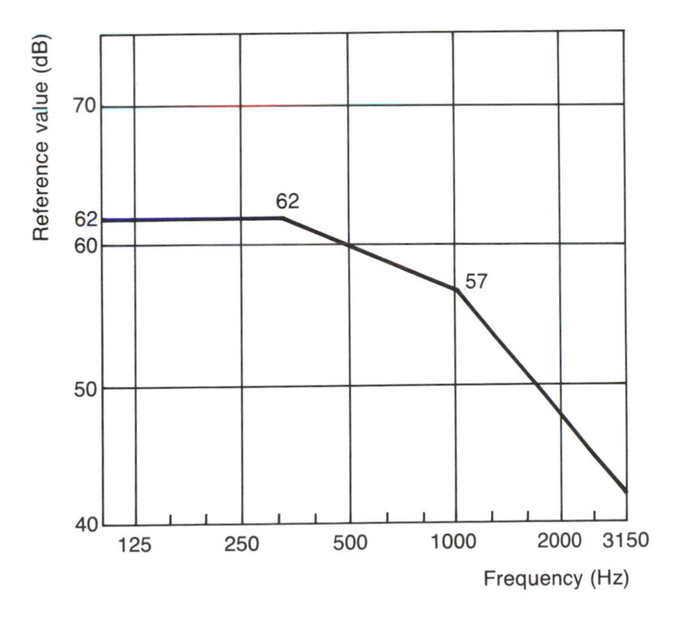

Fig. 4.8 Reference values for impact sound

Table 4.3 Maximum weighted standardized impact sound pressure levels for impact sound in the receiving room

	Individual values (dB)	Test in at least 4 pairs of rooms (dB, mean)	Test in at least 8 pairs of rooms (dB, mean)
Maximum impact sound values	65	61	62

Table 4.4 Reference values for impact sound insulation measurement

Third-octave (Hz)	Corrected measured levels (dB)
100	62
125	62
160	62
200	62
250	62
315	62
400	61
500	60
630	59
800	58
1000	57
1250	54
1600	51
2000	48
2500	45
3150	42

standardized impact sound pressure level of the floor is 63 dB.

4.14 Impact sound insulation rating

It will be appreciated that there is no simple way of stating impact sound insulation. Some floor constructions may

Frequency (Hz)	Corrected levels (dB)	Reference levels (dB)	62 dB curve Ref. + 2 (dB)	Adverse deviation (dB)
100	67	62	64	3
125	66	62	64	2
160	65	62	64	1
200	68	62	64	4
250	68	62	64	4
315	68	62	64	4
400	70	61	63	7
500	69	60	62	7
630	65	59	61	4
800	65	58	60	5
1000	59	57	59	—
1250	56	54	56	—
1600	51	51	53	—
2000	48	48	50	—
2500	46	45	47	—
3150	44	42	44	—
			Total	41

achieve the standards required by the Building Regulations between dwellings, but lesser standards may be acceptable in many other situations. It may be convenient to use a family of reference curves on which to plot corrected measured values (Fig. 4.9). The weighted standardized sound pressure level is the reference curve such that the sum of the unfavourable deviations is as near as possible to 32 dB without exceeding it as in the last example. For further details, see ISO 717 (BS 5821) part 2.

Questions

(1) An elastic mounting for a vibrating machine has a static deformation of 1·13 mm and gives an energy reduction of 9 dB when the machine is running steadily. Calculate the forcing frequency of the machine.

(2) An insulation of 40 dB is required by the use of an elastic mount for a machine vibrating at 2400 rev/min. What static deflection would be needed?

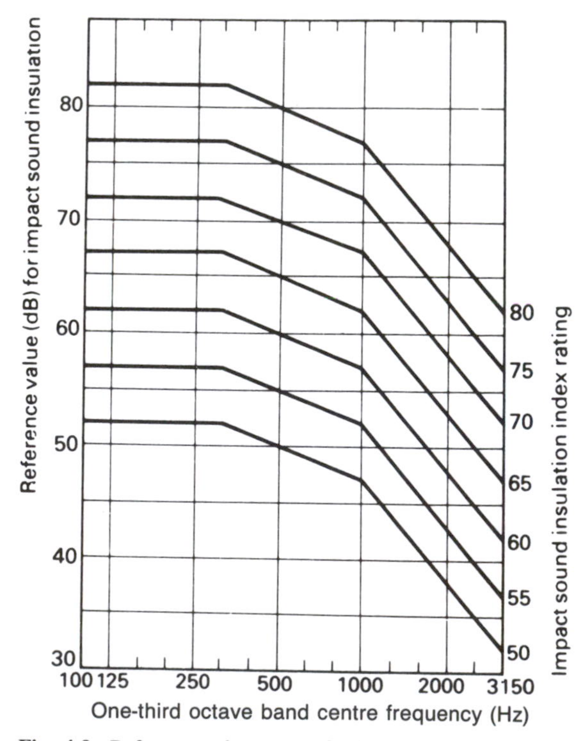

Fig. 4.9 Reference value curves for impact sound insulation measurements

(3) A 50 mm thick concrete panel supported at its edges was found to have a lowest resonant condition at 56 Hz. An ultrasonic pulse was found to take 10 μs to travel through the thickness of the concrete. Find the dimensions of the panel.

(4) A 13 mm thick plasterboard has a critical frequency of 2000 Hz and a density of 1700 kg/m^3. Calculate Young's modulus of elasticity. (V = 333 m/s.)

(5) Third-octave impact insulation measurements were found to give the following results for a particular floor:

Frequency (Hz)	Mean level (dB) re 2×10^{-5} Pa	R.T. of receiving room (s)
100	76	2·35
125	75	1·95
160	70	2·2
200	65	1·95
250	62	1·5
315	59	2·1
400	60	1·8
500	57	1·8
630	50	1·4
800	49	1·1
1000	49	1·1
1250	49	1·1
1600	42	1·0
2000	40	1·0
2500	38	0·8
3150	35	0·75

State whether these results would meet the requirements of the Building Regulations between two dwellings.

(6) Find the velocity of sound in a concrete beam of density 2400 kg/m^3 and with a dynamic modulus of elasticity (E) of 38 400 N/mm^2.

(7) (a) By means of diagrams and notes explain how the sound insulation properties of a flooring system may be measured in an acoustic testing laboratory.

(b) Briefly discuss the reasons why it is possible for a flooring system to give acceptable results in such a test and yet fail when incorporated in actual constructions.

(8) (a) Outline the principal requirements for a mounting to act as a vibration isolator.

(b) What precautions should be taken to ensure that a design for a piped hot-water central heating system is as quiet as possible in operation? Describe the ways in which these precautions will reduce noise.

(9) (a) Outline the method for testing the impact sound insulation of a party-floor separating two flats. Explain how the measurements are made and how the results are processed.

(b) Describe in principle how a recognized single-figure rating may be obtained from the results.

(c) List measures that could be taken to increase the impact sound insulation of a party-floor.

(*IOA*)

5 Airborne sound

The term airborne sound is conveniently taken to mean sound that is transmitted mainly, but not necessarily exclusively, through the air. For instance, sound transmitted from one room to another may have to excite the adjoining wall into vibration. This chapter looks at the propagation of airborne sound and the methods which may be used to reduce it.

5.1 Propagation of sound in open air

Point source

It was shown in Chapter 1 that air cannot sustain a shear force and only a longitudinal type of sound wave is possible. Sound will radiate spherically in a free field so that the sound intensity will decrease with the square of the distance from the source (see Fig. 5.1).

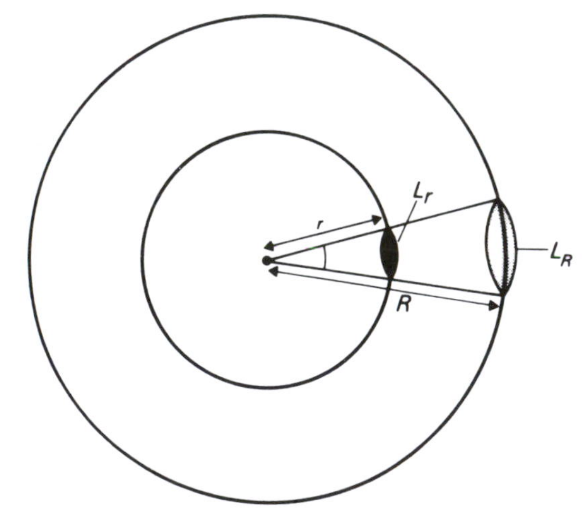

Fig. 5.1 The surface area of a sphere is proportional to the square of its radius. The sound intensity will therefore decrease in proportion to the square of the distance from the point source

$\therefore$ I is proportional to $\dfrac{1}{d^2}$

$$\therefore \frac{I_r}{I_R} = \frac{R^2}{r^2}$$

where I_r = intensity at a distance r from the source
I_R = intensity at a distance R from the source

If L_r = sound pressure level in decibels at distance r from the source

$$\text{Then } L_r - L_R = 10 \log_{10} \frac{I_r}{I_R}$$

$$= 10 \log_{10} \frac{R^2}{r^2}$$

$$= 20 \log_{10} \frac{R}{r}$$

Thus the reduction in sound pressure level for sound from a point source will be 6 dB for each time the distance is doubled:

$$L_r - L_{2r} = 20 \log_{10} \frac{2r}{r}$$

$$= 20 \times 0{\cdot}3010$$

$$\simeq 6 \text{ dB}$$

If the sound power level in picowatts (L_W) had been given, then the formula becomes:

$$L_r - L_W = 20 \log_{10} \frac{1}{r} - 10{\cdot}9 \quad (r \text{ in metres})$$

$$\text{or} \quad L_r = L_W - 20 \log_{10} r - 10{\cdot}9$$

If the sound field is confined to a hemisphere because the source is on the (perfectly reflecting) ground, then the intensity is doubled or the sound pressure level is increased by 3 dB.

$\therefore L_r = L_W - 20 \log_{10} r - 7{\cdot}9$

Example 5.1

The sound power level of a certain jet plane flying at a height of 1 km is 160 dB (re 10^{-12} W). Find the maximum sound pressure level on the ground directly below the flight path assuming that the aircraft radiates sound equally in all directions.

$$\begin{aligned} L_r &= 160 - 20 \log_{10} 1000 - 10{\cdot}9 \\ &= 160 - 20 \times 3 - 10{\cdot}9 \\ &= 160 - 60 - 10{\cdot}9 \\ &= 89{\cdot}1 \text{ dB (say 89 dB)} \end{aligned}$$

It should be noted that the assumption that sound is radiated equally in all directions is not correct although the figures given are of the correct order. Directionality is discussed in section 8.8.

Line source

When the sound source is a line instead of a point, such as is the case near a busy motorway, the sound will radiate in the form of a cylinder (Fig. 5.2).

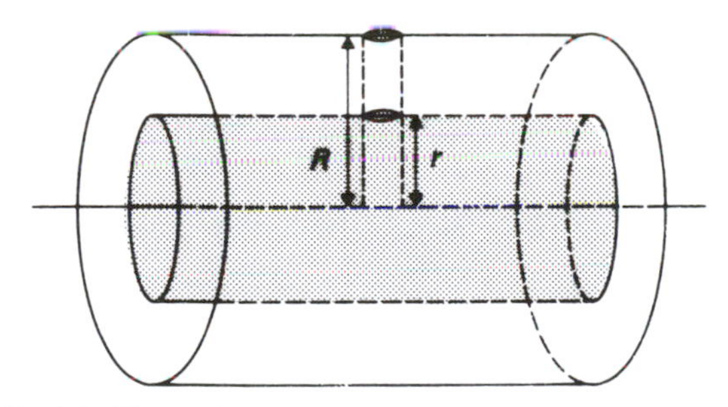

Fig. 5.2 The surface area of a cylinder is proportional to its radius. Sound intensity will therefore decrease directly with distance from a line source

Thus $\dfrac{I_r}{I_R} = \dfrac{R}{r}$

$\therefore L_r - L_R = 10 \log_{10} \dfrac{R}{r}$

If $R = 2r$

$$\begin{aligned} \text{Then } L_r - L_R &= 10 \log_{10} 2 \\ &= 10 \times 0{\cdot}3010 \\ &= 3 \text{ dB} \end{aligned}$$

It can be seen that there is only a 3 dB reduction for a doubling of the distance.

Example 5.2

A police officer measures the sound pressure level at a distance of 7·5 m from the line of traffic on a road and finds it to be 80 dB. What would the level be at a distance of 75 m from the line of traffic if the officer's reading was from:

(a) an isolated vehicle?
(b) a continuous line of closely packed identical cars? (Assume that the ground is flat and unobstructed.)

(a) The isolated vehicle constitutes a point source

$$\begin{aligned} \therefore \text{reduction in dB} &= 10 \log_{10} \left(\frac{75}{7{\cdot}5}\right)^2 \\ &= 20 \log_{10} 10 \\ &= 20 \times 1 \\ &= 20 \text{ dB} \end{aligned}$$

The sound pressure level at 75 m from the vehicle would therefore be $80 - 20 = 60$ dB.

(b) The continuous line of closely packed cars could be taken as a line source.

$$\begin{aligned} \therefore \text{reduction in dB} &= 10 \log_{10} \left(\frac{75}{7{\cdot}5}\right) \\ &= 10 \log_{10} 10 \\ &= 10 \text{ dB} \end{aligned}$$

The level at 75 m from the line of cars would be $80 - 10 = 70$ dB

5.2 Reduction of noise by walls

The reduction of sound by means of a wall or fence is only effective where the barrier is large compared with the wavelength of the noise. It has been shown that approximately

$$x = \frac{H^2}{\lambda D_S}$$

Where x, H and D_S are shown in Fig. 5.3, and where x is related to the sound reduction in dB as shown in Fig. 5.4, provided that $D_L \gg D_S$ and $D_S > H$.

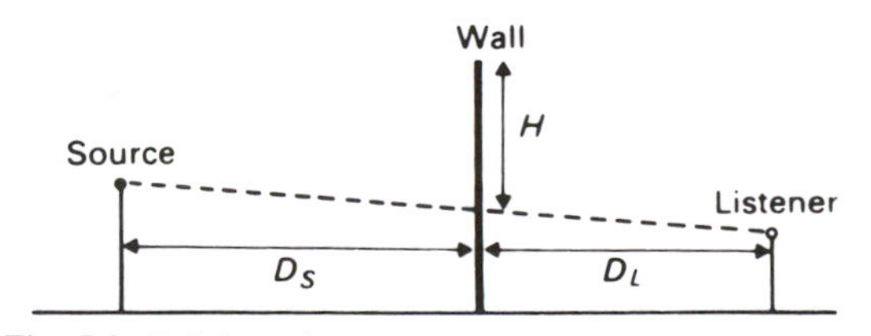

Fig. 5.3 Relation of source and listener to wall

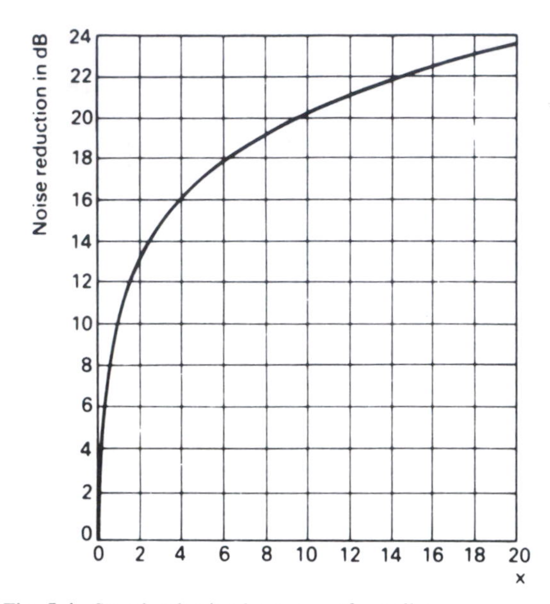

Fig. 5.4 Sound reduction by means of a wall or screen

Example 5.3

A school is situated close to a furniture factory with a flat roof on top of which is a dust extractor plant as shown in Fig. 5.5. The highest window in the school facing the factory is at the same height as the noise source, which can be taken as 1 m above the factory roof. The extractor plant is 6 m from the edge and produces a note of 660 Hz. Find the minimum height of wall to be built on the edge of the factory roof to give a reduction of 15 dB.

From Fig. 5.4, the value of x required to give 15 dB reduction is 3

$$\therefore \quad 3 = \frac{H^2}{(V/f)D_S} = \frac{H^2}{\frac{330}{660}6} = \frac{H^2}{3}$$

$$\therefore \quad H^2 = 9$$

$$H = 3 \text{ m}$$

Total height of wall = 4 m

If the distance D_L from the wall to the listener is not large compared with D_S,

$$\text{then} \quad x = \frac{2}{\lambda}\left\{D_S\left[\sqrt{\left(1 + \frac{H^2}{D_S^2}\right)} - 1\right] + D_L\left[\sqrt{\left(1 + \frac{H^2}{D_L^2}\right)} - 1\right]\right\}$$

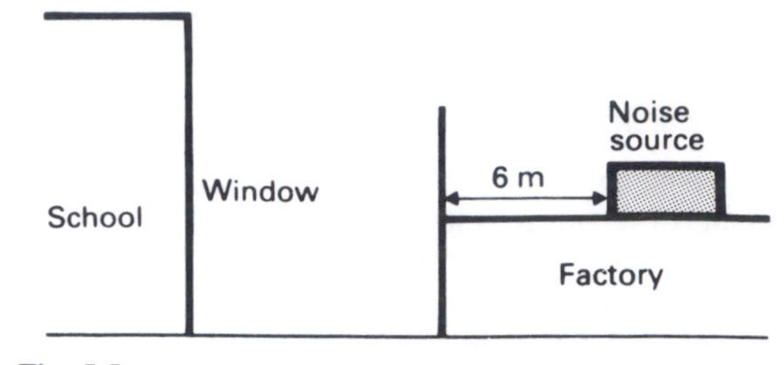

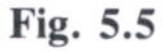

Fig. 5.5

These calculations are based on the amount of sound diffracted around the barrier. It is essential that the barrier itself does not transmit sound. In effect this means that the insulation of the barrier must be greater than the screening effect needed. Gaps in a fence would make the screen useless although it can be quite light in weight. A mass of 20 kg/m^2 is usually sufficient.

5.3 Maekawa method

A formula based on work by Maekawa can be used to give the approximate attenuation produced by a thin rigid barrier between the source of sound and the receiver.

Attenuation, $E_b = 10 \log_{10} (3 + 40\,\delta/\lambda)$ dB

where λ = wavelength of the sound

$\delta = a + b - c$, the difference in path lengths of sound travelling over the barrier and directly (Fig. 5.6)

The path difference may be obtained by scale drawing or calculation.

Example 5.4

A compressor is to be used 30 m away from a house and a long, 2 m high barrier is to be erected 5 m from the compressor to reduce the noise (Fig. 5.7). The ground between the compressor and the house is hard, flat and level. The centre of the compressor is 0·5 m above the ground. Calculate the noise reduction produced by the barrier at a height of 1·5 m above the ground at the house. The velocity of sound is 330 m/s.

$$a^2 = 5^2 + (2 - 0{\cdot}5)^2 = 25 + 1{\cdot}5^2$$

$$a = \sqrt{27{\cdot}25} = 5{\cdot}220$$

$$b^2 = 25^2 + 0{\cdot}5^2$$

$$b = \sqrt{625{\cdot}25} = 25{\cdot}005$$

$$c^2 = 30^2 + 1^2$$

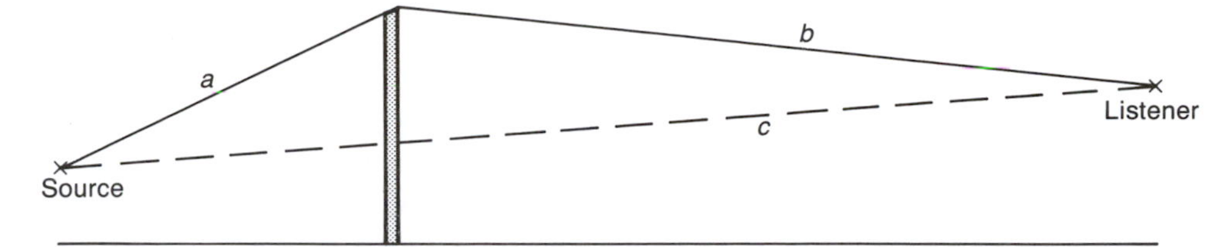

Fig. 5.6 Maekawa method for attenuation by a thin rigid barrier

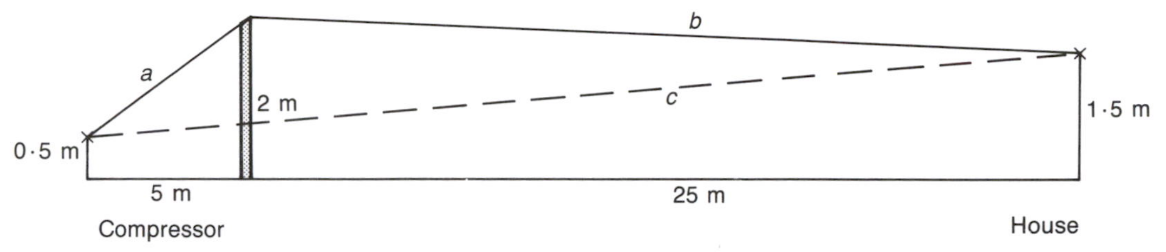

Fig. 5.7 Diagram of compressor and house for Example 5.4

$$c = \sqrt{901} = 30{\cdot}017$$

$$\delta = 5{\cdot}220 + 25{\cdot}005 - 30{\cdot}017$$
$$= 0{\cdot}208 \text{ m}$$

At 500 Hz the wavelength, $\lambda = \dfrac{330}{500} = 0{\cdot}66$ m

$$\text{Attenuation, } E_b = 10 \log_{10}\left(3 + \frac{40 \times 0{\cdot}208}{0{\cdot}66}\right)$$
$$= 10 \log_{10} (3 + 12{\cdot}8)$$
$$= 12 \text{ dB}$$

At other frequencies the attenuation may be calculated in the same way.

Octave band (Hz)	125	250	500	1000	2000	4000
Attenuation by barrier (dB)	8	10	12	15	17	20

5.4 Bushes and trees

The noise attenuation achieved by shrubs and trees can only be marginal. Measurements made in jungle conditions allowing for the normal loss by distance alone are shown approximately in Fig. 5.8.

Considering the marginal improvement, trees are not an economic means of achieving sound insulation. However, there is some absorption effect which would not be present if the ground were paved. If trees are used they must be of a leafy variety and be thick right to ground level. A point sometimes forgotten is that the rustle of leaves may occasionally produce a noise level as high as 50 dB(A).

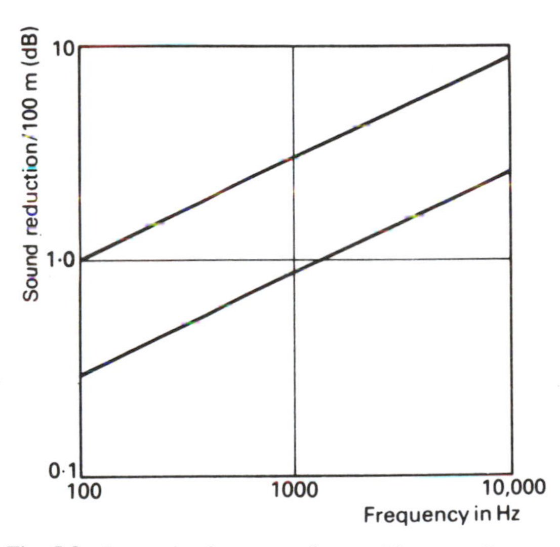

Fig. 5.8 Attenuation by means of trees. The upper line represents maximum attenuation under conditions of very dense growth of leafy trees and bushes where penetration is only possible by cutting. The lower line shows the attenuation under conditions where penetration is easy and visibility up to 100 m, but still very leafy undergrowth

5.5 Air absorption

Energy is absorbed as a sound wave is propagated through the air. These losses are due to a relaxation process and depend upon the amount of water vapour present. They are approximately proportional to the square of the frequency.

The attenuation per metre, α, has been shown to be such that:

$$\alpha = kf^2 + \alpha_2$$

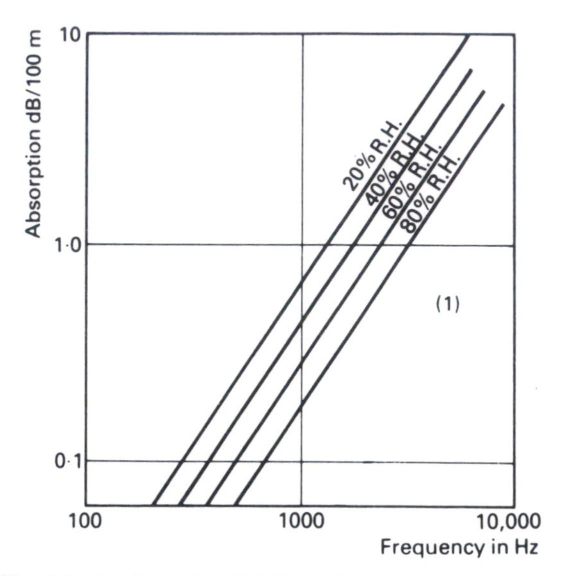

Fig. 5.9 Air absorption dB/100 m. Curve (1) shows the absorption due to the kf^2 part of the equation. The other lines show the total contributions

where $k = 14{\cdot}24 \times 10^{-11}$

f = frequency in Hz

α_2 is humidity dependent.

Typical values are about 3 dB per 100 metres at 4000 Hz dropping to 0·3 dB per 100 metres at 1000 Hz. Air attenuation becomes very important for ultrasonic frequencies and is greater than 1 dB/m at 100 kHz, but is of comparatively little significance in architectural acoustics (see Fig. 5.9).

5.6 Velocity gradients

For propagation close to the ground, sound velocity gradients have a big influence on the levels received from a distance. These velocity gradients can be caused by wind or temperature.

Friction between the moving air and the ground results in a decreased velocity near ground level. This causes a distortion of the wave front. Downwind from the source, sound rays are refracted back towards the ground and the received level is not affected. Upwind the sound is refracted up and away from the ground, causing acoustic shadows in which the level is considerably reduced (Fig. 5.10).

The velocity of sound in air is proportional to the square root of the absolute temperature. A temperature lapse condition will cause the sound rays to be bent up, causing a symmetrical shadow around the source (see Fig. 5.11).

A motorway built downwind (east) of a town would have an advantage of about 10 dB extra attenuation over one to the west.

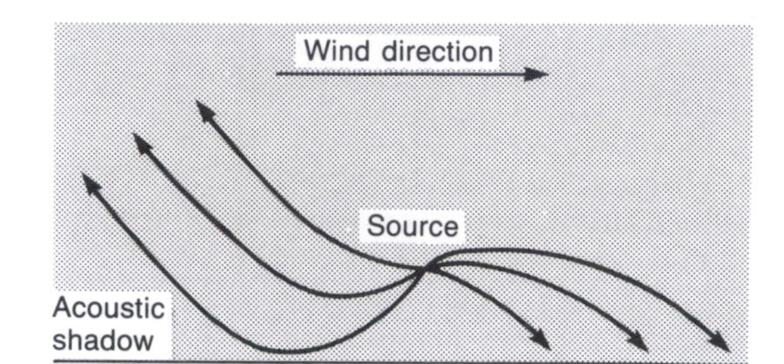

Fig. 5.10 Wind-created velocity gradients

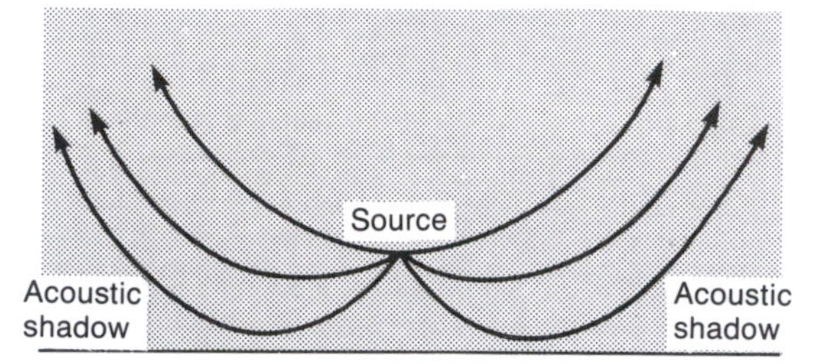

Fig. 5.11 Temperature lapse-created velocity gradients

5.7 Ground absorption

Hard surfaces such as concrete appear to have virtually no absorption properties, whereas there may be a maximum attenuation over grassland of 25 to 30 dB in the 200 to 400 Hz octave band along a distance of about 1000 metres. It seems that there is a preferential absorption in this frequency range for sound produced near the ground.

5.8 Sound insulation by partitions

Sound can be transmitted into a room by some or all of the methods shown in Fig. 5.12.

1. Airborne sound in the source room excites the separating partition into vibration which directly radiates the sound

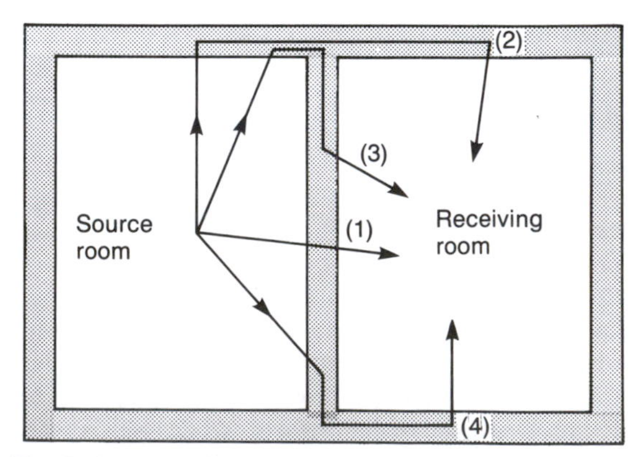

Fig. 5.12 Four different transmission paths from source room to receiving room

into the receiving room. The amount of attenuation will depend upon the frequency of sound, the mass, fixing conditions and thus the resonant frequencies of the partition.

2. Airborne sound in the source room may excite walls other than the separating one into vibration. The energy is then transmitted through the structure and re-radiated by some other partition into the receiving room.
3. Any wall other than the separating one may be excited. The sound is transmitted to the separating wall and then re-radiated by it.
4. Sound energy from the separating partition is radiated into the receiving room by some other wall.

This means that a laboratory measurement of the sound insulation provided by a partition, which is mounted in massive side walls, may give results quite different from those achieved in actual buildings. There is a limit to the insulation obtained by improving only the adjoining partition. Where a partition has a low insulation value of 35 dB or less, flanking transmission is of little consequence, but when partition values of 50 dB are reached, further improvement is limited by the indirect sound paths.

The sound transmission properties of a partition can be divided into three distinct regions (see Fig. 5.13).

Region 1, where resonance and stiffness conditions are important.

Region 2 which is mass controlled and where the partition can be considered as a large number of small masses free to slide over each other.

Region 3 where again the partition becomes stiffness-controlled above a certain frequency known as the critical frequency.

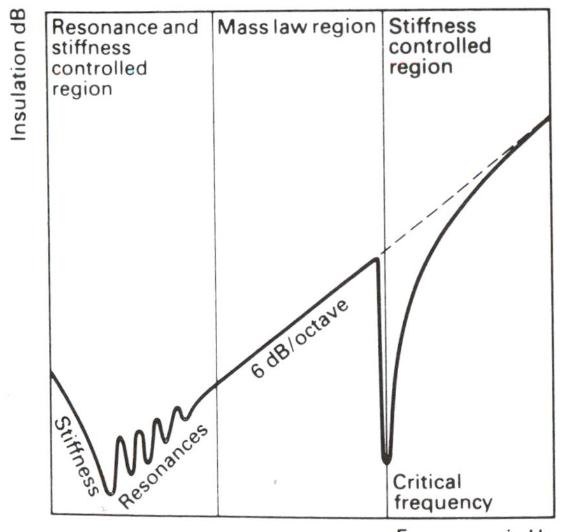

Fig. 5.13 Three distinct regions showing the way a panel will react to different frequency sounds

Table 5.1

Material	Mass/unit area × coincidence frequency × 10^3 kg/m^2 × Hz
Plywood	17
Glass	35
Plasterboard	50
Concrete	100
Steel	150

Most building materials fall into the middle category. Typical values of coincidence frequency can be obtained from Table 5.1. A 50 mm reinforced concrete panel of mass per unit area 120 kg/m^2 can be expected to have a coincidence frequency at about 800 Hz.

5.9 Measurement of airborne sound insulation of panels

The field measurement consists of producing a suitable sound on one side of the panel and measuring the reduction in sound pressure level at the other side. Ideally, a completely diffuse sound field is needed in the source room. To attempt to eliminate the variations due to room modes for each frequency a band of noise is used in the form of white noise.

If a white noise is used then the measurements may be made in one-third, octave intervals, the centre frequencies being:

For one-third octave:

100, 125, 160, 200, 250, 315, 400, 500, 640, 800, 1000, 1250, 1600, 2000, 2500, 3150 Hz

For octave:

Starts at 100 Hz and goes up to 3150 Hz

Suitable means of producing the sound can be by the use of a randon-noise generator for white noise.

The measurement is made by finding the average sound pressure level difference between source and receiving rooms. This may conveniently be done by using two microphones (whose calibration has been checked) connected to a switchbox through filters to a level recorder as shown in Fig. 5.14. By this means the difference may be read off directly. The average is found over the entire room excepting those parts where direct radiation or boundary reflection are of significance. The average level, L, should be taken as:

$$L = 10 \log_{10} \frac{p_1^2 + p_2^2 + p_3^2 + \ldots + p_n^2}{nP_0^2}$$

$$= 10 \log_{10} \left(\frac{10^{L_1/10} + 10^{L_2/10} + \ldots + 10^{L_n/10}}{n} \right)$$

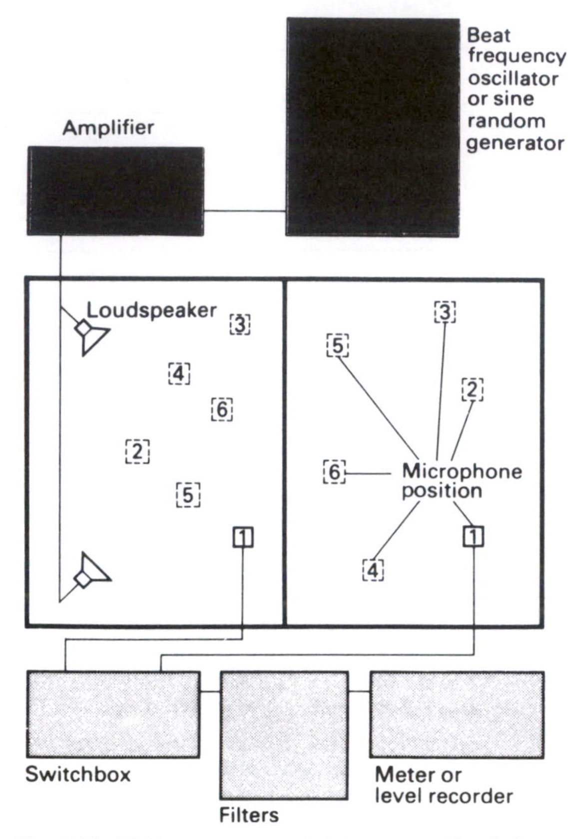

Fig. 5.14 Field measurement of airborne sound insulation

where $p_1, p_2, \ldots, p_n$ = RMS levels at n different points in the room

p_0 = reference sound pressure

However, in practice so long as the values of $p_1, p_2, \ldots, p_n$ do not differ greatly it is satisfactory to find the average level, L, from:

$$L = \frac{L_1 + L_2 + L_3 + L_4 + \ldots + L_n}{n}$$

where $L_1, L_2, \ldots, L_n$ = measured pressure levels at n different points in the room. The average sound pressure level difference, D, is then found from:

$$D = L_S - L_R$$

where L_s = average sound pressure level in the source room

L_R = average sound pressure level in the receiving room

Six different positions are usually sufficient in each room up to 500 Hz and three each above this frequency. If the measured levels at any frequency in either room differ by more than 6 dB then more measurements are needed so that the dB average does not differ from the pressure average by more than ±1 dB.

Variations in results for a particular partition could be produced by changes in the reverberation time of the receiving room. It is necessary for comparison purposes to normalize the result to a particular standard. The standard method adjusts the value to that which would have been obtained if the receiving room had 10 m^2 of absorption units. Hence:

$$D_n = L_S - L_R + 10 \log_{10} \frac{10}{A}$$

where D_n = standardized level difference

A = measured absorption in the receiving room in m^2

In Britain it has been found that for domestic dwellings the reverberation time seldom varies much from 0·5 s. A slightly different method of standardizing is used:

$$D_n = L_S - L_R + 10 \log_{10} \frac{T}{0{\cdot}5}$$

Where T = reverberation time of the receiving room in seconds.

The Building Regulations require that for a particular type of construction, not less than four pairs of rooms be tested. The separating partitions should be at least 7 m^2 and the rooms should each have a volume at least 25 m^3. In practice this size requirement can present problems in some small new dwellings. However, results do show variation particularly at low frequencies for smaller rooms. The field measurement enables a result to be obtained for the normalized level difference.

The laboratory test enables the sound reduction index to be obtained. The method is essentially similar except that a partition is inserted between two reverberant rooms, each having a volume of more than 50 m^3. The laboratory measurement virtually eliminates the flanking transmission. It is likely that there will be differences between results obtained from laboratory and field measurements. While this may not be significant in most cases, there have been instances where flanking transmission has lowered the overall reduction very significantly.

5.10 Airborne sound insulation standards

Section E of the Building Regulations in the Approved Document on Sound outlines the standards for walls and floors. There are essentially two ways of achieving the standards. Firstly, by copying one of the examples of suitable construction methods shown in the approved

Table 5.2 Reference values of transmisson loss for airborne sound between dwellings

Frequency (Hz)	Reference value (dB)
100	33
125	36
160	39
200	42
250	45
315	48
400	51
500	52
630	53
800	54
1000	55
1250	56
1600	56
2000	56
2500	56
3150	56

document. Secondly, by showing via tests that the construction meets the approved standards. In this second case, if such a construction has already been built, it is clearly safer to use test results from that structure before embarking on the unknown.

Both British and ISO standards are based on known practical constructions. Standards are quantified via reference values in BS 5821:Part 1 1984 for airborne sounds as shown in Table 5.2 and Fig. 5.15. Measurements of airborne sound insulation are made as described earlier in one-third octave bandwidths and a graph drawn as shown in Fig. 5.15. The reference curve is then adjusted up or down in 1 dB intervals until the average adverse deviation of the measured graph is a maximum of 2 dB. The resulting position of the reference curve at 500 Hz gives a single figure for the insulation value. The minimum acceptable values are shown in Table 5.3. It should be noted that the curve in Fig. 5.15 is a reference. If a particular set of measurements were identical to that curve then the reference curve could be moved up by 2 dB and the standardized level difference at 500 Hz would be 54 dB, not 52 dB.

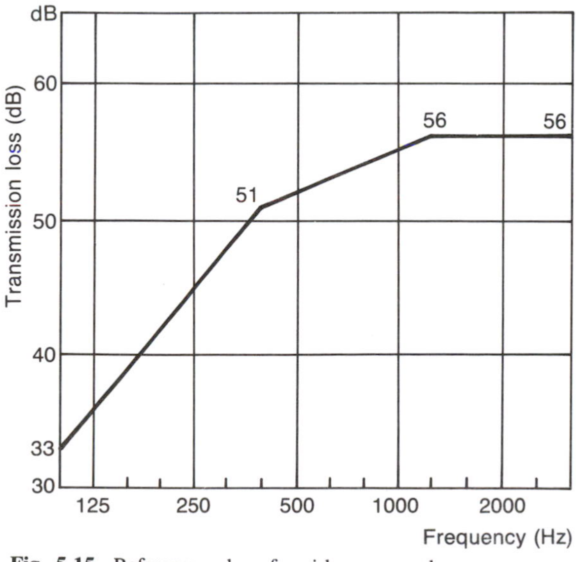

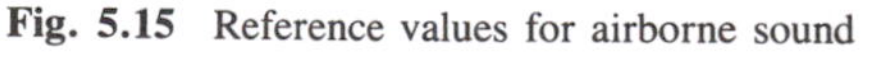
Fig. 5.15 Reference values for airborne sound

Example 5.5

Measurements were made between 10 pairs of rooms in separate dwellings with the same type of separating wall construction. The standardized level differences are shown in Table 5.3. Find the weighted standardized level difference for this construction and whether or not the construction is shown to meet the requirements of the Building Regulations.

The adverse differences for the construction measured may either be calculated or found from a graph.

Frequency (Hz)	Mean standardized level difference (dB)	Reference value (dB)	Adverse deviation (dB)
100	33	33	—
125	32	36	4
160	38	39	1
200	43	42	—
250	44	45	1
315	45	48	3
400	49	51	2
500	51	52	1
630	53	53	—
800	53	54	1
1000	54	55	1
1250	55	56	1
1600	55	56	1
2000	56	56	—
2500	58	56	—
3150	60	56	—
		Total	16

Note that only the unfavourable deviations from the reference are counted, so the favourable figures at 200 Hz, 2500 Hz and 3150 Hz do not offset poor results. The total unfavourable deviation is 16 dB. The standard allows an average of 2 dB or 32 dB in total. Hence, the reference curve could be moved up by 1 dB giving a 500 Hz reading of 53 dB. Thus the standards are achieved with this

Table 5.3 Minimum values of weighted standardized level differences for airborne sound between two habitable rooms

	Airborne sound tests in up to 4 pairs of rooms		Airborne sound tests in at least 8 pairs of rooms	
	Mean value (dB)	Individual value (dB)	Mean value (dB)	Individual value (dB)
Walls	53	49	52	49
Floors	52	48	51	48

construction and it is possible that this could have been shown with no more than four measurements.

Some examples of party wall and floor construction are given in Figs 5.16 and 5.17.

5.11 Sound insulation of composite partitions

The sound reduction of a partition measured in dB will depend upon the proportion of sound energy transmitted.

$$\text{Reduction in dB} = 10 \log_{10} \frac{1}{t}$$

where t is the transmission coefficient.

When calculating the sound insulation of a partition consisting of more than one part (e.g. a 115 mm brick wall with a door) it is first necessary to find the transmission coefficient of each. From this the average transmission coefficient may be calculated using this formula:

$$t_{AV} \times S = t_1 \times S_1 + t_2 \times S_2 + t_3 \times S_3 + \ldots$$

where t_{AV} = average transmission coefficient

S = total area of the partition

t_1, t_2 = transmission coefficients of each section

S_1, S_2 = areas of each part

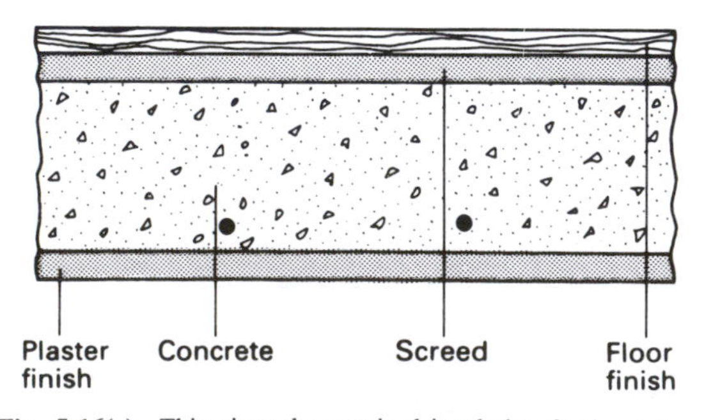

Fig. 5.16(a) This gives the required insulation for impact and airborne sound if: concrete + plaster + screed weighs more than 365 kg/m^2 and a soft floor finish is used. If a hard floor finish is used airborne insulation is still adequate, but the floor fails for impact sound

Example 5.6

A partition of total area 10 m^2 consists of a brick wall plastered on both sides to a total thickness of 254 mm and contains a door of area 2 m^2. The brickwork gives a mean sound reduction of 51 dB and the door 18 dB. Calculate the sound reduction of the complete partition.

Brickwork

$$51 = 10 \log_{10} \frac{1}{t_B}$$

where t_B = transmission coefficient of the brick

$$\therefore \quad \frac{1}{t_B} = \text{antilog}_{10}\, 5{\cdot}1$$

$$= 10^{5\cdot 1}$$

$$t_B = 10^{-5\cdot 1}$$

$$= 0{\cdot}000\,008$$

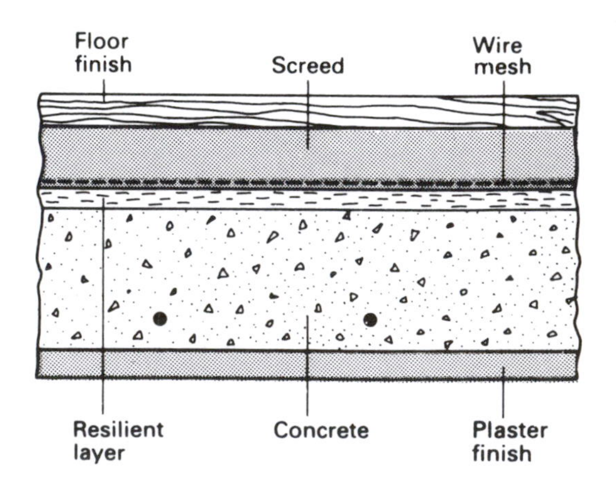

Fig. 5.16(b) This gives the required insulation for airborne and impact sound if: concrete + plaster weigh more than 220 kg/m^2

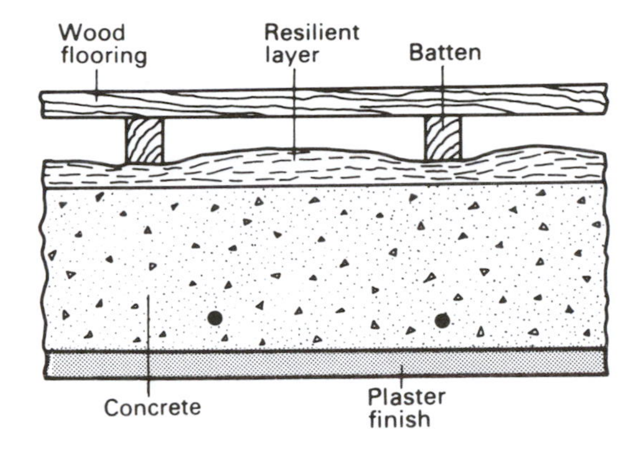

Fig. 5.16(c) This gives the required insulation for airborne and impact sound if: concrete and plaster weigh more than 220 kg/m^2

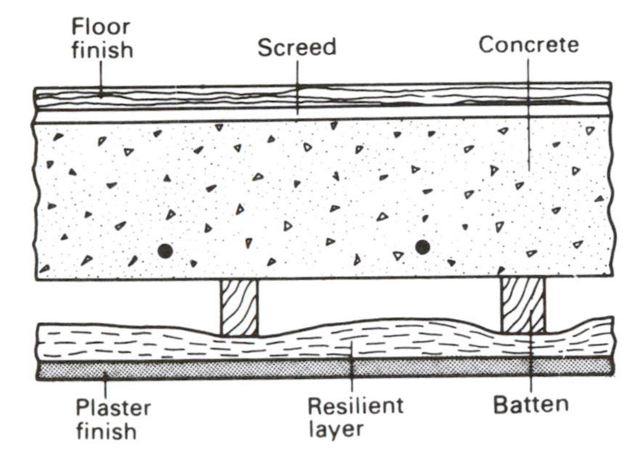

Fig. 5.16(d) This gives the required insulation for airborne and impact sound if: concrete + screed weigh more than 220 kg/m^2 and a soft floor finish is used. If a medium or hard floor finish is used, airborne insulation is still adequate but the floor fails for impact sound

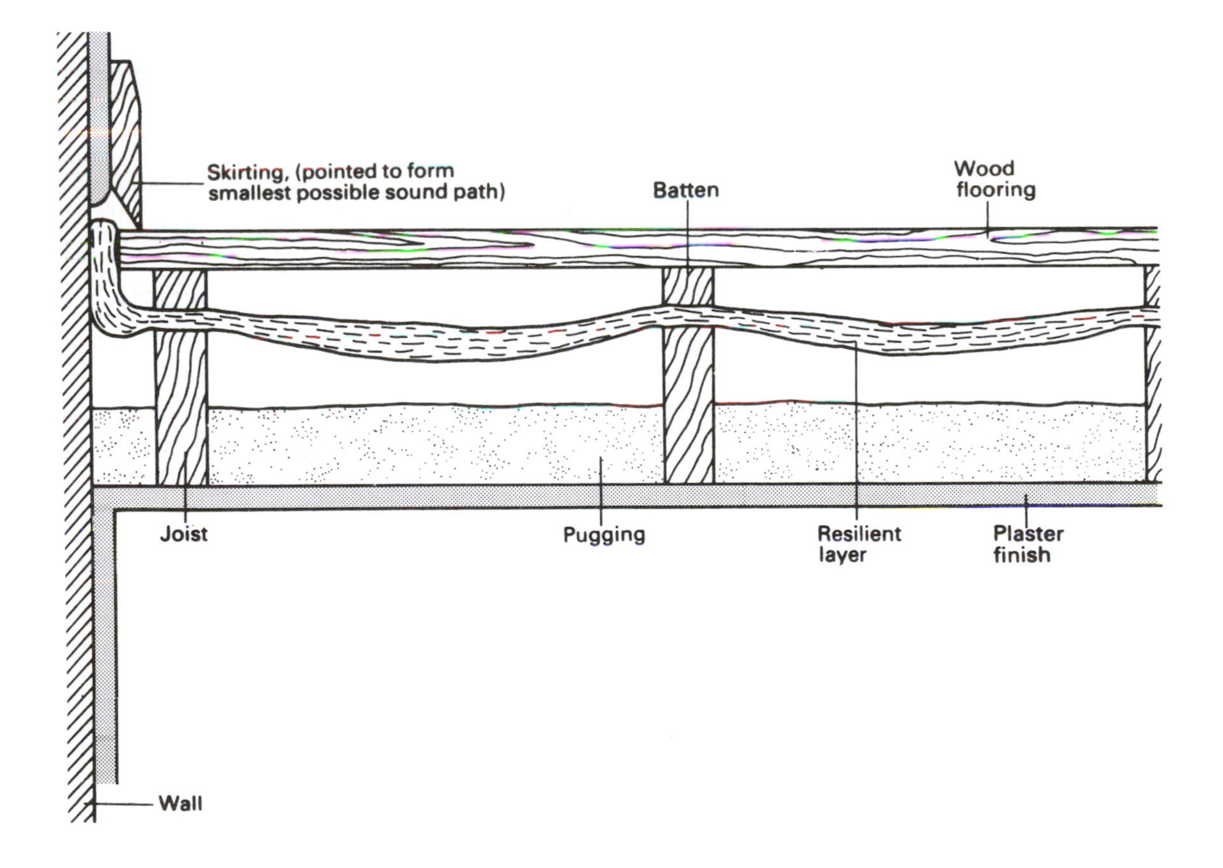

Fig. 5.16(e) Adequate sound insulation, possible for both airborne and impact sound if: heavy lath and plaster with 80 kg/m^2 of pugging or heavy lath and plaster with 15 kg/m^2 of pugging on heavy walls or heavy lath and plaster with no pugging but with very heavy walls or plasterboard and one coat plaster with 15 kg/m^2 pugging and very heavy walls

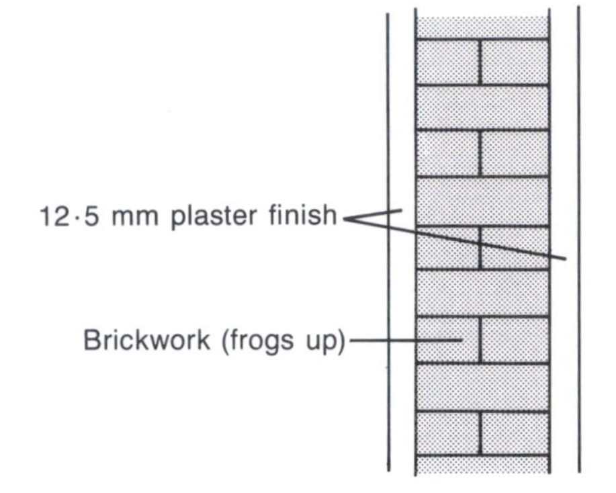

Fig. 5.17(a) Solid brick wall, frogs upwards, plastered both sides to at least 12·5 mm each with a minimum mass of 375 kg/m^2

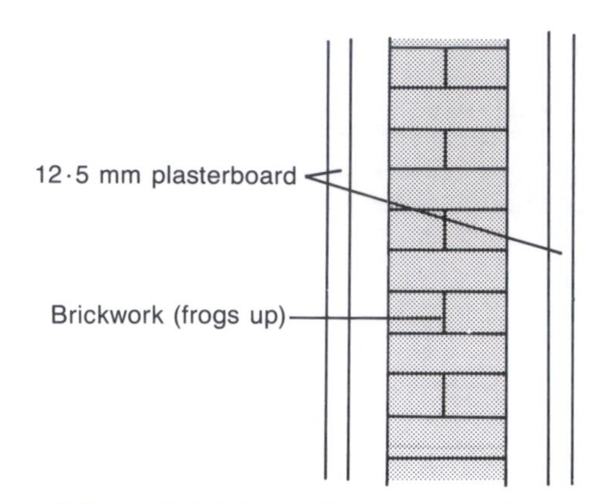

Fig. 5.17(b) Solid brick wall, frogs upwards, plasterboard both sides at least 12·5 mm each with a minimum mass of 375 kg/m^2

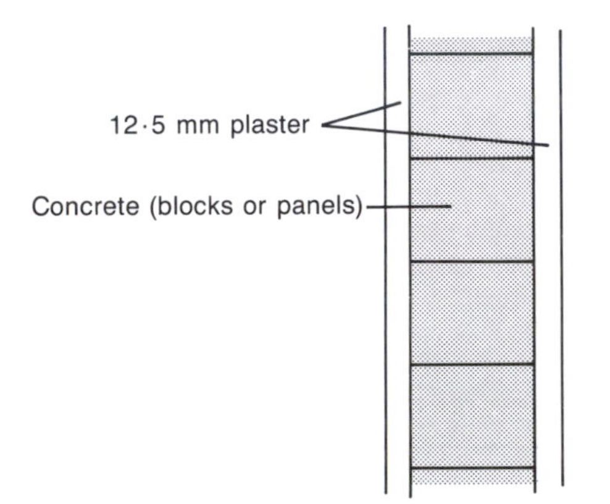

Fig. 5.17(c) Solid concrete plastered both sides with at least 12·5 mm each and minimum mass of 415 kg/m^2

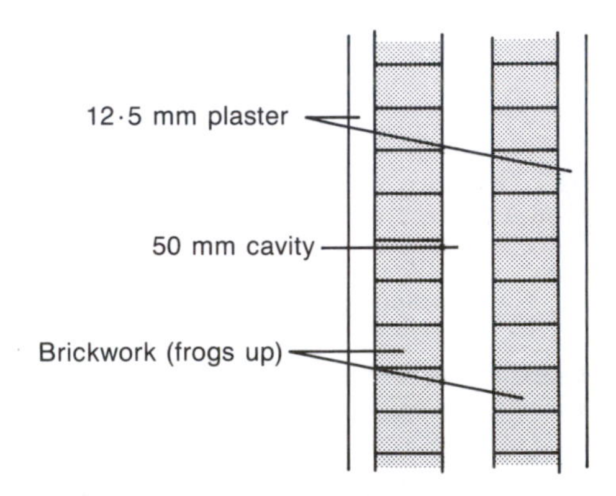

Fig. 5.17(d) Cavity brickwork plastered to at least 12·5 mm each side with a total mass of 415 kg/m^2

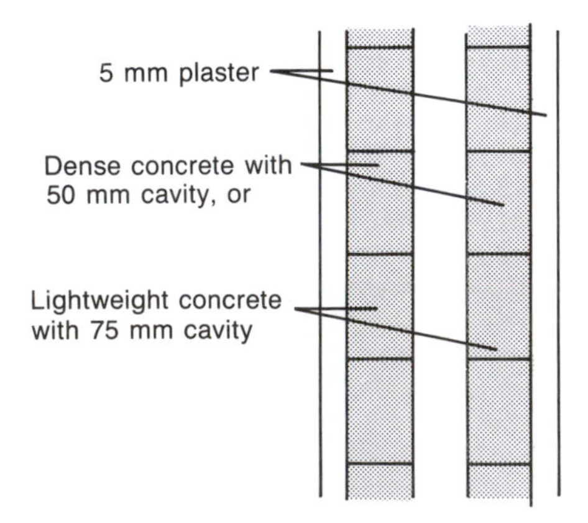

Fig. 5.17(e) Two leaves of concrete plastered both sides. Minimum mass: 415 kg/m^2 for dense concrete with 50 mm cavity; 250 kg/m^2 for lightweight concrete with 75 mm cavity

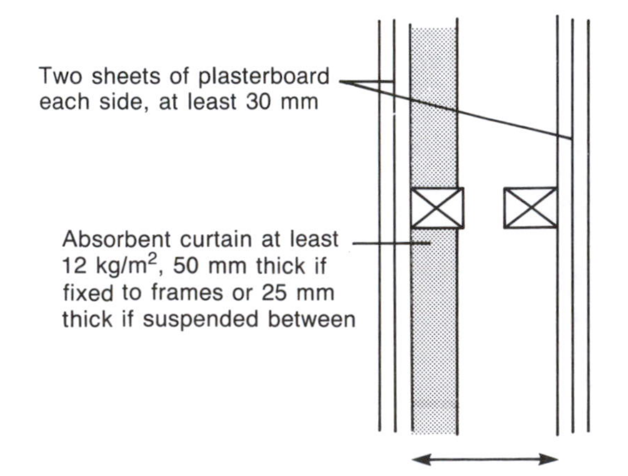

Fig. 5.17(f) Timber frame with two layers of plasterboard each side and an absorbent curtain between; cavity at least 200 mm

Door

$$18 = 10 \log_{10} \frac{1}{t_D}$$

where t_D = transmission coefficient of the door

$$\therefore \quad \frac{1}{t_D} = \text{antilog}_{10} 1{\cdot}8$$

$$= 10^{1\cdot 8}$$

$$t_D = 10^{-1\cdot 8}$$

$$= 0{\cdot}015\ 85$$

Now:

$$t_{AV} \times 10 = 0{\cdot}000\ 008 \times 8 + 0{\cdot}015\ 85 \times 2$$

$$t_{AV} = \frac{0{\cdot}000\ 064 + 0{\cdot}031\ 700}{10}$$

$$= 0{\cdot}003\ 176\ 4$$

Actual sound reduction, dB

$$= 10 \log_{10} \frac{1}{t_{AV}}$$

$$= -10 \log_{10} 0{\cdot}003\ 176\ 4$$

$$= 25 \text{ dB}$$

It can be seen that the poor insulation of the door of small area reduces the overall insulation very considerably. If the door had fitted badly the insulation would be even lower.

5.12 Requirements to achieve good sound insulation

1. Mass

The insulation from a single partition is approximately

$$R_{AV} = 10 + 14{\cdot}5 \log_{10} m$$

where R_{AV} = average sound reduction in dB

m = mass/ unit area in kg/m^2

The greater the mass, the larger the insulation provided by a partition. A one-brick wall (approximately 230 mm thick) has a mass of 415 kg/m^2 and gives an average insulation of about 50 dB. The use of mass up to that of the brick wall is often the most economic method of providing sound insulation. Above an average of 50 dB other methods must be considered. In some cases, structural considerations may prevent the use of mass to provide even moderate sound insulation.

2. Completeness

Example 5.6 showed that the actual insulation of the composite partition was much nearer the value given by the poorer part than that of the brickwork. A wall which might normally have an insulation of 50 dB would have this reduced by means of a hole of only $\frac{1}{100}$ of the total area to about 20 dB. The first consideration must be to try to raise the insulation of the poorest parts, which means that air gaps around doors and windows should be eliminated. Care must be taken where a partition is taken up to the underside of a perforated false ceiling (see Fig. 5.18). This may be a very difficult problem to overcome because of the services above the ceiling. The result can easily be a reduction in the insulation of the partition of 10 dB or more.

3. Multiple or discontinuous construction

This is not likely to be economic where the insulation of each leaf of the construction is already high. It would usually then be cheaper to use the same total mass in the form of a single partition. Any improvement in such a case would normally only be in the higher frequency region, where an improvement is less important. Thus a 280 mm cavity brick wall has little advantage over a 230 mm solid brick wall for sound insulation.

Where the panel is light in weight, a double partition may give a definite improvement provided that:

(a) the gap is large, below 50 mm having no advantage;
(b) the two panels are of different superficial weight;
(c) the gap contains sound-absorbent material;
(d) there are no air paths through the panels;
(e) the panels are not coupled together by the method of construction. It will be fairly clear that if it is necessary to tie the panels together, then to get the greatest impedance mismatch, very light ties should be used with heavy panels, and massive ties with light panels.

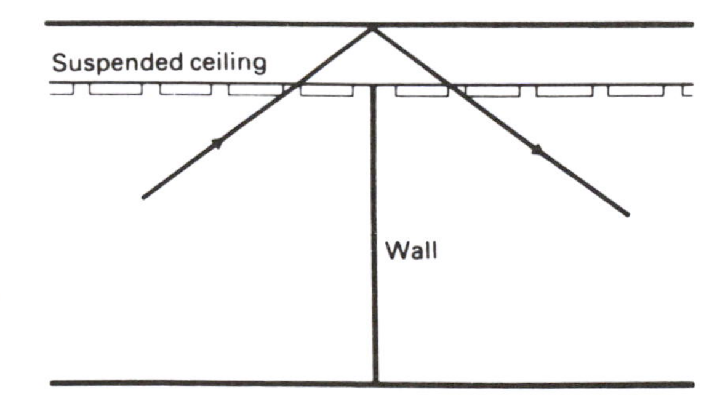

Fig. 5.18 An example of incomplete construction where a good sound path is present above a perforated false ceiling

Table 5.4 Sound reduction in dB for windows

Sound reduction index (dB)	Frequency (Hz)						
Construction	125	250	500	1000	2000	4000	Mean
1. 2·5 mm glass about 500 × 300 mm in metal frames, closed openable sections	23	12	18	22	29	27	21
2. Same as 1, but sealed	18	12	20	27	30	28	23
3. Same as 2, but wood frames	14	17	22	26	29	30	23
4. Same as 1, but 6 mm glass in wood frames	16	20	24	29	29	36	26
5. 25 mm glass 650 × 540 mm in wood frames	28	25	30	33	41	47	34
6. Same as 1, but double wood frames in brickwork, 50 mm cavity between glazing	19	19	27	37	47	52	33
7. Same as 6, but 100 mm cavity	22	20	35	43	51	53	37
8. Same as 6, but 180 mm cavity	28	25	36	43	50	53	39
9. Same as 8, but with acoustic tiles on reveals	25	34	41	45	53	57	42

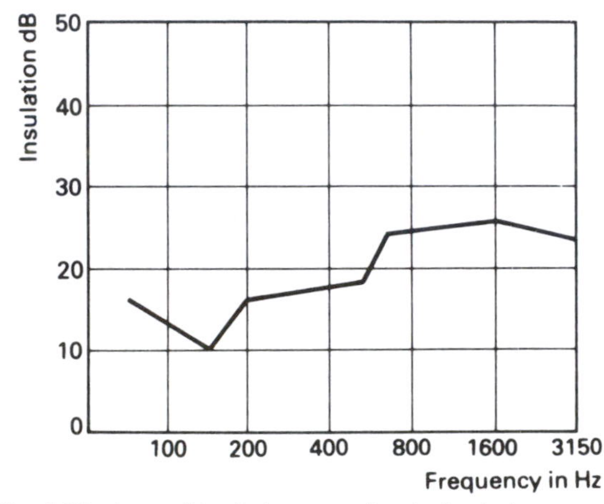

Fig. 5.19 A sound insulation curve for single glazing

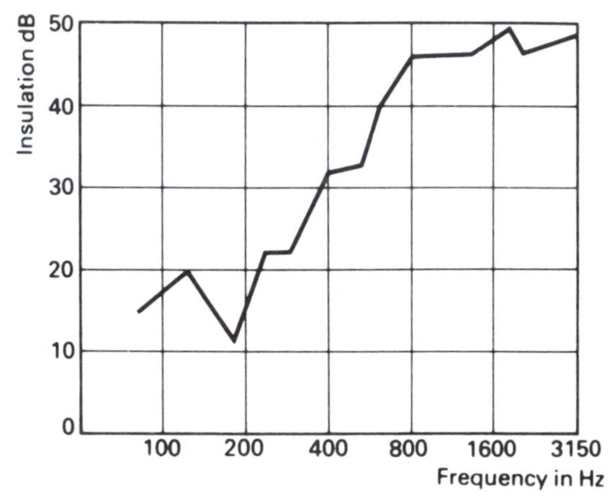

Fig. 5.20 A sound insulation curve for double glazing

The most common use of double construction is for windows. Typical insulation curves for single and double windows are shown in Figs 5.19 and 5.20, and sound reduction values are given in Table 5.4.

4. Apparent insulation by the use of absorbents

The necessity for a correction to be applied to the measured value of sound insulation shows that the sound pressure level in a room is to some extent dependent upon the reverberation time of that room. The improvement that can be achieved by the use of absorbent materials in dB = 10 $\log_{10} (A + a)/A$

where A = mumber of m² of absorption originally

a = number of m² of absorption added

The improvement in a highly reverberent room can be considerable, but in a room which already has a large amount of absorption, little improvement is likely to be possible.

Example 5.7

Using the Sabine formula calculate the area of absorption that needs to be added to a room of volume 100 m³ whose reverberation time was originally 2·00 s to give:

(a) 3 dB reduction in the sound pressure level
(b) 6 dB reduction in the sound pressure level
(c) 10 dB reduction in the sound pressure level

Amount of absorption originally:

$$2{\cdot}00 = \frac{0{\cdot}16 \times 100}{A}$$

$$\therefore A = 8 \text{ m}^2$$

$$\text{(a) } 3 = 10 \log_{10} \frac{8 + a_1}{8}$$

$$\therefore 0{\cdot}3 = \log_{10} \frac{8 + a_1}{8}$$

$$\therefore \frac{8 + a_2}{8} = 2$$

$$a_1 = 8 \text{ m}^2$$

$$\text{(b) } 6 = 10 \log_{10} \frac{8 + a_2}{8}$$

$$\therefore \frac{8 + a_2}{8} = 4$$

$$a_2 = 24 \text{ m}^2$$

$$\text{(c) } 10 = 10 \log_{10} \frac{8 + a_3}{8}$$

$$1 = \log_{10} \frac{8 + a_3}{8}$$

$$\therefore \frac{8 + a_3}{8} = 10$$

$$8 + a_3 = 80$$

$$a_3 = 72 \text{ m}^2$$

It can be seen that a tenfold increase is needed for 10 dB reduction. For 3 dB the amount of absorption must be doubled.

Absorbent materials can be useful near a source of sound because the amount needed can be fairly small. They can often be a way of localizing noise in factories. This is not a means of insulation and the improvement is not nearly as good but nevertheless can be useful. In a large factory it can improve safety because a machine operator can hear his own machine more clearly than others (see Fig. 5.21).

The use of absorbents in ducts and corridors can prevent noise reaching quite rooms.

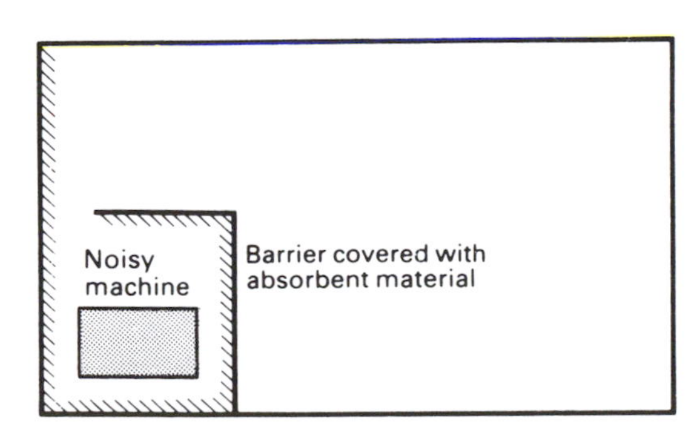

Fig. 5.21 Use of absorbent covered screen can help to localize noise. Absorbent material is necessary to prevent sound being reflected back to the operator of the machine raising the level there even more

5.13 Attenuation in ducts

Noise from ventilator fans or from other rooms can easily be transmitted by means of ducts. By lining the duct with absorbent an attenuation per metre, R_1, is obtained, given approximately by:

$$R_1 = \left(\frac{P}{S}\right) \alpha^{1 \cdot 4}$$

where P = perimeter of the duct in metres

S = cross-sectional area in m^2

α = absorption coefficient

This equation is accurate enough for most practical purposes provided that the area of the duct does not exceed about 0·3 m^2 for ducts which are more nearly square than 2:1.

Increased attenuation can be obtained by means of duct splitters running along all or part of the duct effectively making it a number of small parallel ducts. The chief disadvantage of this is that it cuts the total cross-sectional area with the result that the air speed must be increased.

5.14 Grille attenuation

Reduction of sound at the opening is dependent upon the area of the grille and the total sound absorption within the room.

$$R_2 = 10 \log_{10} \frac{A}{S}$$

where A = room absorption in m^2 units

S = area of the grille in m^2

As the absorption within a room may vary with frequency it may be necessary to make a number of calculations for grille attenuation.

Example 5.8

A fan for a hot-air heating system is found to produce the following noise:

Frequency (Hz)	125	250	500	1000
Level (dB)	90	85	90	75

The duct is 0·2 m × 0·4 m internally with a lining of absorbent whose coefficient is:

Frequency (Hz)	125	250	500	1000
Absorption coefficient, α	0·1	0·2	0·5	0·85

Calculate the level in a room of volume 500 m^3 with a reverberation time of 2·00 s if the duct is 4 m long and straight with a grille the same area as the duct.

Sabine's formula: $T = \dfrac{0\cdot 16V}{A}$

$$A = \frac{0\cdot 16 \times 500}{2\cdot 00}$$

$$= 40 \text{ m}^2 \quad \left(\text{Hence } 10 \log_{10} \frac{A}{S}\right)$$

Frequency	Fan level (dB)	α	$\alpha^{1\cdot 4}$	$4\frac{P}{S}\alpha^{1\cdot 4}$	$10 \log_{10}\frac{A}{S}$	Room level (dB)
125	90	0·1	0·04	2·4	27	60
250	85	0·2	0·10	6	27	52
500	80	0·5	0·38	22·8	27	30·2
1000	75	0·85	0·80	48	27	0

5.15 Attenuation due to bends

The amount of attenuation at a bend depends upon frequency, size of duct and whether it is lined or unlined. An approximate relation is given in Table 5.5.

Table 5.5 Attenuation at right-angled bend for lined square ducts

Octave band centre frequency (Hz)	0·1m^2 Area (square)	1 m^2 Area (square)
63	0	1
125	1	3
250	3	8
500	6	16
1000	8	17
2000	16	18
4000	17	18
8000	18	18

5.16 Sound power levels of ventilation fans

Information about the sound power level in each octave band will normally be supplied by the manufacturer. Alternatively, measurements may be made using a sound level meter and microphone equipped with a windshield. In the absence of the necessary data it is possible to estimate the values approximately from the formula:

$$L_W = 61 + 10 \log_{10} W$$

where L_W = sound power level in dB re 1 picowatt

W = power of the motor in watts

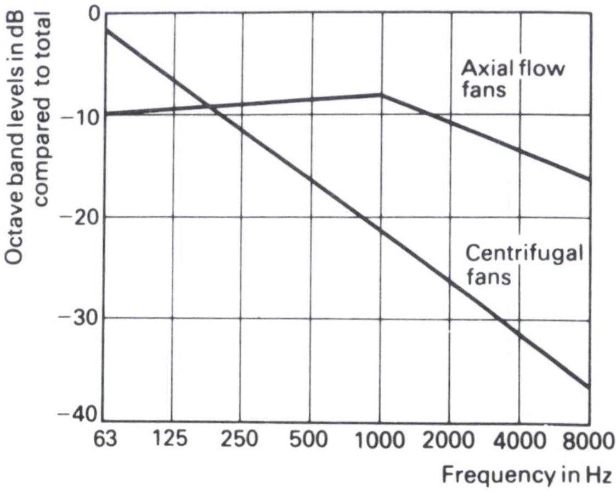

Fig. 5.22 Spectral distribution for axial and centrifugal fans

This formula assumes that the motor is run at full power. If a fan is run at half its rated power then a reduction of 6 dB will result. If a larger motor is used twice the power the formula shows that an increase of 3 dB is expected.

The spectra of sound depend upon the type of fan being used. It can be seen from Fig. 5.22 that centrifugal fans produce most of their noise in the low frequencies whereas axial flow fans have a far more even distribution.

The sound pressure level in a room may be calculated from:

$$\text{SPL} = L_W - 10 \log_{10} \left(\frac{A}{4}\right)$$

where A = area of room absorbent in m^2

When the sound power level is known, the distribution through the duct system can be calculated taking into account that where ducts split, the energy will be divided. If the maximum permissible level in a room is known, the attenuation needed within the system can be found. Further details of noise in HVAC systems are given in Chapter 8.

Questions

(1) The noise level from a sewing machine was found to be 96 dB (linear) at a distance of 1 metre. Find the sound power of the machine in picowatts. (Assume it is in a free field.)

(2) A lighthouse has a foghorn with a sound power output of 100 watts. If it radiates over a quarter of a sphere, at what distance would the level be 100 dB? Why would a low-frequency horn be better than one of higher frequency?

(3) Calculate the sound output of a man whose loud

speech produces a sound pressure level of 80 dB(C) at a distance of 1 metre.

(4) (a) Explain what is meant by each of the terms 'decibel' and 'sound reduction index'.

(b) An external wall, of area 4 m by 2·5 m, in a house facing a motorway is required to have a sound reduction of 50 dB. The construction consists of a 280 mm cavity wall containing a double-glazed (and sealed) window. The sound reduction indices are 55 dB for the 280 mm cavity wall and 44 dB for the sealed double-glazed window with a 150 mm cavity. Calculate to the nearest 0·1 m^2 the maximum size of window to achieve the required insulation.

(5) (a) Explain how sound is transmitted from one room to another in a building and show how it can be reduced.

(b) A partition across a room 6 m by 5 m includes a timber doorway of area 2 m^2. Determine by how much the partition will be a better insulator of sound if constructed in 115 mm brickwork instead of 100 mm building blocks, each plastered both sides, at the particular frequency where the sound reduction indices of brick, building block and timber are respectively 48 dB, 34 dB and 26 dB.

(6) (a) Name four principles to be observed when considering sound insulation and say whether they are effective against airborne or structure-borne sound.

(b) Show how to apply these principles in dividing a room into a consultant's office and waiting room.

(c) Calculate the average sound reduction factor of a partition made of 24 m^2 of 115 mm brickwork and 6 m^2 of plate glass where the sound reduction factors of brickwork and glass at a certain frequency are 45 dB and 30 dB respectively.

(7) A room of volume 200 m^3 is supplied with air from a centrifugal fan of 400 watts. Find the amount of attenuation needed in each octave band. The reverberation times of the room and the maximum acceptable levels at the different frequencies are:

Octave band centre frequency (Hz)	63	125	250	500	1000	2000	4000
Reverberation time (s)	5	3	1·5	0·6	0·5	0·5	0·6
Maximum acceptable levels (dB)	66	59	52	46	42	40	38

(8) Calculate the attenuation provided by a 0·3 m square duct 3 m long and lined with material having the following absorption coefficients:

Octave band centre frequency (Hz)	63	125	250	500	1000	2000	4000
α	0·10	0·37	0·47	0·70	0·80	0·85	0·80

(9) Calculate the sound reduction of a partition of total area 20 m^2 consisting of 15 m^2 of brickwork, 3 m^2 of windows and 2 m^2 of door. The sound reduction indices are 50 dB, 20 dB and 26 dB for the brickwork, windows and door respectively.

(10) A workshop of volume 1000 m^3 has a reverberation time of 3 seconds. The sound pressure level with all the machines in use is 105 dB at a certain frequency. If the reverberation time is reduced to 0·75 seconds what would the sound pressure level be?

(11) Find the reduction in sound pressure level of a 165 Hz frequency by a barrier of height 4 m alongside a motorway if the noise source is 500 mm above the base of the barrier and 5 m away from it. The listener is 1 m below the base and 10 m away from it.

(12) If an average sound reduction of 20 dB is required by a barrier, what is the approximate minimum mass per unit area required?

(13) (a) Discuss the problems associated with field measurement of airborne sound insulation as opposed to laborarory measurements and explain what corrections are necessary to achieve valid comparative results.

(b) In a field test carried out to assess the transmission loss through a partition separating two adjoining dwellings, the following results were obtained: At a frequency of 250 Hz the sound pressure levels at six positions in the source room were respectively 99, 100, 95, 95, 97 and 98 dB. In the receiving room the levels were 60, 58, 59, 63, 59 and 61 dB. The measured reverberation time in the receiving room was 1·53 seconds. Apply the appropriate correction factor and calculate the normalized level difference between the two rooms.

(14) (a) Explain what is meant by the term 'absorption coefficient'.

(b) A factory workshop with hard uniform surfaces have dimensions 30 m by 15 m by 6 m high and a dB(A) weighted reverberation time of 3 seconds. The noise level in the workshop is 98 dB(A). Predict the approximate new noise level if the ceiling is covered with acoustic absorbent, having a dB(A) weighted absorption coefficient of 0·8.

(15) (a) In the context of noise on construction and demolition sites explain what is meant by L_{eq}.

(b) The sound pressure level of noise from a stationary item of plant on a construction site was 81 dB(A) at a distance of 10 m.
The window of a flat is 40 m from this item. It is of single-glazed construction and of area 1 m^2 with a sound reduction index of 19 dB. If the total area of exposed external brick wall of a room is 5 m^2 find the level in the room. (Sound reduction index for the brickwork may be taken as 51 dB.)

(16) (a) Describe the method of measurement of the impact sound insulation provided by a floor between two flats.

(b) In a measurement of the impact and airborne sound insulation provided by a floor, the following results were obtained at 500 Hz using one-third octave filters:

Airborne sound:	Source room level	105 dB
	Receiving room level	58 dB
	Receiving room reverberation time	1·2 s
Impact sound:	Receiving room level	65 dB

Find the correct airborne sound insulation and the octave-band sound pressure level for the impact sound. The Building Regulations give figures of 48 dB for sound reduction and 66 dB for octave-band sound pressure levels, at that same frequency. State whether the floor is likely to be satisfactory at this frequency. *(HND)*

(17) Discuss the principles underlying the 'mass law' for sound reduction by a partition. Describe how practical partitions deviate from this law.
A floor of 20 dB sound reduction index in the 2000 Hz octave band is 2 m high by 1 m wide. It is set in a wall 6 m long by 2·5 m high, which has a sound reduction of 50 dB in the same frequency band. Estimate the sound reduction of the composite partition if:
(i) the door fits perfectly;
(ii) there is a gap of 3 mm around the edge of the door. *(HND)*

(18) The typists of a large typing pool have complained about the level of noise generated within the room in which the furnishings are somewhat austere.
Explain how the level of noise can be reduced and suggest suitable treatments. *(IOB)*

(19) Using annotated diagrams illustrate the various routes by which sound may be transmitted into a dwelling unit in a multi-storey block, and the measures that may be taken to reduce the nuisance value of such sound transmission. *(IOB)*

(20) (a) Define sound reduction index, sound transmission coefficient and normalized sound level differences as applied to the acoustic properties of a partition.

(b) Describe fully with the aid of diagrams the instrumentation required for field measurements of airborne sound transmission in buildings.

(c) Explain the procedures which are adopted to give an accurate measurement of sound level difference. *(IOA)*

(21) A new office building, 25 m high, is to be erected close to an urban motorway. Describe, with the aid of sketches where appropriate, details of how noise levels in the building from the following sources could be reduced:
(a) external noise from the motorway;
(b) internal noise due to office machinery;
(c) internal noise from the mechanical services, including an airconditioning plant room. *(IOB)*

(22) What do the building regulations specify concerning noise control in dwellings?
The sound insulation of a wall is found to fail the requirement for party walls in the building regulations. The insulation has a total adverse deviation of 60 dB mainly below 800 Hz, but with some adverse deviation about 2 kHz.
The wall is of a dry construction having 10 mm plasterboard mounted on both sides of 100 mm timber studs.
Suggest steps to bring the sound insulation up to the standard for party walls. *(IOB)*

(23) Which components of a ventilation system attenuate the sound power by a fan which is used to ventilate a room?
Give a brief description of the mechanism involved in each case and state the frequency range in which you would expect it to be most effective.
Describe the principal features of a duct attenuator. How is the insertion loss of such an attenuator measured under standardized conditions?
What procedures can be adopted to reduce noise at a grille or diffuser through which air is being discharged into a room. *(IOA)*

(24) The safe floor-loading in a building places an upper limit on the permissible weight per unit length of any wall dividing the space. What steps would you propose to obtain the greatest sound level difference that can be achieved between two adjacent rooms constructed in the space?
A length of ventilation ductwork of cross-section 0·3 m × 0·15 m passes through a dividing wall, which is 5 m wide and 3 m high, and opens into each room.
The following data applies to the system.

Frequency (Hz)	125	500	2000
Sound reduction index of wall (dB)	25	40	55
Total sound absorption in each room (m^2)	5	10	15

Calculate the least value of the sound level difference that you might expect between the rooms at each frequency.
Indicate a reason why the estimate might prove to be pessimistic.
Calculate (to the nearest decibel) the insertion loss through the duct to ensure that the maximum achievable level difference is not degraded by more than one decibel at any frequency. *(IOA)*

(25) There are a number of schemes throughout the UK for home insulation against environmental noise.
Discuss the differences in insulation requirements to protect the householder from traffic noise and aircraft noise.
Would you expect an external noise barrier to be effective in each case? *(IOA)*

(26) A motorway is planned to pass close to a cottage in a rural area. Discuss the parameters which determine the traffic noise environment at a distance of 1·0 metre from the nearest façade of the cottage.
Illustrate your answer by describing the manner in which these parameters are accounted for in traffic noise prediction methods.
Describe the steps which could be taken to reduce the traffic noise to an acceptable level at the nearest façade if your predictions suggest that the level is likely to be unacceptable. *(IOA)*

6 Criteria

The optimum noise environment must depend upon both subjective and economic limitations. To find people's requirements necessitates a consideration of three main factors: the noise level, the individual and his activity. It is impossible to make allowances for individual differences and criteria must be stringent enough to ensure that nearly everyone is satisfied. There may be multiple activities within one part of a building, such as a hospital ward. Noise requirements are activity dependent, so a noisy data-processing office is not going to have the same criteria as an executive office, for example.

There are three main bases for the determination of criteria: hearing damage, speech interference and nuisance. The first two are fairly easy to determine, whereas nuisance is far less clear-cut.

6.1 Deafness risk criteria

A sound pressure level of about 150 dB would result in instant hearing damage. People must not be exposed to noises of this magnitude, and in fact it is unwise to exceed 140 dB even for noises of under one minute duration. Fortunately, noise of this level very rarely occurs in buildings. Of far greater importance are the long-term effects of noise, particularly in industry. Table 2.4 gave a suggested limit for occupational noise 8 hours a day, 5 days a week for 40 years. An increase in these levels is possible if the period of exposure each day is reduced, whereas a decrease would be necessary if the sound was of very narrow bandwidth, as may sometimes occur in industry.

Where it is necessary to exceed the criteria, hearing conservation measures need to be taken. These include the use of earplugs or earmuffs, shorter exposure and regular audiometric tests (see Fig. 6.1).

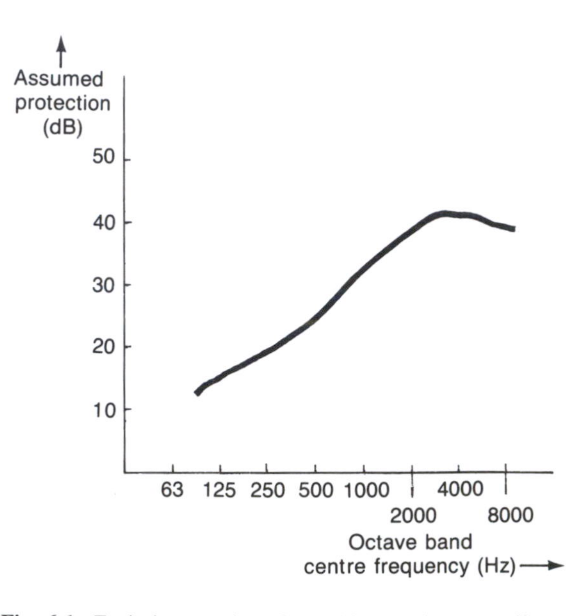

Fig. 6.1 Typical attenuation of sound by wearing earmuffs; maximum attenuation is in the high frequency region

6.2 Performance specification of ear protectors

The attenuation of noise provided by ear protectors (earmuffs or earplugs) is measured according to BS 5108 Part 1:1991, in a method which involves measuring the hearing threshold of test subjects with and without wearing the protectors. Since the attenuation varies with a number of factors, including how well the protector fits the particular individual, an average attenuation is specified, from tests with several subjects, together with a standard deviation indicating the variability from the average. Only half the population of wearers will enjoy the degree of

Table 6.1 Data for calculating the assumed protected level

(1) Octave-band centre frequency (Hz)	(2) Octave-band SPL (dB)	(3) A weighting value (dB)	(4) A-weighted band SPL (dB)	(5) Assumed protection (dB)	(6) Assumed protected A-weighted band SPL (dB)
63	91	−26	65	11	54
125	104	−16	88	13	75
250	100	−9	91	17	74
500	105	−3	102	20	82
1000	108	0	108	25	83
2000	110	+1	111	29	82
4000	107	+1	108	32	76
8000	103	−1	102	29	73

protection indicated by the average attenuation values, with the other half being less well protected. Noise Guide 5, issued under the Noise at Work Regulations 1989, specifies that the mean value minus one standard deviation should be taken as the assumed protection, as shown in Table 6.1. Typical values of assumed protection for muffs and plugs are given in Fig. 6.1, which shows that the attenuation tends to increase with frequency. These values are used to calculate the assumed protected level in dB(A) at the wearer's ear.

Example 6.1

An employee works at a location where the octave-band sound pressure levels, from 63 Hz to 8 Hz, are as follows:

91 dB, 104 dB, 100 dB, 105 dB, 108 dB, 110 dB, 107 dB, 103 dB

Calculate the assumed protected level, in dB(A), at the employee's ear when he or she is wearing ear protectors having the assumed protection shown in Table 6.1.

The procedure is as follows. Column 4 is obtained by subtracting column 3 from column 2, and column 6 is obtained by subtracting column 5 from column 4.

The final answer, the assumed protected level (APL) in dB(A) is obtained by combining the decibel values of column 6, either by using the chart method (Chapter 1) or using a calculator:

$$\text{APL} = 10 \log_{10} (10^{5.4} + 10^{7.5} + 10^{7.4} + 10^{8.2} + 10^{8.3} + 10^{8.2} + 10^{7.6} + 10^{7.3}) = 88 \text{ dB(A)}$$

The level of the noise at the unprotected ear is 115 dB(A), obtained by combining the levels of column 4, so that in this case the protectors have provided a reduction of 27 dB(A). A different noise spectrum would produce a different reduction, because the attenuation varies with frequency, so it is not possible to quote the attenuation as a simple dB(A) value.

An alternative graphical method is provided in Noise Guide 5 for converting octave-band levels into dB(A). The octave-band spectrum is plotted on a chart against a set of approximate dB(A) contours. The dB(A) value is that of the highest contour touched by any of the octave-band values.

The Noise at Work Regulations require that the ear protectors provide sufficient attenuation to reduce noise exposure levels to below the second action level of 90 dB(A). In cases where the noise level is just a few decibels above 90 dB(A), most plugs and muffs will give adequate protection, and a major consideration in selecting the type of protector will be the comfort of the wearer, to ensure that the protection is always worn. But if the noise level is above 100 dB(A), a calculation similar to this example must be performed to ensure that the protection is adequate. Noise Guide 5 gives information about the various types of protectors which are available and other factors to be considered in their selection. The manufacturers of ear protectors also provide much useful information.

If the variability of the attenuation from the mean follows a Gaussian statistical distribution, the above calculation gives the protection afforded to 68% of the wearer population. If attenuation values of mean minus two standard deviations were used, this percentage would increase to 95%. The assumed protected level would then be 85 dB(A).

6.3 Speech interference criteria

Necessity for speech communication will set a more severe limit on the maximum levels of ambient noise than hearing damage. Clarity of speech in the presence of masking noises is dependent not only on their magnitude, but also upon the loudness of the speaker's voice and the hearing acuity of the listener. Little allowance can be made for the variations in the last two and an average is taken. The noise

Table 6.2 Speech interference levels

Distance from speaker to hearer (m)	Maximum level of background noise to ensure speech intelligibility (dB) Normal voice	Raised voice	Very loud voice	Shouting
0·1	73	79	85	91
0·2	69	75	81	87
0·3	65	71	77	83
0·4	63	69	75	81
0·5	61	67	73	79
0·6	59	65	71	77
0·8	56	62	68	74
1·0	54	60	66	72
1·5	51	57	63	69
2·0	48	54	60	66
3·0	45	51	57	63
4·0	42	48	54	60
8·0	35	41	47	53
16·0	29	35	41	47

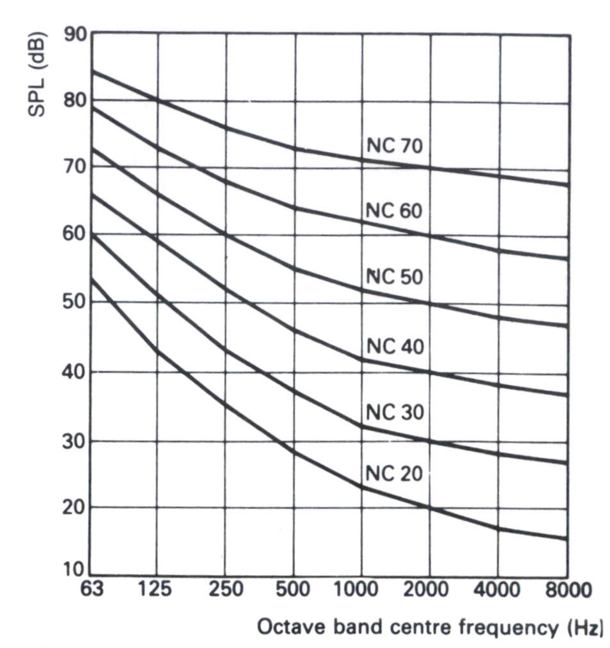

Fig. 6.2 Noise criteria curves (NC)

spectra will also determine the interfering effect. One method involves taking the arithmetic mean of the sound pressure levels in the three octave bands with centre frequencies of 1 kHz, 2 kHz and 4 kHz. This average is known as the *speech interference level* (SIL). Distance between speaker and listener is important. Table 6.2 gives permissible speech interference levels for different distances. These are for a normal voice, and if the level is raised an increase in the speech interference level is permissible.

It is possible for the speech interference level to be below the levels shown in Table 6.2 but for the noise to interfere because of the presence of excessive low frequency components. To overcome this difficulty the *loudness level* in phons is also calculated. This is defined as the sound pressure level of a 100 Hz note that sounds equal in loudness to the noise concerned. If the difference in the loudness level and the speech interference level is less than 22 there will be general satisfaction with the noise environment provided the speech interference level is below the appropriate maximum. However, if the loudness level minus the speech interference level is greater than 30 there will be general dissatisfaction even if the speech interference level is low enough. This is due to the presence of low frequency noise.

Noise criteria (NC) curves are used to specify the maximum speech interference level with the added fact that the difference between loudness level and the speech interference level is equal to 22 (see Fig. 6.2). If lower standards for the aural environment are acceptable then the alternative noise criteria (NCA) curves may be used for specification. These allow the presence of more low frequency noise and are calculated on the basis of the loudness level minus the speech interference level being equal to 30 (see Fig. 6.3).

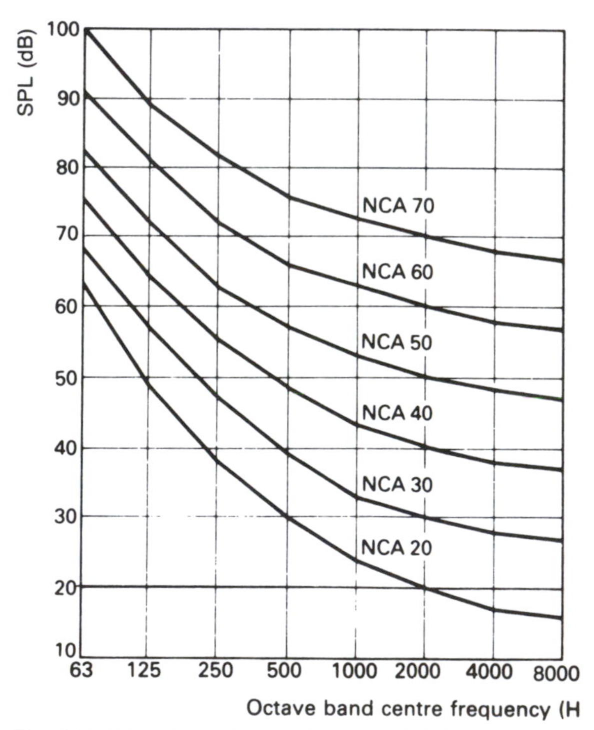

Fig. 6.3 Alternative noise criteria curves (NCA)

Example 6.2

Calculate the sound insulation needed to achieve the

(a) NC 25 curve
(b) NCA 25 curve

in a classroom if there is an external noise of:

Octave band centre frequency (Hz)	63	125	250	500	1000	2000	4000	8000
Level (dB)	75	71	70	69	65	62	61	60

(a) To achieve NC 25 level

Octave band centre frequency (Hz)	Level (dB)	NC 25 (from Fig. 6.2)	Insulation (dB)
63	75	57	18
125	71	47	24
250	70	39	31
500	69	32	37
1000	65	28	37
2000	62	25	37
4000	61	22	39
8000	60	21	39

(b) To achieve NCA 25 level

Octave band centre frequency (Hz)	Level (dB)	NCA 25 (from Fig. 6.3)	Insulation (dB)
63	75	66	9
125	71	53	18
250	70	42	28
500	69	34	35
1000	65	28	37
2000	62	25	37
4000	61	22	39
8000	60	21	39

Each of these standards could be achieved by means of a 115 mm brick wall.

Example 6.3

The recommended maximum sound level for the sleeping area of a house is the NC 35 curve. The external bedroom wall of a house is 4 m by 2·5 m in area, facing a motorway at a distance of 30 m from the line of traffic. An octave-band analysis at 7·5 m from the line of traffic gave the following results:

Octave band centre frequency (Hz)	Sound level (dB)
63	99
125	95
250	93
500	92
1000	93
2000	90
4000	88
8000	85

The construction consists of a 280 mm cavity wall containing a double-glazed (and sealed) window. The sound reduction indices are 55 dB for the 280 mm cavity wall and 44 dB for the sealed double window, both at 1000 Hz. Assume for the purpose of this question that the insulation of both the window and brickwork improves by 5 dB for each doubling of the frequency. Calculate, for 1000 Hz, the maximum area of window to the nearest 0·1 m^2. State what the insulation is expected to be at the other seven octave values and whether they are likely to be adequate.

The level of the sound will drop 3 dB for each doubling of the distance for noise from a line source. In this example the distance has been doubled twice so that 2 × 3 = 6 dB must be subtracted to find the level immediately outside the window.

Octave band centre frequency (Hz)	63	125	250	500	1000	2000	4000	8000
Motorway (dB)	99	95	93	92	93	90	88	85
Outside window (dB)	93	89	87	86	87	84	82	79

To find the necessary insulation, the maximum recommended level given by the NC 35 curve must be subtracted from the values outside the window. Hence:

Octave band centre frequency (Hz)	63	125	250	500
Outside window (dB)	93	89	87	86
NC 35	63	55	47	42
Insulation needed (dB)	30	34	40	44

Octave band centre frequency (Hz)	1000	2000	4000	8000
Outside window (dB)	87	84	82	79
NC 35	37	35	33	32
Insulation needed (dB)	50	49	49	47

Thus it can be seen that the necessary insulation at 1000 Hz must be 50 dB.

Now the reduction in dB $= 10 \log_{10} (1/t)$ where t is the transmission coefficient

$$\therefore\ 50 \text{ dB} = 10 \log_{10} \frac{1}{t_{av}}$$

(t_{av} = average transmission coefficient for the window and brickwork combined)

$$\therefore\ 5 = \log_{10} \frac{1}{t_{av}}$$

$$= -\log_{10} t_{av}$$

$$\therefore\ \bar{5}\cdot 0000 = \log_{10} t_{av}$$

$$\text{av} = 0\cdot 000\ 01$$

For the 280 mm cavity wall:

$$55 \text{ dB} = 10 \log_{10} \frac{1}{t_B}$$

(t_B = transmission coefficient for the brickwork)

$$\therefore\ 5\cdot 5 = -\log_{10} t_B$$

$$\bar{6}\cdot 5 = \log_{10} t_B$$

$$t_B = 0\cdot 000\ 003\ 162$$

Similarly for the window:

$$44 = 10 \log_{10} \frac{1}{t_W}$$

(t_W = transmission coefficient for the window)

$$\therefore\ 4\cdot 4 = -\log_{10} t_W$$

$$\bar{5}\cdot 6 = \log_{10} t_W$$

$$t_W = 0\cdot 000\ 039\ 81$$

Now:

$$(\text{Total area}) \times t_{av} = (\text{area of brick}) \times t_B + (\text{area of window}) \times t_W$$

$$\text{Let the window area} = a \text{ m}^2$$

$$\text{Total area} = 10 \text{ m}^2$$

$$\text{Area of brick} = (10 - a) \text{ m}^2$$

$$\therefore\ 10 \times 1\cdot 0 \times 10^{-5} = (10 - a) \times 3\cdot 162 \times 10^{-6} + a \times 3\cdot 981 \times 10^{-5}$$

$$1\cdot 0 \times 10^{-4} = 3\cdot 162 \times 10^{-5} - 3\cdot 162 \times 10^{-6} a + 3\cdot 981 \times 10^{-5}$$

$$\therefore\ 6\cdot 838 \times 10^{-5} = 3\cdot 665 \times 10^{-5} a$$

$$a = \frac{6838}{3665}$$

$$= 1\cdot 9 \text{ m}^2 \text{ to the nearest } 0\cdot 1 \text{ square metre}$$

A 5 dB improvement in insulation for each doubling of frequency also means a reduction of 5 dB for each halving of the frequency below 1000 Hz. Thus the expected insulation would be:

Hz	Insulation (dB)
2000	55
4000	60
8000	65
500	45
250	40
125	35
63	30

Comparing these values with those from the NC curve the insulation should be adequate for all frequencies.

6.4 Noise rating curves

The spectra of the noise rating curves (Fig. 6.4) is essentially very similar to those of the noise criteria curves in that the octave-band levels are highest at the lowest frequencies and decrease progressively with increasing frequency; less annoyance is caused by low frequency noise than by a higher frequency noise at the same sound pressure level. Hence higher levels may be tolerated for low frequency noise. The curves may be used for several purposes including the assessment of the acceptability of

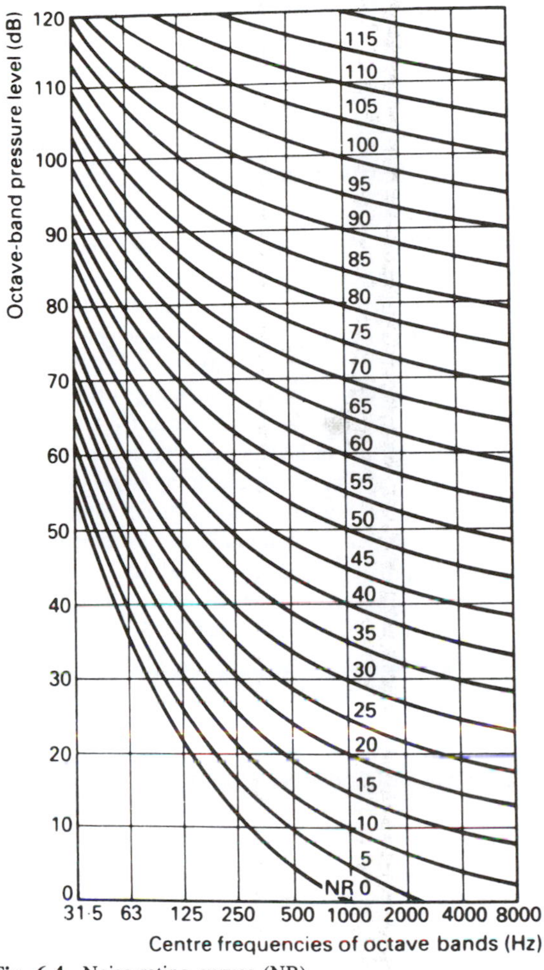

Fig. 6.4 Noise rating curves (NR)

a noise to ensure preservation of hearing, speech communication, and to avoid annoyance.

Corrections need to be applied to noise ratings for broad band continuous noise of −5, if it is pure tone, intermittent or impulsive. dB(A) values can be taken to be between 4 and 8 units above the equivalent NR number.

6.5 Criteria for offices

Speech interference levels were developed to give office criteria. The use of an NC or NCA curve is therefore convenient. Selection of a particular NC or NCA curve is dependent upon the type and use of the office, combined with the economic requirements. NCA curves should not be used unless cost considerations prevent the use of NC curves. Levels of background noise from NC 20 for the best executive offices to NC 55 in offices containing typewriters or other machinery are suitable.

Table 6.3 Maximum levels of background noise in L_{Aeq} to ensure speech intelligibility

Distance from speaker to hearer (m)	Maximum noise level L_{Aeq} (dB)	
	Normal voice	Raised voice
1	57	62
2	51	56
4	45	50
8	39	44

It seems that for certain types of offices, higher levels would be acceptable where speech is not important. In data-processing offices levels of 79 dB(A) for machine operators, 72 dB(A) for punch operators and 64 dB(A) for other clerical staff appear to be satisfactory. The main difference in the latter cases is that the noise is under the control of the operators and is not an external sound.

BS 8233:1987 gives criteria for good speech communication which uses the equivalent continuous A-weighted sound pressure level (L_{Aea}), Table 6.3. Hence, in private offices and small conference rooms it suggests a maximum of 40–45 dB (L_{Aeq}), and in large offices a maximum of 45–50 dB.

6.6 Audiometric rooms

In order that audiometric rooms can measure accurately the amount of hearing loss, if any, that a person suffers, it is important that there is no masking noise. A criterion must be set at such a level that it is possible to measure the threshold of hearing for people with normal hearing (zero hearing loss). Figure 6.5 shows the maximum permissible levels measured in octave bands for this to be achieved. If the levels are measured in half- or third-octave bands the criteria must be set correspondingly lower.

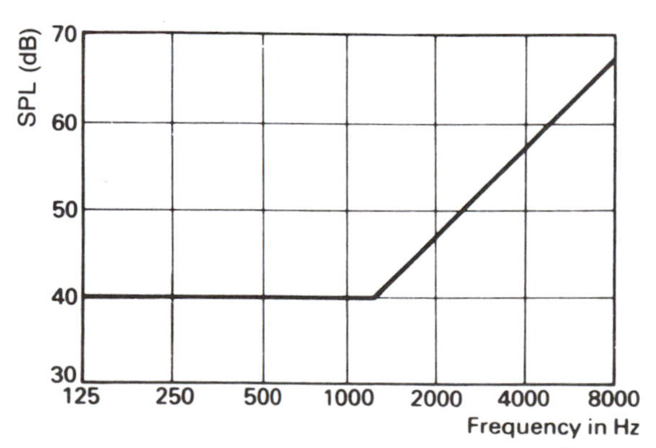

Fig. 6.5 Maximum sound pressure levels for no masking above zero hearing loss

The criteria for third-octave bandwidth measurements at each frequency are $10 \log_{10} 3 = 10 \times 0{\cdot}4771 \simeq 5$ dB lower, and for half octaves, $10 \log_{10} 2 = 3$ dB lower.

If narrowband sound is present then the criteria given can be satisfied, but masking may be caused at the audiometric measuring frequencies. It is necessary, therefore, to check that there is no narrowband sound at frequencies near the test frequency. For further information on this the reader is referred to bibliographical reference, Chapter 2 references 25 and 36.

Where the audiometric room is only to be used for testing people with some hearing loss, higher levels of background noise are permissible. This means that for testing people with a 10 dB hearing loss, the criteria can be relaxed by approximately 10 dB at each frequency, and similarly 20 dB higher for a 20 dB hearing loss. It also applies, however, that if the room is to be used to measure the hearing acuity of people with better than average hearing, a more severe criterion would be needed. A maximum room noise level of 30 dB(A) is recommended in practice.

6.7 Hospitals

There are no standards at present relating to noise or noise control in hospitals. There are reports from various countries suggesting criteria. The NC 30 curve would appear to be in general agreement with these recommended maximum criteria for hospital wards (see Table 6.4).

Hospitals probably represent one of the most difficult environments for which to obtain satisfactory criteria. Of necessity they must have a considerable quantity of mechanical aids. At the same time people are sleeping while others are working. Some internal ward noise is inevitable. If internal noise is reduced to too low a level by means of insulation, the masking effect it provides is lost and internal noises are more noticeable. In large wards there is an additional problem, in very quiet conditions, of patients hearing doctors discussing their diagnosis of other patients. Taking this need for masking into account, daytime levels from all sources within the range 45–50 dBA have been provisionally recommended.

Night-time noise may represent a severe problem at far lower levels than those recommended. In the absence of external masking noises, structure-borne noises can become far more serious. Care in placing machinery in the planning stage is needed to prevent this becoming disturbing to patients. Machinery should be selected with due regard to its noise level as well as to cost.

Table 6.4 Suggested criteria for hospitals

Bedrooms	NC 30
Day rooms	NC 35
Treatment rooms	NC 35
Bathrooms	NC 35
Toilets	NC 35
Kitchens	NC 40

6.8 Domestic dwellings

Ideally the level in the sleeping area of homes should not exceed the NC 25–35 curves. One problem in dwelling-houses will be the noise from neighbours. A correctly specified minimum wall insulation should make certain that, with the exception of particularly noisy neighbours, the desired reduction is achieved. Criteria for impact and airborne sound insulation were given in Fig. 4.8 and Table 4.3, and Fig. 5.15 and Table 5.2, respectively.

To achieve a greater insulation would be uneconomic and probably undesirable as people like to be aware of a small amount of neighbourhood noise during the daytime. In any case noise produced by equipment within an individual dwelling will provide some masking effect.

It has been suggested that the noise criteria for bedrooms depend upon a number of factors, and Table 6.5 shows how this may be calculated using the NR curves. These criteria may be given approximately in dB(A) for bedrooms as shown in Table 6.6.

Table 6.5 Noise rating (NR) criteria for bedrooms and living rooms

	Noise rating (NR)
Basic criteria for living rooms	30
Basic criteria for bedrooms	25
Corrections:	
Pure tone easily perceptible	−5
Impulsive and/or intermittent	−5
Noise only during working hours	+5
Noise only 25% of time	+5
6%	+10
1·5%	+15
0·5%	+20
0·1%	+25
0·02%	+30
Economic tie	+5
Very quiet suburban	−5
Suburban	0
Residential urban	+5
Urban near some industry	+10
Area of heavy industry	+15

Table 6.6 Criteria for the maximum noise levels in dB(A) in bedrooms

Country	25
Suburban	30
Urban	35
Busy urban	40

6.9 Classrooms and lecture rooms

The problem is primarily one of speech interference. Ideally a level no greater than the NC 25 curve should be achieved for continuous noise. It appears that higher levels can be satisfactory, provided they do not exceed the NC 35 curve. Where the noise is intermittent and infrequent, such as is the case with railway and aircraft noise, lower standards may be tolerable. When the noise occurs every few minutes or more frequently, the interruption can have serious effects on the rate of learning.

A recent development which should cut down intermittent noise involves windows which are automatically closed when the external noise level exceeds a predetermined amount. This system is basically a microphone located on the roof of the school controlling a hydraulic servo-mechanism to close all the double-glazed windows. The advantage of this method is that ventilation problems are overcome. The alternative of fixed double-glazed windows requires a system of forced ventilation which will itself produce noise exceeding the NC 25 level.

Music room requirements are similar to those of classrooms, but present a problem if situated close to other teaching rooms. The level of sound produced even by a small school orchestra can be fairly high. If it is necessary to situate a music practice room near to a teaching room, then increased insulation is advisable, equivalent to a 230 mm brick wall.

Present Department for Education (DFE) recommendations for the maximum background noise levels (those not generated by class activities) are given in terms of the equivalent continuous A-weighted sound pressure level as shown in Table 6.7.

Table 6.7 Maximum background noise levels in educational buildings

Type of space	Maxiumum $L_{\mathrm{Aeq},T}$ dB
Circulation spaces, workshops, practical areas	50
Teaching groups of <15 within a communication distance <4 m Resource areas, individual study spaces, libraries	45
Teaching groups of 15–35 within a communication distance ≤8 m Small lecture rooms Medical inspection rooms Seminar rooms	40
Teaching groups >35 Large lecture rooms Language laboratories	35
Music and drama spaces	30

6.10 Hotels

The requirements for hotel bedrooms are similar to those for domestic dwellings with a recommended level of around NC 25. This is quite impossible economically in certain places, and slightly higher levels up to about NC 35 or 40 should be acceptable. Restaurants should be satisfactory if they achieve NC 45 or NR 45.

The standards of insulation between hotel bedrooms should normally be similar to that between private dwellings for both walls and floors. Lower standards are acceptable where radios and televisions are not used. However, if new buildings are constructed to lower standards, problems may arise during future upgrading.

6.11 Broadcasting and recording studios

Different research workers have suggested various criteria including those lying below the threshold of hearing at low frequencies, and below the level of self-generated noise of a high quality condenser microphone at higher frequencies. (This exceeds the threshold of hearing at high frequencies.) The levels given in Table 6.8 would be more reasonable. A noise rating criterion of NR 15 has been suggested. This is similar to the values in Table 6.8 for Drama.

Table 6.8 Maximum permissible background noise levels in studios

Octave band centre frequency (Hz)	Sound only Light entertainment (dB)	Drama (dB)	Other sound All television (dB)
63	55	45	50
125	45	36	40
250	38	27	32
500	32	23	27
1000	27	18	22
2000	23	14	17
4000	20	10	14

6.12 Concert halls

The best concert halls should have a background noise no higher than the threshold of hearing for continuous noise. This level is shown approximately in Table 6.9. In many cases this standard is too costly and a lower specification

Table 6.9 Maximum background noise levels for concert halls

Octave band centre frequency (Hz)	63	125	250	500	1000	2000	4000	8000
Sound pressure level (dB)	53	38	28	18	12	11	10	22

must be accepted, but not worse than the NC 20 curve. Opera-houses and large theatres with more than 500 seats should achieve NC 20. These figures are in line with those recommended in BS 8233:1987 which suggest 20–30 dB(A) for concert halls, opera-houses, large theatres and auditoria and 30–35 dB(A) for small auditoria. Clearly the level of insulation needed to reduce external noise is dependent upon location. However, most major concert halls are situated in noisy urban areas where road, rail and air traffic noise is at its worst. Hence average sound insulation of the order of 55–60 dB will be needed in many situations.

6.13 Churches

Traditional massive construction in Britain has provided most churches with very high standards of insulation. Increase in road traffic combined with lighter construction is making it more difficult to achieve the desirable noise level in modern buildings. Lighter construction with larger windows has increased the ventilation problem, necessitating open windows and further reducing the sound insulation. The NC 25 curve should represent a suitable level if there is no means of amplification. On a busy main road site this will be difficult to achieve.

6.14 Industrial and commercial noise affecting residential areas

The main problem with much industrial noise has already been discussed in relation to hearing loss. However, there is an additional problem where noise from an industrial process will be heard by people in their homes. A limit must be set to the disturbing effect of such a noise which would result in complaints. BS 4142:1990 aims to predict whether complaints can be expected. In general complaints arise when the level of a particular noise exceeds that of the background. The number and seriousness of the complaints rises with the difference. Noises with a particular characteristic, such as a whine, hiss, screech, hum, bang, clatter or thump, are more disturbing than continuous steady noises. Full details of the rating method, with examples, are given in BS 4142:1990 to which the reader should refer before carrying out any actual ratings. The following is a summary of the method:

1. Determine by measurement and/or calculation the equivalent continuous A-weighted sound pressure level of the specific noise over the time interval concerned near the dwellings. If the noise already exists, or a similar noise exists elsewhere, measurements can be made. Sometimes no calculations may be needed, but the measured figures usually need to be adjusted to take account of measuring time or predictions will be needed for new noises.
2. The reference time used is 5 min at night-time and 60 min during the day. Hence it is the highest L_{eq} over such a time during day or night. The term night is not defined precisely, but is the normal adult sleeping time (e.g. 10:00 PM to 7:00 AM might be chosen).
3. The equivalent continuous A-weighted sound pressure level over the time interval ($L_{Aeq,T}$) for night or day is calculated from:

$$L_{Aeq,T} = 10 \log_{10} \left(\frac{1}{T} \sum_i T_i \cdot 10^{0.1 L_{Aeq,T_i}} \right)$$

where $T = \Sigma T_i$

T_i = time intervals for which the equivalent continuous A-weighted SPL (L_{Aeq,T_i}) occurs

4. Measurement positions should be 1 m from the façade of the dwelling.
5. The background level, $L_{A90,T}$ should be measured. This is the A-weighted sound pressure level which is exceeded for 90% of the time (5 min at night or 60 min during the day).
6. For the measurement to be valid the specific noise needs to be at least 10 dB above the background level. If not, a correction needs to be applied as shown in Table 6.10.
7. If the specific noise contains a discrete continuous noise (whine, hiss, screech or hum) or discrete impulses (bangs, clicks, clatters or thumps) then add 5 dB to the specific noise level to obtain the rating.
8. To assess the expectation of complaints, subtract the background noise level from the specific noise rating level, then use the following approximation:

$>$ 10 dB complaints likely
$\cong$ 5 dB marginal significance
$<$ 5 dB complaints unlikely

6.15 Construction sites

Construction-site noise can have a number of effects including deafness to workers, speech interference and nuisance to neighbours. The first two have been dealt with

Table 6.10 Corrections when the specific noise is less than 10 dB above the background level

$L_{Aeq,T} - L_{A90,T}$ (dB)	Correction to be subtracted from the noise level reading (dB)
10	0
6–9	1
4–5	2
3	3
<3	Calculation to be used

earlier but the question of nuisance has special implications for contractors because of provisions within the Control of Pollution Act 1974. This gives local authorities power to deal with noise from construction and demolition sites. These can be before or after works commence. It may be to the contractor's advantage to discover the local authority's noise requirements at the tender stage. The aim of any local authority will be to avoid or if that is not possible to minimize noise nuisance. This may affect the contract price and the plant to be used.

Criteria used by local authorities may include L_{eq} limits and perhaps L_{10} to avoid excessive peak levels. In addition limits to hours of noise-producing work may be given. It is thus advisable for contractors to predict likely noise levels. Details of calculation methods are given in BS 5228.

Certain construction operations besides producing noise may result in high levels of vibration. It has been known for local authority noise limits to be applied but for neighbouring dwellings to be damaged by vibration.

It can be appreciated that the criterion is set by local authorities to avoid nuisance to those activities taking place near to the construction site. A guide to the particular situations can be obtained from the earlier paragraphs. It is unlikely that a serious noise nuisance will be produced on contracts under about £1 000 000.

6.16 Motor vehicle noise criteria

In 1963 the Committee on Problem of Noise suggested maximum sound levels for new vehicles. Since then legislation has imposed maximum levels both for new vehicles where type approval, after testing, may be given, and all vehicles in use (Table 6.11). It will be noticed that the type approval figures are generally at least 3 dB(A) lower than for the same vehicle in use. There are of course standard methods for testing.

These restrictions have been introduced to try to prevent the continuous increase in the total traffic noise which in the early 1970s was about 1 dB(A) per year. Most of this increase was due to the rapid expansion in the number of vehicles on the road. The situation had arisen that road traffic was causing more noise nuisance than any other source. More than 50 per cent of the population of the United Kingdom are now distributed by traffic noise.

With more vehicles came new and improved roads. In many cases these new roads may take very large volumes of traffic close to residential property. As there is a practical limit to the reduction economically possible to individual vehicles, alternative means must be used to minimize noise nuisance caused by these new roads. Various units of traffic noise levels have been suggested. All must be time-based due to the large fluctuations during any 24 h period. The simplest are either based on L_{eq} or some combination of L_n values.

Traffic noise index or TNI was suggested, being

$$TNI = 4 \times L_{10} - 3 \times L_{90} - 30$$

where the measurements are made 1 m from the façade of the building.

L_{10} = the level in dB(A) exceeded for 10% of the 24 h period.

L_{90} = the level in dB(A) exceeded for 90% of the 24 h period.

If measurements were made prior to the building being erected then the levels would all be 3 dB(A) lower and the TNI would be calculated from:

$$TNI = 4 \times L_{10} - 3 \times L_{90} - 27$$

The number 30, or 27, is subtracted merely to give a convenient numerical scale. It can be seen that the TNI is a unit combining the peaks (roughly L_{10}) and the background levels (roughly L_{90}). A criteria of a TNI of 74 was suggested so that only 1 person in 40 would be dissatisfied. Using this index as the criterion the minimum distance of buildings from a motorway can be determined.

Table 6.11 Maximum sound level for vehicles

	Maximum sound levels in dB(A)		
Type of vehicle	New vehicles (1st registered after 1 April 1970)	Vehicles in use (1st registered before 1 Nov 1970)	Vehicles in use (1st registered after 1 Nov 1970)
1. Motor cycle up to 50 c.c.	77	80	80
2. Motor cycle 50 to 125 c.c.	82	90	85
3. Motor cycle above 125 c.c.	86	90	89
4. Motor car	85	88	88
5. Passenger vehicle for more than 12 passengers	89	92	92
6. Any other passenger vehicle	84	87	87
7. Heavy goods vehicle	89	92	92
8. Tractor, locomotive works truck or engineering plant	89	92	92
9. Any other vehicle	85	92	89

As explained in Chapter 5, sound levels decrease by 6 dB(A) for an isolated source and by 3 dB(A) for a line source each time the distance is doubled. It can be shown that TNI should decrease by 15 for each doubling of distance. In practice it is slightly lower at 14. If the TNI is measured at a certain distance d_0 from the road, the distance d_1 at which an acceptable TNI of 74 would be achieved can be calculated from the formula:

$$\text{required reduction in TNI} = 45 \log_{10} \frac{d_1}{d_0}$$

L_{10} (18 h) In practice the use of TNI has been very limited due to the large amount of measurement needed to obtain the data. It has been shown that there is good correlation between certain values of L_{10} and likelihood of nuisance. The value used is the L_{10} (18 h) being the sound level in dB(A) exceeded for 10 per cent of the time between 6:00 AM and midnight. In practice the hours from 12:00 midnight to 6:00 AM are unimportant for two reasons: most people are already asleep (noise may prevent one going to sleep but seldom awakes one unless it is unusual), and in any case traffic noise almost always drops down during the early hours of the morning.

The L_{10} (18 h) criterion suggested was originally 70 dB(A). However, because of limitations in accuracy of measurement at that time 68 dB(A) was used. A measurement accuracy of ±2 dB(A) being specified. This could reasonably be obtained using a sound level meter, tape recorder, analysis technique. Present requirements are for accuracy of ±1·5 dB(A) which require precision equipment and expert operation. It is not necessary to make continuous measurement throughout the 18 h period. The minimum time, in minutes, is related to the number of vehicles and can be calculated from:

$$t_{min} = \frac{4000}{q} + \frac{120}{r}$$

where q = number of vehicles per hour

r = number of measurements per minute > 5

t_{min} = minimum measurement time in minutes > 5

The measurement position should normally be 1 m from the façade of the building. The L_{10} (18 h) is then calculated from the arithmetic mean of the 18 measurements.

The measurement procedure may be shortened by making three L_{10} measurements in consecutive hours between 10:00 AM and 5:00 PM on a normal working day.

Then, L_{10} (18 h) $\simeq L_{10}$ (3 h) − 1 dB(A)

The advantage of the L_{10} (18 h) unit for the measurement of traffic noise is that it can be done with very simple equipment. In practice even a sound level meter alone can be used by taking a large number of 'instant' readings.

For new major road schemes the L_{10} (18 h) can be predicted, the aim being for 15 years after its completion. For more detail the reader is referred to the HMSO publication *Calculation of Road Traffic Noise*.

Questions

(1) The recommended internal 10 per cent noise level for a courtroom is 35 dB(A). What average insulation is required to achieve this if the octave band spectrum of traffic noise outside is:

Frequency (Hz)	63	125	250	500	1000	2000	4000	8000
SPL (dB)	81	79	76	70	66	65	58	46

(2) A church is to be built near a busy road intersection at which the noise levels at weekends are:

Frequency (Hz)	63	125	250	500	1000	2000	4000	8000
SPL (dB)	81	84	84	82	75	70	66	58

Find the insulation required to achieve NC 25 inside.

(3) An office is to be built in part of a factory and the insulation required is such that the SIL for a normal male voice at 1 m is achieved. The levels in the factory are:

Centre frequency (Hz)	63	125	250	500	1000	2000	4000	8000
SPL (dB)	50	53	67	74	72	87	78	64

What average insulation is required at the SIL frequencies? If the insulation increases by 5 dB each time the frequency is doubled, what insulation is obtained at each frequency if it was just sufficient in the 1200–2400 Hz band? Explain why the appropriate NC curve normally has an advantage over the SIL.

(4) A recording studio is situated in a street which is partly residential. It is intended to be able to continue recording sessions until late at night without disturbance to neighbours who may be sleeping. The bedrooms of local residents are fitted with double-glazed windows giving the following insulation:

Frequency (Hz)	125	250	500	1000	2000	4000	8000
Insulation (dB)	19	19	27	37	47	52	53

If the highest level of sound at any frequency during a recording session is estimated to be 100 dB, what insulation is needed in the studio to achieve NC 30 level in surrounding houses?

(5) A small new factory is to be built in an urban area. The SPL inside with all machines running is expected to be 105 dB(A) and is a continuous noise with no particular tonal characteristics. Domestic dwellings are situated close to one side of the factory. What is the minimum weight of wall to make it unlikely that there will be complaints from householders if:
(a) the factory is only in use during the day?
(b) the factory is to be used at night?
(Assume reduction in dB $= 10 + 14{\cdot}5 \log_{10} m$, where m is mass/unit area in kg/m^2.)
If the factory was designed for use (a) and it was later changed to use (b), what area of material with an absorption coefficient of 0·8 would be needed? The volume of the factory is 1000 m^3 and its original reverberation time was 3·2 s.

(6) At a particular site 66 dB(A) was exceeded by road traffic for 90 per cent of the time and 81 dB(A) for 10 per cent of the time. The measurements were made 10 m from the roadway and clear of any obstructions. How far from the road should a building be erected so that the TNI at its face was 74?

(7) (a) Predict the sound pressure levels at positions 1, 2, 3 and 4 shown in the site sketch, Fig. 6.6, from a 37 kW excavator and 3 T dumper with (i) dumper only working; (ii) both working.

Site	To dumper (m)	To excavator (m)
1	38	33
2	16	11
3	43	39
4	18	90 (NB excavator has moved)

The sound power level of 37 kW excavator is 108 dB(A) and for the 3 T dumper 94 dB(A). It may be assumed that there are no other buildings nearby.

(b) If side 4 were 1 m from the bedroom window of a house, what would be the predicted sound pressure level? Hence, what would be the SPL inside the room? The external wall consists of 8 m^2 brick (sound reduction index 50 dB) and 2 m^2 double glazing (sound reduction index 35 dB).

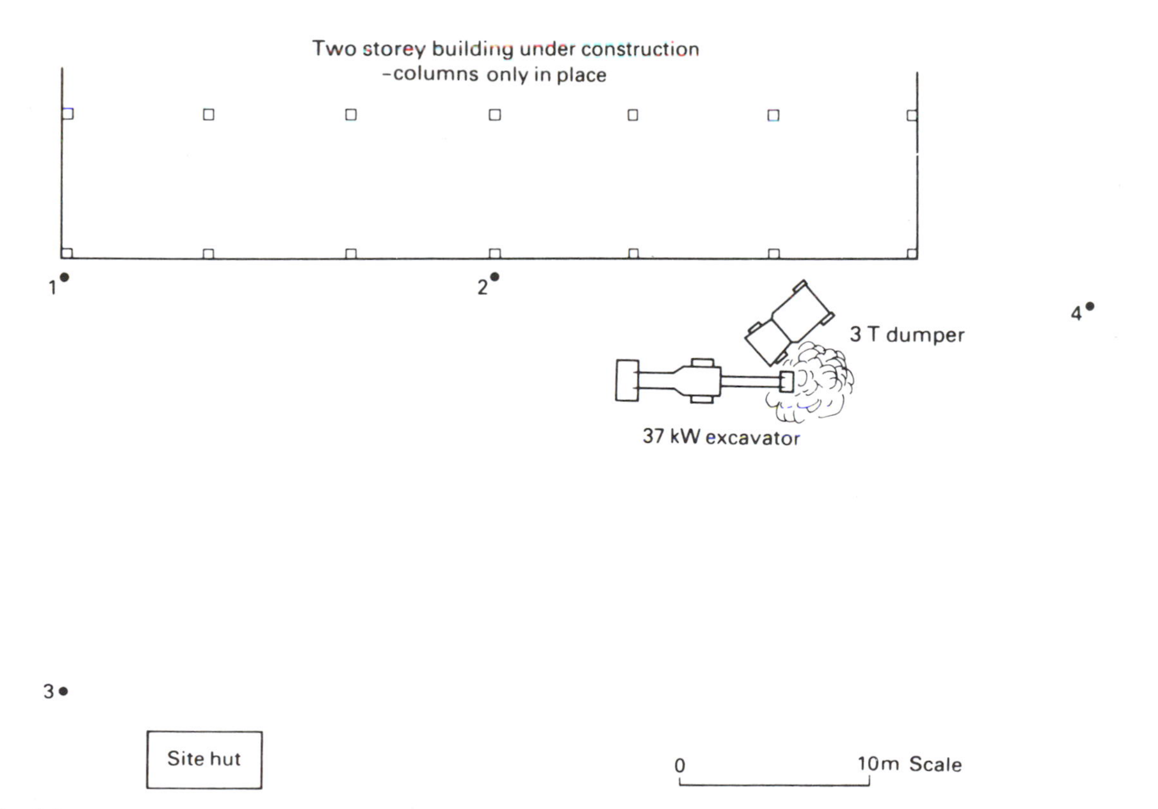

Fig. 6.6

(8) (a) A church is to be built near a busy road intersection at which the noise levels at weekends are:

Frequency (Hz)	63	125	250	500	1000	2000	4000	8000
SPL (dB)	81	84	84	82	75	70	66	58

Find the insulation required to achieve NC 30 inside. The construction is such that the brickwork gives 50 dB insulation and double-glazed windows 38 dB insulation, both at 500 Hz. Find the maximum percentage of windows in the building to achieve the desired insulation at that frequency.

(b) If the church is to be designed to seat 250 people, suggest a suitable volume and reverberation times for low, medium and high frequencies. Assuming that costs must be kept down, explain how you arrive at these values.

Assume that there will on average be 200 people in the congregation; find how many absorption units must be incorporated into the building for the medium frequencies. (Each person contributes $0{\cdot}4\ m^2$ absorption units.) It is essential that speech is heard with the greatest possible clarity.

(9) Discuss the problem of noise (defined as unwanted sound) in particular in relation to nuisance, speech interference levels and physiological effects.

(10) An auditorium is to be sited in a noisy area with a mean road near by and will require a projection room and the provision of light meals for delegates during protracted discussions. The room is to be well lit by daylight and electric lighting and fully air-conditioned. It is to be used for conferences and discussions.

Discuss in detail the acoustic and noise factors that must be checked when reviewing the plans for the auditorium.

(11) (a) Discuss the parameters involved in the assessment of noise in terms of annoyance and nuisance.

(b) Discuss briefly and compare the methods commonly used in the UK to assess community reaction to noise from industrial premises, making reference to their advantages and disadvantages.

(12) Define the term L_{eq}.

Three machines operate intermittently, in the open air, at fixed distances from a central point. Assume that, at the distances involved, each machine acts as a point sound source standing on an acoustically reflecting plane surface which is clear of all obstructions. From the details given in Table 6.12 of the operating cycle for each machine, calculate:

(a) the L_{eq} value for each of the three measurement periods; and

(b) the 24 h L_{eq} value.

Suggest reasons why the calculated values may not coincide with those obtained by direct measurement.

(13) Define and discuss the merits of three different types of noise exposure rating schemes commonly used in the United Kingdom and elsewhere as applied to different modes of transportation.

Describe how the impact of night noise is assessed.

(14) (a) What is meant by daily personal exposure level, $L_{EP,d}$?

(b) What are the duties of employers and employees to fulfil the requirements of the Noise at Work Regulations 1989?

(c) A worker is exposed intermittently throughout an eight-hour working day to the steady noise levels indicated in Table 6.13 from a particular machine, and for the rest of the day is exposed to 75 dB(A). If the worker wears the hearing protection provided while attending to the machine, calculate the maximum daily attendance time at the machine that can be permitted, if the worker's $L_{EP,d}$ is to be 85 dB(A). *(IOA)*

Table 6.12

	Machine A	Machine B	Machine C
Distance to the measurement point	200 m	200 m	100 m
A-weighted sound power level	150 dB(A)	140 dB(A)	140 dB(A)
Operating cycle 7 am to 4 pm	20 minutes on 40 minutes off	10 minutes on 50 minutes off	20 minutes on 40 minutes off
Operating cycle 4 pm to 10 pm	Not operating	5 minutes on 15 minutes off	30 minutes on 30 minutes off
Operating cycle 10 pm to 7 am	10 minutes on 50 minutes off	5 minutes on 5 minutes off	Not operating

Table 6.13

Centre frequency (Hz)	63	125	250	500	1000	2000	4000	8000
Machine noise levels (dB)	106	111	115	112	102	95	89	87
Mean muff attenuation (dB)	15·6	11·9	15·7	25·1	34·6	36·2	36·5	31
Standard deviation (dB)	3·9	3·8	2·7	3·7	3·6	4·0	2·0	5·1
A-weighting values (dB)	−26	−16	−9	−3	0	+1	+1	+1

Table 6.14

Frequency (Hz)	63	125	250	500	1000	2000	4000	8000
A-weighted factory noise level (dB)	57	79	100	100	92	87	76	62
Mean attenuation of earmuffs (dB)	13	9	13	19·5	27	31·5	36	32
Standard deviation (dB)	3·5	3·5	3·7	3·2	2·8	2·8	2·2	3·7

(15) Discuss the UK approach to the prevention of hearing damage due to industrial noise exposure, with particular reference to the Noise at Work Regulations 1989.

A worker is required to work in a factory area where he will be exposed to the noise levels indicated in Table 6.14. He is provided with hearing protection (earmuffs) having the performance characteristics specified.

Note: To assist you in the calculation, the factory noise levels have already been A-weighted.

Calculate:

(i) the total dB(A) reduction provided by the earmuffs.

(ii) the maximum daily exposure time that should not be exceeded when using the hearing protection provided. (*IOA*)

(16) A pump running at 1800 rpm is located in a plant room above a conference room in which a pronounced low frequency hum causes disturbance. The sound pressure level in the conference room (L_{p2}) is measured and is found to exceed the required NR 30 (see table). The room measures 5 m × 5 m × 2·4 m high and has a reverberation time of 0·5 s (ideal for speech intelligibility). The reverberant sound pressure level inside the plant room (L_{p1}) and the sound reduction index of the party floor (SRI) are given in the table below.

Calculate the sound pressure level in the conference room due to breakout of noise from the plant room. Diagnose the cause of the problem by comparing this level to the measured level (L_{p2}).

Recommend means of noise control.

If it is desired to achieve NR 30, are the necessary noise reductions practicable to achieve? Quantify your answers. (*IOA*)

Frequency (Hz)	31·5	63	125	250	500	1000
Location						
Plant room L_{p1}	100	90	88	84	81	80
Floor SRI	33	39	41	45	49	57
Measured L_{p2}	85	55	48	40	33	24
NR 30	76	59	48	40	34	30

(*IOA*)

(17) (a) Briefly describe the rating principles contained in BS 4142.

(b) What advice or guidance does the British Standard give on limits to the applicability of the rating method?

(c) A firm wishes to introduce a new item of equipment at an existing site. It is intended that the equipment will only operate during the normal working day, and apart from shutdowns for maintenance will be in continuous operation. There is a sensitive residential façade on a line of sight with, and approximately 30 metres from, the proposed location of the new equipment.

What data are required in order to carry out an assessment in accordance with BS 4142? How would you obtain the necessary data?

(d) What information should be reported regarding subjective impressions? (*IOA*)

(18) There have been many attempts to develop indices which relate noise exposure to human response.

(a) Discuss in detail the factors that need to be considered when developing such an index to quantify the exposure, and the standard expressed in that index for pollution control. Include in your discussion particular reference to the difficulties in establishing the dose/response relationship and the factors which need to be considered when determining an index. Include in your discussion the rôle that economics play in the standard.

(b) Outline, and discuss with reference to the issues and factors mentioned in (a) above, any index currently in use in the United Kingdom.

(*IOA*)

(19) Outline the principal provisions of current UK legislation in relation to:

(i) the protection of the hearing of workers exposed to noise at work,

(ii) annoyance to residents by noise from industrial premises.

The working day for supervisory staff in an industrial firm involves:

- 2 hours in a factory space where the noise level is 85 dB(A).
- 3 hours in a press shop where the background noise is 82 dB(A), with impact noise from the presses occurring at 5 s intervals. The single noise event level, SEL, of the noise from each press impact has been separately measured at 98 dB(A).
- 3 hours in the drawing office where the noise level is approximately 60 dB(A).

Estimate the daily personal noise exposure of the supervisors.

What would your advice be to the firm concerning compliance with current legislation? (*IOA*)

(20) An air diffuser is mounted in a wall close to the junction with the ceiling (Fig. 6.7). The design engineer has estimated that the sound power level spectrum emitted at the diffuser will be as follows.

Octave band centre frequency (Hz)	63	125	250	500	1000	2000	4000	8000
PWL (dB re 10^{-12} W)	34	35	36	32	30	25	20	18

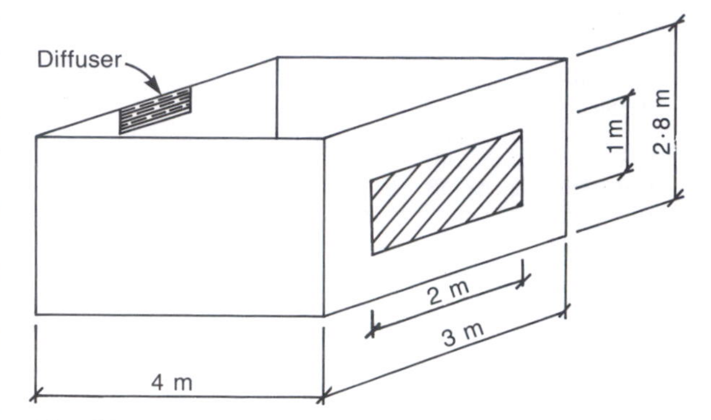

Fig. 6.7 Room with air diffuser: diagram for Question 20

The room that the diffuser serves is a conference room, and the designer hopes to achieve NR 25. Use the room data given in Fig. 6.7 to determine whether this criterion will be reached. (Include the NR chart attached with your script.)

When the system is installed, the engineer discovers that the sound pressure level measured in the 2 kHz octave band is 5 dB higher than predicted. What is the most likely cause of the increased noise in this octave band, and how could it be alleviated?

Assume that any door has similar acoustic properties to the wall fabric.

(*IOA*)

	Absorption Coefficient at the given frequency (Hz)							
	63	125	250	500	1000	2000	4000	8000
Floor: thick pile carpet	0·05	0·15	0·25	0·50	0·60	0·70	0·70	0·65
Ceiling: suspended plasterboard	0·20	0·20	0·15	0·10	0·05	0·05	0·05	0·05
Walls: 'acoustic plaster'	0·05	0·10	0·15	0·20	0·25	0·30	0·35	0·35
Window: 6 mm plate glass	0·08	0·15	0·06	0·04	0·03	0·02	0·02	0·02

7 Vibration

7.1 Introduction

A vibration is a type of motion in which a particle or body moves to and fro, or oscillates, about some fixed position. The distance of the particle or body from its fixed, or reference position is called the displacement, which may be positive or negative. The average displacement of a vibrating object is usually zero, so that the fixed position is also called the mean position. Any motion which results in a change in the mean position is called a translation. Any vibration may be described by a graph showing how the displacement varies with time (Fig. 7.1).

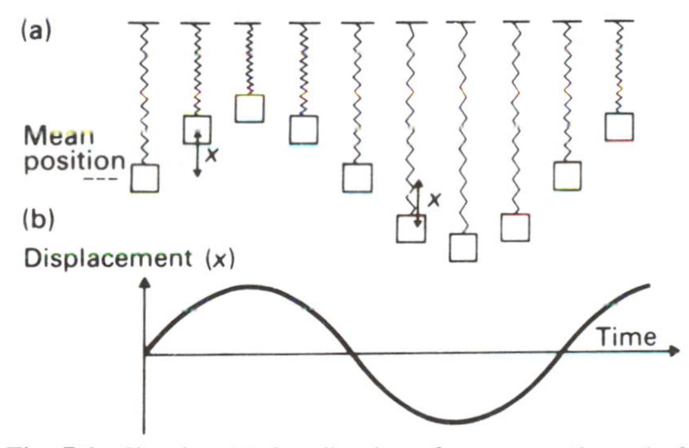

Fig. 7.1 Showing (a) the vibration of a mass on the end of a spring at ten instants equally spaced in time and (b) the graphical representation of the vibration

7.2 Types of vibration — periodic, random and transient

Vibrations may be classified as either periodic, random or transient. In a **periodic** vibration the motion repeats itself exactly, after a time interval called the period. For the simplest type of periodic vibration, called simple harmonic motion, the displacement–time graph is a sine wave. A motion of this type can be described in terms of a single frequency — examples are the motion of the prongs of a tuning fork, or of the bob of a simple pendulum. More complex periodic motions can be made up from a combination of different sine waves, and so these vibrations contain a number of different frequencies and have a frequency spectrum consisting of a number of lines representing a fundamental frequency and its harmonics. Examples are the motion of a piston in an internal combustion engine, or vibrations produced by regularly repeating forces in rotating machinery, such as motors, generators and fans.

In a **random** vibration the oscillations never repeat exactly. Examples are vibrations produced in structures by wind, or wave forces, or vibrations produced in a motor car as a result of its ride along a bumpy road. Random vibrations contain a little of every frequency, and so the frequency spectrum is a continuous curve, called a broadband spectrum.

A periodic vibration is said to be **deterministic** because its displacement at any time can be predicted. In the case of a random vibration it is not possible to do this, and the vibration can only be described in statistical terms. Random vibrations may be further subdivided into **stationary** and non-stationary types. In a stationary random process the statistical characteristics of the vibration, such as the RMS displacement, for example, remain constant in time. For a non-stationary process this is not so.

Transient vibrations die away to zero after a period of time. Examples are the vibrations in a building caused by the passage of a heavy vehicle, or the vibrations of a plucked violin string. The simplest type would be represented by a decaying sine wave. The frequency content of transient vibrations is complicated, but obviously in the case of the violin string it would contain the fundamental frequency and its harmonics. In the case of a repeated transient such as the impacts between teeth in a gear mechanism, the repetition rate of the impacts, and its harmonics will also be important.

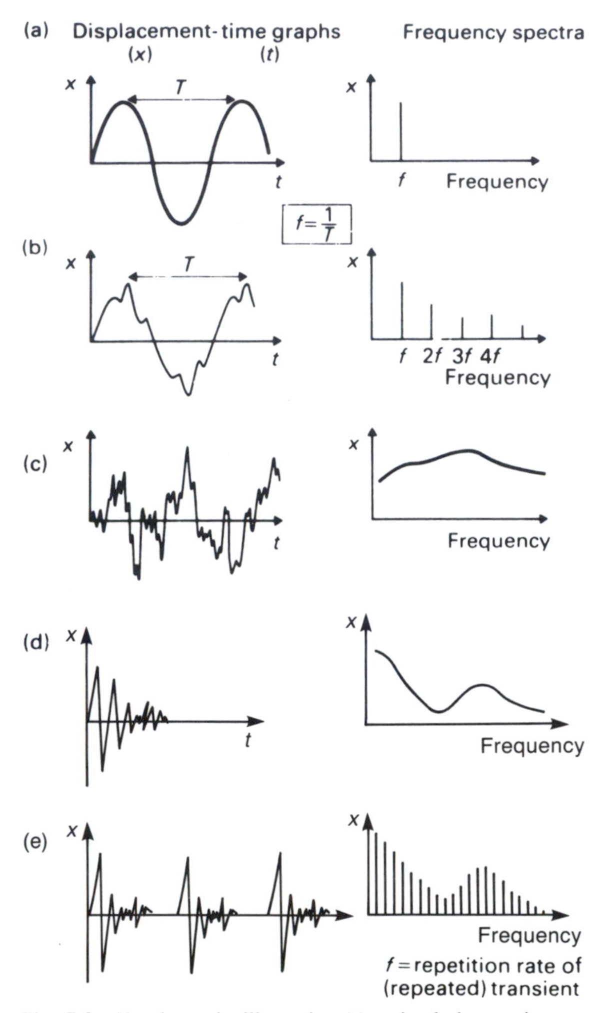

Fig. 7.2 Sketch graphs illustrating (a) a simple harmonic motion (b) a complex periodic vibration (c) a random vibration (d) a transient vibration and (e) a repeated transient vibration

Figure 7.2 shows some examples of displacement–time graphs for various types of vibrations and their frequency spectra.

7.3 Vibrations and waves

Vibrations occur at any point in a medium in which there is a wave motion. However, for a complete description of the wave we need to consider the relationship between the vibrations at different points. It is possible to distinguish between **progressive waves** in which the wave energy is always travelling away from the source (there being no reflections to send the energy back towards the source), and **standing waves** where the wave travels to and fro in a confined space, e.g. a room, or a pipe, or a string. In

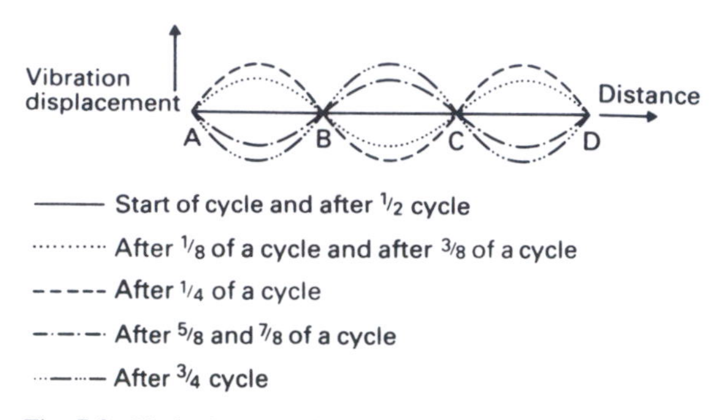

Fig. 7.3 Mode shape graphs, showing how the vibration displacement varies at various points during one cycle of a standing or stationary wave. Points between A and B, and between C and D are all in phase, but 180 degrees (half a cycle) out of phase with points between B and C

a progressive wave the vibrations of adjacent points are slightly out of step, or out of phase; the difference increasing with separation until the two points are a complete wavelength apart, when the phase difference is one complete cycle, i.e. the two vibrations are back in phase. In a standing wave the situation is different; all points between two nodes being in phase, and 180° out of phase with all points between the adjacent pair of nodes (Fig. 7.3).

A vibration, then, takes place at a point, and if the vibration of a body is being considered the implied assumption is that all points in that body are vibrating in phase, or that any phase difference across the body is small. This is another way of saying that the dimensions of the body are small compared with the wavelength of waves in the body.

It is, therefore, a common feature of vibrating objects such as beams, plates, and panels that at low frequencies their motion may be described simply; all parts vibrating in step, whereas at higher frequencies wave propagation in the body means that the description needs to be in terms of a mode shape, indicating the relative magnitudes and phases of different areas.

7.4 Displacement, velocity and acceleration

A vibration may be measured and described in terms of any one of these three quantities. The interrelationship between them may best be explained by considering the simplest sort of vibration — a single frequency, i.e. simple harmonic motion. Lengthy consideration of such a simple example is fully justified because the French mathematician Fourier showed that any periodic motion, however complex, can be made up from a series of simple harmonic motions.

Displacement is usually measured in metres (m) but this unit is too large to be conveniently used for vibrations.

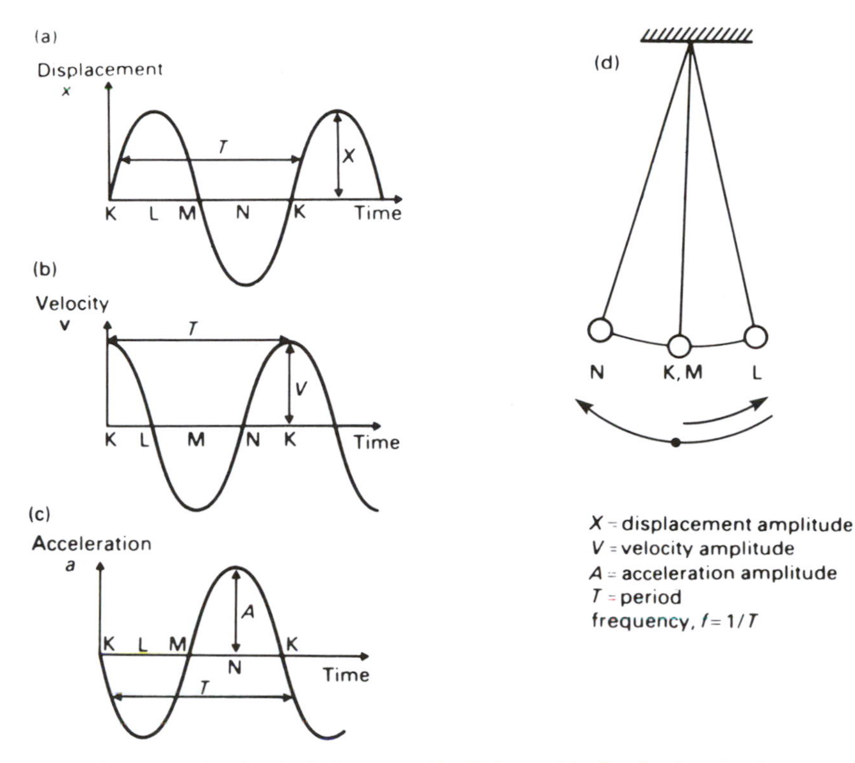

Fig. 7.4 Graphs representing one cycle of a single frequency (simple harmonic) vibrational motion in terms of (a) displacement (b) velocity (c) acceleration; the motion of a single pendulum, illustrated in (d), is an example of such a motion

Therefore either millimetres (mm) or microns (μm) are also used; where 1 μm $= 10^{-3}$ mm $= 10^{-6}$ m. Velocity may be measured in metres per second (m/s or m s^{-1}) or in millimetres per second (mm/s or mm s^{-1}). Acceleration is usually measured in metres per second per second (m/s^2 or m s^{-2}) but g units are also sometimes used, where g, the acceleration due to gravity, is 9·81 m/s^2.

Figure 7.4(a) shows the displacement–time graph for the motion. The amplitude of the vibration X, and its period T are also shown. The frequency f, in hertz, equals $1/T$. The 'size' of the vibration may be defined in terms of its amplitude X (peak to zero) or its root mean square (RMS) value, which for the sine wave is $X/\sqrt{2}$ (0·707X) or even by the peak-to-peak value (2X). Most instruments will give RMS values, although some will also allow measurement of peak values or peak-to-peak values.

Velocity is defined as the rate at which displacement changes with time. It may be found from the slope of the displacement–time curve, a high or steep slope indicating high velocity at that instant, and vice versa (Fig. 7.4(b)). By measuring the slopes at various points along the displacement–time graph it is possible to estimate velocities and construct the velocity–time graph, shown in Fig. 7.4(b). It is also a sine wave, of the same period and frequency, but out of step, or out of phase, with the displacement by a quarter of a cycle. The peak velocity, or velocity amplitude V, will obviously depend on the displacement amplitude X, since X determines the total distance that the vibrating particle has to travel in a period. However, V will also depend upon the period T and therefore on the frequency f, because if for a given amplitude the frequency is increased then the particle will have to travel over the same cycle in a shorter time. It can be shown that the relationship between X (the displacement amplitude), and V (the velocity amplitude) is:

$$V = 2\pi fX$$

where f is the frequency, in hertz.

The same relationship holds between the RMS displacement and RMS velocity.

Example 7.1

A panel is vibrating a frequency of 100 Hz. The RMS vibration is measured at the centre and found to be 1·0 mm/s. Find the RMS and peak displacement by half a cycle. Arguing as before, the peak vibration is sinusoidal.

$$V = 2\pi fX$$

$$\therefore X = \frac{V}{2\pi f}$$

$$\text{RMS } X = \frac{1\cdot 0}{2 \times 3\cdot 142 \times 100} = 1\cdot 6 \times 10^{-3}\ \text{mm}$$

$$= 1\cdot 6\ \mu\text{m}$$

$$\text{Peak } X = \frac{\text{RMS } X}{0\cdot 707} = \frac{1\cdot 6}{0\cdot 707} = 2\cdot 25\ \mu\text{m}$$

Example 7.2

One of the walls of a building is subject to vibrations of 100 Hz frequency. The peak displacement is measured and found to be 0·01 mm. Calculate the peak displacement and RMS velocities of the wall at the measurement point, assuming the vibrations to be sinusoidal.

$$V = 2\pi f X$$

$$\text{peak } V = 2 \times 3\cdot 142 \times 100 \times 0\cdot 01$$

$$= 6\cdot 3\ \text{mm/s}$$

$$\text{RMS } V = 0\cdot 707 \times 6\cdot 3 = 4\cdot 4\ \text{mm/s.}$$

Acceleration is defined as the rate at which velocity changes with time, and it can be found from the gradient of the velocity–time graph, in the same way that velocity is derived from the displacement–time graph. The graph of acceleration against time is shown in Fig. 7.4(c). It is again another sine wave of the same period and frequency, but out of step with velocity by a quarter of a cycle, and with displacement by half a cycle. Arguing as before, the peak acceleration or acceleration amplitude A depends on both the velocity amplitude V and on frequency f. It can be shown that:

$$A = 2\pi f V$$

The same relationship holds between the RMS acceleration and the RMS velocity.

Example 7.3

An annoying whine from a machine is identified as being produced by a sheet-metal cover which is performing resonant vibrations at a frequency of 1000 Hz. The vibrations are measured using an accelerometer, and the RMS acceleration is found to be 5·0 m/s^2. Calculate the RMS velocity and displacement of the cover.

$$A = 2\pi f V$$

$$V = \frac{A}{2\pi f} = \frac{5}{2 \times 3\cdot 142 \times 1000} = 0\cdot 8 \times 10^{-3}\ \text{m/s}$$

$$V = 2\pi f X$$

$$X = \frac{V}{2\pi f} = \frac{0\cdot 8}{2 \times 3\cdot 142 \times 1000} = 1\cdot 3 \times 10^{-4}\ \text{mm}$$

$$= 0\cdot 13\ \mu\text{m}$$

This example illustrates the effect of frequency on the relationship between X, V and A. The acceleration is fairly high and the vibration produces considerable noise, but because of the frequency factor it corresponds to a displacement which is only miniscule. Acceleration and displacement may be directly related by combining equations $V = 2\pi f X$ and $A = 2\pi f V$ to eliminate V:

$$A = 4\pi^2 f^2 X$$

Both peak (amplitude) and RMS values are related in this way. This equation could obviously have been used to calculate X directly from A in the last example. The nomograph in Fig. 7.5 can also be used to interrelate displacement velocity and acceleration.

Example 7.4

The vibration level near to a suspected worn bearing is measured using a displacement meter and is found to be 0·5 mm peak. The vibration is predominantly of one frequency, 20 Hz. The measurement is checked using an accelerometer and the peak acceleration found to be 9·0 m/s^2. Do the two measurements agree?

Either the equation $A = 4\pi^2 f^2 X$ or Fig. 7.5 can be used to find the value of peak acceleration which corresponds to a displacement amplitude of 0·5 mm at 20 Hz. This turns out to be 8·0 m/s^2, so the two measurements are in reasonable but not complete agreement.

This example also illustrates the fact that any vibration can in principle be measured in terms of either x or v or a; the three measurements would be related and not be independent of each other. WARNING Relationships between RMS (or peak) values of x, v and a can only be calculated using the previous equations (or Fig. 7.5) if the vibration is of one single known frequency. It may be extended, approximately, to vibration measurements made in frequency bands, e.g. third-octaves, but could not be used for overall vibration levels containing a wide range of frequencies. However, electrical signals which represent overall acceleration, incorporating a wide range of frequencies, may be converted electrically into a signal which represents velocity or displacement using an integrating circuit or integrator in the measuring instrument.

The relationship between displacement, velocity and acceleration may be illustrated by considering the motion of a simple pendulum (Fig. 7.4(d)). At the central, lowest position, the displacement is zero but the velocity of the pendulum bob is at its maximum. However, the rate at which velocity is changing, i.e. the acceleration, is also zero at this point. At the end positions, displacement is a maximum, and velocity is zero. However, the rate at which velocity is changing, i.e. the acceleration, is a maximum.

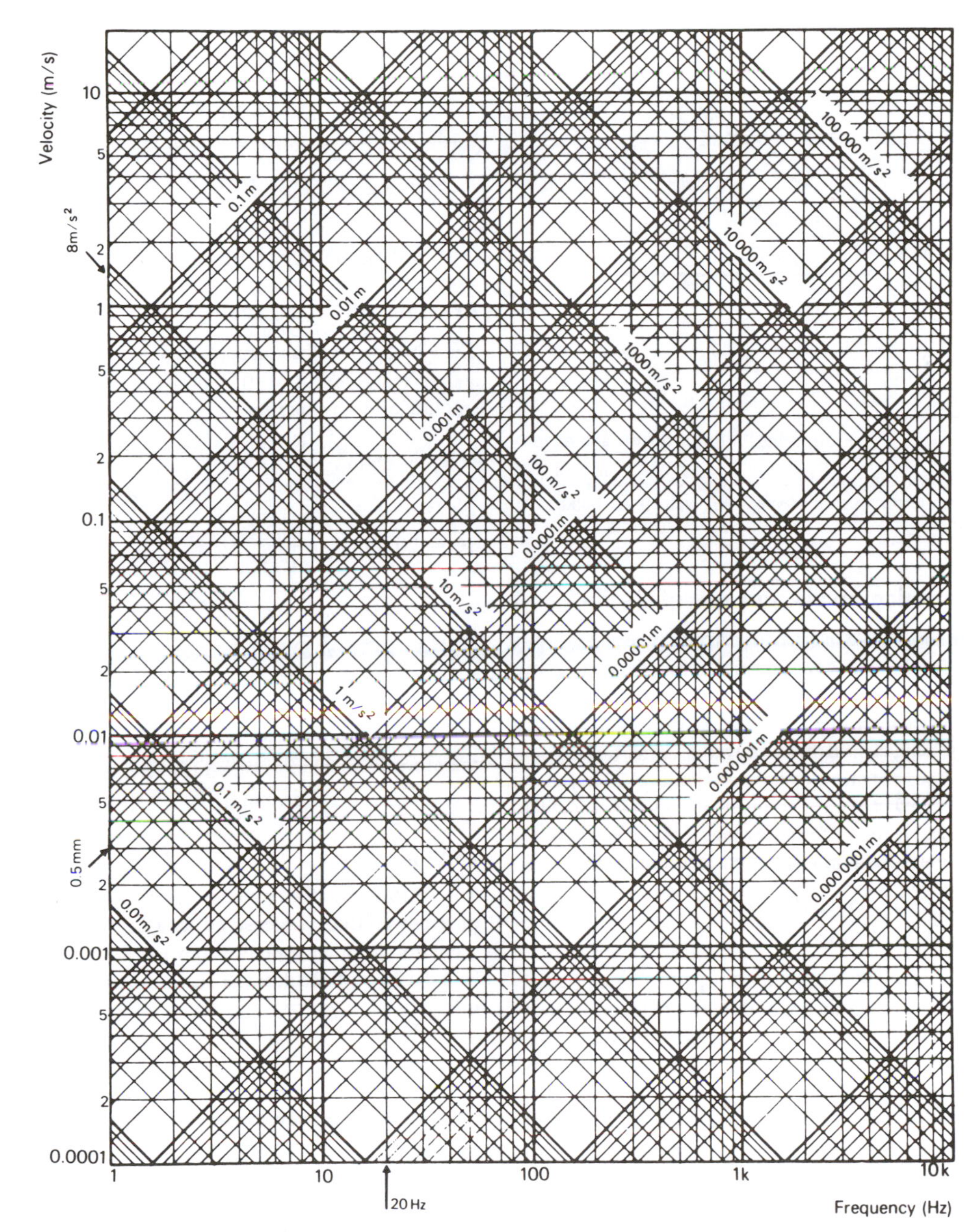

Fig. 7.5 Nomograph relating displacement, velocity, acceleration and frequency for sinusoidal vibrations; the three arrows representing 20 Hz, 0·5 mm and 8 m/s² relate to Example 7.4

7.5 Vibration spectra

Vibration levels are often measured in frequency bands and plotted as a frequency spectrum. When interpreting such spectra the frequency factor inherent in equations $V = 2\pi fX$ and $A = 2\pi fV$ should be taken into consideration. Figure 7.6 shows the spectrum of a body in terms of x, v, a. The vibration displacement spectrum is flat, i.e. the vibration displacement is the same for all frequencies over the range shown. However, for the same vibration the velocity amplitude increases with frequency in terms of decibel changes at a rate of 6 dB per octave. The acceleration spectrum increases even more steeply at a rate of 12 dB per octave.

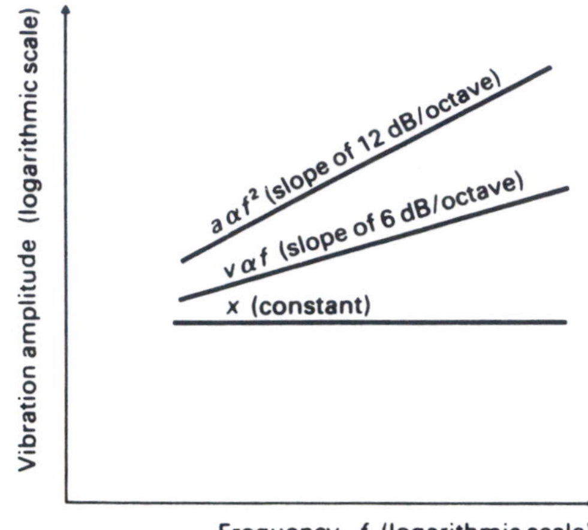

Fig. 7.6 Illustrating the different slopes of vibration spectra expressed in terms of displacement (x), velocity (v) and acceleration (a)

7.6 Reference levels and use of decibels

The logarithmic scale is often used to compare and measure vibration levels. The decibel system may be used in two ways: firstly, simply to compare two levels. For example, when comparing two levels, 1 and 2, we might state that level 2 is, say, 16 dB above level 1. This description, however, does not allow us to know the level 2 absolutely; to do this we must always compare our level to a standard or reference level. The following reference levels for vibration measurement have been recommended:

velocity 10^{-9} m/s, 10^{-6} mm/s
acceleration 10^{-6} m/s^2

It can be shown that the total energy of a vibrating object is proportional to the square of the amplitude (cf. intensity of a sound wave proportional to square of pressure). This applies, of course, whether amplitude is expressed in terms of displacement velocity or acceleration.

Since the decibel system is used to compare the energy, intensity or power of two quantities, the difference, N in decibels between two vibration levels is given by:

$$N\ (\text{dB}) = 10 \log_{10}\left(\frac{A}{A_0}\right)^2 = 20 \log_{10}\left(\frac{A}{A_0}\right)$$

where A is the level being described and A_0 is the reference level. A similar formula is used for displacement and velocity.

Example 7.5

The displacement amplitude of a vibration is measured and found to be 250 μm. Convert this to dB re 10^{-11} m.

$$N = 20 \log_{10}\left(\frac{X}{X_0}\right)$$

$$= 20 \log_{10}\left(\frac{250 \times 10^{-6}}{10^{-11}}\right)$$

$$= 20 \log_{10}(250 \times 10^5)$$

$$= 20 \times 7{\cdot}4$$

$$= 148 \text{ dB (re } 10^{-11} \text{ m)}$$

Example 7.6

The vibrational velocity amplitude of a machine is measured and quoted as 96 dB re 10^{-6} mm/s. Calculate the velocity amplitude in absolute terms.

$$N = 20 \log_{10}\left(\frac{V}{V_0}\right)$$

$$96 = 20 \log_{10}\left(\frac{V}{10^{-6}}\right)$$

$$\log_{10}\left(\frac{V}{10^{-6}}\right) = \frac{96}{20} = 4{\cdot}8$$

$$\frac{V}{10^{-6}} = \text{antilog}_{10}\ 4{\cdot}8 = 10^{4\cdot 8} = 6{\cdot}3 \times 10^4$$

therefore $V = 6{\cdot}3 \times 10^4 \times 10^{-6}$
$= 0{\cdot}063$ mm/s

Example 7.7

The RMS level of vibrational acceleration on the handle of a power tool is measured as 15 dB above that of a standard vibration source whose RMS level is known to be $6{\cdot}94$ m/s^2. Calculate the acceleration on the handle in m/s^2.

$$N = 20 \log_{10}\left(\frac{A}{A_0}\right)$$

$$15 = 20 \log_{10}\left(\frac{A}{A_0}\right)$$

$$\log_{10}\left(\frac{A}{A_0}\right) = \frac{15}{20} = 0{\cdot}75$$

$$\frac{A}{A_0} = \text{antilog}_{10}\ 0{\cdot}75 = 10^{0\cdot 75}$$

$$\frac{A}{A_0} = 5{\cdot}62$$

In this case, $A_0 = 6{\cdot}94$ m/s^2
therefore $A = 5{\cdot}62 \times 6{\cdot}94$
$= 39{\cdot}0$ m/s^2

Example 7.8

The vibration amplitude of a resonating sheet panel is measured and found to be 15·0 m/s^2. After treating the panel with a spray-on proprietary damping compound, the level was reduced to 0·85 m/s^2. Express the reduction in decibels.

$$N = 20 \log_{10}\left(\frac{A}{A_0}\right)$$

$$= 20 \log_{10}\left(\frac{15}{0 \cdot 85}\right) = 20 \log_{10} 17 \cdot 6$$

$$= 25 \cdot 0 \text{ dB}$$

The vibration level has been reduced by 25 dB.

Note: The above examples could also be performed using charts such as Table 7.1 for the conversion of dB to ratio and vice versa.

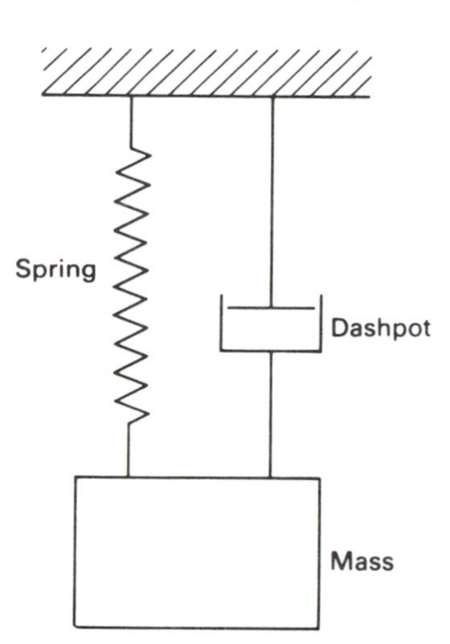

Fig. 7.7 Mass, spring and dashpot model of a simple vibrating system

7.7 Vibration of a mass–spring system

All vibrating objects possess three essential features: **mass** (or inertia), **stiffness** and **damping**. If the object is displaced from its equilibrium position, it is the stiffness which provides a restoring force which is always trying to bring the object back to its starting point. When this happens, the inertia of the object causes it to overshoot the zero mark, and the restoring force is again called in to play, this time acting in the opposite direction. It is this continual interplay between mass and stiffness which causes the vibratory motion at the natural frequency of the system. The vibrating object possesses energy which is shared between that stored in the mass (as kinetic energy of motion), and that stored as elastic or strain energy in the stiffness component. This stored energy is continually being converted from one of these forms to the other. The damping mechanisms in the system cause the vibrational energy to be lost to the surroundings, eventually being turned to heat energy. Therefore in continually maintained, or **forced vibrations**, energy is constantly being supplied to replace that lost by damping. Otherwise, as with **free** or **transient vibrations** the damping eventually causes all the vibrational energy to be dissipated and the oscillations die away.

A mass suspended on the end of a spring (Fig. 7.7) may be regarded as the simplest of all vibrating systems, and the vibrations of real and complicated objects may be studied by modelling them in terms of a series of interconnected masses and springs. This is the so-called **lumped parameter** approximation because the mass, stiffness and damping are lumped together at various discrete points in the model, whereas in a real structure such as a beam or a plate the mass (and also the stiffness and damping) is distributed continuously throughout, i.e. a **distributed parameter** system.

In the mass–spring system the relevant property of the spring is its **stiffness** k, defined as the static force required to produce unit static deflection, and measured in units such as newtons per metre (N/m), or newtons per millimetre (N/mm). **Damping** is represented in Fig. 7.7 by a dashpot or damper, which may be thought of as a pot of viscous liquid through which the mass moves as it vibrates. In the case of viscous damping, implied above, the damping force which acts on the mass is proportional to its velocity. However, there are other forms of damping, such as hysteresis damping or structural damping, in which the damping force is proportional to the displacement and also depends on the frequency. The amount of damping is quantified, in the case of viscous damping, by a **damping constant**, c, defined as the damping force per unit velocity.

7.8 Free (undamped) vibrations

If the mass is set into vibration, either as the result of a transient blow, or by holding it aside and letting it go, then the resulting vibrations are sinusoidal and the frequency depends entirely on the mass and the stiffness of the system and not at all on the initial conditions which set it in motion (the magnitude of the blow or of the displacement). This frequency, called the **free** or **natural frequency** is given by the formula:

$$f_0 = \frac{1}{2\pi}\sqrt{\frac{k}{m}}$$

Where f_0 is the natural frequency in hertz, k the spring constant in N/m, and m the mass in kilogrammes. The angular frequency, ω, is also often used to describe vibrations, where $\omega = 2\pi f$.

Table 7.1 Ratio-to-decibels, conversion to nearest 0·1 ($20 \log_{10}x/x_0$)

Ratio $\frac{x}{x_0}$	0·00	0·01	0·02	0·03	0·04	0·05	0·06	0·07	0·08	0·09
1·0	0	0·1	0·2	0·3	0·3	0·4	0·5	0·6	0·7	0·7
1·1	0·8	0·9	1·0	1·1	1·1	1·2	1·3	1·4	1·4	1·5
1·2	1·6	1·7	1·7	1·8	1·9	1·9	2	2·1	2·1	2·2
1·3	2·3	2·3	2·4	2·5	2·5	2·6	2·7	2·7	2·8	2·9
1·4	2·9	3·0	3·0	3·1	3·2	3·2	3·3	3·3	3·4	3·5
1·5	3·5	3·6	3·6	3·7	3·8	3·8	3·9	3·9	4·0	4·0
1·6	4·1	4·1	4·2	4·2	4·3	4·4	4·4	4·5	4·5	4·6
1·7	4·6	4·7	4·7	4·8	4·8	4·9	4·9	5·0	5·0	5·1
1·8	5·1	5·2	5·2	5·3	5·3	5·3	5·4	5·4	5·5	5·5
1·9	5·6	5·6	5·7	5·7	5·8	5·5	5·8	5·9	5·9	6·0
2·0	6·0	6·1	6·1	6·2	6·2	6·2	6·3	6·3	6·3	6·4
2·1	6·4	6·5	6·5	6·6	6·6	6·6	6·7	6·7	6·8	6·8
2 2	6·8	6·9	6·9	7·0	7·0	7·0	7·1	7·1	7·2	7·2
2·3	7·2	7·3	7·3	7·3	7·4	7·4	7·5	7·5	7·5	7·6
2·4	7·6	7·6	7·7	7·7	7·7	7·8	7·8	7·9	7·9	7·9
2·5	8·0	8·0	8·0	8·1	8·1	8·1	8·2	8·2	8·2	8·3
2·6	8·3	8·3	8·4	8·4	8·4	8·5	8·5	8·5	8·6	8·6
2·7	8·6	8·7	8·7	8·7	8·8	8·8	8·8	8·9	8·9	8·9
2·8	8·9	9·0	9·0	9·0	9·1	9·1	9·1	9·2	9·2	9·2
2·9	9·2	9·3	9·3	9·3	9·4	9·4	9·4	9·5	9·5	9·5
3·0	9·5	9·6	9·6	9·6	9·7	9·7	9·7	9·7	9·8	9·8
3·1	9·8	9·9	9·9	9·9	9·9	10·0	10·0	10·0	10·0	10·1
3·2	10·1	10·1	10·2	10·2	10·2	10·2	10·3	10·3	10·3	10·3
3·3	10·4	10·4	10·4	10·4	10·5	10·5	10·5	10·6	10·6	10·6
3·4	10·6	10·7	10·7	10·7	10·7	10·8	10·8	10·8	10·8	10·9
3·5	10·9	10·9	10·9	11·0	11·0	11·0	11·0	11·1	11·1	11·1
3·6	11·1	11·2	11·2	11·2	11·2	11·2	11·3	11·3	11·3	11·3
3·7	11·4	11·4	11·4	11·4	11·5	11·5	11·5	11·5	11·6	11·6
3·8	11·6	11·6	11·6	11·7	11·7	11·7	11·7	11·8	11·8	11·8
3·9	11·8	11·8	11·9	11·9	11·9	11·9	12·0	12·0	12·0	12·0
4·0	12·0	12·1	12·1	12·1	12·1	12·1	12·2	12·2	12·2	12·2
4·1	12·3	12·3	12·3	12·3	12·3	12·4	12·4	12·4	12·4	12·4
4·2	12·5	12·5	12·5	12·5	12·5	12·6	12·6	12·6	12·6	12·6
4·3	12·7	12·7	12·7	12·7	12·8	12·8	12·8	12·8	12·8	12·8
4·4	12·9	12·9	12·9	12·9	12·9	13·0	13·0	13·0	13·0	13·0
4·5	13·1	13·1	13·1	13·1	13·1	13·2	13·2	13·2	13·2	13·2
4·6	13·3	13·3	13·3	13·3	13·3	13·3	13·4	13·4	13·4	13·4
4·7	13·4	13·5	13·5	13·5	13·5	13·5	13·6	13·6	13·6	13·6
4·8	13·6	13·6	13·7	13·7	13·7	13·7	13·7	13·8	13·8	13·8
4·9	13·8	13·8	13·8	13·9	13·9	13·9	13·9	13·9	13·9	14·0
5·0	14·0	14·0	14·0	14·0	14·0	14·1	14·1	14·1	14·1	14·1
5·1	14·2	14·2	14·2	14·2	14·2	14·2	14·3	14·3	14·3	14·3
5·2	14·3	14·3	14·4	14·4	14·4	14·4	14·4	14·4	14·5	14·5
5·3	14·5	14·5	14·5	14·5	14·6	14·6	14·6	14·6	14·6	14·6
5·4	14·6	14·7	14·7	14·7	14·7	14·7	14·7	14·8	14·8	14·8
5·5	14·8	14·8	14·8	14·9	14·9	14·9	14·9	14·9	14·9	14·9
5·6	15·0	15·0	15·0	15·0	15·0	15·0	15·1	15·1	15·1	15·1
5·7	15·1	15·1	15·1	15·2	15·2	15·2	15·2	15·2	15·2	15·3
5·8	15·3	15·3	15·3	15·3	15·3	15·3	15·4	15·4	15·4	15·4
5·9	15·4	15·4	15·4	15·5	15·5	15·5	15·5	15·5	15·5	15·5

Table 7.1 (continued)

Ratio $\frac{x}{x_0}$	0·00	0·01	0·02	0·03	0·04	0·05	0·06	0·07	0·08	0·09
6·0	15·6	15·6	15·6	15·6	15·6	15·6	15·6	15·7	15·7	15·7
6·1	15·7	15·7	15·7	15·7	15·8	15·8	15·8	15·8	15·8	15·8
6·2	15·8	15·9	15·9	15·9	15·9	15·9	15·9	15·9	16·0	16·0
6·3	16·0	16·0	16·0	16·0	16·0	16·1	16·1	16·1	16·1	16·1
6·4	16·1	16·1	16·2	16·2	16·2	16·2	16·2	16·2	16·2	16·2
6·5	16·3	16·3	16·3	16·3	16·3	16·3	16·3	16·4	16·4	16·4
6·6	16·4	16·4	16·4	16·4	16·4	16·5	16·5	16·5	16·5	16·5
6·7	16·5	16·5	16·5	16·6	16·6	16·6	16·6	16·6	16·6	16·6
6·8	16·7	16·7	16·7	16·7	16·7	16·7	16·7	16·7	16·7	16·8
6·9	16·8	16·8	16·8	16·8	16·8	16·8	16·9	16·9	16·9	16·9
7·0	16·9	16·9	16·9	16·9	17·0	17·0	17·0	17·0	17·0	17·0
7·1	17·0	17·0	17·1	17·1	17·1	17·1	17·1	17·1	17·1	17·1
7·2	17·1	17·2	17·2	17·2	17·2	17·2	17·2	17·2	17·2	17·3
7·3	17·3	17·3	17·3	17·3	17·3	17·3	17·3	17·3	17·4	17·4
7·4	17·4	17·4	17·4	17·4	17·4	17·4	17·4	17·5	17·5	17·5
7·5	17·5	17·5	17·5	17·5	17·5	17·6	17·6	17·6	17·6	17·6
7·6	17·6	17·6	17·6	17·7	17·7	17·7	17·7	17·7	17·7	17·7
7·7	17·7	17·7	17·8	17·8	17·8	17·8	17·8	17·8	17·8	17·8
7·8	17·8	17·9	17·9	17·9	17·9	17·9	17·9	17·9	17·9	17·9
7·9	18·0	18·0	18·0	18·0	18·0	18·0	18·0	18·0	18·0	18·1
8·0	18·1	18·1	18·1	18·1	18·1	18·1	18·1	18·1	18·1	18·2
8·1	18·2	18·2	18·2	18·2	18·2	18·2	18·2	18·2	18·3	18·3
8·2	18·3	18·3	18·3	18·3	18·3	18·3	18·3	18·4	18·4	18·4
8·3	18·4	18·4	18·4	18·4	18·4	18·4	18·4	18·5	18·5	18·5
8·4	18·5	18·5	18·5	18·5	18·5	18·5	18·5	18·6	18·6	18·6
8·5	18·6	18·6	18·6	18·6	18·6	18·6	18·6	18·7	18·7	18·7
8·6	18·7	18·7	18·7	18·7	18·7	18·7	18·8	18·8	18·8	18·8
8·7	18·8	18·8	18·8	18·8	18·8	18·8	18·9	18·9	18·9	18·9
8·8	18·9	18·9	18·9	18·9	18·9	18·9	18·9	19·0	19·0	19·0
8·9	19·0	19·0	19·0	19·0	19·0	19·0	19·0	19·1	19·1	19·1
9·0	19·1	19·1	19·1	19·1	19·1	19·1	19·1	19·2	19·2	19·2
9·1	19·2	19·2	19·2	19·2	19·2	19·2	19·2	19·2	19·3	19·3
9·2	19·3	19·3	19·3	19·3	19·3	19·3	19·3	19·4	19·4	19·4
9·3	19·4	19·4	19·4	19·4	19·4	19·4	19·4	19·4	19·4	19·5
9·4	19·5	19·5	19·5	19·5	19·5	19·5	19·5	19·5	19·5	19·5
9·5	19·6	19·6	19·6	19·6	19·6	19·6	19·6	19·6	19·6	19·6
9·6	19·6	19·7	19·7	19·7	19·7	19·7	19·7	19·7	19·7	19·7
9·7	19·7	19·7	19·8	19·8	19·8	19·8	19·8	19·8	19·8	19·8
9·8	19·8	19·8	19·8	19·9	19·9	19·9	19·9	19·9	19·9	19·9
9·9	19·9	19·9	19·9	19·9	19·9	20·0	20·0	20·0	20·0	20·0
10	20·0	20·0	20·0	20·0	20·0	20·0	20·0	20·0	20·0	20·0
100	40·0	40·0	40·0	40·0	40·0	40·0	40·0	40·0	40·0	40·0
1000	60·0	60·0	60·0	60·0	60·0	60·0	60·0	60·0	60·0	60·0

Example 1 Ratio of 43·5 ie 4·35 × 10 ≡ 12·8 dB + 20 dB = 32·8 dB
Example 2 36 dB = 16 dB + 20 dB ≡ 6·3 × 10 = ratio of 63·0

Therefore high stiffness and low mass lead to a high natural frequency whereas increasing mass and reduced stiffness tend to reduce the frequency. Although the numerical factor $1/\pi$ only applies for this simple system, the proportionality between f_0 and $\sqrt{k/m}$ can be applied to more complicated cases.

Example 7.9

The natural frequency of a thin sheet-steel panel is 1000 Hz. The mass of the panel is 0·8 kg. A paint-on mastic compound is applied in an attempt to reduce the vibrations by increasing the damping. This has the effect of increasing the mass to 1·6 kg. Estimate the new natural frequency, assuming that the stiffness of the panel remains unchanged.

$$f_0 = \frac{1}{2\pi}\sqrt{\frac{k}{m}}$$

therefore $f_0 \propto 1/\sqrt{m}$ because 2π and k are constant.

$$\text{and so } \frac{f_1}{f_2} = \sqrt{\frac{m_2}{m_1}}$$

where f_1 is the natural frequency of a system of mass m_1 and f_2 is the frequency for one of mass m_2. In this case $m_1 = 1\cdot6$ kg, $m_2 = 0\cdot8$ kg and $f_2 = 1000$ Hz.

$$\text{Therefore } \frac{f_1}{1000} = \sqrt{\frac{0\cdot8}{1\cdot6}} = 0\cdot707$$

$$\text{and } f_1 = 1000 \times 0\cdot707 = 707 \text{ Hz}$$

Care has to be taken when estimating changes in natural frequency since many modifications such as increasing the thickness of the panel, or the use of stiffening ribs, will increase both mass and stiffness, maybe in a complicated way.

7.9 Damped free vibrations

In all real systems there will be some damping. If this is small, as in the case of underdamping, the system will still oscillate but the amplitude of the vibration decays exponentially with time; see Fig. 7.8(a). For overdamped systems there is no vibration and the mass returns slowly to its original position, as in Fig. 7.8(b). The critically damped situation is the one in which the system just fails to oscillate, and returns to its original position in the minimum time, as in Fig. 7.8(c). Overdamping and critical damping are of most concern to those involved in the design of suspension systems, e.g. for vehicles, or for the meter needles of electrical instruments, but it is underdamping which is of most relevance to noise and vibration problems.

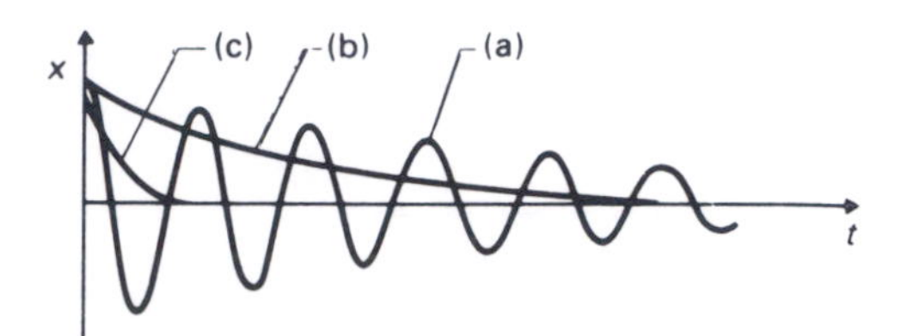

Fig. 7.8 Displacement (x) against time (t) graphs for damped vibration, illustrating (a) underdamping (b) overdamping and (c) critical damping

The amount of viscous damping in a system is expressed as a **damping ratio**, ξ, which compares the damping constant of the system with the critically damped case:

$$\xi = \frac{\text{damping constant of the system}}{\text{damping constant of the system if critically damped}}$$

Sometimes the ratio is expressed as a percentage.

Another way in which the damping of a system is specified, particularly for hysteresis damping, is in terms of a **loss factor**, which for very lightly damped systems can be equated, approximately, to twice the damping ratio. Figure 7.9 shows damped oscillations for various values of damping ratio. Typical values of damping ratio for lightly damped structures are in the range 0·01–0·03 (or 1–3%).

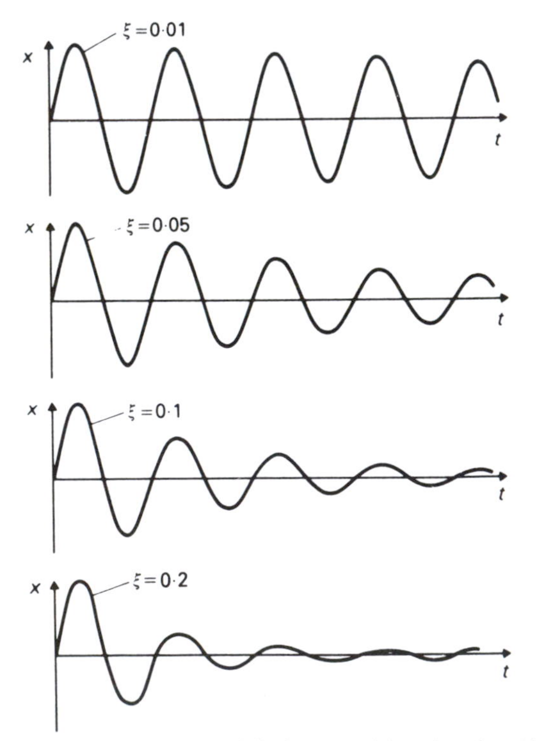

Fig. 7.9 Sketch graphs of displacement (x) against time (t) of damped oscillations for various values of damping ratio ξ

The frequency of the damped natural vibrations is:

$$f = f_0 \sqrt{1 - \xi^2}$$

where f is the damped frequency, f_0 the undamped natural frequency, and ξ is the damping ratio.

For small values of ξ the damped natural frequency is very close to but slightly less than the undamped frequency.

The amount of damping may also be expressed in terms of the rate at which the oscillations die away. The decay of amplitude is exponential and so the ratio of successive amplitudes, expressed logarithmically (natural logarithms) is a constant, called the **logarithmic decrement**, δ.

$$\delta = \frac{1}{n} \log_e \frac{x_1}{x_n}$$

where x_1 and x_n are two amplitudes n cycles apart. The amplitude decay and hence δ can be measured experimentally and related to the damping constant as follows:

$$\delta = 2\pi\xi$$

7.10 Forced vibrations

If the mass (Fig. 7.7) is subjected to a continuous vibratory force, then it will be constrained to vibrate at the forcing frequency. For the sake of simplicity, the response of the system to a sinusoidally varying force will be considered, since more complex oscillating forces can be broken down into a combination of such sine waves.

The amplitude of the forced vibrations depends not only on the magnitude of the oscillatory exciting force, but also on its frequency. Figure 7.10 shows how the amplitude of forced vibration varies as the forcing frequency changes, the force amplitude remaining constant. The frequency is expressed as a ratio, in terms of the resonance frequency of the system, f_0, thus allowing Fig. 7.10 to be generally applicable to all mass–spring systems.

If the force was constant, and not oscillatory, the displacement of the mass would correspond to its static deflection, according to Hooke's law. This is the case of zero frequency. As the frequency increases, the response of the system remains almost constant, increasing slightly until the natural frequency of the system is approached. The amplitude increases greatly in this frequency region: this is the phenomenon of **resonance** and the forced vibration amplitude is a maximum at the **resonance frequency** which for lightly damped systems is almost the same as the free or natural frequency. The amplitude of damping in the system, and measurements made from resonant response curves such as Fig. 7.10 provide another means of assessing damping expressed as a magnification or Q **factor**. Q may be found from the resonance curve in

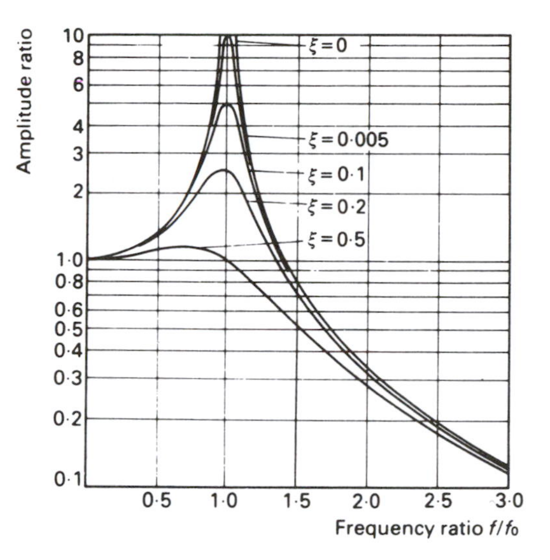

Fig. 7.10 Illustrating the variation with frequency of the forced vibration response of a simple mass–spring system, with viscous damping, for various values of damping ratio ξ. The horizontal axis is the ratio f/f_0 of the forcing frequency to the natural frequency of the system. The vertical axis is the ratio of the forced vibration amplitude to the static deflection produced by the same magnitude of force. The vertical scale is also simply related to receptance: Receptance = amplitude ratio × k; where k is the spring constant

two ways; from the maximum amplitude and from the width of the resonance curve.

$$Q = \frac{\text{displacement amplitude at resonance}}{\text{static displacement produced by the same force}}$$

Therefore a Q factor of 10 means that forced vibrations at resonance have ten times the amplitude of those produced by the same force amplitude at a much lower frequency, well below resonance. Alternatively, Q may be found, for lightly damped systems, from the shape of the curve near resonance, using the following formula:

$$Q = \frac{f_0}{f_1 - f_2}$$

where f_1 and f_2 are frequencies at which the forced vibration amplitude has dropped to $1/\sqrt{2}$ (i.e. 0·707) times its resonant value — the so-called half-power points. In practice this is a better method for measuring Q than the resonant magnification method. The Q factor and the damping ratio are simply related:

$$Q = \frac{1}{2 \times \text{damping ratio}}$$

The phase relationship between the sinusoidal force applied to the system and the sinusoidal response to it

Table 7.2 Interrelationships between the various damping terms, with some typical values

Material or structure	ξ Damping ratio	Loss factor $= 2\xi$	δ Logarithmic decrement $= 2\pi\xi$	Q Dynamic Magnification factor $= 1/2\xi$
Mild steel, Aluminium, Brass	0·01 (or 1%)	0·02	0·063	50
Metal fabricated structures: Welded; Building structures (masonry, concrete); Cast iron	0·02 (or 2%)	0·04	0·126	25
Metal fabricated structures: Riveted, Bolted; Natural rubber, Neoprene; High-damping polymers	0·05 (or 5%)	0·1	0·314	10
Felt, Cork; High-damping Alloys: Copper-manganese	0·1 (or 10%)	0·2	0·628	5
High-damping polymers; High-damping Alloys: Copper-manganese	0·2 (or 20%)	0·4	1·256	2·5

depends on the frequency. Well below resonance, force and displacement are in phase, and at frequencies well above resonance they are 180° (or a half-cycle) out of phase with each other. Near to resonance the phase difference between them changes rapidly with changing frequency, and is 90° (a quarter of a cycle) at resonance. Further discussion of this area is outside the scope of this chapter, but it is worth noting that measurement of the rate of phase change near to resonance can be related to, and provides another method of estimating the damping.

In summary: the damping of a system is most important in controlling the free vibration or ringing of that system, in response to transient forces such as impacts, and in determining its behaviour when subject to resonant forced vibration. It may be measured from the rate of decay of free vibrations, or from the shape of the resonance response curve, or from the phase changes which occur near to resonance. Damping is expressed in several different ways, and the interrelationships between these are summarized in Table 7.2.

7.11 Stiffness, damping and mass controlled frequency regions

The resonance curve can be broadly divided into three regions. **Well below resonance** the behaviour of the system is controlled mainly by its **stiffness**. In this region the frequency response of the system is flat. Microphone diaphragms and accelerometers are both mechanical systems which can be approximated in their behaviour to a mass–spring system. If they are to have a flat frequency response, they must be operated well below their resonance frequency, which should therefore be made as high as possible — well above the audible frequency range — in order to maximize the operating frequency range.

At and near resonance the behaviour of the system is controlled mainly by its **damping**.

At frequencies above resonance the forced vibration amplitude decreases with increasing frequency, eventually falling below the low frequency or static displacement response (Fig. 7.10). In this region, **well above resonance**, the forced vibration is mainly **mass-controlled**. It is because of this that panels and partitions, excited into forced vibration by airborne sound pressure, often obey the mass law of airborne sound insulation, since the sound frequencies of general interest are usually well above the resonance frequency of the panels.

7.12 Response functions

There are several ways in which the vibration response of a structure may be described; all of them relating the exciting force to the vibration amplitude it produces. The response may be measured at the point at which the force is applied, or at some other point, depending upon whether it is the point or the transfer response that is required. The six possible response functions, of which receptance, impedance and mobility are probably most common, are:

receptance = displacement/force
dynamic stiffness = force/displacement
impedance = force/velocity
mobility = velocity/force
apparent mass = force/acceleration
inertance = acceleration/force

Figure 7.10 shows how the receptance of the simple mass–spring system (viscously damped) varies with frequency. Response spectra like this may be measured by exciting the structure with a known force of variable frequency, e.g. by using an electrodynamic vibrator, and measuring the vibration produced. The measured response spectrum can be very useful if the forces exciting the structure in service are known, because the expected vibration spectrum may be predicted by combining the force and response spectra.

7.13 Transmissibility

The response function relates the vibration of a system to the force exciting it. The transmissibility relates the vibration at one point to the vibration at another point in the structure. For a simple mass–spring system the transmissibility may be defined in two ways. First of all, in Fig. 7.11(a) the mass is being excited by a vibratory force applied directly to it. This system is a model of a machine mounted on antivibration mounts (or vibration isolators). The operation of the machine, represented by the mass, produces vibratory forces, and the purpose of the isolator, represented by the spring, is to reduce to a minimum the vibratory force transmitted to the floor.

The force transmissibility, T, is defined as:

$$T = \frac{\text{amplitude of the force transmitted to the base (the floor)}}{\text{amplitude of the force exciting the mass (the machine)}}$$

$$= \frac{F_T}{F_0}$$

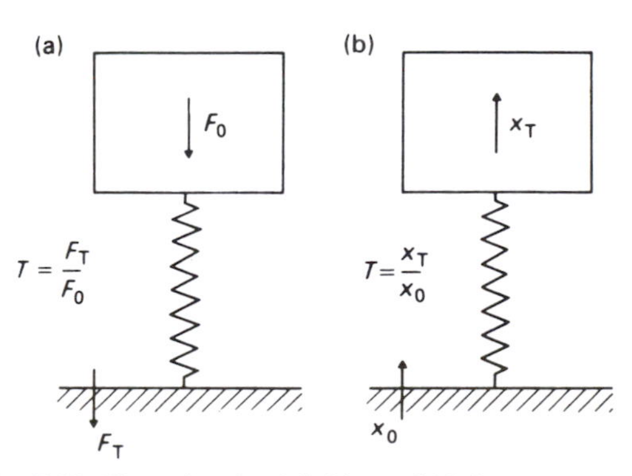

Fig. 7.11 Illustrating the definitions of (a) force transmissibility and (b) displacement transmissibility

T therefore is a measure of the force transmitted from the point of application (the mass) to the surroundings via the isolator. The definition assumes that the exciting and transmitted forces are sinusoidal, with amplitudes F_0 and F_T respectively.

Figure 7.11(b) shows an alternative way in which the mass may be excited into vibration as a result of the vibration of the base. Such a system could again be a model of a machine or piece of equipment (the mass) mounted on vibration isolators, but in this case the purpose of the isolators is to minimize the vibration transmitted from the vibrating floor to the mounted mass. An example might be the need to protect precision instrumentation and machines from damaging external vibrations.

In this case the displacement transmissibility is defined as:

$$T = \frac{\text{amplitude of the transmitted displacement (of the mass)}}{\text{amplitude of the applied displacement (of the base)}}$$

$$= \frac{x_T}{x_0}$$

The definition assumes that the displacement of both the floor and the mass are sinusoidal with amplitudes x_0 and x_T respectively.

For a simple mass–spring system the force and displacement transmissibilities are in fact identical, and Fig. 7.12 shows how T varies with frequency for various amounts of damping. At frequencies well below resonance, $T = 1$, indicating that the mass and base in effect move together, as if rigidly connected. As the resonance frequency is approached, the transmissibility increases greatly, indicating an amplification of the vibrations being transmitted through the structure. The maximum transmissibility at resonance depends on the amount of damping. Above resonance the transmissibility falls until at a frequency of $\sqrt{2} \times f_0$ (i.e. $1{\cdot}414 \times f_0$) it falls to 1. At higher frequencies T is less than 1, indicating that the vibrations are being attenuated, or isolated as they travel through the structure. In this region, well above resonance, it can be seen that the amount of damping still affects the transmissibility; a lightly damped system has a lower T value at the same frequency ratio than a more heavily damped system.

7.14 Vibration isolation — simple theory

A machine such as a motor, fan or engine produces a vibratory force at a particular frequency f, related to its rotational speed. The machine and the isolator form a mass–spring system with a resonance frequency f_0, which is related to the mass of the machine and the stiffness of the isolator. The isolator must be selected so that f_0 is low enough to achieve a value of f/f_0 high enough to produce the required degree of isolation (see Fig. 7.12), where f/f_0

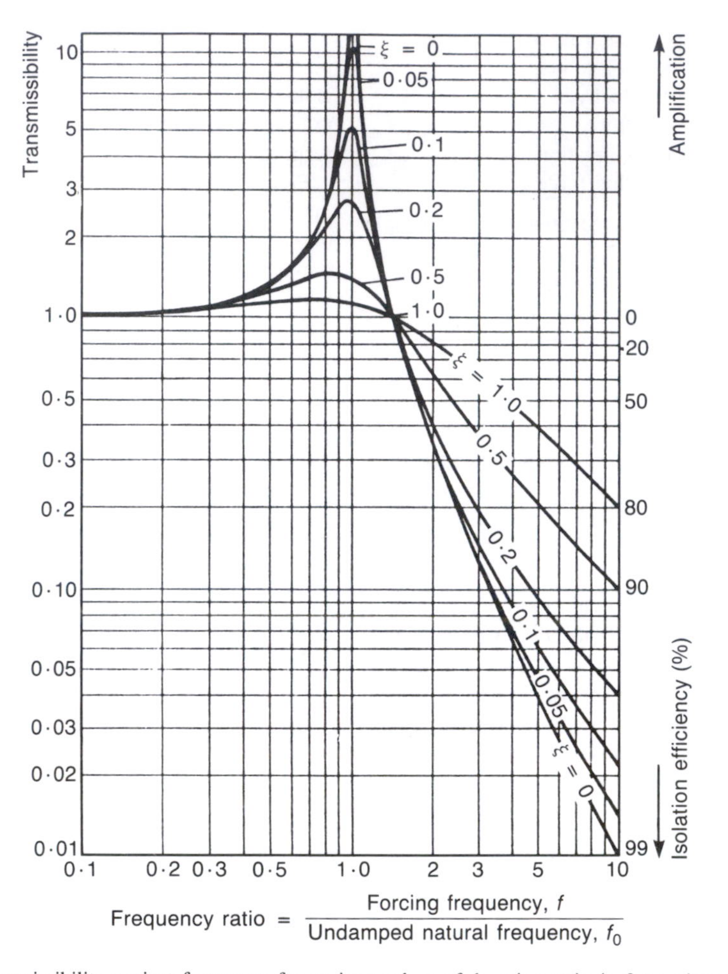

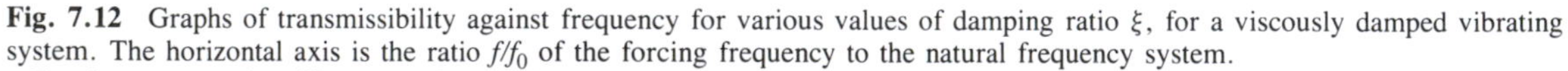

Fig. 7.12 Graphs of transmissibility against frequency for various values of damping ratio ξ, for a viscously damped vibrating system. The horizontal axis is the ratio f/f_0 of the forcing frequency to the natural frequency system.

For frequency ratios f/f_0 greater than 3, the vibration reduction obtained can be found by using Fig. 4.6, which assumes zero damping.

Transmissibility can also be calculated from the formula:

$$T = \sqrt{\frac{1 + 4\xi^2(f/f_0)^2}{[1 - (f/f_0)^2]^2 + 4\xi^2(f/f_0)^2}}$$

Alternatively, assuming the damping is very small the transmissibility can be calculated from the formula

$$T = \frac{1}{(f/f_0)^2 - 1}$$

is the ratio of the driving frequency to resonance frequency.

In theory, the ratio f/f_0 should be as high as possible and this means using isolators of the lowest possible stiffness to achieve a low f_0. However, if the isolator is too soft the static deflection of the isolator will be large, and stability problems can arise. Sometimes the machine may be mounted rigidly on to a large concrete block, usually called an inertia base. The machine and block assembly is then isolated. This arrangement in effect increases the mass of the machine and allows stiffer isolators to be used to achieve the same resonant frequency. The centre of gravity of the assembly is lowered and the stability is improved.

The damping of an isolator should, in theory, be as small as possible to achieve the best reductions at a given frequency ratio. However, in practice, damping can be useful since in many machines the vibratory forces pass through the resonance frequency of the system during start up or run down, and damping helps to limit the vibration amplitude of the machine as it passes through the resonant speed.

7.15 Isolation efficiency, η

This is another way of expressing the effectiveness of an isolation system:

$$\eta = (1 - T) \times 100\%$$

Example 7.10

A motor when rigidly bolted to the floor produces vibrations at a frequency of 40 Hz. The motor is to be mounted on isolators, with an isolation efficiency target of 80% at this frequency. Calculate the transmissibility, the expected vibration reduction in decibels and the resonance frequency when the isolators are installed. The isolators are very lightly damped (assume $\xi = 0$).

Using $\eta = (1 - T) \times 100$, when $\eta = 80\%$, $T = 0{\cdot}2$. Therefore the expected reduction in vibration level is $1/0{\cdot}2 = 5$ times, in terms of vibration amplitude. Therefore the expected vibration reduction $= 10 \log_{10} (5)^2 = 14$ dB.

From the graph of Fig. 7.12, in order to achieve a value of $T = 0{\cdot}2$ at zero damping the frequency ratio must be about 2·5. Hence the resonance frequency

$$f_0 = \frac{40}{2{\cdot}5} = 16 \text{ Hz}$$

7.16 Static deflection

The resonance frequency of the mass (the machine) and the spring (the isolator) system is simply related to the static deflection X_{st}, which the spring undergoes in response to the weight of the mass. The static deflection is related to the mass m (in kg) via the spring constant k (in N/m) by the equation $X_{st} = mg/k$, where g, the acceleration due to gravity, is $9{\cdot}81$ m/s^2. Combining this with the equation $f_0 = (1/2\pi)\sqrt{k/m}$ for the natural frequency f_0, yields the relationship

$$f_0 = 15{\cdot}8\sqrt{1/X_{st}}$$

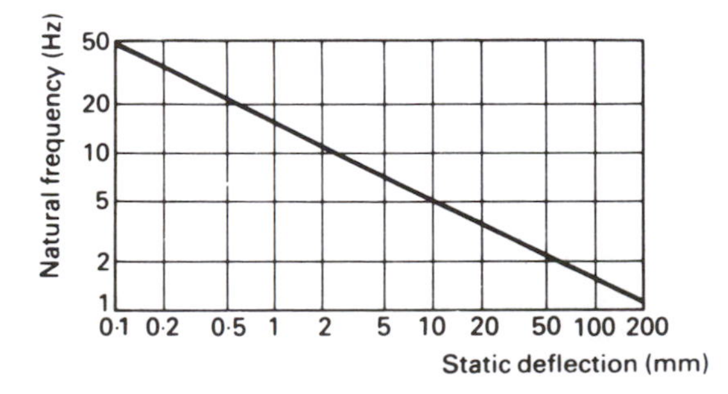

Fig. 7.13 The relationship between the natural frequency and static deflection of a mass–spring system

where X_{st} is the static deflection in millimetres. This result is shown graphically in Fig. 7.13.

The static deflection is therefore a very important property of an isolator since it can be directly related to the resonance frequency and to the isolation efficiency (or transmissibility) obtained. It is common practice to specify the requirements of an isolator in terms of the static deflection it will undergo as an alternative to specifying the required resonance frequency.

7.17 Selection of isolators — simple design procedure

This essentially follows Example 7.10. The steps are:

1. Determine the target frequency — the lowest frequency for which isolation is required. This will be the lowest frequency of the vibratory forces exciting the system, and will depend upon machine running speed.
2. Decide upon the degree of isolation required at the target frequency either in terms of isolation efficiency or transmissibility.
3. Using the transmissibility–frequency graphs (Fig. 7.12) and assuming a value of damping coefficient, find the ratio of forcing frequency to natural frequency which gives the required degree of isolation.
4. Calculate the required natural or resonance frequency of the system from the target frequency and frequency ratio. This may be converted into a static deflection if required.
5. Using the weight of the machine and the number of isolators to be used (maybe they are at each corner of the machine), calculate the weight loading of the isolators.
6. (a) Using the isolator load (from 5) and the required static deflection (from 4) a suitable isolator may be selected from manufacturers product data.
   (b) Alternatively, if non-proprietary isolators are to be designed, the information from steps 4 and 5 may be used to calculate the required static stiffness.

In Example 7.10, which stopped at stage 4 of the procedure, the target frequency was 40 Hz, the required transmissibility 0·2, the required frequency ratio 5, and the required resonance frequency 16 Hz. In order to complete the design (steps 5 and 6), the mass of the machine and the number of isolators to be used must be known.

It can be seen that the first step in the design, the selection of the target frequency, is very important. It must be the lowest possible frequency at which vibration excitation can occur. If this stage of the design is incorrect so that lower forcing frequencies do exist below the target frequency, then it is possible that the selected isolators, while satisfactorily reducing levels at the target frequency, will cause

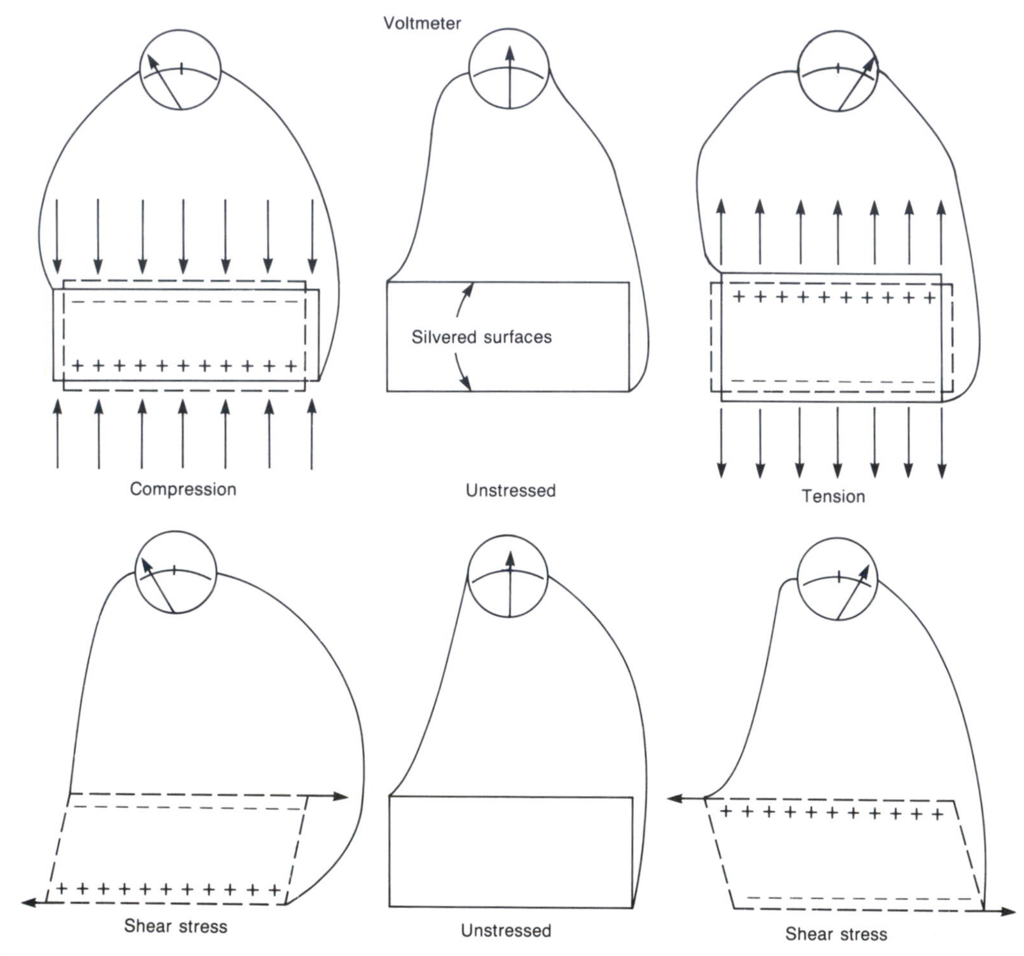

Fig. 7.14 Piezoelectric slab in compression and shear

amplification at the lower frequency. The target frequency may be calculated from the lowest operating speed of the machine and other relevant machine data (e.g. number of blades, teeth, cylinders), or from frequency analysis measurements of vibration caused by the machine at its lowest operating speed.

It must be emphasized that the design procedure outlined above is a very simple one and many other factors may have to be considered. The choice of isolator material may be limited by environmental conditions, such as high temperatures, or the presence of oil or other chemicals, or by cost. The weight distribution of the machine, the positioning of the isolators and any possible stability problems that may arise from their use will have to be investigated. If the machine is to run through a resonance while starting or stopping, then the maximum acceptable amplitude at resonance must be considered and if necessary limited by damping or other means. The effect of the isolators may be reduced if the base or floor on which they are mounted is not perfectly rigid, or if other vibration paths exist which bridge the isolation.

Vibration isolation is discussed again in the next chapter.

7.18 The measurement of vibration

Transducers

A transducer is a device which produces an electrical signal proportional to the physical quantity of interest. A variety of transducers are available for the measurement of mechanical vibrations. These include displacement gauges which work on a capacitative principle, with the vibrating surface forming one plate in a variable air-gap condenser and electrodynamic (coil and magnet type) gauges which give a direct measurement of vibration velocity. However, the most commonly used device for vibration measurement in the audio-frequency range is the piezoelectric accelerometer, which gives an electrical signal which is proportional to the vibration acceleration.

The piezoelectric effect

When a slab of piezoelectric material is subjected to a stress, the material becomes polarized, with electric charge collecting at the stressed faces of the slab (Fig. 7.14). These materials are ceramics, and therefore non-conducting, and in order to utilize the effect the charged slab faces are silvered to create electrodes. The piezoelectric effect then results in an electrical potential difference (i.e. a voltage difference) being set up between the electrodes which is proportional to the stress. If the stress changes from a compression to a tension, then the polarity of the voltage difference also changes, and so it follows that an alternating stress across the slab results in the production of an alternating signal. It is also possible to arrange for piezoelectric materials to respond to shear stresses in a similar way.

Although some naturally occurring crystalline materials such as quartz and tourmaline are piezoelectric, modern transducers are usually made in ceramic form from a combination of lead zirconate and lead titanate, known as PZT.

All piezoelectric materials have a Curie point, which is a temperature above which all piezoelectric properties are permanently lost. In practice, operating temperatures must be limited to well below the Curie point to prevent significant loss of sensitivity. Most accelerometers can be used at up to 250 °C, and certain special types to much higher temperatures.

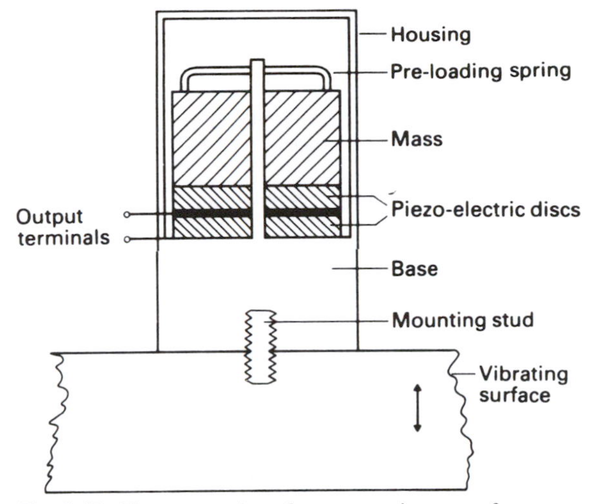

Fig. 7.15 The construction of a compression type of piezoelectric accelerometer

The piezoelectric accelerometer

The compression type of accelerometer (Fig. 7.15) consists of two piezoelectric discs sandwiched between a mass and the base of the device. The base is thick and stiff so that bending of the base (base strain) is minimized. The piezoelectric discs form part of a mass–spring system whose resonance frequency is designed to be well above the

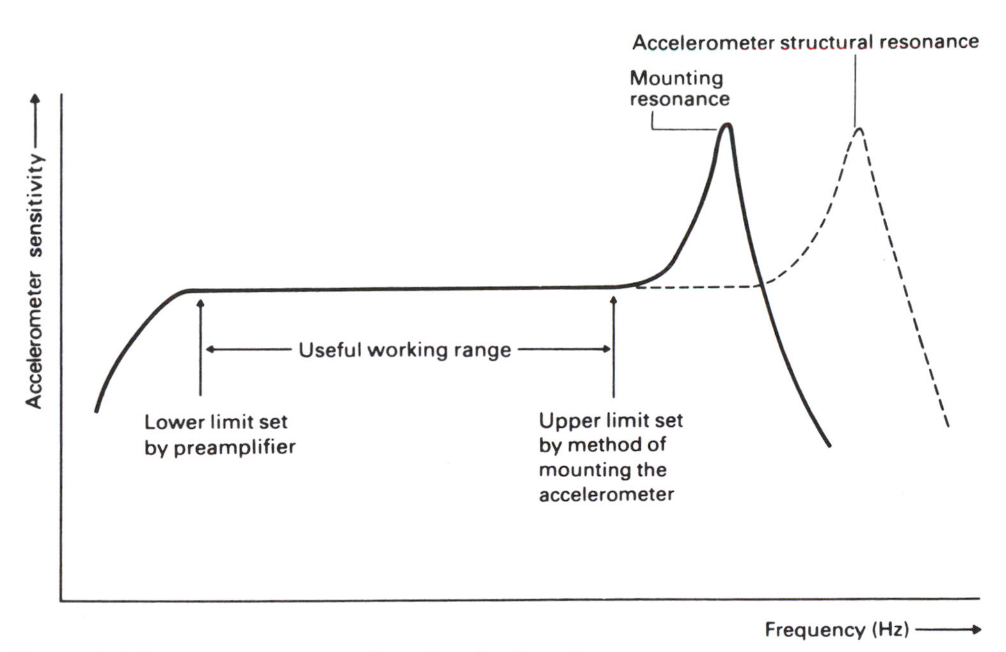

Fig. 7.16 A typical frequency response curve for a piezoelectric accelerometer

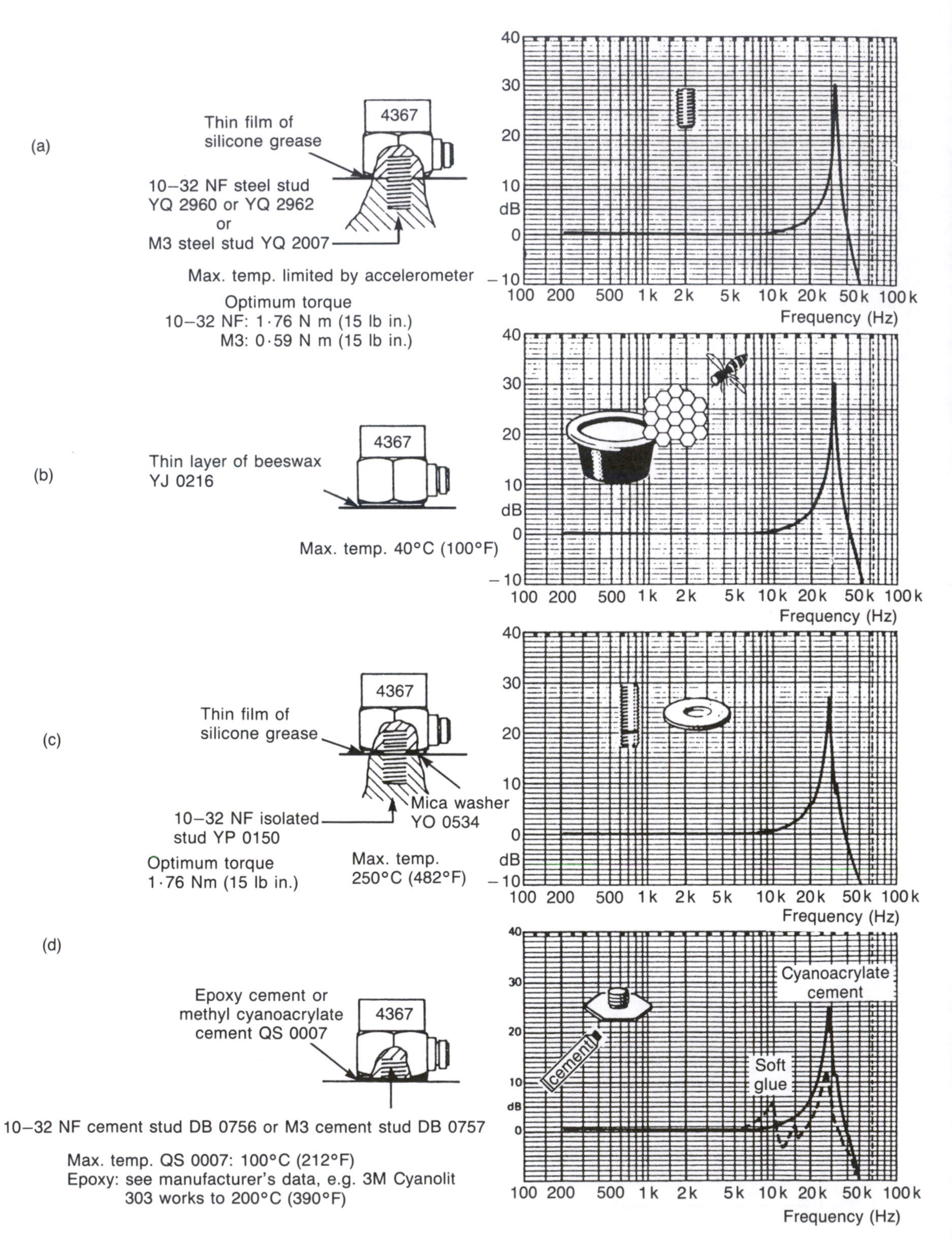

Fig. 7.17 Methods of mounting piezoelectric accelerometers and examples of responses obtained (Courtesy Bruel and Kjaer)

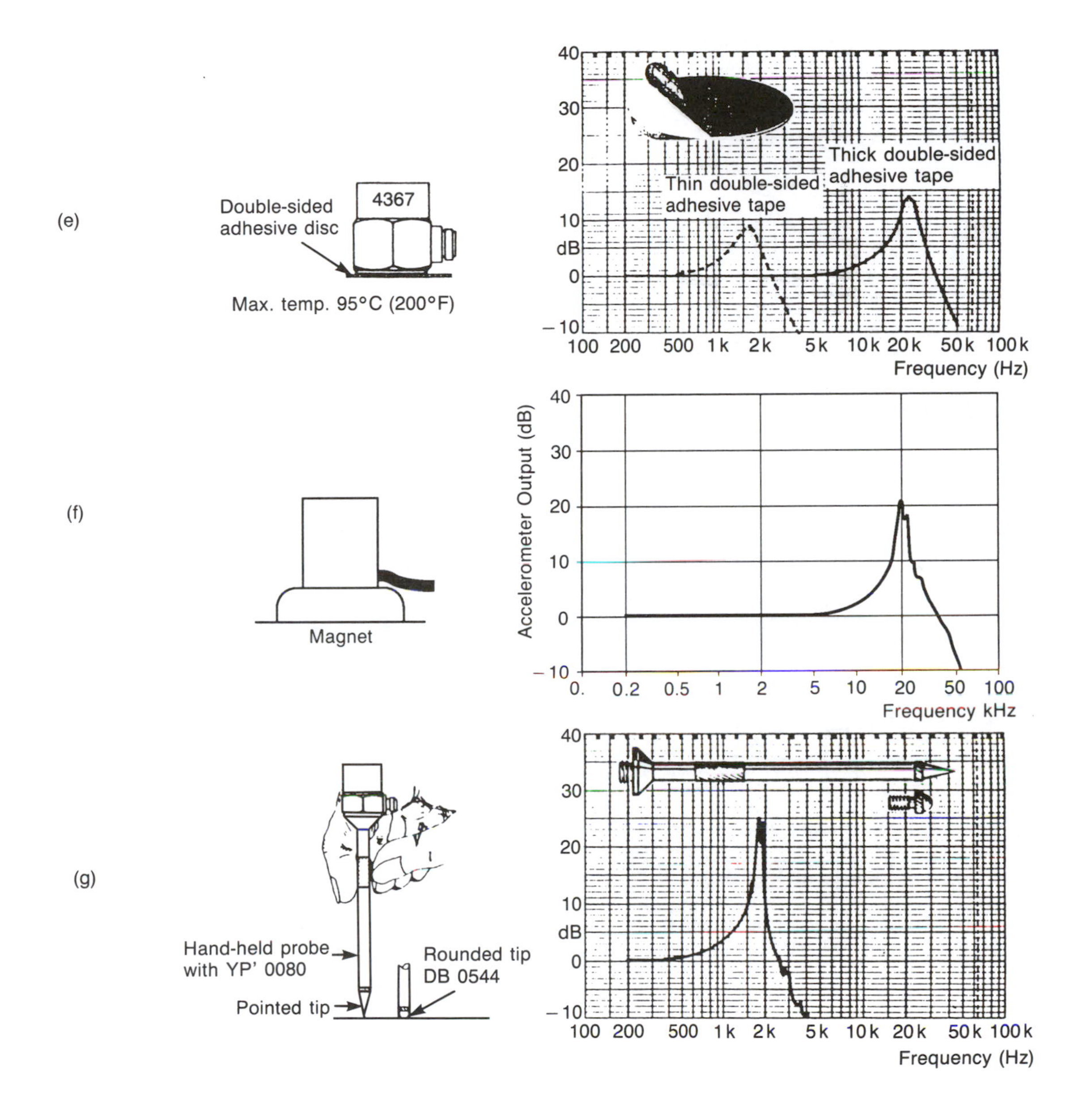

measurement frequency range. The motion of the vibrating surface is transmitted via the base and the discs to the mass. In reaction to this the mass exerts a force on the piezoelectric discs which is proportional to the mass and its acceleration (force = mass × acceleration). The mass, the base and the vibrating surface all move together with the same acceleration, at frequencies well below resonance. Therefore the force on the piezoelectric discs, and the voltage signal it produces, are proportional to the acceleration of the vibrating surface:

Force on piezoelectric discs $\propto$ acceleration of vibrating surface

Voltage signal across piezoelectric discs $\propto$ force on piezoelectric discs

Therefore:

Voltage signal across piezoelectric discs $\propto$ acceleration of vibrating surface

The preloading spring ensures that the discs remain under compression during the entire cycle.

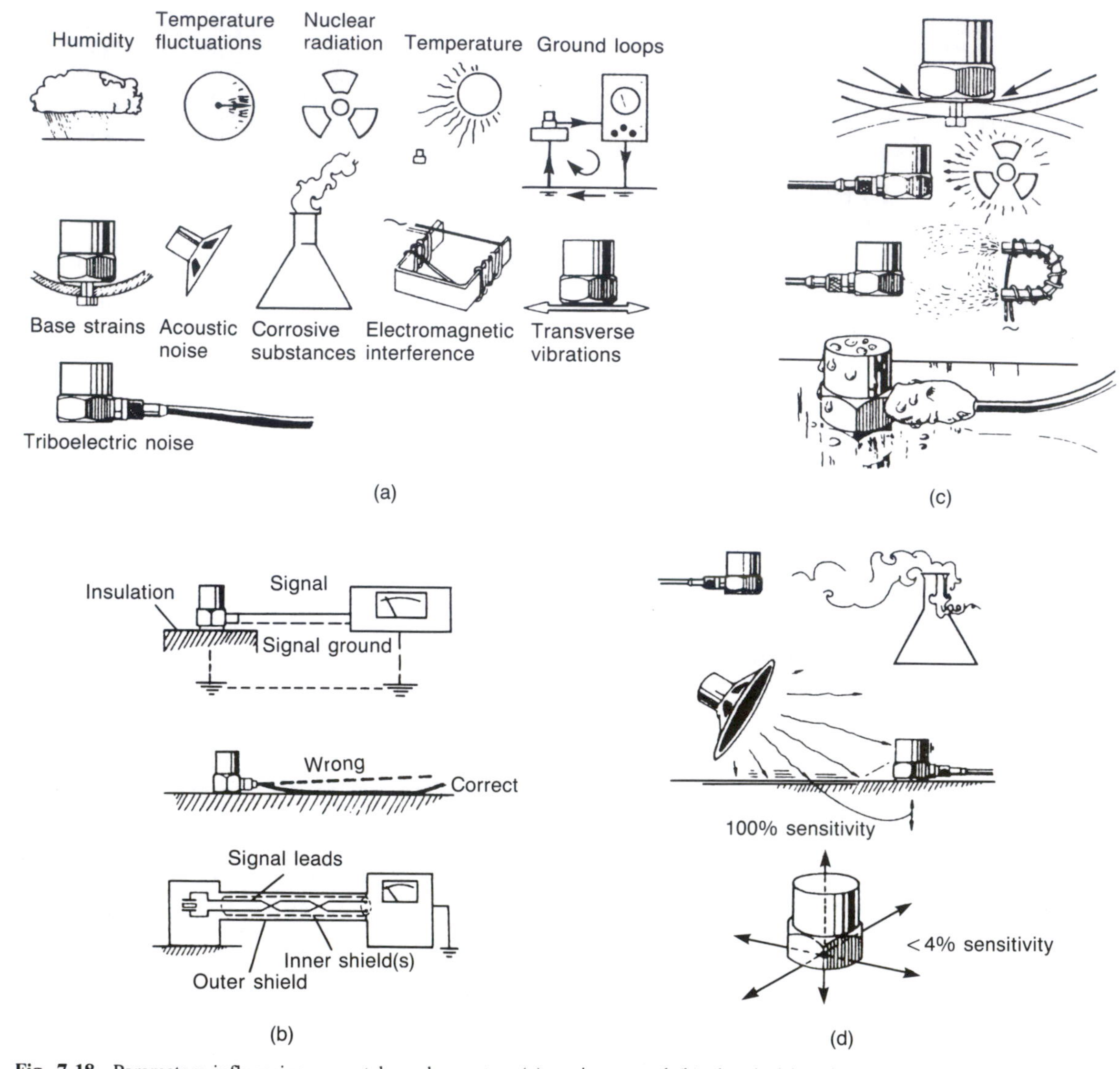

Fig. 7.18 Parameters influencing a crystal accelerometer: (a) environmental (b) electrical interference and ground loops (c) base strain radiation, electromagnetics and moisture (d) chemical, acoustic and cross-axis (Courtesy Bruel & Kjaer)

Accelerometer frequency response This is reasonably flat (to within 12 per cent) for frequencies up to about one-third of the resonant frequency, and this sets the upper frequency limit of the device. The lower frequency limit is set by the electronics of the measurement system (preamplifiers and cables).

Mounting the accelerometer The way in which the accelerometer is attached to the vibrating surface is very important. The method of fixing should be as rigid as possible, since any flexibility between the accelerometer and the surface will produce a 'mounting resonance' which will reduce the upper frequency limit of the device. The best method of mounting is to use a steel stud which screws into both the vibrating surface and the base of the accelerometer. Other methods include the use of studs which screw into the base of the accelerometer and are attached to the vibrating surface by cement, by a thin layer of beeswax or by a permanent magnet. Thin double-sided adhesive tape may also be used. The cement should be as stiff as possible: certain epoxy resins and cyanoacrylates are recommended, but soft setting glues or gums should be avoided. The beeswax method is good provided that the surfaces are clean and smooth, and the wax layer thin and even, in which case measurements may be made up to about 8 kHz. However, a disadvantage is that the wax softens at higher temperatures, and cannot be used above about 40 °C. The magnet method can be used for frequencies up to about 3 kHz. The accelerometer may be attached to a hand-held probe, but although this method is very convenient for carrying out quick vibration surveys it should only be used for frequencies below about 1 kHz because of the low mounting resonance frequency. Figure 7.16 shows a typical frequency response curve of an accelerometer, including the effect of the mounting.

Various mounting methods and their effects are illustrated in Fig. 7.17. Further details are given in BS 7129:1989 (ISO 5348:1987) British Standard Recommendation for mechanical mounting of accelerometers for measuring vibration and shock.

Accelerometer size The sensitivity of an accelerometer depends upon the type of piezoelectric material used, and also on the mass which produces the stress in the discs. Large accelerometers, with a greater mass, are more sensitive than smaller ones. Increased mass, however, reduces the resonance frequency of the device and therefore its upper working frequency limit. Accelerometers are available in a wide range of sizes, with total masses (i.e. the whole device) ranging from 2 g to 100 g giving a corresponding range of resonant frequencies from 80 kHz to 10 kHz.

The size of the accelerometer should also be considered in relation to the surface to which it is to be attached. If it is too large it can significantly alter the vibration level and the resonance frequency of the surface (see Example 7.9). To avoid this, a good rule is that the mass of the accelerometer should be less than one-tenth of the mass of the vibrating surface to which it is attached. This means that only the smallest devices are suitable for measuring the vibration of small, thin sheet-metal panels.

The effect of adverse environments Piezoelectric accelerometers may have to operate at high temperatures or in the presence of corrosive substances, high humidity, nuclear radiation or high acoustic noise levels (Fig. 7.18). The accelerometer has been designed to withstand and be sensitive to these adverse environmental influences, but under extreme conditions special types may have to be used. The manufacturer of the accelerometer should be consulted in these cases.

Sensitivity to base strain, temperature transients, transverse vibrations When very low-level vibration signals are being measured it is important that spurious signals, i.e. those which are unrelated to the acceleration, are kept to a minimum. Such signals can be generated in the accelerometer as a result of transient changes in ambient temperatures, particularly for very low frequencies. Ideally accelerometers should be sensitive only to motion in the axial direction (perpendicular to the base) and should have zero sensitivity in other directions. In practice there is a small sensitivity (a few per cent of the axial value) to vibration in transverse directions. The third way in which the accelerometer can generate signal noise is as a result of flexure of the base of the device.

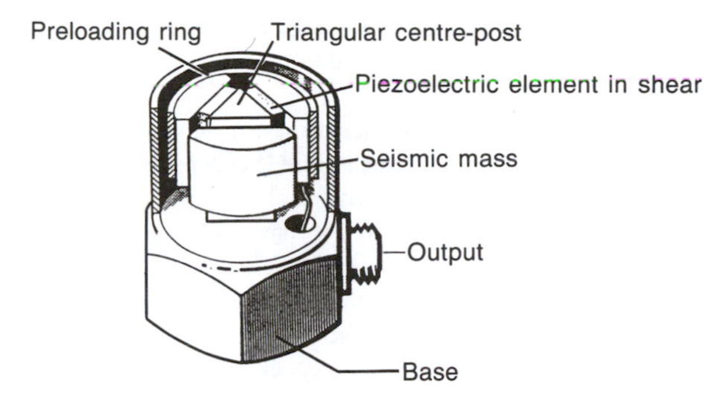

Fig. 7.19 Shear type of piezoelectric accelerometer

Accelerometers which are designed to subject the piezoelectric element to shear strains are usually less sensitive to all of these influences than compression-type accelerometers. A diagram of a shear-type accelerometer is shown in Fig. 7.19.

Cable noise Spurious signals may be generated in the accelerometer cables as a result of cable vibration. These result from the generation of electrical charges caused by relative motion between various internal parts of the cable. They can be reduced by the use of specially designed (graphited) cables but it is always good practice to avoid cable whip by clamping down the cable, with tape or adhesive, as close to the accelerometer as possible, and at other points along its length as well.

Another way in which cable noise can be generated is as a result of ground loops between the separate ground levels of the accelerometer and its associated measuring equipment. This can be minimized by using a mounting method which electrically isolates the accelerometer from the vibrating surface, e.g. by using an insulating washer beneath the accelerometer base.

Sometimes induced currents are generated in the cable as the result of the operation of electrical machinery nearby. The use of properly shielded cables reduces this pick-up which can in extreme cases be further reduced by using special accelerometers and preamplifiers which allow both terminals of the accelerometer to be isolated from the case of the accelerometer, and from instrument ground level. In order to minimize the effect of pick-up the signal should be amplified as early in the measurement chain as possible, preferably before the signal is transmitted down long cables.

The ideal position for the preamplifier is, therefore, as close to the accelerometer as possible.

Dynamic range of acceleration measurements The upper amplitude measurement limit is dependent ultimately upon the accelerometer's structural strength, and also of course on the way it is fixed to the vibrating surface. A typical general-purpose accelerometer can respond linearly to levels of up to about 50 000 m/s^2, assuming the fixing method is satisfactory at these levels. Using the beeswax method the upper limit will be about 100 m/s^2. The lower limit of measurement is governed by some of the factors discussed above, i.e. noise produced in accelerometer and cables, and also by the preamplifier circuitry. A typical lower limit value is about 0·01 m/s^2.

Electronic instrumentation for use with accelerometers

The essential parts of a vibration measuring system, in addition to the accelerometer are a preamplifier, further amplifier stages and filters for signal conditioning and analysis, and some form of read-out, display or recording device, e.g. a meter. A range of specialist vibration meters are available incorporating these and additional features (Fig. 7.20), but in addition some sound level meters can be used for vibration measurement. To do this the microphone of the sound level meter is removed and replaced by the accelerometer and its cable, using a suitable adaptor. The meter scale of the instrument will then read in arbitrary decibel units, but can be calibrated using a known vibration level (Fig. 7.21). Alternatively, some instruments have a variety of removable meter scales and allow vibration levels to be read off directly in appropriate units (e.g. in m/s^2).

Preamplifiers The accelerometer has a very high electrical output impedance, which in effect means that it delivers a signal of very low power, and so the size of the output signal can be reduced by the electrical loading effect of the cables and amplifiers which follow it. The purpose of the preamplifier is not only to amplify the weak signals from the accelerometer, but more importantly to present the signals at a lower, more convenient output impedance to further amplifier stages.

There are two types of preamplifier which are suitable; the voltage preamplifier and the charge preamplifier. One essential difference between the two is the loading effect of the accelerometer cable. With a voltage preamplifier, the sensitivity of the system depends on the length and type of cable being used (in fact, upon the electrical capacitance of the cable). This means that if the cable is changed the sensitivity of the measuring system changes. This may be inconvenient but may be overcome provided a standard vibration source is always available for recalibration. With a charge preamplifier the sensitivity of the system is independent of cable length. The low frequency limit of a charge preamplifier is lower, although with most voltage preamplifiers measurements can be made accurately down to about 2 Hz, which is low enough for most vibration work. The preamplifier in the sound level meter is of the voltage type whereas in specialist vibration meters, charge amplifiers are often used.

Accelerometer manufacturers specify the sensitivity of their devices in two ways to suit either preamplifier, i.e.

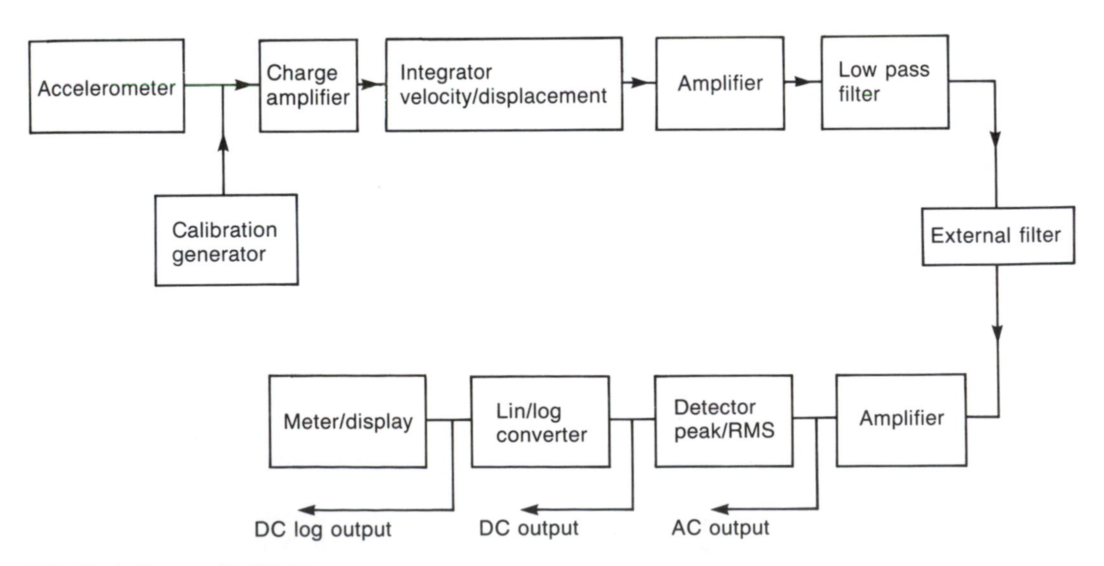

Fig. 7.20 Block diagram of vibration meter

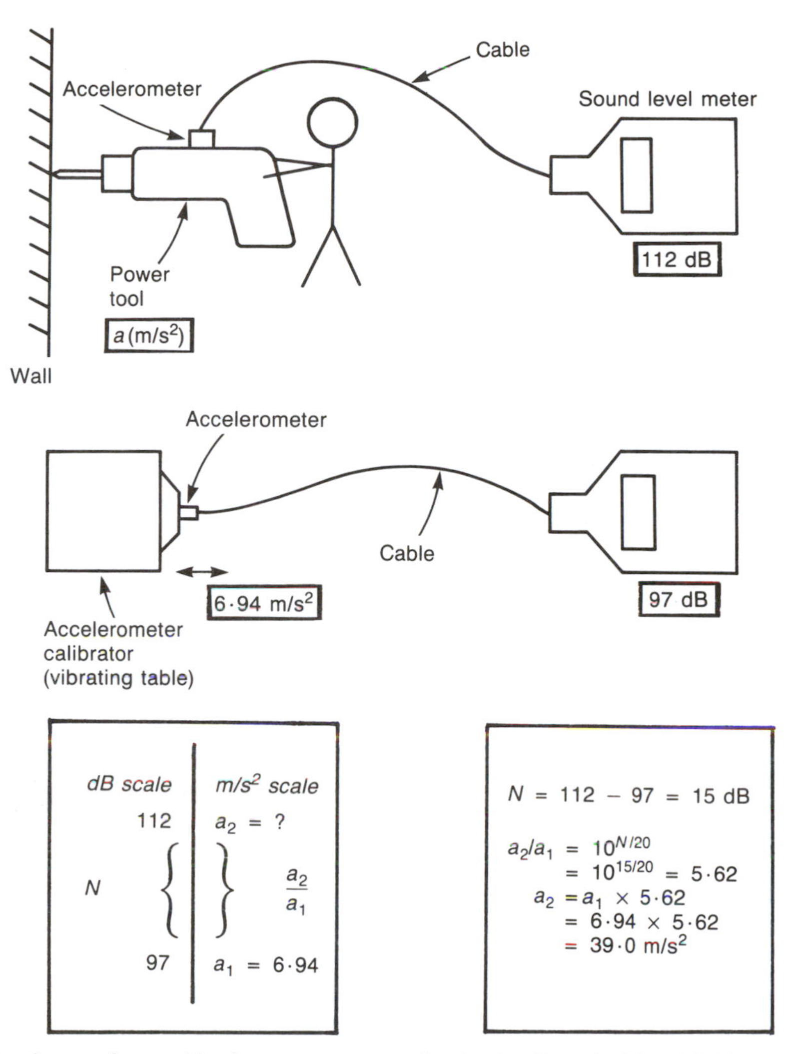

dB scale	m/s² scale
112	a_2 = ?
N	$\frac{a_2}{a_1}$
97	a_1 = 6·94

$N = 112 - 97 = 15$ dB

$a_2/a_1 = 10^{N/20}$
$= 10^{15/20} = 5{\cdot}62$
$a_2 = a_1 \times 5{\cdot}62$
$= 6{\cdot}94 \times 5{\cdot}62$
$= 39{\cdot}0$ m/s²

Fig. 7.21 Illustrating the use of a sound level meter to measure vibration (see Example 7.7 and Table 7.1)

in volts (or millivolts) per m/s² (voltage sensitivity for a given cable length) or in picocoulombs, pC, per m/s² (charge sensitivity). The two sensitivities may be related if the total capacitance of the system (accelerometer, cable and preamp) measured in picofarads, pF, is known, using the relationship:

charge sensitivity = voltage sensitivity × capacitance

[picocoulombs per m/s² = volts per m/s² × picofarads]

Integrating devices These are circuits placed after the preamplifier to electrically integrate the accelerometer signal; once to convert it into the analogue of vibration velocity, and twice to convert it into vibration displacement. The perfect integrator would in effect have to have a frequency response which counterbalances the frequency characteristics shown in Fig. 7.6, i.e. a slope of −6 dB/octave to produce velocity, and −12 dB/octave for displacement. Practical integrating circuits approximate to this above a certain frequency. They are suitable for use with continuous random signals but not for shock signals and transients.

High pass and low pass filters for signal conditioning The purpose of these is to limit measurement errors caused by spurious signals or by signals outside the working range of the accelerometer and its instrumentation. The low pass filter, for example, may be used to cut out high frequency signals from the mounting resonance of the accelerometer. The high pass filter may be used to cut out frequencies below the range of interest which may contain spurious temperature-transient signals or electrical interference, or

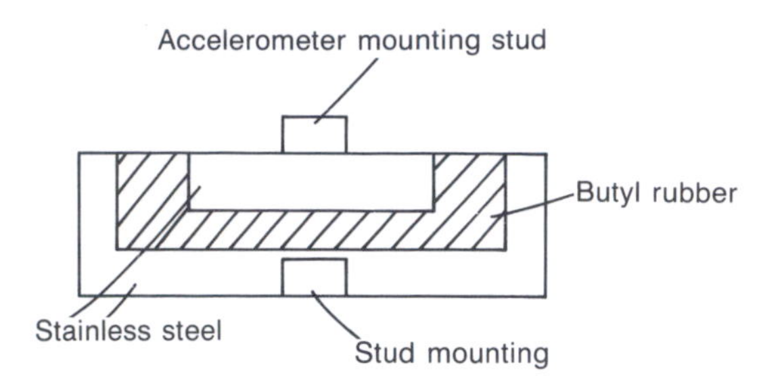

Fig. 7.22 Mechanical low pass filter

to reduce measurement error from frequencies below that of the integrator cut-off.

Mechanical overloading

At very high vibration levels, the accelerometer response becomes non-linear and measurements will then be in error, maybe by large amounts. High amplitude shocks and impacts which may excite the accelerometer resonance can thus give rise to measurement problems, even though the frequencies of interest may be well below those frequencies in the excitation which causes the overload. This is a problem which cannot be overcome by electrical filtering. One possible solution is to use a **mechanical low pass filter** between the accelerometer base and the vibrating surface, to remove the high frequencies before they reach the accelerometer (Fig. 7.22). Such a filter is in effect a layer of suitable resilience. An overload indicator in the input preamplifier can also be used to warn of possible measurement inaccuracies.

Band pass filters The incorporation of band pass filters into the system allows the frequency spectrum of the vibration to be measured either in octave, third-octave or narrow bands. Narrowband analysis can enable the frequency of pure tones in the signal to be measured, and knowledge of the frequency can help to identify the source of the vibration. Special frequency weighting networks can be incorporated into the meter to indicate human response to vibration.

Detector and indicator circuits

Indications of RMS and either peak or peak-to-peak values are usually provided, with peak-hold facilities for impulsive signals. Vibration signals are often of much lower frequency than many noise signals, so much longer averaging times such as 1 s or 10 s are used in order to provide a reasonably steady meter display, instead of the fast and slow responses used in sound level meters. Some meters will have time integration for measuring time-varying vibration signals, either using RMS time averaging to give an energy equivalent value, similar to L_{eq} for noise, or using fourth-power law, root mean quad (RMQ) time averaging to give vibration dose values (explained later in the chapter).

The display may be analogue (i.e. a meter with a pointer) or digital, and will be calibrated directly in vibration units, mm, or m/s. Vibration meters usually also have a variety of output facilities to allow the vibration signal to be connected to an oscilloscope, tape recorder, printer or computer.

Calibration

The best way is to calibrate the entire measurement system by using a vibrational source of known level. Various vibrating tables on to which the accelerometer is mounted are available for this purpose. Some of these operate at one vibration level and frequency, others allow calibration over a range of frequencies and levels. If the vibration is to be tape-recorded for subsequent analysis then the calibration signal should be recorded on the tape and a note made of all instrumentation settings during the recording of the calibration and the subsequent vibration signals.

Calibration of the electronic instrumentation may be performed by feeding an accurately known voltage into the system. Many instruments supply a highly stabilized reference voltage for this purpose. This type of calibration can allow for correction of any changes in the sensitivity of the instrumentation (the electronics) but not for changes in sensitivity of the accelerometer or its cable.

Background noise levels

In this context, *noise* means electronic noise, or unwanted signal in the instrumentation system, arising either from the electronics (e.g. random thermal movement of electrons) or from the accelerometer (e.g. from transient temperature fluctuations) or from the cable (e.g. electromagnetic pick-up, ground loops or triboelectricity).

A good practice when measuring vibration levels, especially very low levels, is to check the background levels produced by the complete measurement system — by measuring the signal produced — i.e. the apparent vibration level, when the accelerometer is mounted on a non-vibrating surface. This should be done as far as is possible with all other environmental conditions the same as for the actual measurement. For reasonable accuracy the background noise level of the instrument system should be less than one-third of (or 10 dB less than) the measured vibration signal.

Measurement of force

If the transfer function of a structure is to be measured,

such as its impedance, then a force transducer must be used in conjunction with a vibration transducer such as an accelerometer. Piezoelectric force transducers are available which give an electrical output proportional to the force transmitted through them. If the point response function is required then force and acceleration are measured at the same point, often using a combined transducer called an impedance head. For transfer functions the force and acceleration are applied at different points in the structure.

7.19 The effects of vibration on people and on buildings

Introduction

Depending upon the level, and a variety of other factors, vibration may affect people's comfort and well-being, impair their efficiency at performing a variety of tasks, or even at very high levels become a hazard to their health and safety. The best known example of the harmful effects of vibration is the white finger syndrome (also known as Reynaud's disease) in which prolonged use of hand-held equipment, such as chain-saws in very cold conditions, produces loss of sensation in the fingers. The vibration produced by the various forms of transportation (e.g. road traffic, trains, aircraft, helicopters, ships and boats) is of great interest for a variety of reasons. First of all, there is concern about the safety and efficiency of the driver or operator subjected to vibration; secondly, there is the effect of vibration levels on the comfort of passengers; and thirdly, there is often great concern among members of the public about vibration produced in buildings, including domestic dwellings adjacent to roads or railway lines, or near to air routes. A great variety of industrial machinery produces vibration which is experienced by people at work. Particular sources which can cause vibration to be experienced by the occupants of nearby buildings and thus often give rise to concern among members of the public include heavy-duty air compressors, forge hammers, pile-driving and quarry-blasting operations.

Factors involved in human response to vibration

The assessment of human response to vibration is made

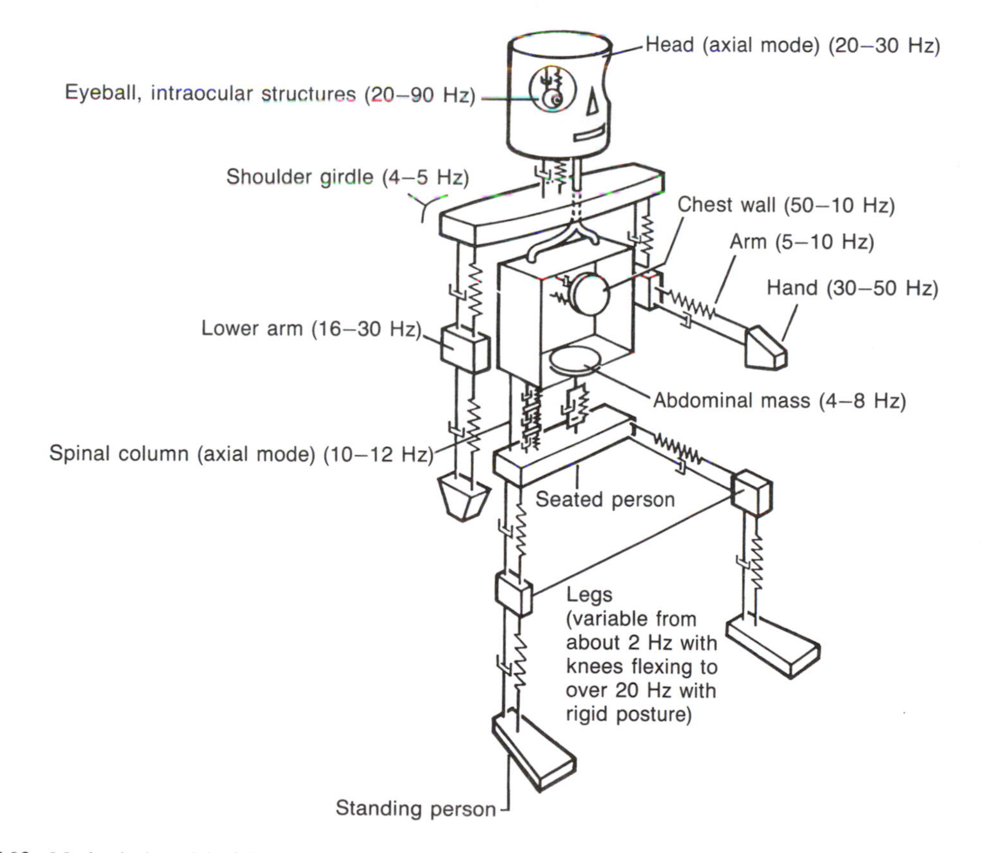

Fig. 7.23 Mechanical model of the human body showing resonance frequency ranges of the various body sections (Courtesy Bruel & Kjaer)

difficult by the fact that there are a large number of factors involved, and because of the great differences between individuals. The main physical factors determining the response to a vibration are the amplitude (or intensity) and frequency, and also the duration (exposure time), point of application and direction of the vibration. Among the configurations which are of interest are the transmission of vibration from the floor through the feet of the standing person and from the seat via the buttocks and possibly the head (through the headrest) of the seated person. In these cases the vibration may be transmitted and felt throughout all parts of the body, and it is the whole-body response which will be required. In other cases the vibration may be applied and sensed at a particular part of the body — the vibration produced in the fingers and hands by power tools being a good example. In this particular case it is the response of the hand–arm system which is important.

As far as vibration transmission is concerned, the human body may be thought of as a complex mass–spring system (Fig. 7.23). It therefore has a complicated frequency response which includes resonances associated with either the whole body or various parts of the body such as the head or the shoulder girdle. These frequencies may vary greatly for different people. Different parts of the body are therefore most sensitive to different frequencies of vibration. Therefore there is not only a difference in individual sensitivity to vibration, but also a difference in individual transmissibility as well. Even the response of any one individual can vary with posture and body tension. The situation may be further complicated by the effects of seats, headrest, gloves, etc., unless great care is taken to measure the vibration level at the exact point of application of the vibration to the body. One more difficulty in the assessment of human response to vibration is in separating it from the response to the high noise levels which are often associated with vibration-causing processes.

Early research into human response: the Reiher–Meister and the Dieckmann schemes for vibration assessment

Many schemes have been developed for the assessment of human response to vibration. One of the earliest, published by Reiher and Meister in 1931, covers the frequency range 1–100 Hz and vibration amplitudes, specified as displacements in the range 1–1000 μm. Using the Reiher–Meister scale it is possible to rate the vibration as belonging to one of six categories ranging from imperceptible to painful. Separate scales are used for vibrations in the vertical and horizontal directions. The scale for vertical vibrations is shown in Fig. 7.24. The threshold of perception corresponds to a velocity amplitude of 0·3 mm/s and the annoyance threshold to 2·5 mm/s.

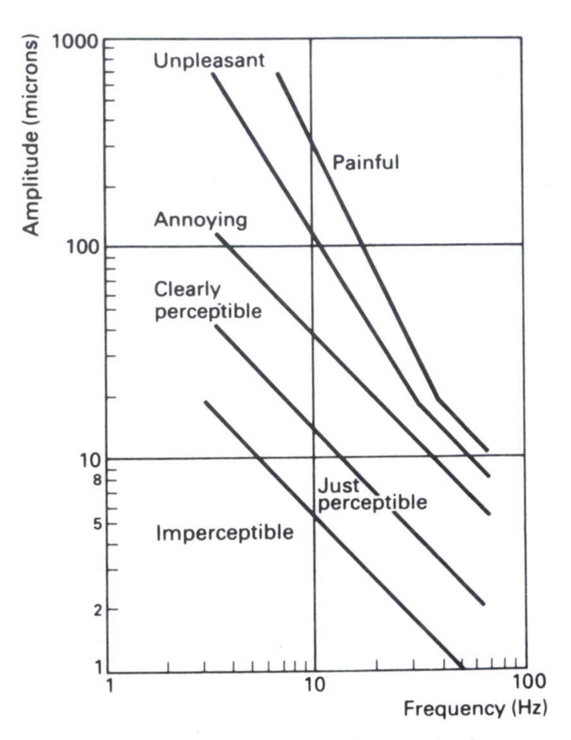

Fig. 7.24 The Reiher–Meister scale for assessing human response to vertical vibrations (persons standing)

Dieckmann (1955) proposed a similar scheme but extending to lower frequencies (down to 0·1 Hz) and higher amplitudes than Reiher and Meister. The vibration level is quantified in terms of *K*-values, ranging from 0·1 to 100, which are related to the intensity. The effect of a vibration can be assessed from its *K*-value:

K = 0·1 — lower limit of perception
K = 1 — allowable in industry for any period of time
K = 10 — allowable only for a short time
K = 100 — upper limit of strain allowable for the average man

The *K*-values may be read off charts of frequency against amplitude, similar to the Reiher–Meister scales, or they may be calculated in terms of displacement amplitude A and frequency f, as shown in Table 7.3.

British and International Standards on human response to vibration

ISO 2631 Evaluation of human exposure to whole-body vibration The introduction to the standard states:

> Various methods of rating the severity of exposure and defining limits of exposure based on laboratory or field data have been developed in the past for specific appli-

Table 7.3 Calculation of Dieckmann K-values for assessing human

Vertical vibrations		Horizontal vibrations	
Below 5 Hz:	$K = 0{\cdot}001\ A.f^2$	Below 2 Hz:	$K = 0{\cdot}002\ A.f^2$
5–40 Hz:	$K = 0{\cdot}005\ A.f$	2–25 Hz:	$K = 0{\cdot}004\ A.f$
Above 40 Hz:	$K = 0{\cdot}2\ A$	Above 25 Hz:	$K = 0{\cdot}1\ A$

(A being amplitude in microns and f the frequency in Hz)

cations. None of these methods can be considered applicable in all situations and consequently none has been universally accepted.

In view of the complex factors determining the human response to vibrations, and in view of the shortage of consistent quantitative data concerning man's perception of vibration and his reactions to it, this International Standard has been prepared first, to facilitate the evaluation and comparison of data gained from continuing research in this field; and, second, to give provisional guidance as to acceptable human exposure to whole body vibration.

Part 1: 1985 General requirements This part of the standard takes into account frequency (in the range 1–80 Hz), vibration amplitude (acceleration), duration (from 1 min to 24 h exposure) and the direction of the vibration relative to the human body. Three different criteria are proposed; working efficiency, health and safety and comfort. These three criteria give rise to three boundaries or limits; the fatigue-decreased proficiency boundary, the exposure limit (for health and safety) and the reduced comfort boundary. Fig. 7.25 shows the limits for the fatigue-decreased proficiency boundary in terms of the amplitude, frequency and duration for a vertical vibration (along the toe-to-head axis). Exposure limits are 6 dB above and reduced comfort values 10 dB below these values, the shape of the contours remaining the same. The human subject is most sensitive to vertical vibrations in the frequency range 4–8 Hz. Above 8 Hz the response contours correspond to constant velocity amplitudes. The ISO standard also allows the effect of broadband vibrations (i.e. containing many frequencies) to be evaluated.

Parts 2, 3 and 4 of ISO 2631 are concerned with the

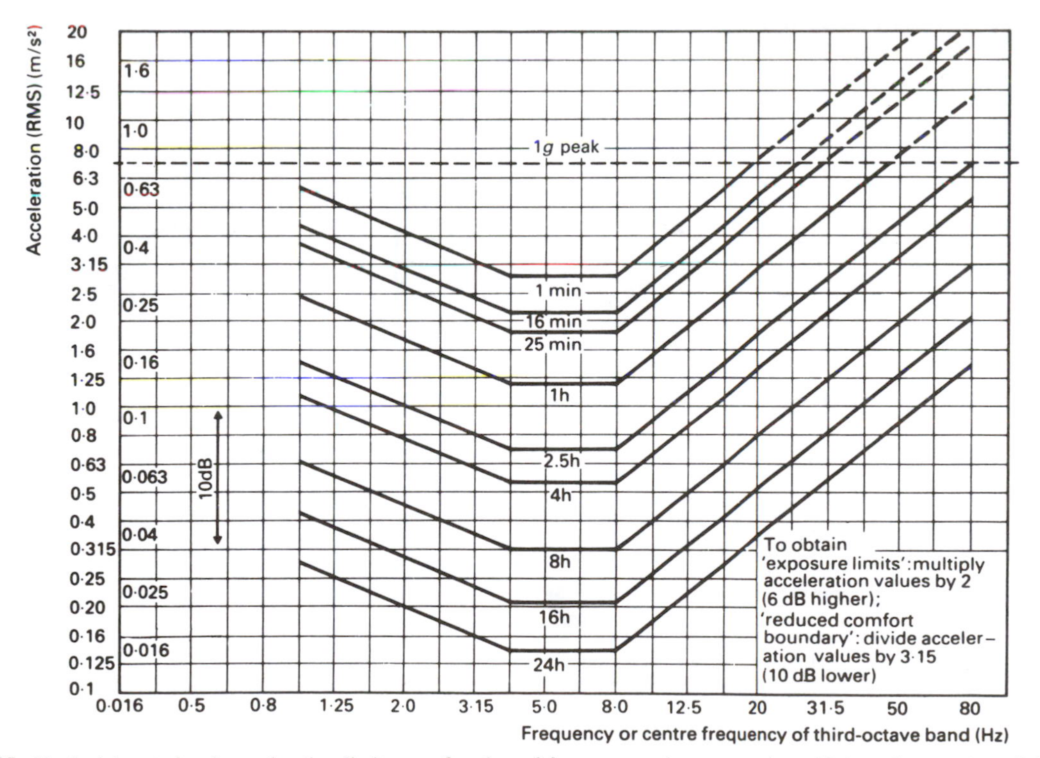

Fig. 7.25 Vertical (toe-to-head) acceleration limits as a function of frequency and exposure time; 'fatigue-decreased proficiency boundary' (Taken from ISO 2631: Part 1: 1985)

vibration of humans in buildings, at low frequencies and on board ships:

- Part 2 1989: Human exposure to continuous and shock-induced vibration in buildings (1 to 80 Hz).
- Part 3 1985: Evaluation of exposure to whole-body z-axis vertical vibration in the frequency range 0·1 to 0·63 Hz.
- Part 4 (draft): Evaluation of crew exposure to vibration on board sea-going ships (1 to 80 Hz).

The subject-matter of Parts 1, 2 and 3 of ISO 2631 are also covered by BS 6841 and BS 6472. Although there are some similarities, there are also some very significant differences between the British and International Standards. One of them is the introduction of the concept of vibration dose value into the British Standards, in order to take account of the effects of impulsive and intermittent vibration.

BS 6841:1987 British Standard Guide to the measurement and evaluation of human exposure to whole-body mechanical vibration and repeated shock The British Standard Guide gives methods for quantifying vibration and repeated shocks in relation to human health, interference with activities, discomfort, the probability of vibration perception and the incidence of motion sickness.

The guide is applicable to motions transmitted to the body as a whole through the supporting surfaces: the feet of a standing person, the buttocks, back and feet of a seated person and the supporting area of a recumbent person.

The four principal effects of vibration considered by the guide are:

- Degraded health
- Impaired activities (e.g. hand manipulation, effects on vision)
- Impaired comfort (and perception)
- Motion sickness

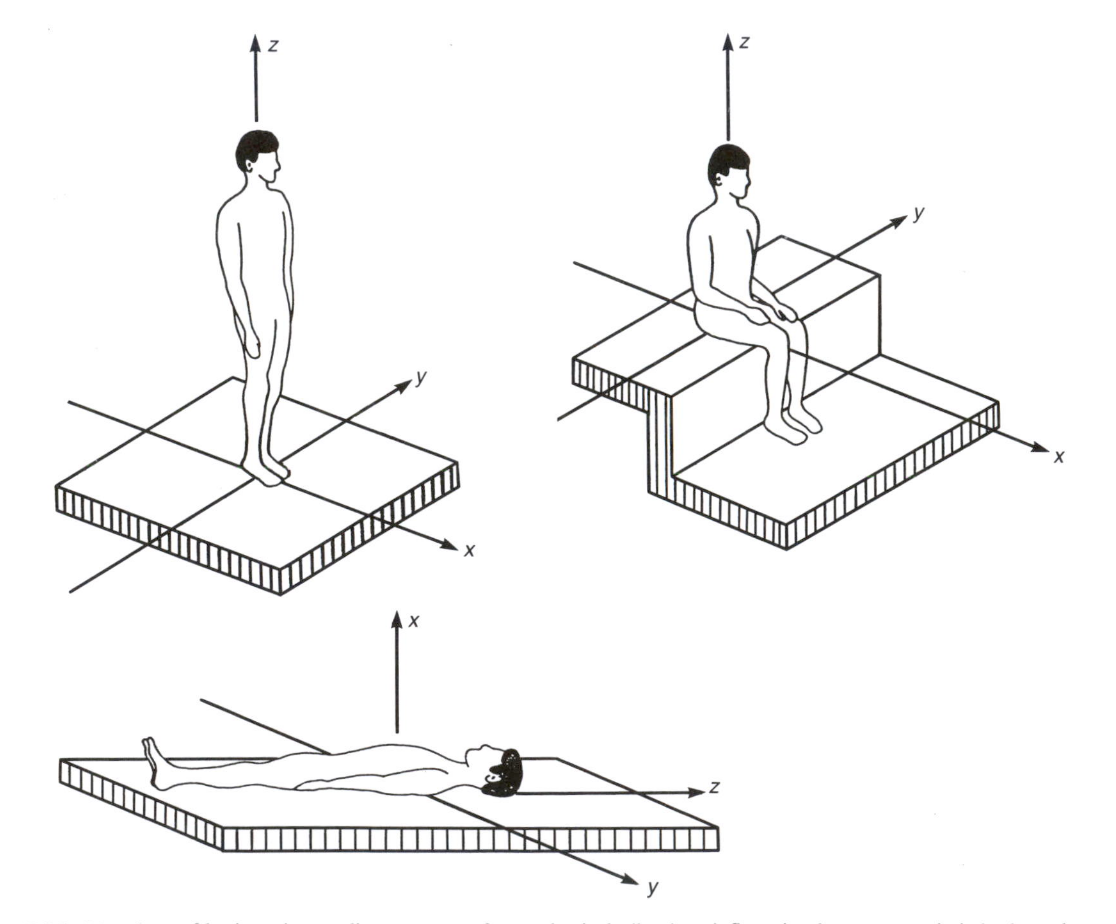

Fig. 7.26 Directions of basicentric coordinate systems for mechanical vibrations influencing humans: x-axis is back to chest; y-axis is right side to left side; z-axis is foot to head

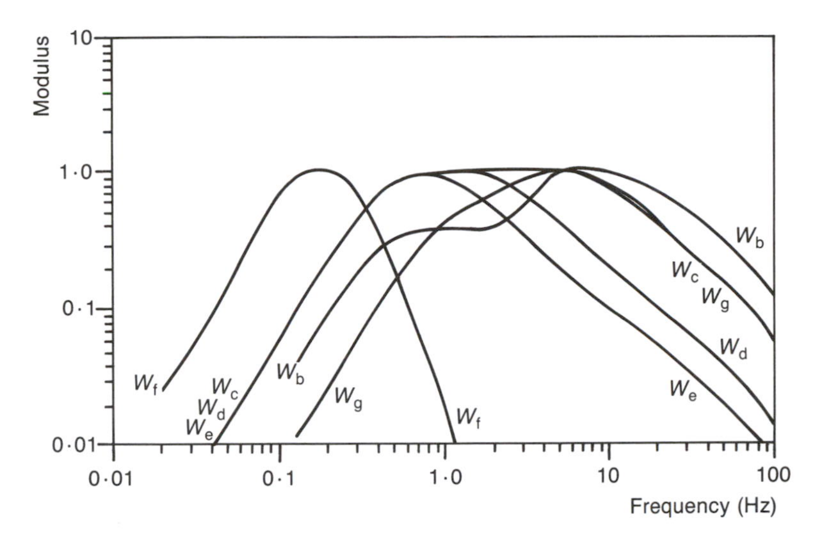

Fig. 7.27 Frequency weighting networks for human response to vibration (BS 6841: 1987)

The guide specifies requirements for:

- Measurement of vibration magnitude
- Measurement of the frequency content of the vibration
- The direction of measurement

The primary quantity for expressing vibration magnitude is the root mean square (RMS) acceleration, in m/s^2 for translational vibration and rad/s^2 for rotational vibration. RMS values give good correlation with human response for steady, continuous vibration. But it has been found that the severity of vibrations which are intermittent, or impulsive, with occasional short duration, high peak values will often be underestimated by RMS values. Therefore the standard also gives alternative methods for measurement and evaluation in these cases, leading to the establishment of a vibration dose value, which will be described later.

The guide specifies measurements in a direction relative to the axes of the human body. The basicentric coordinate system used is shown in Fig. 7.26. Measurements should always be taken as close as possible to the interface between the human body and the source of the vibration.

Human response to vibration depends on the frequency of the vibration, and the guide defines six different frequency weighting networks (rather like the A weighting used for sound). Designated W_b, W_c, W_d, W_e, W_f and W_g (Fig. 7.27), they cover the frequency range from 0·1 Hz to 100 Hz, although frequencies below 0·5 Hz relate only to travel sickness and to the W_f weighting. The choice of the appropriate weighting depends on the different effects of vibration which are being assessed (e.g. health, activity, comfort) and on the direction of vibration measurement (Fig. 7.26). Vibration meters, sometimes called human response vibration meters, are available with these various weightings built in.

BS 6472:1992 Guide to evaluation of human exposure to vibration in buildings (1 Hz to 80 Hz) Experience has shown that complaints regarding vibration in residential buildings are likely to arise from occupants when the vibration levels are only slightly in excess of perception thresholds. For this reason a separate, specific standard, BS 6472, is often used for evaluating disturbance to people in buildings, rather than BS 6841, which is a more general guide with a much wider range of application.

BS 6472 is based on the evaluation of vibration measurements related to the possibility of adverse comments from the occupants, rather than on criteria which are related to health hazard or working efficiency. There are some similarities with BS 6841: the same basicentric coordinate system is used, and vibration magnitudes are specified in RMS acceleration, although there is also the alternative of peak velocity (in m/s). Measurements should be carried out at the point on the building surface where the vibration enters the human body.

The variation in human response with frequency is taken into account by the specification of two base curves, one for the z-axis (i.e. toe to head) and the other for the x and y axes (i.e. back to chest, right side to left side). These base curves (Fig. 7.28) are used to evaluate the measured vibration levels. The z-axis curve shows that in this direction the human body is most sensitive in the frequency range 4–8 Hz, with a base value of 0·005 m/s^2. The curve rises below 4 Hz; above 8 Hz, the rate of rise with increasing frequency is 6 dB per octave, which corresponds, in effect, to constant velocity.

The standard requires that broadband vibrations are measured using a frequency weighting network based on the shape of the base curve. The results of frequency-weighted measurements are then described in terms of

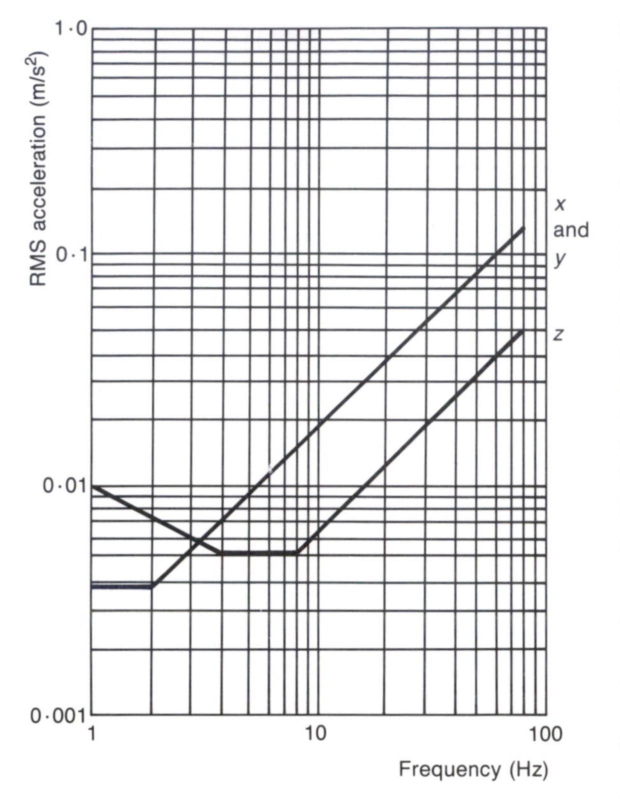

Fig. 7.28 Human response to vibration in buildings: *x*, *y* and *z* base curves (BS 6472: 1984)

multiples of base curve values, and these form the basis for the evaluation of the vibration. Table 7.4, taken from the standard, shows multiplying factors of base values which may be used to specify 'satisfactory' magnitudes of building vibration, i.e. where adverse comment from the occupants is unlikely. The table takes into account different locations and types of buildings, and variations in sensitivity to daytime and night-time exposures.

The sensitivity of subjects to vibration in buildings is such that the standard suggests that a doubling of the vibration magnitudes of Table 7.4 will result in the possibility of adverse comments from occupants, and a further doubling will mean that adverse comments will become probable.

The vibration magnitudes indicated in Table 7.4 are for exposures to continuous vibration over a period of 16 h during the daytime, or 8 h at night-time. The standard indicates how these multiplying factors should be adjusted for intermittent or impulsive vibration. In the case of vibration produced by blasting activities and where there are more than three occurrences in the daytime, a special method of determining the multiplying factor is used, based on the number of events, the duration of each event and the type of floor (wooden or concrete) in the affected building. In all other cases of intermittent, transient or impulsive vibration, the evaluation is based on the calculation of the vibration dose value, which is also used in BS 6841.

Vibration dose value

It seems reasonable to assume that the disturbance caused by a vibration will depend on its duration (t), as well as on its magnitude (a). The question then arises about the relationship between these two variables, a and t, in determining the amount of disturbance. The use of RMS averaging for the measurement and evaluation of steady, continuous vibration implies a relationship of the form a^2t = constant, i.e. that two different vibration events having magnitudes a_1 and a_2, and durations t_1 and t_2, will produce the same degree of disturbance if:

$$a_1^2 t_1 = a_2^2 t_2$$

The power of a vibratory motion is proportional to a^2, and energy is the product of power and time, so this a–t model is, in fact, an equal energy model, i.e. a model that assumes the amount of disturbance produced depends upon the amount of energy in the vibration event. The analogous p^2–t model for noise, where p is the acoustic pressure, is the basis of the use of L_{eq} for the measurement and assessment of time-varying noise.

Although root mean square (RMS) time-averaging seems to work well for the assessment of human response to steady, continuous vibration, it has been shown to underestimate the subjective effects produced by high peak levels of short duration contained in impulsive types of vibration. The vibration dose value, used in BS 6841 and BS 6472, involves the use of root mean quad (RMQ) time-averaging, instead of RMS, in order to take better account of short duration peaks.

The RMQ value of a vibration event is the fourth root of the mean fourth power value, analogous to RMS, which is the square root of the mean square value:

$$\text{RMQ} = \left[\int_0^T \frac{a^4(t)dt}{T} \right]^{0\cdot25}$$

where T is the duration of the event in seconds and $a(t)$ is the frequency-weighted acceleration, which is a function of time (t). The frequency weighting will be one of the possible weightings given in either BS 6841 or BS 6472. Changing the index from 4 to 2, or from 0·25 to 0·5, in the above expression will give the definition of RMS value. Like the RMS, the RMQ is an average value, so it will have the units of acceleration, m/s^2. For a sine wave of amplitude A, the RMS value is $0\cdot7071A$ and the RMQ value is $0\cdot7825A$.

Table 7.4 Multiplying factors for base levels (BS 6472), used to specify satisfactory magnitudes of building vibration with respect to human response

Place	Time	Mulltiplying factors (see notes 1 and 5)	
		Exposure to continuous vibration: 16 h day, 8 h night (see note 2)	Impulsive vibration excitation with up to 3 occurrences (see note 8)
Critical working areas (e.g. hospital operating theatres, precision laboratories (see notes 3 and 10)	Day	1	1
	Night	1	1
Residential	Day	2 to 4 (see note 4)	60 to 90 (see notes 4 and 9)
	Night	1·4	20
Office	Day	4	128 (see note 6)
	Night	4	128
Workshops	Day	8 (see note 7)	128 (see notes 6 and 7)
	Night	8	128

Note 1: Table 7.4 leads to magnitudes of vibration below which the probability of adverse comments is low (any acoustical noise caused by structural vibration is not considered).
Note 2: Doubling of the suggested vibration magnitudes may result in adverse comment and this may increase significantly if the magnitudes are quadrupled (where available, dose–response curves may be consulted).
Note 3: Magnitudes of vibration in hospital operating theatres and critical working places pertain to periods of time when operations are in progress or critical work is being performed. At other times magnitudes as high as those for residences are satisfactory provided there is due agreement and warning.
Note 4: Within residential areas people exhibit wide variations of vibration tolerance. Specific values are dependent upon social and cultural factors, psychological attitudes and expected degree of intrusion.
Note 5: Vibration is to be measured at the point of entry to the subject. Where this is not possible then it is essential that transfer functions be evaluated.
Note 6: The magnitudes for vibration in offices and workshop areas should not be increased without considering the possibility of significant disruption of working activity.
Note 7: Vibration acting on operators of certain processes such as drop-forges or crushers, which vibrate working places, may be in a separate category from the workshop areas considered elsewhere in BS 6472. The vibration magnitudes specified in relevant standards would then apply to the operators of the exciting processes.
Note 8: Appendix C of BS 6472 contains guidance on assessment of human response to vibration induced by blasting.
Note 9: When short-term works such as piling, demolition and construction give rise to impulsive vibrations it should be borne in mind that undue restriction on vibration levels can significantly prolong these operations and result in greater annoyance. In certain circumstances higher magnitudes can be used.
Note 10: In cases where sensitive equipment or delicate tasks impose more stringent criteria than human comfort, the corresponding more stringent values should be applied. Stipulation of such criteria is outside the scope of this standard.

The vibration dose value (VDV) is the fourth root of the integral of the fourth power of vibration value with respect to time. The mathematical definition of VDV is given below:

$$\text{VDV} = \left[\int_0^T a^4(t)dt\right]^{0\cdot 25}$$

The vibration dose value represents an amount or quantity of vibration. The equivalent definition, based on RMS averaging, with the powers 4 and 0·25 replaced, respectively, by 2 and 0·5 in the above expression, would have the units of energy. The units of VDV are m/s$^{1\cdot 75}$ because, dimensionally, it is composed as follows:

$$\left[\left(\frac{\text{m}}{\text{s}^2}\right)^4 \times \text{s}\right]^{1/4} = \left(\frac{\text{m}^4}{\text{s}^7}\right)^{1/4} = \frac{\text{m}}{\text{s}^{1\cdot 75}} = \text{m/s}^{1\cdot 75}$$

If the vibration magnitude does not vary with time but remains constant, so that $a(t) = a$, then the above expression simplifies to:

$$\text{VDV} = aT^{0\cdot 25}$$

In both BS 6841 and BS 6472 the VDV concept is used to evaluate the cumulative effects of bursts of intermittent vibration and of impulsive vibration. The cumulative effects can be estimated by combining the VDVs of individual events according to the fourth-power law:

$$V_{\text{T}} = [V_1^4 + V_2^4 + \ldots + V_N^4]^{0\cdot 25}$$

where V_{T} is the total VDV and $V_1, V_2, \ldots, V_N$ are the VDVs of the individual events.

If there are N identical events, then:

$$V_{\text{T}} = [NV^4]^{0\cdot 25} = VN^{0\cdot 25}$$

Example 7.11

The vibration exposure at a particular location consists of

three different events having individual VDVs of 0·02, 0·03 and 0·04 $m/s^{1.75}$. Calculate the total vibration dose value, V_T:

$$V_T = [(0.02)^4 + (0.03)^4 + (0.04)^4]^{0.25}$$

$$= 0.043 \text{ m/s}^{1.75}$$

The example shows how the fourth-power law emphasizes the contribution of the largest of the three individual values.

Example 7.12

The VDV of an event is 0·04 $m/s^{1.75}$. Calculate the total vibration exposure if the event occurs 16 times during the day:

$$V_T = V \times N^{0.25} = 0.04 \times 16^{0.25} = 0.08 \text{ m/s}^{1.75}$$

The example illustrates the emphasis placed upon amplitude, as compared to duration, in the at^4 relationship. The effect of doubling the amplitude of an event would be the same as repeating the event 16 times.

Estimated vibration dose value (eVDV) Accurate measurement of VDV and RMQ values for impulsive signals requires specialist equipment with fourth-power time-averaging capabilities, not usually available in sound level meters and general-purpose vibration meters. It has been shown, however, that estimated VDVs, which are good approximations to the true VDV may be obtained from RMS measurements, provided that the crest factor of the signal is less than 6.

The crest factor of a signal is the ratio of its peak to its RMS value. It is a measure of the peakiness of a signal, so that impulsive signals, with high peak levels of short duration, will have high crest factor values. The crest factor of a sine wave signal is 0·7071.

The estimated VDV for a vibration event of constant RMS magnitude a m/s^2 and duration T s is given by:

$$\text{eVDV} = 1.4aT^{0.25}$$

The factor 1·4 has been determined empirically. For a sinusoidal motion of duration T and amplitude A:

$$\text{VDV} = 0.7825AT^{0.25}$$

$$\text{eVDV} = 0.9899AT^{0.25}$$

Thus the eVDV overestimates the VDV for sinusoids, but underestimates it for impulsive signals with high crest factors.

BS 6841 and BS 6472 require the crest factor of the frequency-weighted signal, i.e. with both the peak and RMS being weighted values.

The concepts of VDV and eVDV may be used to evaluate the effect of non-continuous vibration events against the criteria of BS 6841 and BS 6472. This requires that the base values and their multiples (from Table 7.4) must be turned into eVDVs, by assuming constant levels over the 16 h daytime or 8 h night-time period. The eVDV of a steady vibration level of 0·005 m/s^2 (i.e. base level) of 16 h duration is:

$$\text{eVDV} = 1.4 \times 0.005 \times (16 \times 60 \times 60)^{0.25}$$

$$= 0.108 \text{ m/s}^{1.75}$$

The equivalent value for the 8 h night-time period is 0·0908 $m/s^{1.75}$.

By multiplying these base-value eVDVs by the appropriate factors and by applying the guidance on the effect of doubling and further doubling of magnitudes, Table 7.4 may be transformed into eVDVs for residential buildings (Table 7.5). Similar tables may be constructed for other building types.

Example 7.13

Sixteen trains pass close to a particular site in the daytime, and twice during the night-time. A proposal to build a residential property on the site is being considered. Vibration measurements are carried out at the site while trains are passing. Each event lasts for 20 s and the RMS level, which is constant during each event, is 0·1 m/s^2. Calculate the daytime and night-time eVDVs and assess the situation using BS 6472.

In the daytime:

Table 7.5 Vibration dose values ($m/s^{1.75}$) above which various degrees of adverse comment may be expected in residential buildings

Place	Low probability of adverse comment	Adverse comment possible	Adverse comment probable
Residential buildings, 16 h day	0·2–0·4	0·4–0·8	0·8–1·6
Residential buildings, 8 h night	0·13	0·26	0·51

$$\text{eVDV} = 1{\cdot}4 \times 0{\cdot}1 \times (16 \times 20)^{0{\cdot}25}$$
$$= 0{\cdot}59 \text{ m/s}^{1{\cdot}75}$$

This value lies between the limits of 0·4 and 0·8 $\text{m/s}^{1{\cdot}75}$ (see Table 7.5), so adverse comment is possible.

At night-time:

$$\text{eVDV} = 1{\cdot}4 \times 0{\cdot}1 \times (2 \times 20)^{0{\cdot}25}$$
$$= 0{\cdot}35 \text{ m/s}^{1{\cdot}75}$$

This value lies between the limits of 0·26 and 0·52 $\text{m/s}^{1{\cdot}75}$ (see Table 7.5), so adverse comment is probable.

Note that, in general, it may be necessary to include a multiplying factor (i.e. a transfer function) to relate the levels of vibration in the ground to those in the proposed building.

Example 7.14

If in the previous example the number of trains per day could be reduced, how many would be allowed if the probability of adverse comment were to remain low?

Table 7.5 indicates that for low probability of adverse comment during the daytime, the eVDV should remain below 0·4 $\text{m/s}^{1{\cdot}75}$, so:

$$0{\cdot}4 = 1{\cdot}4 \times 0{\cdot}1 \times (N \times 20)^{0{\cdot}25}$$
$$= 0{\cdot}30 \times N^{0{\cdot}25}$$

where N = the number of trains allowed per day

From which, $N^{0{\cdot}25} = 1{\cdot}35$ and $N = 3{\cdot}4$, i.e. three trains per day would be allowed.

Example 7.15

In the previous examples it may be possible to site the proposed residential property further from the railway line, where the vibration levels from the trains are lower. What reduced RMS vibration level would ensure that the probability of adverse comment remains low for 16 trains per day and 2 at night-time?

Daytime: as in the last example, the eVDV must remain below 0·4 $\text{m/s}^{1{\cdot}75}$; the required RMS acceleration, a, is given by:

$$0{\cdot}4 = 1{\cdot}4 \times a \times (16 \times 20)^{0{\cdot}25}$$
$$a = 0{\cdot}68 \text{ m/s}^2$$

Night-time: Table 7.5 shows that the eVDV must remain below 0·13 $\text{m/s}^{1{\cdot}75}$, so:

$$0{\cdot}13 = 1{\cdot}4 \times a \times (2 \times 20)^{0{\cdot}25}$$
$$a = 0{\cdot}04 \text{ m/s}^2$$

Example 7.16

The vibration exposure in an office close to a building site arises from three different types of construction activities or events. The duration of each type of event, and the RMS acceleration levels during events are constant. This information, and the numbers of the different types of event, are given below. Calculate the total eVDV due to all these events.

	Event type 1	Event type 2	Event type 3
RMS acceleration (m/s²)	0·02	0·04	0·08
Event duration (s)	10	4	2
Number of events	20	12	6

Step 1: Calculate the component eVDV for each type of event:

For event type 1: $\text{eVDV} = 1{\cdot}4 \times 0{\cdot}02 \times (10 \times 20)^{0{\cdot}25} = 0{\cdot}105 \text{ m/s}^{1{\cdot}75}$

For event type 2: $\text{eVDV} = 1{\cdot}4 \times 0{\cdot}04 \times (4 \times 12)^{0{\cdot}25} = 0{\cdot}147 \text{ m/s}^{1{\cdot}75}$

For event type 3: $\text{eVDV} = 1{\cdot}4 \times 0{\cdot}08 \times (2 \times 6)^{0{\cdot}25} = 0{\cdot}208 \text{ m/s}^{1{\cdot}75}$

Step 2: Combine the components to give the total eVDV, V_T:

$$V_T = [(0{\cdot}105)^4 + (0{\cdot}147)^4 + (0{\cdot}208)^4]^{0{\cdot}25}$$
$$= 0{\cdot}223 \text{ m/s}^{1{\cdot}75}$$

The solution can be obtained in one stage, using the formula:

$$\text{eVDV} = [(1{\cdot}4 \times a_1)^4 \times n_1 \times t_1 + (1{\cdot}4 \times a_2)^4 \times n_2 \times t_2 + (1{\cdot}4 \times a_3)^4 \times n_3 \times t_3]^{0{\cdot}25}$$
$$= [(1{\cdot}4 \times 0{\cdot}02)^4 \times 10 \times 20 + (1{\cdot}4 \times 0{\cdot}04)^4 \times 4 \times 12 + (1{\cdot}4 \times 0{\cdot}08)^4 \times 2 \times 6]^{0{\cdot}25}$$
$$= 0{\cdot}223 \text{ m/s}^{1{\cdot}75} \quad \text{(as before)}$$

This method, although more direct, does not give the contributions to the total from the three different types of event. The previous method shows that the events of type 3 provide the major contribution to the total because of their higher magnitude. Their higher magnitude more than outweighs the effects of their shorter duration and lower number.

Example 7.17

A sample of the vibration exposure in a workshop is measured over a period of 45 min. The VDV of the sample is 0·06 $m/s^{1\cdot75}$. Calculate the exposure over the total working day, of 8 h duration, assuming the sample to be representative of the entire day:

$$\text{Total VDV} = 0\cdot06 \times \left(\frac{8}{0\cdot75}\right)^{0\cdot25} = 0\cdot11 \text{ m/s}^{1\cdot75}$$

BS 6842:1987 Measurement and evaluation of human exposure to vibration transmitted to the hand

This standard provides guidance on methods of measuring and assessing hand-transmitted vibration exposure. It states: 'Habitual use of many vibration hand tools may affect blood vessels, nerves, bones, muscles and connective tissues of the hand and forearm.' It specifies that measurements should be of RMS acceleration (m/s^2) in the frequency range 8–1000 Hz in octave or third-octave bands, or using the frequency weighting defined in the standard (Fig. 7.29). Measurements should be made in one of three orthogonal directions defined relative to the hand–arm system (Fig. 7.30).

The vibration transducer, e.g. an accelerometer, should be attached to the surface in contact with the hand, such as the handle of a vibrating tool. The measurement of vibration on handles with resilient covers can be accomplished using a specially designed adaptor, which might take the form of a suitably shaped, small, light, rigid block, with a mounting arrangement for the accelerometer, which could be held between the fingers and against the handle. Where the vibration levels are affected by the pressure of the operator's grip, the measurements should be taken for normal operating conditions.

When the acceleration is time-varying, a frequency-weighted energy equivalent RMS acceleration, $(a_{h,w})_{eq}$, should be measured. This is defined in a similar way to L_{Aeq} for noise and requires instrumentation with a similar time-integrating capability. In order to facilitate comparisons between different durations of exposure, the daily exposure should be expressed in terms of the 8 h energy equivalent frequency-weighted acceleration $A(8)$:

$$A(8) = \left\{\frac{1}{8}\int_0^T\left[a_{h,w}(t)\right]^2 dt\right\}^{1/2}$$

where $a_{h,w}(t)$ = the instantaneous value of the weighted acceleration, in m/s^2, transmitted to the hand, which varies with time, t

T = the total duration of the working day, in hours

The energy equivalent acceleration, $A(T)$, over any other period, T h, may be converted to $A(8)$ using the 'a^2t' law:

$$A(8) = \left(\frac{T}{8}\right)^{1/2} \cdot A(T)$$

Example 7.18

If $A(6) = 10$ m/s^2, then:

$$A(8) = \left(\frac{6}{8}\right)^{1/2} \times 10 = 8\cdot66 \text{ m/s}^2$$

Example 7.19

The frequency-weighted accelerations for exposure times of 1 h, 3 h and 5 h are, respectively, 15 m/s^2, 12 m/s^2 and 10 m/s^2. Calculate the equivalent accelerations over the total 9 h period, and the $A(8)$ value.

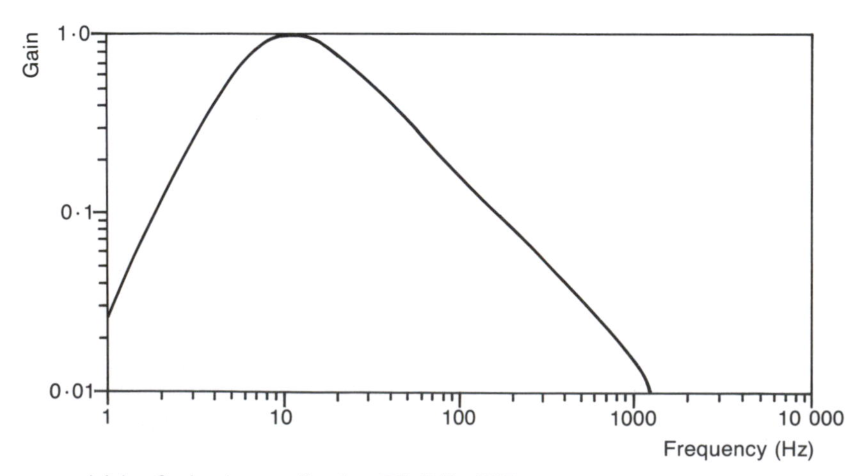

Fig. 7.29 Frequency weighting for hand–arm vibration (BS 6842: 1987)

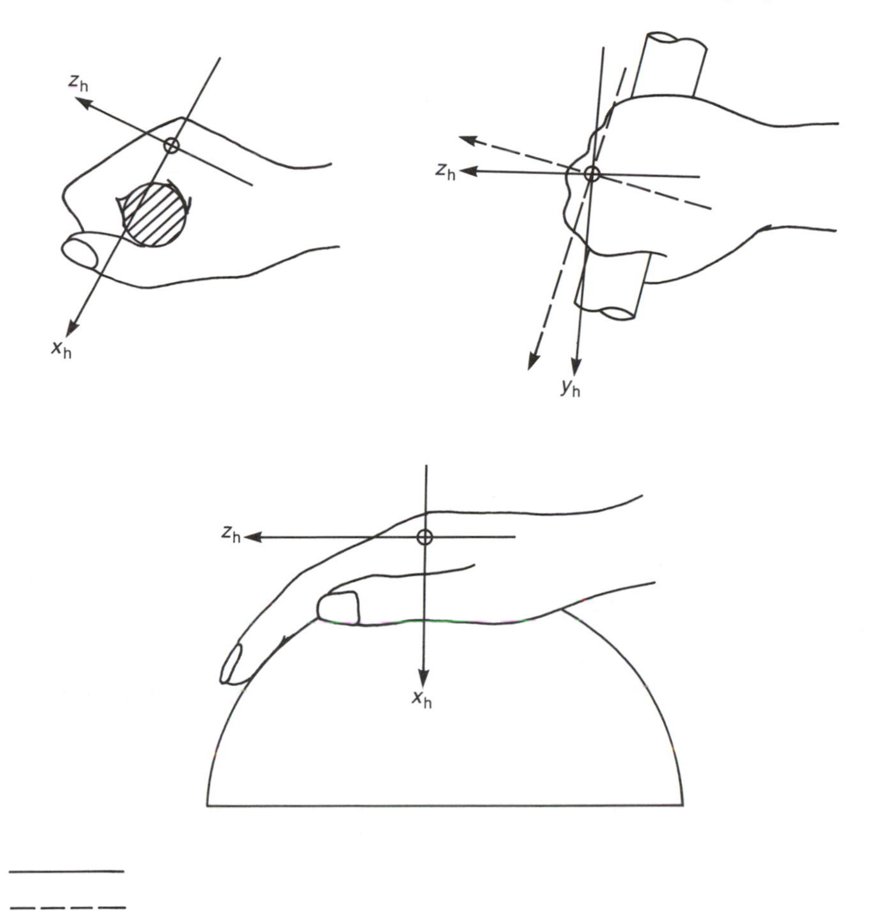

Fig. 7.30 Hand–arm coordinate system (BS 6842: 1987). Solid lines indicate the anatomical coordinate system; broken lines indicate the basicentric coordinate system: (a) handgrip position showing a standardized grip on a cylindrical bar of radius 2 cm; (b) flat palm position showing a hand pressed onto a ball of radius 5 cm. The origin of the anatomical system is deemed to lie in the head of the third metacarpal and the z-axis (i.e. hand axis) is defined by the longitudinal axis of the bone. The x-axis projects forwards from the origin when the hand is in the normal anatomical position (palm facing forwards). The y-axis passes through the origin and is perpendicular to the x-axis. In cases where the hand is gripping a cylindrical handle the system will be rotated so that the y_h-axis is parallel to the handle axis

$$A(9) = \left[\frac{(15^2 \times 1) + (12^2 \times 3) + (10^2 \times 5)}{9}\right]^{1/2}$$

$$= 11{\cdot}34 \text{ m/s}^2$$

$$A(8) = \left(\frac{9}{8}\right)^{1/2} \times 11{\cdot}34 = 12{\cdot}03 \text{ m/s}^2$$

According to the standard, much of the information about the effects of vibration on the hand–arm system relate to vascular symptoms characterized by finger blanching. The probability of a vibration-exposed individual developing finger blanching depends on several other factors besides vibration exposure, including individual susceptibility, tool type, tool condition and method of use plus environmental conditions such as temperature.

The standard gives guidance on how vibration exposure may be related to the prevalence of finger blanching symptoms. It appears that, with normal tool usage, symptoms do not usually occur if the frequency-weighted acceleration is below about 1 m/s^2 RMS. The standard states: 'There is some evidence suggesting that, if a tool having a frequency-weighted vibration magnitude of about 4 m/s^2 RMS were to be used regularly for 4 hours a day, there may be an occurrence of symptoms of blanching in about 10% of the vibration-exposed population after about 8 years.' Alternative combinations of acceleration, daily exposure time (hours) and lifetime exposure (years) which will also produce a 10% prevalence may be obtained using the equal-energy principle and are tabulated in the standard (Table 7.6).

Table 7.6 Hand–arm vibration exposure table (BS 6842: 1987): frequency-weighted vibration acceleration magnitudes (m/s^2 RMS) which may be expected to produce finger blanching in 10% of persons exposed

Daily exposure	Lifetime exposure					
	6 months	1 year	2 years	4 years	8 years	16 years
8 h	44·8	22·4	11·2	5·6	2·8	1·4
4 h	64·0	32·0	16·0	8·0	4·0	2·0
2 h	89·6	44·8	22·4	11·2	5·6	2·8
1 h	128·0	64·0	32·0	16·0	8·0	4·0
30 min	179·2	89·6	44·8	22·4	11·2	5·6
15 min	256·0	128·0	64·0	32·0	16·0	8·0

Note 1: With short duration exposures the magnitudes are high and vascular disorders may not be the first adverse symptom to develop.
Note 2: The numbers in the table are calculated and the figures behind the decimal points do not imply an accuracy which can be obtained in actual measurements.
Note 3: Within the 10% of exposed persons who develop finger blanching there may be a variation in the severity of symptoms.

The effect of vibration on buildings

It is almost inevitable that high noise and vibration levels experienced by the occupants of a building should give rise to concern about the possible effects that these may have on the building. However, cases in which even minor damage to a building can be attributed directly to the effects of vibration alone are very rare. Usually many other factors are involved as well, such as ground settlement or movement caused by changes of moisture content. It is generally accepted that the vibration levels in a building would become absolutely intolerable to the human occupants long before they reached a level at which there was danger of damage to the building. In cases where minor damage does occur, and in which vibration is alleged to play a part, the most common occurrences are damage to unsound plaster, cracking of glass, loosening of roof tiles and cracks to masonry. However, it is very likely that existing minor damage may be noticed for the first time by an occupant whose attention and concern has been aroused by the disturbance caused by a new source of vibration.

Heavy vibrating machinery located at high levels in a building can produce intense vibrations in the horizontal direction, and these are more likely to be damaging than vertical vibration. Bells situated in church towers can also produce high levels in the building structure. Sustained sources of vibration may produce resonances in buildings or parts of buildings. However, this is less likely to cause problems if the vibrations are transient. The natural frequency of a building will depend mainly on its height and base dimensions, typical values ranging from 10 Hz for a low building to 0·1 Hz for a very tall building. The natural frequency is of interest since this type of vibration may be excited by wind loading of the building. Floors, ceilings and windows also have their own natural frequencies. Typical values for floors are in the range 10–30 Hz depending on size and type of construction. People in buildings are more aware of vibrations transmitted via the floor than from any other part of the structure. It is important, therefore, that the natural frequency of the floor does not coincide with the range of maximum sensitivity (4–8 Hz) for vertical vibrations of the human body at which whole-body resonance occurs. The maximum amplitude usually occurs in the centre of the floor. Modern long-span floors are likely to cause an increase in floor vibration amplitudes. Natural frequencies of windows range from 10 to 100 Hz depending on the size and thickness of the glass, and for plaster ceilings typical values range from 10 to 20 Hz.

Many investigations have been carried out in an attempt to define threshold limits for the occurrence of vibration induced damage to buildings. The evidence from these investigations has been fully reported and discussed in a Building Research Establishment report by R. Steffens, *Structural Vibration and Damage*, published in 1974. In one investigation on the effects of blasting on buildings, it was shown that buildings can withstand peak amplitudes of about 400 μm. In contrast, typical levels produced in buildings by nearby road traffic often lie in the range 5–25 μm (10–30 Hz). It is often found that internal sources of vibration in a building, such as footsteps, door slamming, furniture moving, washing machines and vacuum cleaners, will produce levels comparable with or even greater than the external source which is the subject of complaint (road traffic, compressor, pile driver, etc.). Typical levels produced by footsteps and by door slamming can be in the region of 50–150 μm.

Other investigations have suggested that limits for damage are best expressed in terms of peak vibration velocity, and various values have been suggested ranging from 50 to 230 mm/s depending on the type of building and the degree of damage. For comparison it is interesting to note that a level of 75 mm/s corresponds to a Dieckmann K-value of 670, which would be extremely unpleasant, and is well into the painful zone of the Reiher–Meister scale (assuming frequencies in the range 5–40 Hz).

Yet another method for rating vibration, based on energy considerations and developed by Zeller in Germany, has been used for assessing possible damage to buildings. This involves the acceleration and the frequency of the vibration, the Zeller power being given by the acceleration squared divided by the frequency, and measured in mm^2/Hz^3.

BS 7385 Evaluation and measurement for vibration in buildings

Part 1 Guide for measurement of vibrations and evaluation of their effects on buildings This standard, which is identical to ISO 4866:1990, discusses some of the principles involved in the measurement and evaluation of building vibration, and gives guidance on information to be recorded.

The following factors are considered:

- The characteristics of vibration (type of signal, range of magnitudes and frequencies) produced by different types of source, such as traffic, blasting, pile-driving and machinery.
- Type of building within 14 different classes, taking into account the different types of construction, types of soil and foundations and a political importance factor.
- Selection of measurement parameters, equipment, transducers, measurement positions, data collection and analysis.

Most ground-borne vibration entering buildings from artificial sources is in the frequency range from 1–150 Hz. Natural sources, such as wind and earthquakes, produce significant amounts of vibrational energy at frequencies down to 0·1 Hz.

Vibration-induced damage to buildings is classified into three categories: cosmetic, minor and major.

Part 2:1993 Guide to damage levels from ground-borne vibration The preferred method of measurement is simultaneously to record unfiltered time histories of the three different orthogonal components (e.g. x, y and z) of particle velocity. The (total) particle velocity may then be found by taking the root mean square value of the three components and its peak value obtained. The peak values of the individual components should also be measured because it is this type of data which has usually been presented in the various case histories used to develop the limits in the standard.

The case history data suggests that the probability of damage tends towards zero at levels below 12·5 mm/s peak component particle velocity. The limit for cosmetic damage varies from 15 mm/s at 4 Hz to 50 mm/s at >40 Hz for measurements taken at the base of the building. Different low frequency limits (<40 Hz) are given for two different types of buildings. The limits for cosmetic damage should be doubled for minor damage, and doubled again for major damage. The limits for cosmetic damage are shown in Fig. 7.31 and Table 7.7.

BRE Digest 353:1990 Damage to structures from ground-borne vibration This digest reviews various methods for measuring and assessing building damage caused by vibration, including German, Swiss and Swedish standards. German standard DIN 4150:Part 3 1986, which has been widely used, adopts a similar approach to BS 7385 Part 2. Guideline values of peak component particle velocity (mm/s) are given for three different types of building structure. Different limits are given for the frequency ranges: less than 10 Hz, 10–50 Hz and 50–100 Hz.

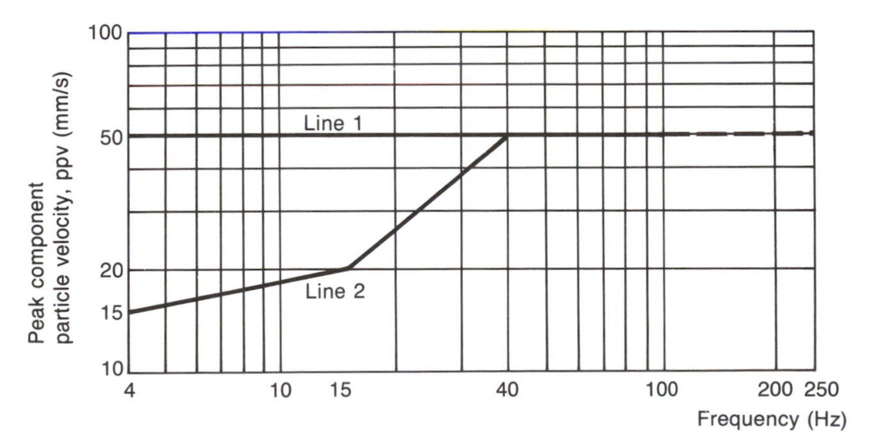

Fig. 7.31 Transient vibration guide values for cosmetic building damage (BS 7385:Part 2 1993)

Table 7.7 Transient vibration guide values for cosmetic building damage (BS 7385: Part 2 1993)

Line	Type of building	Peak component particle velocity (mm/s) in frequency range of predominant pulse	
1	Reinforced or framed structures Industrial and heavy commercial buildings	50 at 4 Hz and above	
2	Unreinforced or light framed structures Residential or light commercial type buildings	4 Hz to 15 Hz 15 at 4 Hz increasing to 20 at 15 Hz	15 Hz and above 20 at 15 Hz increasing to 50 at 40 Hz and above

7.20 Vibration testing and condition monitoring

These are two very important applications of vibration technology which deserve mention for the sake of completeness in any chapter on vibration.

In vibration testing the machine or component to be tested is attached to a vibrator and subjected to a controlled programme of vibration. The purpose is to simulate in a fairly short period of time the vibration exposure which the component might experience in its service lifetime.

In condition monitoring (or machine health monitoring) the vibration from a working machine is continually monitored. The frequency spectrum of the machine can be used as a diagnostic tool to give an indication of the machine's condition. Advance warning of excessive wear or some other fault allows planned maintenance to be carried out before failure occurs.

Further discussion of either of these topics is beyond the scope of this chapter.

Questions

(1) The vibration acceleration amplitude of the floor of an office situated next to a workshop is $0 \cdot 3$ m/s^2 at a frequency of 30 Hz. Find the velocity and displace amplitudes.

(2) The vibration displacement amplitude measured on the ground floor of a house near to a building site is found to be $0 \cdot 002$ mm at a frequency of 50 Hz. What are the corresponding amplitudes of velocity and acceleration?

(3) The maximum vibration level allowed in a certain working area is 10 mm/s. Find the corresponding maximum permitted displacement and acceleration levels for vibrations of 10 Hz frequency.

(4) Two vibration levels are measured and the results quoted in terms of decibels relative to 1 g ($9 \cdot 81$ m/s^2). The first level is +7 dB and the second is −5 dB. Convert both levels into absolute values (m/s^2) and into decibels relative to 10^{-6} m/s^2.

(5) The acceleration vibration level is measured on the casing of a power tool, using a sound level meter adapted for vibration measurements. The level is found to be 84 dB, in arbitrary decibel units. The meter is then calibrated using a vibrating calibration table which produces a level of $7 \cdot 07$ m/s^2. This produces a reading of 97 dB on the meter. What is the level on the casing in m/s^2?

(6) The vibrational velocity of a surface is given as 103 dB (re 10^{-6} mm/s). Express this in absolute terms (i.e. in mm/s).

(7) The resonance frequency of a thin machine panel has been estimated as 1200 Hz, and its effective mass as 120 g. The vibration levels at the centre of the panel are to be measured using an accelerometer of mass 80 g. Calculate the resonance frequency of the panel with the accelerometer attached.

(8) An electric motor is to be mounted on antivibration mounts in an attempt to reduce the transmission of vibration to the floor beneath. The vibration level at the floor is $15 \cdot 0$ m/s^2 at 50 Hz before the isolators are installed and the aim is to reduce this to $6 \cdot 0$ m/s^2.
   (a) Estimate the necessary reduction both in decibels and as a transmissibility required of the isolators.
   (b) Assuming a damping ratio of $0 \cdot 2$, estimate the natural frequency of the motor–isolator system, and the static deflection required.

(9) The displacement–frequency response curve of Fig. 7.10 is flat (horizontal) well below resonance, and well above resonance it falls off at a rate of 12 dB/octave. Using Fig. 7.6 as a guide, sketch the response curve in terms of velocity and of acceleration.

(10) Assess the severity of human response to the vibration levels of Questions 1, 2 and 3 against the Reiher–Meister, Dieckmann and ISO 2631 schemes. Assume that the vibrations are in the vertical direction.

(11) Compare and contrast the use of an adapted sound level meter and a purpose-built vibration meter for measuring vibration.

(12) Discuss the properties required of a transducer for

measuring vibration. Describe the construction and method of operation of a piezoelectric accelerometer.

(13) Piezoelectric accelerometers are available in a variety of sizes. Discuss the relationship between the size of an accelerometer and its performance: sensitivity, dynamic range, frequency response, etc. And discuss any other factors which may affect the choice of accelerometer size.

(14) Describe and explain the frequency response of a piezoelectric accelerometer. Explain the various methods of attaching the accelerometer to a surface, and indicate how the fixing method may affect the frequency response.

(15) Discuss the effects of vibration on people, and explain the various factors which can influence human response to vibration.

(16) Explain the term *vibration dose value (VDV)* and show how VDV may be used in the assessment of human response to vibration. What is the estimated vibration dose value (eVDV) and how does it relate to VDV?

(17) Discuss the factors which should be considered when assessing whether a source of vibration may cause damage to a building.

(18) Briefly describe the effects that vibration can have on the human hand–arm system, and discuss the factors involved in determining them. Compare and contrast the techniques and methods used to assess whole-body and hand–arm vibration.

(19) Calculate the estimated vibration dose value produced by 20 vibration events occurring within a 16 h daytime period. Each event lasts for 10 s during which the constant RMS vibration level is 0·05 m/s^2. If the events were measured in a residential property, evaluate the likely effect using BS 6472.

(20) It is planned to build an office block close to a railway line where 40 heavy goods trains pass by each day. Each train event lasts for 16 s during which the RMS acceleration level is constant. Estimate the limit for the RMS acceleration in the office, using BS 6472, if the probability of adverse comment is to remain low.

(21) There is concern that occupants in a new office building may be disturbed by vibration from three types of event on a nearby industrial site. The VDVs of each type of event have been measured, and their values are tabulated along with their numbers. Calculate the total VDV and use BS 6472 to evaluate the likely effects.

	Type 1	Type 2	Type 3
VDV of event (m/s$^{1·75}$)	0·06	0·04	0·05
Number of events	4	20	10

(22) Vibration from a demolition site affecting nearby residential properties occurs in 20 s bursts, during which the constant RMS acceleration measured at the nearest property is 0·02 m/s^2. How many bursts may occur before the eVDV reaches 0·2 m/s$^{1·75}$?

(23) A centrifugal fan running at 600 rpm and having a total mass of 1080 kg is to be supported on a solid concrete slab which has negligible deflection under the imposed load. The weight of the machine is to be carried on six vibration isolators each carrying an equal load. The isolators are required to provide an installed isolation efficiency of 99%. The transmissibility T of the vibration system is given by

$$T = \left[\frac{1}{(f/f_0)^2 - 1}\right] \times 100\%$$

Using the above information, calculate:

(a) The required natural frequency of the mass–spring system.

(b) The static deflection in mm of the steel springs required to achieve the calculated natural frequency.

(c) The required stiffness in kN m^{-1} of the steel springs.

(*IOA 1992*)

(24) Describe the frequency response of a piezoelectric accelerometer. When a piezoelectric accelerometer is calibrated, the output voltage is 96 mV RMS for a sinusoidal input of 9·81 m s^{-2} at 80 Hz. The accelerometer is then placed on an object vibrating sinusoidally at 50 Hz. The output voltage is now 54 mV RMS. Calculate the peak velocity.

(*IOA 1991*)

(25) (a) Describe using sketched graphs, how the amplitude of oscillation of a damped oscillatory system varies with frequency. How does the response of the system change with increased damping? Indicate, with some explanation, the frequency range over which the system is:

- stiffness controlled
- damping controlled
- mass controlled

(b) Distinguish between force and displacement transmissibility of antivibration systems. What are the advantages and disadvantages of incorporating damping into the design of practical antivibration systems?

An antivibration system has a static deflection of 4 mm. Over what frequency range will the isolation efficiency exceed 95%?

$$T = \frac{1}{\left(\frac{f}{f_0}\right)^2 - 1} \qquad f_0 = 15·8\sqrt{\frac{1}{X_{st}}}$$

(*IOA 1990*)

(26) A large machine is vibrating at a frequency of 850 Hz. Measurement of the vibration level indicates an RMS acceleration of $12 \cdot 3$ m s^{-2}. Estimate the sound pressure level radiated from the panel at a position close to the panel. Explain the assumptions made in your calculation. (Take the specific acoustic impedance of air to be 410 rayls.) (*IOA 1990*)

(27) A sound level meter is to be used as a vibration detector with an accelerometer connected to its input amplifier. Calibration of the accelerometer at 10 m s^{-2} gave a meter reading of 92 dB(linear). When the accelerometer was attached to a vibrating surface, the reading was 75 dB(linear). Determine the acceleration of the surface in dB re 10^{-6} m s^{-2}. If the vibration frequency is 30 Hz, calculate the displacement of the surface. (*IOA 1988*)

(28) A microphone amplifier is calibrated to read dB re 20 μPa when used with a microphone of sensitivity 50 mv/Pa.
 (i) Calculate the RMS acceleration if the meter reads 78 dB when the microphone is replaced by an accelerometer of sensitivity 2 mV/m s^{-2}.
 (ii) If the vibration frequency is known to be 50 Hz calculate the corresponding RMS velocity and displacement. (*IOA 1985*)

(29) Describe, by reference to appropriate diagrams, the essential features of:
 (i) A piezoelectric accelerometer.
 (ii) A vibration meter capable of use for measuring vibratory acceleration, velocity and displacement.

A vibration meter calibrator produces a harmonic vibration of acceleration $g = 9 \cdot 81$ m/s^2 peak at a frequency of 80 Hz. Calculate:
(a) the RMS velocity;
(b) the displacement peak amplitude; and
(c) the vibratory velocity level re 10^{-8} m/s.
(*IOA 1984*)

8 Noise-control engineering

8.1 Introduction

The technology exists to solve many noise problems provided that the costs can be met to implement the noise-control engineering solution. However, many problems could be solved more economically and more effectively if adequate planning had anticipated possible noise problems, and steps had been taken to avoid them.

Therefore adequate methods of predicting noise levels in advance also form an important part of any noise-control programme. Day-to-day management is also important in ensuring that noise-control measures (such as the provision of enclosures around machines or of personal hearing protection) are properly used and maintained, and that personnel always adhere strictly to noise-reducing working procedures.

The formulation and implementation of a noise-control policy therefore involves the techniques of planning, management and economics as well as knowledge of the engineering principles. The aim of this chapter is to give a brief overview of the many different aspects of noise control. More detailed and practical treatments are given in the reference list.

8.2 The steps in a noise-control programme

It is important to adopt a logical approach to noise-control problems. The first step for an existing noise problem is the measurement of the noise levels and the diagnosis of the most important sources, where several exist together. In the case of proposed sources, the levels must be predicted.

The next step is the establishment of a target noise level for the particular situation. The amount of noise reduction which is required can then be estimated. The target level or standard will be set to meet a particular criteria relating for example to employee health and safety, customer acceptability or public annoyance. The required noise reductions may necessitate modifications to an existing noise source or may have to be built in to the proposed source at the planning stage.

Having established the required reduction, the next stage is the application of noise-control engineering principles to design the reduction treatment. At this stage several other factors have to be considered, such as the cost, safety and fire-hazard implications of the proposed noise-reduction measures and their possible effects on working procedures and production processes. Any adverse environmental conditions likely to be encountered by the materials to be used, such as the presence of dust and grit, corrosive chemicals, or high temperatures, must be taken into account at this stage.

The next stage is the assessment of the effectiveness of noise-reduction measures when installed. If the required reduction is not achieved, the process has to be repeated, starting at the design stage. It must be emphasized that many noise sources may contribute to a particular problem, or there may be many transmission paths associated with one particular source. This means that the effective solution of a noise problem may involve more than one noise-control measure (e.g. absorption as well as screening, or isolation as well as insulation). Therefore if a particular treatment does not give the required reduction, it does not necessarily mean that the diagnosis and design has been completely wrong — it may just have been incomplete and require other additional measures as well.

The above sequence is shown schematically in Fig. 8.1.

8.3 The noise chain: source–path–receiver

When considering various noise-control options it is often convenient to think of the noise problem as consisting of a three-link chain: the noise source, the transmission path and the receiver. The noise-reduction measures may then

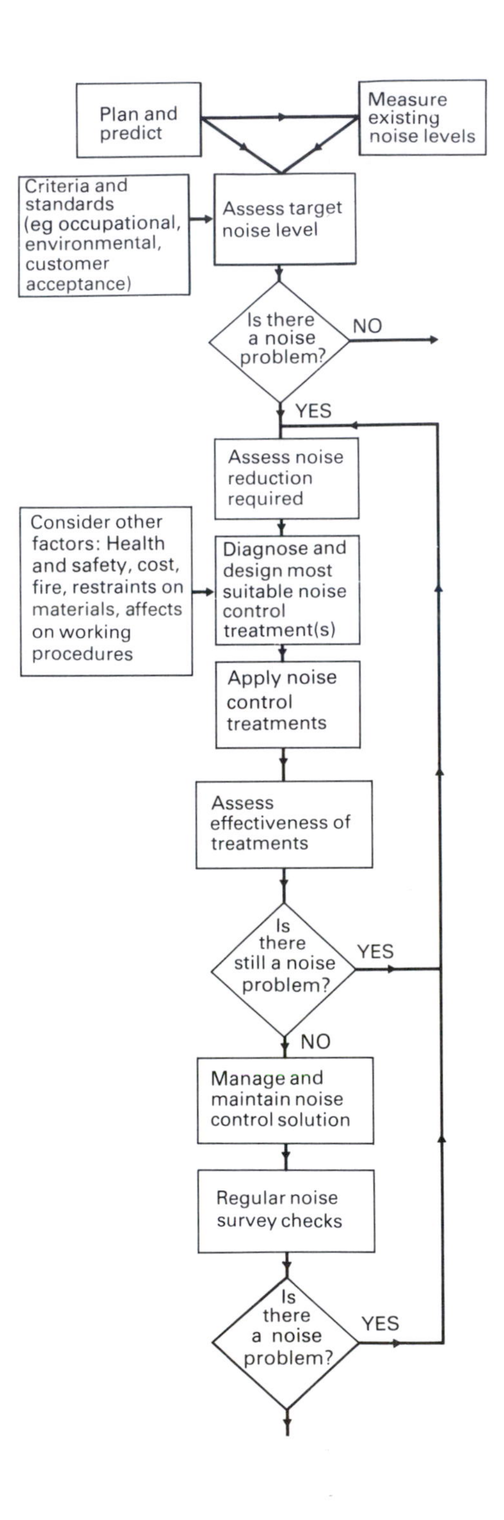

Fig. 8.1 The steps in a noise-control programme

be applied to either one or more of these links. Noise control at source is obviously desirable if it is practical. In extreme cases this might mean the complete removal of the noise-producing process and its substitution by a quieter one. An alternative might be the re-siting of the source (e.g. a plant room) to a less noise-sensitive area. In cases where such a complete solution is inappropriate it might be possible to reduce the noise at source.

Noise sources mainly fall into three broad categories — first of all, noise emitted from vibrating surfaces, e.g. machine panels, and secondly 'aerodynamic noise sources' in which there is a direct disturbance of the air (or other medium, such as water) by, for example, the presence of a fan or a jet, or a pump. The third type of noise source is noise produced by impacts — both the noise from the impact itself and the ringing of components which subsequently occurs.

The noise is transmitted from the source to the receiver by one or more transmission paths (Fig. 8.2). This might be by airborne sound transmission, in which case the use of sound-absorbing and sound-insulating materials will be useful as noise-reduction measures. Alternatively, the path may be via a solid structure, in which case techniques appropriate to vibration reduction, such as isolation and damping, may be the appropriate noise-control measures. The transmission path for fan noise will be via the ductwork as well as direct radiation from the fan, and possibly vibration transmission from both fan and ductwork.

As a last resort, noise-control measures may be applied at the reception point. This might for example be the provision of hearing protection for a worker in industry or the provision of double glazing for a household near to a main road or airport.

Figure 8.3 shows schematically the various stages in the transmission of noise from source to receiver and lists some of the standard noise-control measures appropriate to each stage.

8.4 Radiation of noise from vibrating surfaces

Let us investigate the relation between radiated noise levels and surface vibration amplitude. The simplest example to consider is a large, flat vibrating plate radiating plane sound waves. Theoretically it is necessary that the surface area of the plate is infinitely large (i.e. a vibrating plane) in order that perfectly plane waves will be radiated, but in reality it is sufficient that the dimensions of the plate are large compared with the wavelength of sound in air at the frequency of vibration. It is also assumed that all points on the plate are vibrating together, in phase at the same amplitude. In this simple case, illustrated in Fig. 8.4, the air particles will be set into vibration with exactly the same amplitude as the plate itself. It is the vibration amplitude (or RMS value) expressed as a velocity which is most useful

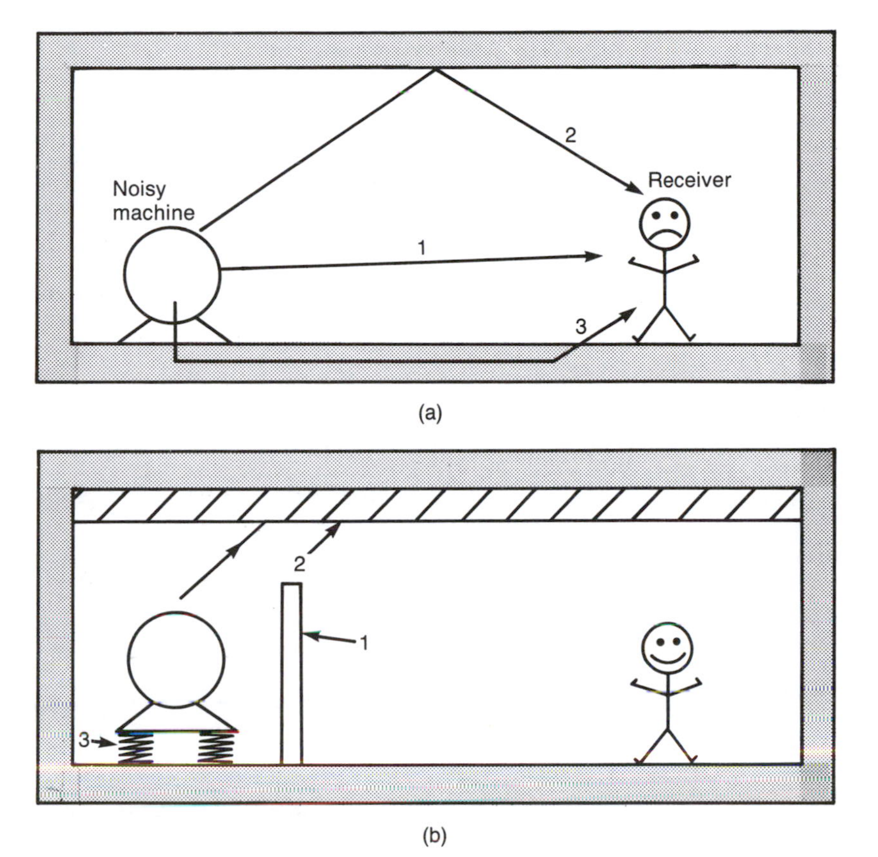

Fig. 8.2 (a) Noise transmission paths and (b) their remedies: (1) direct airborne transmission is remedied by a sound-insulating screen; (2) airborne transmission via reflections from walls, ceiling and other surfaces is remedied by sound-absorbing linings or hangings; (3) structure-borne sound transmission, e.g. via the floor, is remedied by springs or resilient pads to isolate the machine from the structure. All three treatments may be necessary before substantial noise reduction is achieved.

for predicting sound pressures, because in a plane wave the acoustic pressure p is related to acoustic particle velocity V via the relationship

$$p = ZV$$

where Z is the specific acoustic impedance of the medium, which in the case of air at a temperature of 20 °C and standard atmospheric pressure is 415 SI units (kg/m^{2}s), also called rayls. The acoustic pressure p is in N/m^2 and the particle velocity V in m/s (p and V may both be peak values, or may both be RMS values). Therefore if the vibration velocity amplitude of the plate is known the acoustic pressure in the sound wave can be calculated, and from it the sound pressure level.

Example 8.1

The RMS vibration velocity of a metal panel is 2·0 mm/s. Calculate the RMS acoustic pressure and the sound pressure level of sound from the panel assuming plane-wave radiation.

$$p = ZV$$
$$V = 2{\cdot}0 \text{ mm/s} = 2{\cdot}0 \times 10^{-3} \text{ m/s}$$

$$\text{therefore } p = 415 \times 2{\cdot}0 \times 10^{-3} = 830 \times 10^{-3} \text{ N/m}^2$$

The sound pressure level (SPL) in decibels relative to the reference pressure of 2×10^{-5} N/m^2 is given by

$$\text{SPL} = 20 \log_{10}\left(\frac{p}{2 \times 10^{-5}}\right)$$

$$= 20 \log_{10}\left(\frac{830 \times 10^{-3}}{2 \times 10^{-5}}\right)$$

$$= 92{\cdot}4 \text{ dB}$$

If the vibration amplitude is given in terms of displacement or acceleration, this must be first converted to velocity as described in Chapter 7. The frequency of the vibration must be known in this case.

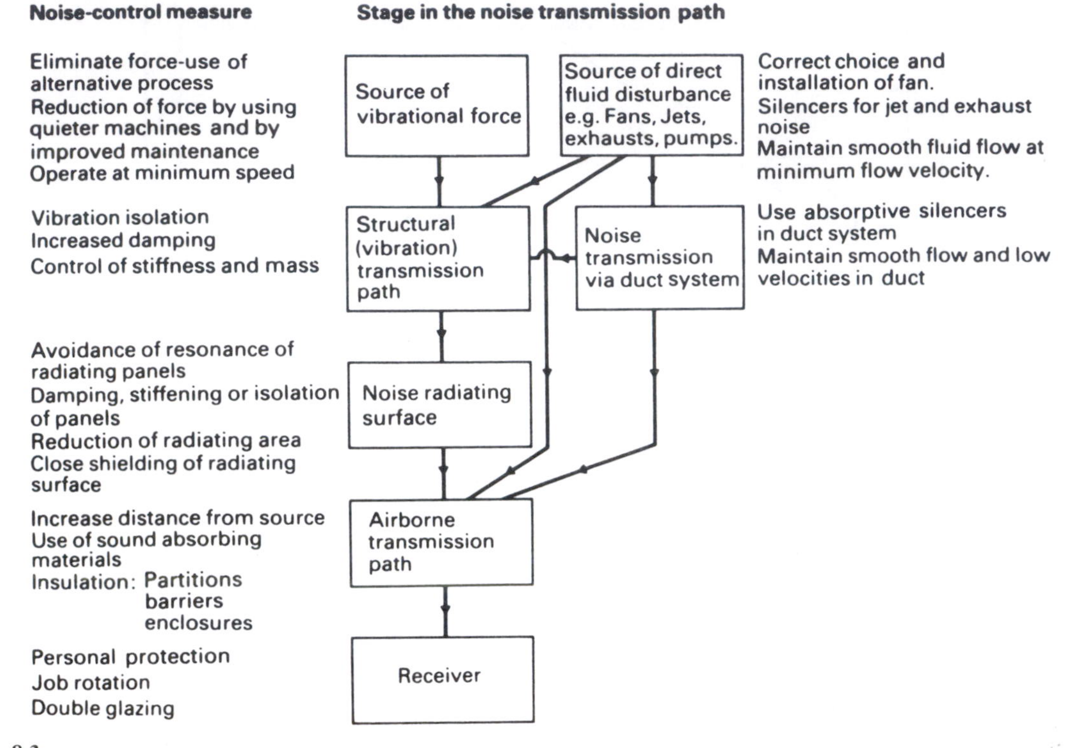

Fig. 8.3

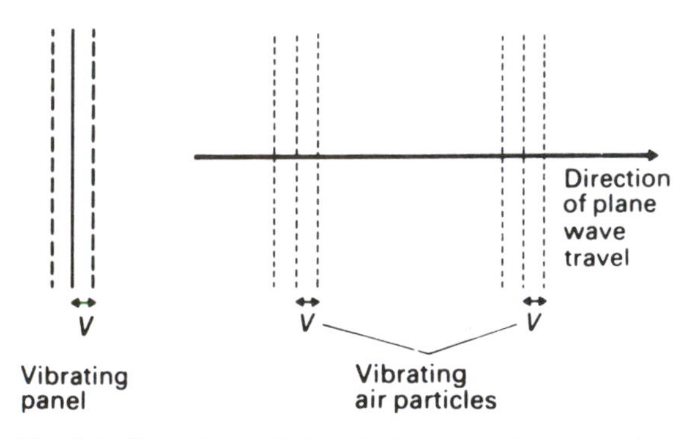

Fig. 8.4 Illustrating radiation of plane wave from a panel. All parts of the panel vibrate in phase, with the same vibration amplitude V. The amplitude of vibration of the air particles is the same as that of the panel

8.5 Radiation efficiency

Real plates are not of course infinitely large and neither do they necessarily vibrate with all points in phase or at the same amplitude. The sound radiated from real panels is related to the simple theory used in Example 8.1 by defining a radiation factor or radiation efficiency σ for the panel or plate. This is the acoustic intensity radiated by the plate divided by the intensity that would be radiated by an infinite plate vibrating (with all points in phase) with the same RMS velocity. This has the effect of modifying the radiated sound pressure level by a factor of 10 log σ decibels. If in fact the radiation efficiency of the panel of Example 8.1 had been 0·2 (instead of 1·0) the change would be 10 log (0·2) = −7 dB, and the SPL would be reduced to 85·4 dB.

For a finite but rigid vibrating plate (i.e. a vibrating piston) the radiation efficiency depends upon the diameter/wavelength ratio. Vibrating sources with diameters small compared with the wavelength of sound in air (at the vibration frequency) behave as simple spherical sources of sound, and σ is low. The radiation efficiency rises as the diameter/wavelength ratio increases until the panel approximates to a plane-wave radiator with σ close to 1·0 when the panel size is approximately equal to or greater than the wavelength (Fig. 8.5). Therefore for small panels (or low frequencies and large wavelengths) the sound is radiated equally in all directions, whereas for large panels (or high frequencies and small wavelengths) the radiation of sound is much more directional, approaching plane-wave conditions.

Real plates and panels do not, however, vibrate like rigid pistons with all points having the same amplitude and phase; they almost invariably vibrate as a result of panel flexure. Plate or panel resonances occur at frequencies which depend

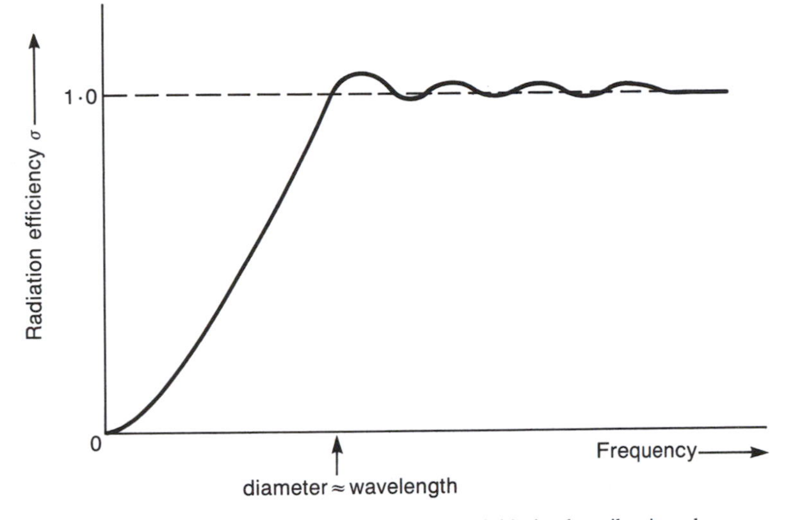

Fig. 8.5 Sketch graph to show variation of radiation efficiency for a rigid circular vibrating plate

upon the panel dimensions and the speed of flexural waves in the panel, which in turn depends on the panel density and its flexural stiffness (thickness and Young's modulus). These resonances can be thought of as standing-wave patterns formed by the interference of flexural waves moving backwards and forwards across the panel. The panel boundaries act as reflectors of flexural waves.

Examples of some simple panel resonance mode shapes are shown in Fig. 8.6. They may be compared with the standing wave patterns on a vibrating string which were illustrated in Fig. 7.3, but in this case the vibrations are in two dimensions. The panel can be divided into areas over which the amplitude varies but the vibrations are in phase. At any instant the particles of the panel, in some areas, marked +, will be moving upwards out of the page in Fig. 8.6, whereas in other areas, marked −, the particles are moving downwards, into the page. Therefore the noise radiated by + areas will be partially cancelled by radiation from − areas.

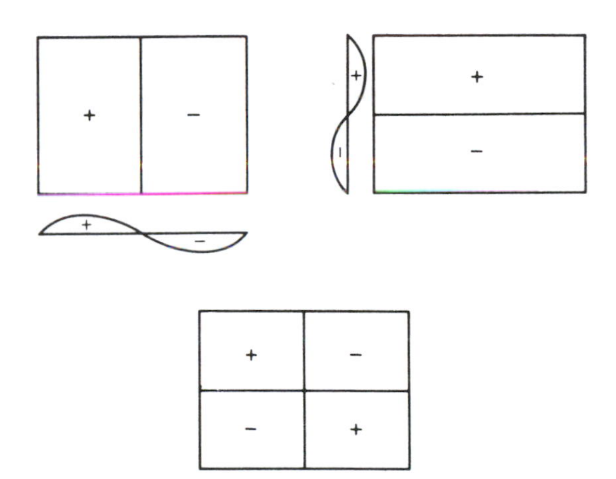

Fig. 8.6 Plan views of a vibrating rectangular panel, clamped at the edges illustrating the mode shapes of three of the simplest modes of vibration. For comparison with Fig. 7.3, edge-on views are shown for two of the modes

The critical frequency for the panel is of central importance in determining its efficiency as a noise radiator. Below the critical frequency the speed of sound in air is greater than the speed of flexural waves in the panel, and these panel waves are called *acoustically slow*. Above the critical frequency the flexural waves in the panel move faster than the sound waves in air (*acoustically fast* panel waves). At the critical frequency the two velocities are the same.

Sound waves in air impinging on a panel cause flexural vibrations in the panel. This process is very effective at and above the critical frequency, when small-amplitude sound waves can produce large-amplitude panel vibrations. This is because of the coincidence effect explained in Chapter 4, and is the reason for the coincidence dip in the sound reduction index curve for typical panels.

The interaction between sound waves and panels is a two-way process, and panels which are easily and effectively excited into vibration by airborne sound waves are also efficiency radiators of sound. It follows that the radiation efficiency of panels is very high at or above the critical frequency, but is low for frequencies well below critical.

The main factors determining the efficiency of panel radiation are therefore the frequency of vibration (as compared to the critical frequency) and the panel dimensions (as compared with the airborne sound wavelength). Large, thin, floppy sheet-steel panels tend to be best at radiating low frequencies, whereas smaller, thicker and stiffer panels are better at radiating high frequencies.

The velocity of airborne sound remains constant and independent of frequency, but the velocity of flexural panel waves increases with frequency. Therefore at low frequencies (below critical) the panel waves are acoustically slow, but as the frequency is increased they 'catch up' until at critical frequency the velocities are the same. The velocity of the panel waves increases with increasing panel stiffness. Therefore if the stiffness is increased this has the effect of reducing the catching-up frequency, i.e. reducing the critical frequency. Therefore the effect of stiffening a panel, perhaps in an attempt to reduce its vibration amplitude, may be to make it into a more efficient radiator of noise at some frequencies.

8.6 The reduction of noise from sound radiating surfaces

The most direct form of noise control is to reduce the forces acting on the surface and causing it to vibrate, as discussed later in the chapter. Given that certain forces are acting on a panel, the aim should be to minimize the amplitude of panel vibration, and its radiating surface area and radiation efficiency. The possible noise reduction treatments to the panel are stiffening, damping, isolating or shielding (i.e. covering or acoustic lagging).

Increasing the stiffness can increase the radiation efficiency, so the most effective treatment for a thin, floppy panel might be to increase its damping rather than to stiffen it. For panels which are already fairly stiff, damping treatments may not be very effective because the mechanism of damping requires a large vibration amplitude; the damping forces can then convert vibration energy into heat. If the panel is already fairly stiff, it may already be an efficient sound radiator; further stiffening will reduce vibration amplitudes and radiated noise levels. The best noise-control treatment for very stiff panels is to try to isolate them from the surrounding framework in order to prevent the vibration-producing forces from the framework reaching the panel. A last resort is to close-shield or cover the panel with an acoustic lagging material. The principles of damping, isolating and close shielding are described later in this chapter.

8.7 Real noise sources — near field and far field

The radiation of sound from an idealized point source was discussed in Chapter 5. The sound intensity from such a source varies with distance according to the inverse square law, which means that there is a drop of 6 dB in the measured sound pressure level from the source, every time the distance is doubled. The sound energy from a simple point source is radiated equally in all directions, provided it is located in free space away from any reflecting surfaces.

Real noise sources such as machines with vibrating surfaces differ from the simple point source in two important ways. Firstly, the source has a definite size and is not a point; and secondly, it may radiate different amounts of sound energy in different directions. However, provided the distance from the source is great enough, real noise sources do behave like simple sources inasmuch as the sound pressure level in any direction (in the open) will decrease at a rate of 6 dB per doubling of distance away from the source.

Each portion of a vibrating machine surface radiates sound. Near to the machine the contributions of the various portions combine in a complex way and it is difficult to predict local variations in sound pressure near to the

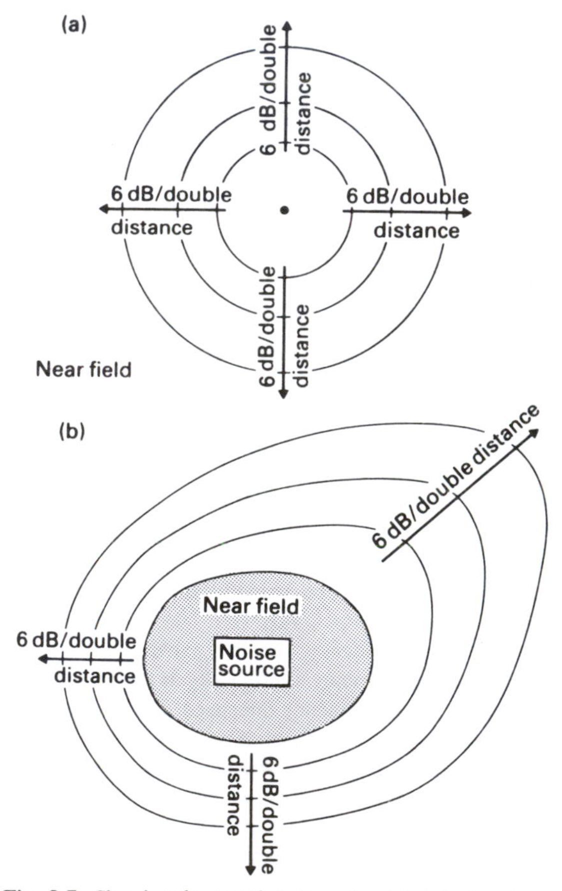

Fig. 8.7 Showing the sound pressure level distribution from (a) a point source and (b) a real source. The contours are of equal sound pressure level. Along each arrowed line the sound pressure level drops at a rate of 6 dB per doubling of distance from the source. The near-field region is shown shaded in (b)

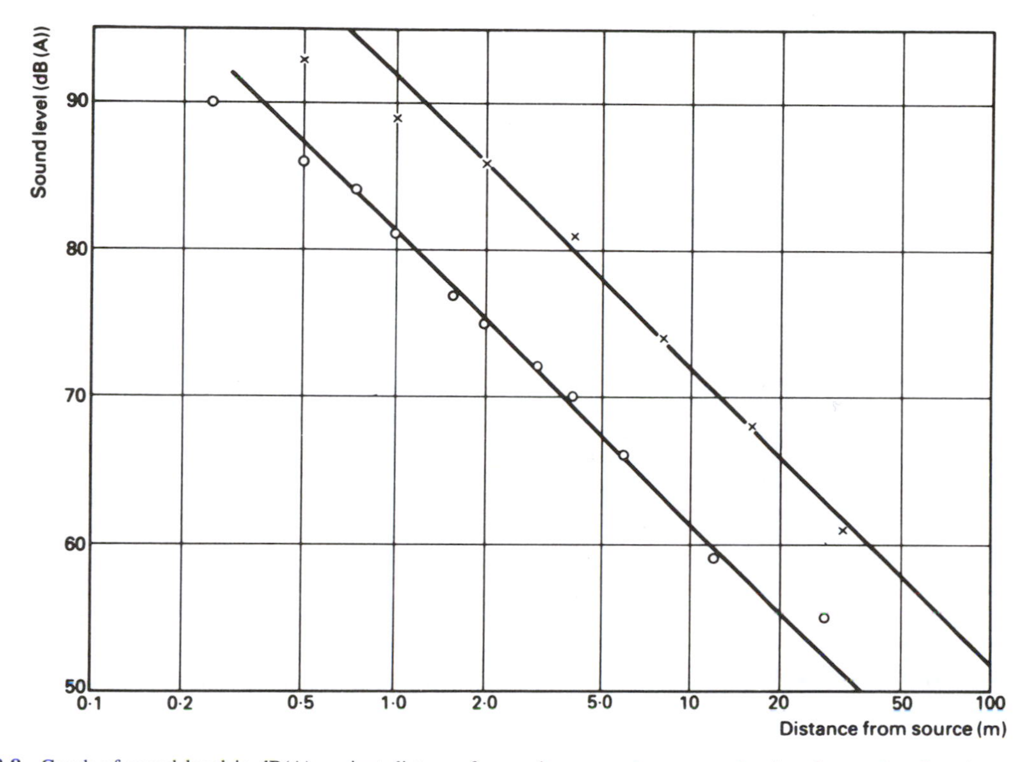

Fig. 8.8 Graph of sound level in dB(A) against distance from noise source in metres, showing the results of outdoor measurements of two small noise sources: (x) lawnmower and (o) electric drill; background level = 54 dB(A). The two lines are of slope equivalent to 6 dB per doubling of distance

machine. This region close to the machine is called the near sound field or simply the **near field**.

Further away from the machine the noise contributions from the various parts of the machine coalesce smoothly, and we can think of the machine radiating 'as a whole'. This region is called the far sound field or simply the **far field**. It is in this far-field region that the sound pressure distribution varies smoothly with distance according to the inverse square law. The distributions of sound pressure levels around a real machine, and around a point source for comparison, are shown in Fig. 8.7.

The extent of the near-field region depends on the dimensions of the machine, and on the wavelength of sound being radiated. There is, of course, no sharp division between the two regions, but in order to ensure far-field conditions measurement of machine noise should ideally be taken at distances of not less than several wavelengths or several 'machine dimensions' from the source, whichever is the greater. A minimum distance of one or two wavelengths and one or two machine lengths away is sometimes quoted. Figure 8.8 shows the results of noise measurements (dB(A) levels) taken at various distances from two machines situated in the middle of a grass-covered field, well away from buildings. The machines were a petrol-engined lawnmower and an electric drill. The noise radiated from the drill is mainly at high frequencies, 1000 Hz and above. The relevant wavelengths will therefore be about a third of a metre or less. The maximum dimensions of the drill are about 30 cm. Therefore, because of the short wavelengths and small dimensions, measurements even as close as 1·0 m from the drill will be in the far field. The lawnmower is bigger, but the engine which is the main noise source is not very much bigger than the drill. However, the noise is predominantly at low frequencies and it is the wavelength criterion which determines the near–far field boundary in this case. The results show that the drop-off of SPL with distance starts to become uniform at about 0·75 m for the drill and at 1·5 m for the lawnmower.

The practical consequences of all this are that far-field measurements may be used as the basis for predicting noise levels at greater distances from the noise source, whereas near-field measurement will give unreliable results if used for this purpose. However, there will be many occasions when near-field measurements have to be taken, particularly if the source is a machine indoors. Far-field measurements may be impossible for large machines in small rooms, or if noise levels have to be taken close to the machine in order to avoid background noise interference from other nearby machines. It may be that the near-field measurement is the most appropriate, as when the noise at the machine operator position is measured for purposes of occupational noise exposure assessment.

8.8 Directionality of noise sources

The directionality of a noise source may be described in terms of a *directivity factor Q*, defined as follows:

$$Q = \frac{\text{the sound intensity produced by the source in a certain direction, at a given distance}}{\text{the average sound intensity (over all directions) at that same distance}}$$

The average intensity is the intensity which would be produced at that distance if the total energy of the source were to be equally distributed in all directions. If, for example, measurements are made at six points evenly spread over a sphere surrounding the source, giving intensities I_1, I_2, I_3, I_4, I_5 and I_6 then

$$\text{average intensity} = I_{av} = \frac{I_1 + I_2 + I_3 + I_4 + I_5 + I_6}{6}$$

and the directivity factor in a given direction, say direction 5, is given by

$$Q_5 = \frac{I_5}{I_{av}}$$

However, measurements are more conveniently made in terms of sound pressure levels in decibels, and the equivalent of the above procedure is to average levels measured in various directions (at the same distance) and compare the level in any direction with the average to get the *directivity index D*:

$$D = \text{SPL} - \text{SPL}_{av}$$

where SPL is the sound pressure level in the direction of interest, and SPL_{av} is the true, logarithmic average of the levels measured at the same distance in various directions around the source.

Directivity factor Q and directivity index D are simply related:

$$D = 10 \log_{10} Q$$

The directivity factor Q is a ratio, which may be greater than one (in directions where more than average power is radiated) or less than one.

The directivity index, in decibels, may be positive (for directions in which more than average power is radiated) or — negative (for directions of less than average radiation).

Directionality of radiation may arise because the noise source itself is inherently directional, as for example a machine which radiates more noise from the front than the back. However, sources which would be omnidirectional in free space become directional to a certain extent if the radiation is restricted to less than a complete spherical space, by the positioning of the source with respect to surfaces such as walls, floors and ceilings. For example, a noise source situated on open but hard reflecting ground is only free to radiate into a hemispherical space, and half the power which would have travelled downwards is reflected upwards. According to the above definitions, such a source has a directivity factor (Q) of 2 and a directivity index (D) of +3 dB. If the ground is soft and completely sound-absorbing so that reflection does not occur then $Q = 1$ and $D = 0$. For a source which is only free to radiate into a quarter of a sphere, $Q = 4$ and $D = +6$ dB. An example might be a source radiating from near to the base of the façade of a large building, which is in the angle made by the façade and the ground adjacent to the building.

This reasoning can be applied to small sources in rooms. The position of the source, for example a ventilation grille radiating noise from a duct into the room, will determine its directivity, as shown in Table 8.1.

The sound pressure level (SPL) at a given distance (r) from a source can be related to the sound power level of the source (L_W) and its directivity factor (D) as follows:

$$\text{SPL} = L_W - 20 \log_{10} r - 11 + D$$

Table 8.1 The directivity of an omnidirectional source when positioned at various points in a room

Position of source	Part of sphere into which source can radiate	Directivity factor Q	Directivity index D
Centre of room	Whole sphere	1	0
Centre of wall, floor or ceiling	Half	2	+3 dB
Junction of two planes, e.g. Wall and ceiling, wall and floor	Quarter	4	+6 dB
Corner (junction of three planes)	Eighth	8	+9 dB

For an omnidirectional source on hard ground, $D = +3$ and the numerical factor becomes 8 instead of 11 (for soft ground).

The situation can become a little confusing if in addition to being situated on hard ground the source is inherently directional. It must be remembered that in the above equation in which the numerical factor of 11 appears, the definition of directivity index refers to a sound pressure level averaged over a whole sphere. It is hoped that the following example will help to clarify the situation.

Example 8.2

Noise measurements are made at 4 m from a small machine situated on hard ground in the open. The measurements are made in the half space above the ground in four equally spaced directions around the machine (north, south, east, west) and directly above the machine (up). The noise levels are:

North (N) 76 dB(A)
South (S) 78 dB(A)
East (E) 80 dB(A)
West (W) 82 dB(A)
Up 84 dB(A)

Calculate the sound power level (A-weighted) of the machine and the directivity index in each of the five directions. Also calculate the A-weighted sound pressure level in the north direction at a distance of 10 m from the machine.

First of all average the five readings (logarithmic average):

Average SPL

$$= 10 \log_{10} \left(\frac{10^{7\cdot 6} + 10^{7\cdot 8} + 10^{8\cdot 0} + 10^{8\cdot 2} + 10^{8\cdot 4}}{5} \right)$$

$= 80{\cdot}9$ dB(A)
$= 81$ dB(A) to the nearest half decibel

(Note that the arithmetic average of the five levels gives 80 dB(A), in error by 1 dB(A) as compared with true logarithmic averaging.) This is the average over a hemisphere, above the ground. The average over a complete sphere is 3 dB(A) less, giving a value of SPL_{av} = 78 dB(A). Next, calculate the directivity indices:

$$D = SPL - SPL_{av}$$

North, $D_N = 76 - 78 = -2$dB

South, $D_S = 78 - 78 = 0$ dB

East, $D_E = 80 - 78 = +2$ dB

West, $D_W = 82 - 78 = +4$ dB

Up, $D_U = 84 - 78 = +6$ dB

Note that these directivity factors include both the inherent directionality of the source, and the directivity (+3 dB) imposed by its position on hard ground. The sound power level is obtained by careful use of the equation given above, and now rearranged:

$$L_W = SPL + 20 \log_{10} r + 11 - D$$

Using the values for the north direction,

$$L_W = 76 + 20 \log_{10} 4 + 11 - (-2)$$
$$= 76 + 12 + 11 + 2$$

$\therefore$ A-weighted $L_W = 101$ dB

Exactly the same result is obtained using the data for each of the other four positions. Alternatively, the average SPL values may be used, i.e. 81 dB(A) with a value of $D = +3$ dB, or 78 dB(A) with a value of $D = 0$.
To find the SPL at 10 m in the north direction,

$$SPL = L_W - 20 \log_{10} r - 11 + D_N$$

$$SPL = 101 - 20 \log_{10} 10 - 11 - 2$$
$$= 101 - 20 - 11 - 2$$
$$= 68 \text{ dB(A)}$$

Alternatively, since the level at 4 m in the north direction is known, the level at 10 m can be calculated directly from the inverse square law without using the sound power level: Difference in levels at distances r_1 and r_2

$$= 20 \log_{10} \left(\frac{r_2}{r_1} \right)$$

Difference in levels at 4 m and 10 m

$$= 20 \log_{10} \left(\frac{10}{4} \right)$$

$= 8$ dB

Therefore level at 10 m = level at 4 m − 8 dB
= 68 dB(A), as before.

8.9 Sound distribution in rooms — direct and reverberant fields

The sound which arrives at a point in a room from a noise source in that room may be considered to consist of two parts. First there is the **direct** sound which arrives from source to receiver without having been reflected from any of the room surfaces. This is followed by first reflections from the walls, ceiling and floor and any other prominent reflectors in the room, and then by innumerable further reflections. The sum total of all these reflections is called the **reverberant** sound. If the noise is transient, the direct and reverberant sound arrive at the receiver at different times, and the reverberant sound decays with a reverberation time given by Sabine's formula (see Chapter 3).

If the noise source is continuous the total sound pressure level is the combination of the direct sound pressure level and the reverberant sound pressure level.

The region close to the source, where direct sound predominates, is called the direct sound field or **direct field**. Further away from the source the reverberant sound may be the dominant component. This region is called the reverberant sound field or **reverberant field**.

The direct SPL from a source depends on the distance from the source and the sound power level and directivity of the source. This was discussed in the previous section and in Example 8.2.

The reverberant component of the sound builds up to a level at which the rate of supply of sound energy to the room from the source is equal to the rate at which it is being absorbed by the room surfaces. A simple theory assumes that the reverberant sound is distributed uniformly throughout the room. The reverberant sound pressure level depends on the sound power level of the source and the amount of acoustic absorption in the room, expressed as a room constant R_c, according to the following equation

$$\text{Reverberant SPL} = L_W + 10 \log_{10} \left(\frac{4}{R_c}\right)$$

$$\text{The room constant } R_c = \frac{S\bar{\alpha}}{(1 - \bar{\alpha})}$$

where S is the surface area of the room in m^2, and $\bar{\alpha}$ is the average absorption coefficient in the room. R_c is measured in m^2 units. The average absorption coefficient $\bar{\alpha}$ may be calculated from the areas of different types of absorbing surfaces in the room. Alternatively, the reverberation time of the room may be measured and $\bar{\alpha}$ calculated using either Eyring's formula or Sabine's formula. For values of $\bar{\alpha}$ small enough to allow Sabine's formula to be used instead of the more accurate but more complicated Eyring's formula, then the room constant R_c can be approximated by $S\bar{\alpha}$, which is equal to A, the amount of acoustic absorption in the room. If $\bar{\alpha} = 0{\cdot}2$ then the approximation of $R_c = A$ makes an error of about 1 dB in the calculated reverberant sound pressure level.

The total sound pressure level is found by adding up the two components logarithmically, i.e., using the usual rules of decibel addition. Alternatively, the following formula may be used:

$$\text{Total SPL} = L_W + 10 \log_{10} \left(\frac{Q}{4\pi r^2} + \frac{4}{R_c}\right)$$

where Q is the directivity factor of the noise source in the direction of the receiver, and r is the distance between source and receiver. The first term in the brackets relates to the direct component and contains information about the source (its directivity and its position), whereas the second term relates to the reverberant component and depends only on the room.

Example 8.3

A small machine is situated on the floor (assumed to be reflecting) in the centre of a room of dimensions 10 m × 6 m × 4 m high. The noise from the machine is predominantly in the 500 Hz band and the sound power level in this band is 87 dB. Calculate the sound pressure level in this band at a distance of 2 m from the machine if the reverberation time of the room is 0·8 seconds. Assume that the machine radiates omnidirectionally.

Volume of room, $V = 10 \times 6 \times 4 = 240$ m^3

Surface area of room, S

$= 2 \times (10 \times 6 + 6 \times 4 + 10 \times 4)$
$= 248$ m^2

Sabine's formula:

$$T = \frac{0{\cdot}16 \times V}{A} \quad (T = \text{reverberation time})$$

∴ Area of acoustic absorption, A

$$= \frac{0{\cdot}16 \times V}{T} = \frac{0{\cdot}16 \times 240}{0{\cdot}8}$$

$= 48{\cdot}0$ m^2

$$\text{Average absorption coefficient } \bar{\alpha} = \frac{A}{S} = \frac{48}{248}$$

$$= 0{\cdot}194$$

$$\text{Room constant } R_c = \frac{S\bar{\alpha}}{1 - \bar{\alpha}} = 59{\cdot}5 \text{ m}^2$$

$$\text{Total SPL} = L_W + 10 \log_{10} \left(\frac{Q}{4\pi r^2} + \frac{4}{R_c}\right)$$

Using $Q = 2$, $r = 2$, $L_W = 87$,

$$\text{total SPL} = 87 + 10 \log_{10} (0{\cdot}040 + 0{\cdot}067)$$
$$= 87 - 9{\cdot}7$$
$$= 77{\cdot}3 \text{ dB}$$

Alternatively, by computing the components separately it can be shown that the direct SPL is 73 dB, and the reverberant SPL is 75·3 dB.

Figure 8.9 shows how the sound pressure level in a typical room varies with the distance from the noise source in the room. Close to the source, where the sound pressure level falls with increasing distance, the direct sound is

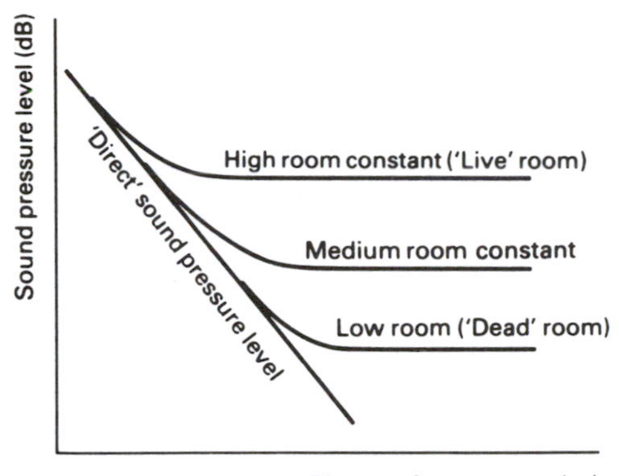

Fig. 8.9 Sketch graph showing the variation of sound pressure level with distance from a noise source in a room, for different room constants

predominant. The graph levels off at greater distances when the reverberant component becomes dominant. The level of the reverberant component and the distance at which the graph levels off depends upon the room constant. For a real source there is the additional complication that the direct sound field of the machine consists of near-field and far-field regions. Sometimes it may be impossible to achieve far-field noise measurements of a machine indoors because of the presence of the reverberant sound field.

A few quick measurements of noise levels at varying distances from a machine can give a rough idea of the relative importance of direct and reverberant sound at any position. This information will be useful in deciding on measures to reduce noise at that point. If the problem is caused mainly by direct sound from a particular machine, a screen or barrier may provide the answer if treatment at source is not possible. If the problem is caused by reverberant sound, a barrier will be ineffective, but noise reductions may be achieved by the use of additional sound absorption in the room, either by treating walls, floors or ceiling or by using space absorbers. The reduction achieved depends on the amount of existing absorption in the room as well as the extra amount to be added. The amount of noise reduction achieved may be calculated as in Example 5.5. In practice reductions will be limited to 10 dB at maximum.

In some cases where the reverberant and direct levels are similar, neither treatment on its own will be effective, but the combination of absorbers to control reverberant sound will then enable a screen to be used to reduce direct sound.

The simple ideas outlined above, in which there is a constant reverberant level, seem to be satisfactory for rooms of 'normal' proportions and for halls and theatres. However, it has been found that in large open-plan offices the propagation of sound is more complicated, and this also appears to be true for large factory spaces. In the case of the large office, the height of the room is small compared to the other two dimensions and so reflections from the ceiling plays a relatively more important part than it does in rooms of more normal proportions. This may also be true in factories where there is the additonal complication that the ceiling or roof shape and construction may be very different from those in a normal room or office. Machines on the factory floor act as barriers to direct sound from other machines, and as scatterers and absorbers of sound. All of these complications mean that the distribution of sound within the room may be different than expected on the assumption of a constant reverberant sound level.

8.10 The specification of noise from machinery — sound pressure level and sound power level

In principle the sound pressure level produced by a sound source at any distance can be predicted from its sound power level, although the calculations may, in general, be more complicated than for the simple free-field and reverberant field situations.

Therefore the output from a sound source may be described either by its sound power level, or by specifying the sound pressure level at a certain distance from the source. There are advantages and disadvantages to each method.

The main advantage of using sound power levels is simplicity — L_W is a single-figure number which depends only on the sound source, not on any other factor such as distance or details of the acoustic environment. This is the idea behind various EC Machinery Noise Directives which require certain classes of machinery to be labelled with their L_W values, so that users and purchasers can easily compare the noise produced by different types of machine.

But there are several disadvantages. The measurement of sound power level requires specialist test rooms or equipment. Also, the L_W value itself has no significance, subjectively, unlike the sound pressure level, which can be related to what is actually heard. Although, in principle the sound pressure level in any situation may be predicted from the L_W value, it is in practice only possible to predict far-field values; near-field values, close to the source, cannot be predicted.

The main disadvantage of specifying the sound pressure level at a particular distance from the source is that it can lead to ambiguity and confusion, for users, when trying to compare the noise outputs from machines which are described in different ways, e.g. if different distances are used, or if the tests are carried out in different acoustic environments. But this method does have the advantage that sound pressure levels may be easily measured, and can be related to what is heard.

8.11 The measurement of sound power levels

There are, in principle, four different test methods:

1. **Free-field methods** require that the measurement of sound radiated from the source is not affected by reflected sound. Such an acoustic environment may sometimes be achieved by performing tests outdoors, or in an anechoic or semi-anechoic test room.
2. **Sound intensity methods** use a sound intensity meter, which is sensitive to the direction of flow of sound energy; they measure sound power levels in non-anechoic environments.
3. **Reverberant methods** require the specialist test environment of a reverberant room.
4. **Substitution methods** require a calibrated sound power source.

8.12 Free-field methods

An anechoic room is a specially constructed chamber in which the walls, floor and ceiling are covered with sound-absorbing material, usually in the form of wedges, which are effective in eliminating reflected sound. Thus measurements of sound pressure levels produced by a noise source in such a room are of the direct sound only, and can be used to compute the sound power level of the source. The directivity of the source can also be measured. The method is convenient for small portable machines but the size of the room limits the size of machines which can be tested since measurements should be made in the far field.

A semi-anechoic room has absorbent walls and ceiling but a hard sound-reflecting floor. Example 8.2 illustrated the method by which sound power is calculated under semi-anechoic conditions although in the example the measurements were made outdoors. In the ISO standard method, sound pressure levels are measured at various positions over a surface surrounding the noise source. The sound power level, L_W, is related to the area of the measurement surface (S in m^2) and the average sound pressure level over the surface (SPL) by the equation.

$$L_W = \text{SPL} + 10 \log_{10} S$$

If the measurement surface is a hemisphere then $S = 2\pi r^2$ and 10 log S become 20 log $r + 8$, giving the 'hard-ground' version of the direct field equation.

8.13 Sound intensity methods

The measurement of sound power level from a noise source using a sound intensity meter is similar in principle to the free-field method using an anechoic or semi-anechoic room. The sound intensity is measured over a test surface, usually a hemisphere or parallelepiped surrounding the noise source. The sound power level (L_W) is obtained from the average sound intensity over the test surface (L_I) and its area (S):

$$L_W = L_I + 10 \log_{10} S$$

A necessary requirement of the free-field method, in which a sound level meter is used, is that sound energy only travels outwards from the source through the test surface. The great advantage of the sound intensity meter is that it allows measurements to be carried out in an ordinary room, where the presence of reflecting surfaces means that some reflected sound energy will pass through the test surface in the opposite, inwards direction. However, provided that there is no acoustic absorption inside the test surface, all of this reflected sound energy will eventually pass out through the test surface again, either directly or maybe after reflection or scattering by the surfaces of the noise source. The sound intensity meter indicates the net flow of acoustic energy through the test surface, so the value of the average sound intensity measured over the entire surface will correspond only to that provided by the outward flow of energy from the source. The reflected sound passing through the surface cancels out, making a net zero contribution to the flow of acoustic energy through the surface.

For similar reasons, measurements can also be taken in the presence of background noise from another source, provided that it produces a constant level and is located outside the test surface, so that it also produces a net zero flow of sound energy through the test surface.

BS 7703 gives details of the measurement method, including recommendations for the size and shape of the test surface, number of measurement points to be used, and checks on the adequacy of the equipment and measurement procedures.

8.14 Reverberant methods

A truly reverberant room is one in which the energy from a noise source in a room is diffused completely throughout the room, so that ideally the sound pressure level would be the same everywhere. In practice, measurements are taken at several points in the room and the average sound pressure level is obtained. This is related to the sound power level of the source L_W and to the room constant, as described in the last section. Expressing the room constant in terms of the reverberation time T (which can be measured directly), by using Sabine's formula the following relationship is obtained:

$$L_W = \text{SPL} + 10 \log_{10} V - 10 \log_{10} T - 14$$

where V is the room volume in m^3. A modified and more accurate version of the above formula is given in the ISO standard. Only sound power levels can be obtained from reverberant-room methods — the directivity information can only be obtained from free-field (anechoic and semi-anechoic) measurements.

8.15 Substitution methods

In the substitution method of sound power level measurement, a reference noise source of known sound power level is used. Comparisons are made of the noise levels produced in the room by the machine under test and the standard source. This type of test may be carried out in a special reverberant room, or with lesser accuracy in the actual location of the test machine. The reference source should be capable of being calibrated and should be omnidirectional, as far as possible, over the frequency range of interest.

Example 8.4

Measurements were made in a reverberant room of the noise produced by an electric drill. The average sound pressure level in the 1000 Hz octave band was 85 dB. When the drill was replaced by a calibrated noise source which has a sound power level of 100 dB, in this octave band, the average level produced was 89 dB. Calculate the sound power level of the drill in the 1000 Hz band.

The principle of the method is that the difference in the two sound pressure levels is the same as the difference in their two sound power levels.

Sound power level of drill − 100 = 85 − 89

therefore sound power level of drill = 100 − 4
= 96 dB

Further details of the ISO methods for the measurement of sound power levels are given in Appendix 4.

8.16 Control of noise by good planning and management

Much can be done to avoid noise problems arising by good noise-control management, practised on a day-to-day basis, and by giving careful consideration, in advance, to situations which can possibly give rise to noise problems. Some of the possibilities are listed below.

1. By selecting the quietest procedure for an operation or process, where alternatives exist.
2. By including noise level requirements in the specification for new machines and equipment.
3. By careful consideration of the layout and siting of noisy processes and equipment, in relation to noise-sensitive areas and personnel.
4. By trying to arrange that such processes and equipment do not operate at the most noise-sensitive times (e.g. evenings and weekends).
5. By arranging for regular maintenance of machines, and of noise-control equipment.
6. By arranging for regular noise surveys to be carried out, by properly trained personnel.
7. By education and information, so that people are aware of the purpose of noise-control equipment, and how it should be used. This might simply be a matter of ensuring that noise-shielding hoods or enclosures are always replaced after removal for inspection or maintenance, or that vibration isolation is not bridged inadvertently.

8.17 Noise surveys — identification and diagnosis of noise sources

Noise surveys

The first step is to gather information about existing noise levels. Measurements should be carried out carefully and methodically by someone trained in the proper use of a sound level meter. Careful observation, both visual and aural, can add much to the value of such a survey. A well-designed measurement report sheet should be used to serve as a check-list to ensure that essential routine procedures such as battery checks, use of windshield over microphone, and calibration of the meter are not forgotten, and that meter settings (e.g. fast/slow/impulse/peak, etc.) and background levels are noted, as well as the actual noise level readings. Other details to be recorded include information about the noise source, the instrumentation, measurement positions, the acoustic environment and the subjective impression of the noise. Appendix 4 includes a list of information to be recorded when measuring sound power levels of machines. Much of this is relevant to basic noise surveys, and could be incorporated into the design of the noise measurement report sheet.

Identification and diagnosis of noise sources

A correct diagnosis of the situation is obviously an essential step in any noise-control programme. The problem might be in deciding which of several machines in a workshop are the important contributors to the overall noise level. The best thing to do, if possible, would be to turn them all off and then to make noise level measurements of each one, switched on in turn. Because of the way decibels 'add up' this would be preferable to the alternative procedure of switching each one off, in turn. However, it may not be possible to switch off any of the sources, in which case the diagnosis will have to be based on noise level measurements taken close to each source.

Sometimes it is necessary to decide which are the most important noise-radiating areas of a particular machine. One technique for doing this is to cover all possible areas with a close-fitting shield, or cladding, of thin lead sheet lined with foam or mineral wool, or using a commercial sound barrier mat. Noise measurements are then made with each noise-radiating area uncovered in turn. If this sort of test is to be effective, great care must be taken in accurate fitting of the noise covers, so that there are no holes or gaps, and so that a good initial noise reduction, of at least 10 dB(A), is achieved, with all surfaces covered. Another possibility is to take vibration measurements (octave or third-octave) on each noise-radiating surface, and, based on its vibration levels and area, to estimate the sound power radiated by each surface, usually assuming a radiation efficiency of one. Noise level measurements made with the microphone positioned very close to each surface can also be used (using the average level over the surface).

An insight into the noise-producing mechanisms at work in a machine can often be obtained by an intelligent and systematic investigation of the effect (on the noise) of varying machine parameters such as load, speed, feed rate, etc.

Sometimes the source of the noise is obvious — a fan, for example — and the diagnosis problem is in deciding the relative importance of the various transmission paths from source to receiver. This is particularly difficult with low frequencies, the sources of which are anyway difficult to locate because the radiation is usually omnidirectional, and in addition there is the possibility of standing waves being set up in the room. One of the greatest difficulties can be in deciding whether the received noise is airborne or structure borne.

Specialist signal analysis techniques can be very useful in diagnosis. An analysis of the vibration of a structure can find the shape of the important modes of vibration. Analysis of the acoustic data can be carried out either in the 'time domain' or in the 'frequency domain'. A study of the time history of the noise or vibration, using an oscilloscope, can help to elucidate the sequence and relative importance of noise-producing events in a machine, such as mechanical impacts and valves opening and closing. Correlation techniques can be used to compare the noise signal at the reception point with signals emitted by various possible sources, in order to assess their relative importance. Narrowband frequency analysis can be a useful tool for identifying the particular source of repetitive events produced in rotating machines (e.g. in gears, fans and bearings) which produce pure tones in the frequency spectrum of the noise.

8.18 Reduction of vibration-producing forces

The first step is to eliminate all avoidable impacts, squeaks and rattles, whether caused by human thoughtlessness (e.g. door slamming), poor maintenance, insufficient lubrication or by loose, worn or defective parts. The next step is to try to substitute the existing process for a quieter one, or to look for a quieter version of the existing machine.

All rotating machines are sources of noise-producing forces such as the cyclic mechanical forces produced at bearings and gears, the electromagnetic forces in motors and alternators, and the inertial forces caused by out-of-balance rotating parts. There are also the aerodynamic and hydrodynamic forces in fans and pumps. In all cases the noise depends strongly on the rotational speed and one way of achieving noise reduction is to reduce speed where possible.

Many of the noise-producing forces in machines are of course vital to the operation or process involved. Examples of such working forces are the combustion forces in engines, magnetic forces in motors, impact forces in punching and hammering operations, and cutting forces in many metal-working processes. In these cases, and with rotating machines, complete elimination of the noise-producing force is impossible. However, a great deal of research has been carried out into all these mechanisms, and many noise-reducing solutions are available. Examples include helical cutter blades for woodworking machines, specially designed punch and press tools and highly damped circular saw blades. Quieter gears, pumps and bearings are also available. In many cases the quieter design involves better quality engineering (for example, to produce closer tolerances between moving parts or better balancing of rotating parts) and is therefore of course more expensive.

Processes which produce impulsive forces, such as hammering, punching and pressing operations, are a very important class of noise producers. The spectrum of an impulsive force contains a wide variety of frequencies, and can usually excite almost any of the natural frequencies of a machine structure. The most effective method of control is to replace the impulsive force by a steadily applied force which will achieve the same effect. Examples are in 'hush' piling (as opposed to impact pile driving), in the quiet breaking of concrete (where a bending action is used, rather like that used in breaking a bar of chocolate) and in the use of squeezing forces rather than hammering to insert rivets into sheet metal work.

Where the impact process cannot be replaced, the noise can sometimes be lessened by reducing either the mass or the velocity (or both) of the impacting parts, or by introducing some resilience into or in between the impacting parts. An obvious example of the latter is the lining of chutes and hoppers with a soft material such as rubber. Another example is the use of plastic gears to reduce the noise-producing effects of the impacts between teeth. Modifications to the shape of cams, gears and cutting tools can reduce the 'sharpness' of the forces, and reduce noise.

Finally, the effect of the force can be minimized by feeding it into a stiff and massive part of the structure, and by avoiding amplification effects caused by structural and acoustic resonances.

8.19 Reduction of vibration

As a result of the transmission of vibration through a structure, the vibratory force which is the cause of a noise problem may be quite remote from the surface which radiates the noise. Vibration reduction treatments can be applied at any point in the transmission path but most attention is usually paid to the beginning of the path, where the force enters the structure, and at the end, where the vibrational energy is fed into the noise-radiating surfaces.

The methods available for the reduction of vibration are:

1. The control of resonances
2. The control of stiffness
3. Vibration isolation
4. Increase of damping

Control of resonances

The vibration amplitude of a noise-radiating panel is considerably increased if it has a natural frequency which coincides with the frequency of the vibrational force exciting the structure. If this happens the natural frequency of the panel can be changed, to avoid resonance, by altering either its mass and/or its stiffness. Such a 'detuning' process can sometimes be effective in reducing vibration. An alternative approach which is sometimes possible is to change the machine running speed, which usually controls the frequency of the force. However, there are often very many different natural frequencies in the transmission path, and the force often has components at several frequencies as well, and so in practice it may become difficult to avoid the resonant condition altogether.

Control of stiffness

A general increase in the stiffness of a structure will reduce vibration levels, except where the result of stiffening a panel is to move its natural frequency into coincidence with the forcing frequency. Increasing the stiffness of panels can also cause an increase in their radiation efficiency. The points at which vibration is fed into the structure should be as rigid as possible, and so care should be taken to ensure that vibration sources such as pumps, motors, vibrating pipework, etc., are connected to stiff parts of the structure, and not to the flexible areas such as the centre of thin plates and panels.

The drastic reduction of stiffness, i.e. the provision of resilience at some point in the vibration path can be an effective means of vibration reduction, and is really an example of vibration isolation. The use of a resilient layer in floor constructions is illustrated in Chapter 5.

Vibration isolation

Machines can be mounted on isolating springs to reduce the transmission of vibration to the floor and into the building structure. The theory of vibration isolation was discussed in the previous chapter.

Materials A wide variety of materials may be used as isolators of vibration. At the 'high frequency' end of the range, at 25 Hz and above, where static deflections are small, cork, cork composites, felt, foamed plastic and foamed rubber may be used in the form of pads or mats. All these materials derive part of their springiness from the air they contain and they should be used in compression.

In the intermediate frequency range, 5–35 Hz, rubber and elastomer materials are used, in a wide variety of shapes. They are used either in compression or in shear. Natural rubber can only operate over a limited temperature range and is attacked by oil. These and other disadvantages may be overcome by using one of a wide variety of synthetic rubbers. An important property of rubber, relevant to its use as an isolator, is that its great compressibility arises from its change of shape, hence the frequent use of rubber in shear isolators. A solid rubber cube squeezed between two opposite faces will only be compressible if the other four faces are allowed to bulge outwards. Therefore consideration of the shape of rubber isolators is very important.

Metal springs are used in the lowest frequency range, 2–15 Hz, where static deflections are greatest. Advantages of metal springs are that they can be designed and fabricated into a variety of shapes and configurations to give any required stiffness, and they are able to withstand high loads. They have very good resistance to adverse environments, being unaffected by oil and high temperatures. A disadvantage is that although they are effective in isolating low frequency vibration, high frequencies travel along the coils of metal springs and are not isolated. For this reason metal springs are often fitted with a pad of neoprene or similar material to reduce the transmission of the high frequencies. Another disadvantage of metal springs is that they have very low damping compared with the other materials mentioned above. This can cause problems when a machine passes through its resonant speed during starting and stopping. To reduce these problems a damping mechanism using some form of friction, or a viscous dashpot can be built into the metal spring isolator. Another solution is to use some physical stop mechanism (called a snubber) to limit the resonant vibration amplitude. Yet another solution is to ensure that the machine is accelerated rapidly through resonance on starting up, and braked sharply on switching off, so that resonant vibrations do not have time to build up.

Pneumatic isolators in which an air cushion provides the springiness, may also be used to isolate vibrations of the very lowest frequencies.

Installation and loading Whichever type of antivibration mount is selected, it is important that it is fitted according to the manufacturer's instructions, and in particular that the load it bears lies within the recommended range.

Resilient bases The simple theory outlined in the previous chapter assumed that the machine was to be mounted on a rigid base. If this is not so, the chosen isolators may be much less effective and could even cause an increase in vibration. A resilient floor can itself be represented as a mass–spring system and so the result of mounting an isolated machine on it is to create a more complicated vibrating system which consists of two masses and two springs, and which has two resonant frequencies. It is in theory possible to analyse such a system and choose isolators accordingly, if the effective mass and stiffness of the resilient floor are known. This is not often done — it is better to try to arrange for the machine to be mounted on a more rigid base.

Bridging of isolation The effectiveness of correctly designed and installed isolators can be greatly reduced if alternative vibration paths from the machine to the building structure 'bridge' the isolation. Possible bridging paths are ducts, conduits and pipes which are attached rigidly to the machine. These should all be isolated by flexible connection to the machine in order to prevent vibration transmission.

Applications The use of vibration isolation is not restricted solely to the fitting of mounts under large machines. Smaller machines such as fans, motors and pumps can be isolated from the structure of larger machines to prevent radiation of noise from much larger surface area. Ducts and pipes can be isolated from floors, walls and ceilings, and machinery panels, guards and covers can be isolated from the main frame of the machine. Floors, ceilings and even whole rooms and buildings can be isolated from their surroundings. In all cases, however, the following points are important:

1. The isolators must be carefully designed to give the required reduction. They must be installed correctly.

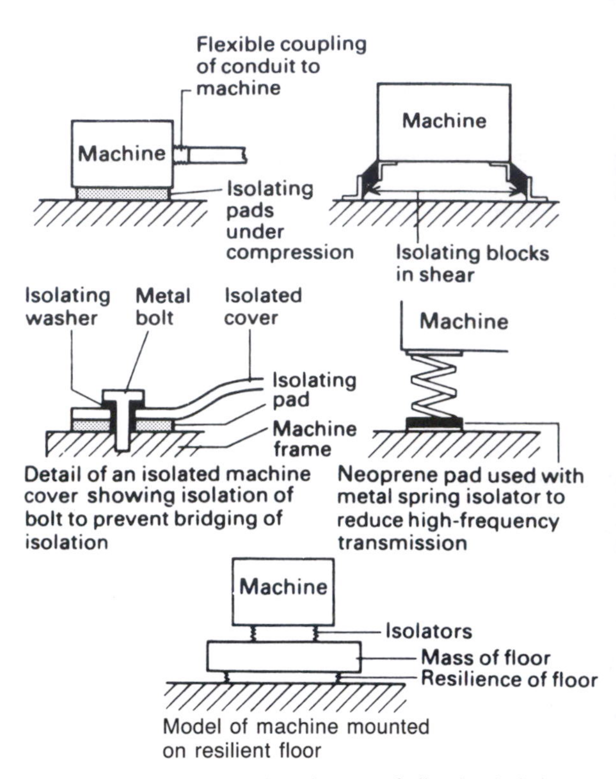

Fig. 8.10 Illustrating various features of vibration isolation systems

2. Isolators should be attached to bases which are as rigid as possible.
3. Care must be taken to avoid bridging or bypassing the isolation with alternative transmission paths.

Figure 8.10 illustrates some of the above points relating to vibration isolation.

Damping

An increase in damping is most effective in reducing levels of sustained vibrations when applied to resonant systems, or in reducing transient vibrations (ringing) of structures which are subjected to repeated impulsive forces. Damping mechanisms remove vibrational energy from the transmission path, converting it to heat. Some of the damping occurs within the material and some at joints between different parts of the structure. Structures which contain riveted or bolted joints tend to be more highly damped than welded or one-piece constructions. Damping treatments are very effective when applied to large areas of thin sheet-steel panels because the inherent damping of the steel itself is very low, and because the undamped vibration amplitudes can be high. Such treatments are not effective when applied

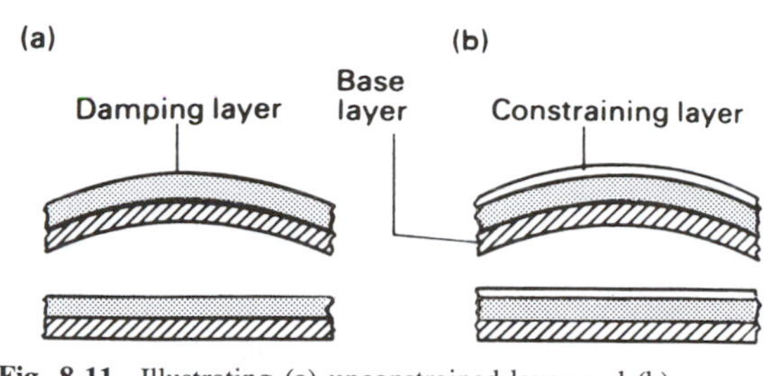

Fig. 8.11 Illustrating (a) unconstrained-layer and (b) constrained-layer damping, and showing the deformation of the damping layer

to stiff panels — it might be better to stiffen them even more, or to isolate them.

The principle of damping treatments applied to panels is to create a highly damped layer, often of a viscoelastic material, next to the sheet metal. The layer is made to vibrate, following the motion of the base layer, and much of its vibrational energy is abstracted by the damping processes. The aim is to create as much relative motion as possible between the molecules of the damping layer, to allow the 'internal friction' of the material to get to work.

Damped panels may be of two types, depending upon whether or not the damping layer is constrained. Figure 8.11 illustrates the difference. Unconstrained-layer damping may be created by the application of damping materials in a variety of ways, including mastic treatments which can be sprayed or painted on to the sheet-metal base layer, or in a sheet form which is attached with adhesives. Several sandwich materials are available in sheet form which exploit constrained-layer damping. The purpose of the constraining layer is to create shear strain, i.e. more relative motion, in the damping layer.

Constrained-layer damping is usually more efficient than unconstrained-layer damping. Applied damping treatments can be very effective even if applied to only part of the panel surface. It is not necessary to cover the whole surface with damping treatment.

The mechanical handling, transport and storage of materials offer many opportunities for use of increased damping. Chutes, hoppers, trays and storage bins can radiate a lot of noise as a result of impacts with objects thrown into or on to them, and this noise can be considerably reduced by damping sheet-metal areas.

The machining of hollow metal castings or the riveting of metal cans and drums creates a great deal of 'ringing' noise. Fairly simple 'clamping and damping' of the workpiece can sometimes produce considerable noise reductions. The use of sand fillings as a damping treatment can also be very effective.

8.20 The transmission of sound through walls and partitions

Sound reduction index/transmission loss

The basic property of a partition which determines its effectiveness as a sound insolator is the *sound reduction index R*, also called the *transmission loss*, in dB.

$$\text{Sound reduction index } R = 10 \log_{10}\left(\frac{1}{t}\right) \text{ decibels}$$

where t, the transmission coefficient of the partition, is the fraction of the sound energy incident on the partition which is transmitted through it. As an example, if $t = 0{\cdot}001$ then $R = 30$ dB, and if $t = 10^{-5}$ then $R = 50$ dB. If a partition is made up of different materials, as for example a brick wall with a door and window, the sound reduction index of the total façade may be calculated as in Example 5.6.

The sound reduction index of a partition is measured in a laboratory under specialized conditions. Two adjacent reverberant rooms are required and the test partition is built into the party wall separating them. The reverberant suite is constructed to a very high standard of sound insulation so that flanking paths (see Fig. 5.12) become unimportant and the transmission of sound from one room to the other is by airborne sound transmission through the test partition.

When using sound reduction index values to predict noise levels in actual field (i.e. non-laboratory) situations, the following points should be borne in mind:

1. The published R-value does not take into account any flanking paths which may exist.
2. The way in which the partition was fixed into position can affect its R-value. The fixing method may be different from that used in the laboratory test.
3. The size of the partition may affect its R-value — for example, the size can affect resonance frequencies.
4. The laboratory test was carried out under truly reverberant conditions which may not prevail in actual field situations. The sound reduction index depends upon the angle of incidence of the sound striking the partition. In the laboratory test a random distribution of incidence angles is achieved — this may not be the case in less reverberant field situations.

8.21 Factors affecting the sound reduction index of partitions and panels

The sound reduction index of a single-leaf partition may be calculated, approximately, using the mass law, in which case R increases by 5 or 6 dB for every doubling of mass and for every doubling of frequency. The mass law assumes that only the mass of the partition is significant in determining the R-value, i.e. the effect of stiffness and damping are ignored. Resonances of the partition, at low frequencies, and the coincidence effect at high frequencies cause departure from the mass law. Stiffness and damping are important in both these frequency regions. These points are discussed in Chapters 4 and 5. The requirements of the ideal single-leaf partition, as far as achieving good sound insulation is concerned, are that it should have a high mass combined with low stiffness and should be well damped.

Composite or laminated partitions consisting of two or more sheets bonded together by well-damped viscoelastic material, e.g. rubber or polymer, are often fairly effective sound insulators, when compared with single-sheet constructions, because the composite construction achieves increase in mass with relatively little increase in stiffness, and improved damping, compared to single sheets of the same total thickness.

Double leaf constructions, i.e. with a cavity between the two leaves, can achieve a bigger increase in R-value over the single-leaf value than is achievable simply by doubling the thickness, and the mass, of the single leaf, provided that:

1. The width of the cavity is at least 50 mm, to avoid the mass–spring–mass resonance of the construction (caused by the entrapped air acting as a spring with the two leaves as masses) falling inside sensitive frequency ranges.
2. The cavity is filled with sound-absorbing material, to prevent the build-up of reverberant sound in the cavity.
3. The leaves are isolated from the building structure, and from each other, to prevent flanking transmission.

Using these design principles, lightweight double-skin partitions can achieve a similar overall sound insulation performance to much heavier masonry constructions, except perhaps at the lowest frequencies. The performance of double-glazed windows is illustrated in Chapter 5.

Transmission of sound between two adjacent rooms

In addition to the sound reduction index R, the amount of sound energy transmitted through a partition will depend on the area of the partition, S. For a given amount of sound energy transmitted through the partition, the level of reverberant sound created in the receiving room will depend upon the amount of absorption in the receiving room, A. It is for this reason that level difference measurements are 'normalized' as discussed in Chapter 5.

The level difference between the two rooms is given by

$$SPL_1 - SPL_2 = R - 10 \log_{10} S + 10 \log_{10} A$$

where SPL_1 = average level in the source room, in dB

SPL_2 = average level in the receiving room, in dB

S = area of partition in m^2

A = absorption in receiving room, in m^2

This equation, which is used in the laboratory measurement of R, may be used for predicting noise levels in real situations where the sound distribution in each room is fairly reverberant. It does not include effects of flanking transmission.

Example 8.5

The noise level in a room used for a discotheque is 88 dB in the 1000 Hz octave band. Calculate the level expected in the same octave band in an adjacent room if the partition wall is of a type of construction which has a published value of $R = 47$ dB in this band. The dimensions of the partition are 10 m × 5 m. The amount of absorption in the receiving room, obtained from measurements of its reverberation time is 32 m^2.

$$SPL_1 - SPL_2 = R - 10 \log_{10} S + 10 \log_{10} A$$
$$= 47 - 10 \log_{10} 50 + 10 \log_{10} 32$$
$$= 45 \text{ dB}$$

therefore, level in receiving room = 88 − 45 = 43 dB

Similar calculations for the other octave bands will allow the overall level dB(LIN), the A-weighted level or the NR or NC value for the receiving room to be estimated.

Transmission of sound from indoors to outside

In this case the sound energy transmitted through the partition is radiated into open space, where there is no reverberant field. The sound pressure level close to the partition on the outside is given by

$$SPL_2 = SPL_1 - R - 6$$

where SPL_1 = reverberant sound pressure level on the source side of the partition (indoors)

SPL_2 = sound pressure level close to the partition on the outside

R = sound reduction index of the partition

The factor of 6 dB arises from the fact that the sound energy incident on the partition on the room side is diffuse, whereas on the open side it is not. The relationship between acoustic pressure and acoustic intensity is different for directional and diffuse sound, by a factor of 4. This in turn shows up as a difference in the relationship between the sound intensity level (in dB re 10^{-12} W/m^2) and the sound pressure level (in dB re 2×10^{-5} Pa). In a plane wave these two are numerically equal, whereas in a diffuse field the factor of 4 means that the intensity level is 6 dB less than the sound pressure level. This is the origin of the 6 dB in the above expression.

The sound power levels of surfaces and façades

If the average sound intensity flowing through a surface is 1 W/m^2 and its area is S m^2 then the surface can be considered as a source of sound whose power P in Wm^2 is given by

$$P = IS$$

The logarithmic equivalent of this equation relates the average sound pressure level SPL at or close to the surface to the sound power level L_W of the surface acting as a noise source:

$$L_W = \text{SPL} + 10 \log_{10} S$$

This relies on the assumption of plane-wave propagation from or through the surface so that as mentioned above sound pressure level and intensity level will be numerically equal. Therefore the above equation should be used as a rough guide only. The sound power level may be used to predict the sound pressure level produced by the surface some distance away, provided that the distance between the surface and the receiving point is several times the dimensions of the radiating surface, to achieve far-field conditions. The radiation may have a directionality depending on the dimensions of the radiating surface and the wavelength of the sound, but approximate calculations may be made assuming omnidirectional propagation.

Example 8.6

A light workshop contains several noise sources, but it is suspected that the noise level at a certain workbench position is mainly produced by the vibration of a sheet-metal hood designed to extract dust and chippings from a woodworking machine. The vibration levels of the hood are measured and the average vibration velocity is 2·0 mm/s in the octave band of the offending noise. The area of the hood, which is situated in the centre of the workshop 2 m from the workbench, is 0·4 m^2. Estimate the sound pressure level produced by the hood at the workbench position. The room constant for the workshop has been estimated as 150 m^2.

The vibration velocity is the same as for Example 8.1, and so the sound pressure level close to the hood is 92·4 dB.

$$\text{Sound power level } L_W = \text{SPL} + 10 \log_{10} S$$
$$= 92{\cdot}4 + 10 \log_{10} 0{\cdot}4$$
$$= 88{\cdot}4 \text{ dB}$$

To calculate the SPL at a point in the room:

$$\text{SPL} = L_W + 10 \log_{10} \left(\frac{Q}{4\pi r^2} + \frac{4}{R_c} \right)$$
$$= 88{\cdot}4 + 10 \log_{10} (0{\cdot}02 + 0{\cdot}027)$$
$$= 88{\cdot}4 - 13{\cdot}3 = 75{\cdot}1 \text{ dB}$$

Example 8.7

The noise level inside a parish hall during a social function is 83 dB(A). One side of the hall contains an open window facing a house 30 m away. The dimensions of the window are 1·5 m × 2 m. Calculate the sound level outside the house. (Assume $R = 0$ for the open window.)

$$\text{SPL}_2 = \text{SPL}_1 - 6$$

Level just outside window = 83 − 6 = 77 dB(A)

$$L_W = \text{SPL} + 10 \log_{10} S$$

Sound power level of window (A-weighted)

$$= 77 + 10 \log_{10} 3$$
$$= 81{\cdot}8 \text{ dB}$$

$$\text{SPL} = L_W - 20 \log_{10} r - 8$$
$$= 81{\cdot}8 - 20 \log_{10} 30 - 8$$

Sound level outside the house = 44·3 dB(A)

Example 8.8

A small factory building is to be constructed with one façade of dimensions 12·5 m × 4 m facing a dwelling 80 m away. It is estimated that the maximum octave band SPL inside the factory will be 88 dB in the 500 Hz band. Calculate the sound reduction index required of the façade in this frequency band if the level at the façade is not to exceed 40 dB.

First we can calculate the sound power level produced by the façade which would give 40 dB at the dwelling.

$$L_W = \text{SPL} + 20 \log_{10} r + 8$$
$$= 40 + 20 \log_{10} 80 + 8$$
$$= 86 \text{ dB}$$

Next we can calculate the sound pressure level just outside the façade which will produce this sound power level.

$$L_W = \text{SPL} + 10 \log_{10} S$$

$$\text{SPL} = L_W - 10 \log_{10} S$$
$$= 86 - 10 \log_{10} 50$$
$$= 69 \text{ dB}$$

Finally we can estimate the R-value to produce this level just outside the façade, when the level inside is 88 dB.

$$\text{SPL}_2 = \text{SPL}_1 - R - 6$$
$$69 = 88 - R - 6$$
$$R = 13 \text{ dB}$$

8.22 Radiation of sound from plane areas

The last three examples illustrate a very simple, idealized approach to noise radiation from vibrating surfaces. First of all, a plane-wave radiation model is assumed in order to calculate the sound power level of the surface from the sound pressure level close to it, then the surface is assumed to behave like a point source, in order to calculate sound pressure levels far from it.

Such an approach will only be justified for distances which are large compared to the dimensions of the plane surface, when the rate of reduction of sound pressure level with distance will approach the rate of 6 dB per doubling of distance, as expected from a point source. At distances which are very small compared to the surface dimensions (and if these dimensions are large compared to the sound wavelength) the radiation pattern will approximate to plane-wave conditions, i.e. with no reduction of level with distance. At distances which are intermediate between these two extremes, the rate of reduction with distance will vary between zero and 6 dB per doubling of distance. In particular, if one of the surface dimensions is very much larger than the other, the radiation pattern may approximate to that from a line source with a 'drop-off' rate of 3 dB per doubling of distance.

An approximate method for predicting sound pressure levels at any distance from a rectangular plane surface, of dimensions a and b ($b > a$), assumes plane-wave radiation for distances up to a/π, line source radiation for distances between a/π and b/π, and point source radiation for distances greater than b/π, as shown in Fig. 8.12.

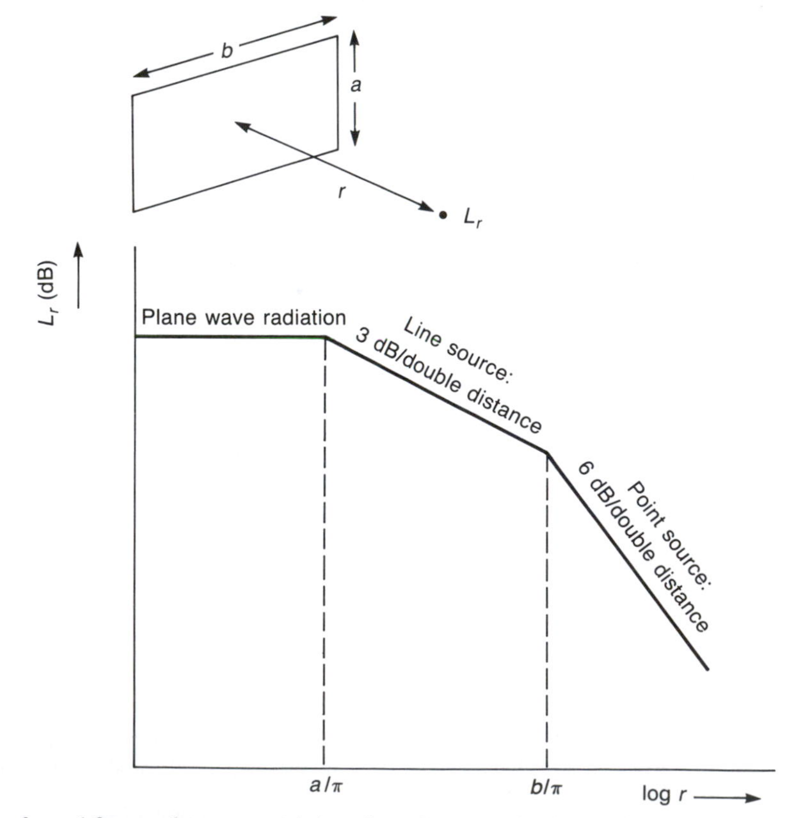

Fig. 8.12 Radiation of sound from a plane area: variation of sound pressure level with distance

Example 8.9

The average sound pressure level close to the rectangular façade of a building, of dimensions 12·5 m and 4 m, is 69 dB. Estimate the sound pressure level opposite the centre point of the façade, at distances of 1 m, 2 m, 4 m, 8 m and 80 m.

The two cut-off points are $a/\pi = 4/\pi = 1{\cdot}3$ m and $b/\pi = 12{\cdot}5/\pi = 4{\cdot}0$ m.

Between 0 and 1·3 m: plane-wave radiation with SPL = 69 dB, i.e. SPL at 1 m is 69 dB.

Between 1·3 m and 4·0 m: line source radiation with

$$\text{SPL at 2 m} = 69 - 10\log_{10}(2{\cdot}0/1{\cdot}3) = 67 \text{ dB}$$
$$\text{SPL at 4 m} = 69 - 10\log_{10}(4{\cdot}0/1{\cdot}3) = 64 \text{ dB}$$

Beyond 4·0 m: point source radiation with

$$\text{SPL at 8 m} = 64 - 20\log_{10}(8/4{\cdot}0) = 58 \text{ dB}$$
$$\text{SPL at 80 m} = 64 - 20\log_{10}(80/4{\cdot}0) = 38 \text{ dB}$$

Using the simpler method of Example 8.8 gives an SPL of 37 dB at 80 m, assuming spherical radiation, i.e. soft ground conditions, or 40 dB assuming hard ground conditions (as in Example 8.8). The two methods give similar results for large distances, but the point source method will be inaccurate close to the source.

Sound transmission from outside to indoors

Sound incident upon the outside wall of a room is transmitted through the wall, which then acts as a noise source radiating sound into the room. It is useful to be able to predict the sound pressure level inside the room, from the outside level. There are, however, two complications. First of all, the sound incident on the outside surface may be directional, arriving from a single source of noise, such as a fan. In this case the angle of incidence will be important. Alternatively, the external sound may be diffuse, arriving at the wall from many different directions, as for example in a busy street. The second complication concerns the way in which the external sound pressure level, at the wall, is specified. Suppose that the sound is directional and is incident normally on the surface, i.e. from a noise source directly opposite. There are two possible approaches: either the level at the required point, in the absence of the wall, could be predicted from the sound power level of the noise source, or the level close to the wall could actually be measured. The difference between these two sound pressure levels would be 6 dB, the measured value being the higher one. In the case of diffuse incident sound the difference would be 3 dB.

The level inside the room, assumed to consist of reverberant sound only is:

$$\text{SPL}_{in} = \text{SPL}_{out} - R + 10\log_{10} S - 10\log_{10} A + K$$

where SPL_{in} = sound pressure level inside the room

SPL_{out} = sound pressure level outside, close to the wall

R = sound reduction index of wall

S = area of wall through which the sound is transmitted in m^2

A = amount of acoustic absorption in room in m^2

K = a numerical factor

For direct normal incidence sound $K = +6$ if the outside level is predicted (in the absence of the building), and $K = 0$ if the value measured close to the wall is used. If the sound strikes the wall with random incidence then $K = 0$ if a predicted outside level is used (in the absence of the building) and $K = -3$ if the level measured close to the wall is used.

It should of course be remembered that the sound reduction index of a surface may be different for normal and randomly incident sound, and the appropriate value should be used.

8.23 Acoustic enclosures

The use of an enclosure made of a good sound-insulating material to reduce the noise from a machine is an obvious and attractive solution to a noise problem. With very careful design and use of materials with a high sound reduction index, large noise reductions outside the enclosure can be achieved. However, there are disadvantages, since enclosures can create problems of access to the machine, inconvenience to the operator and create cooling problems; they take up floor space and can be expensive. All of these factors have to be considered by the acoustics engineer, as well as the acoustic parameters such as required noise reductions, and the frequency content of the noise source. In many straightforward situations modest reductions of, say, 10 to 20 dB may be achieved in the medium to high frequency ranges by careful choice of material and design. However, if much larger reductions are required, and particularly at low frequencies, great engineering skill and attention to detail are required.

Principles

The principles of design are the same as those for good insulation given in Chapter 5, which are:

1. The material of the enclosure must have a high enough *sound reduction index* at all frequencies to give the

required noise reductions.

2. Any '*weak links*' in the enclosure (as far as sound insulation is concerned) can seriously reduce the amount of noise reduction achieved. These could include gaps, cracks and holes, poor sealing around windows and doors, and openings for access, ventilation and services.
3. *Flanking paths* for the transmission of sound must be avoided or minimized. This means that if there is any possibility of structure-borne sound being transmitted then the machine must be effectively isolated from the floor and from the enclosure.

Practice

In practice this means that all pipes and ducts (for water, air, fuel, etc.) should be mechanically isolated from the machine, using flexible connectors. Where these pass through the enclosure, care should be taken so that access holes are as small as possible, and air-sealed with flexible, mastic type of material. The aim should be to minimize the leakage of airborne sound and yet to prevent transmission of vibration. Similarly it is important that any gaps around doors and windows in the enclosure are well sealed. If the air-cooling requirements of the machine are such that either natural or forced ventilation is needed, then the air can be led through absorbent lined ducts, preferably with one or two 90° bends to help minimize the escape of airborne sound from the enclosure. Sometimes access for materials entering and leaving the machine can be through absorbent lined tunnels. The ends of the tunnels can be covered with push-aside flaps of sound barrier mat which close automatically. The number of practical problems which may be encountered, and the ways of solving them, are endless, but it is attention to these details, as outlined above, which is essential if the simple concept of a box around the machine is to be turned into a practical working solution, and if the noise-reduction performance of the enclosure is to approach a value which is theoretically achievable for the material being used for the enclosure. The sketch diagram of Fig. 8.13, illustrates some of the features of a good enclosure.

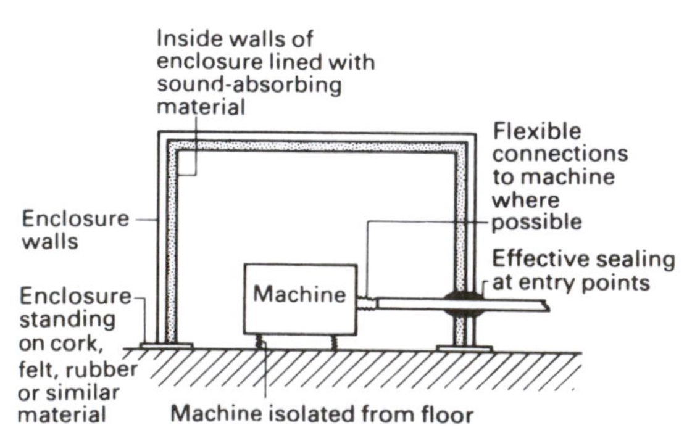

Fig. 8.13 Illustrating some of the features of an acoustic enclosure

Use of absorption inside enclosure

Although an enclosure may be successful in reducing noise levels outside, noise energy is 'bottled up' within the enclosure and noise levels within may be much higher than they would be in the absence of the enclosure. To reduce this high level of reverberant sound it is common practice to line the inside walls of enclosures with sound-absorbing materials. The benefits of reducing noise levels inside the enclosure in this way are obvious for large enclosures which contain working personnel, such as the machine operator. However, even when this is not the case it is still useful to use absorption in this way since a reduction of noise within the enclosure leads to a corresponding reduction in levels outside. Although these reductions are modest compared with the total reduction, the use of sound-absorbing material in this way does provide a useful supplement to the noise reduction produced by the insulation of the enclosure walls. The use of sound-absorbing materials also helps to reduce the effect of small gaps and leaks in the enclosure.

Prediction of enclosure performance

Enclosures which are much larger than the machines they surround may be treated as 'rooms' and the ideas of the previous paragraphs can be used to estimate noise levels both inside the enclosure and at various positions outside it. The noise level inside the enclosure may be assumed to be reverberant and the sound power radiated by the enclosure walls will depend on the internal reverberant sound pressure level, the area of the walls, and the sound reduction index of the walls.

If the enclosure is situated within a larger room, in which the sound is mainly reverberant, and caused predominantly by transmission through the enclosure walls, then the situation is rather similar to that of two rooms separated by a partition. In this case the area of the partition is the total area of the enclosure (four walls and roof). The difference in levels of reverberant sound inside and outside the enclosure may then be calculated from the equation

$$\mathrm{SPL_{in}} - \mathrm{SPL_{out}} = R - 10 \log_{10} S + 10 \log_{10} A$$

where R is the sound reduction index of enclosure walls

S is the total area of enclosure walls

A is the area of absorption in the receiving room, i.e., the room outside the enclosure

SPL_{in} = sound pressure level within the enclosure

SPL_{out} = reverberant sound pressure level outside the enclosure (caused by transmission through the enclosure walls)

This level difference, sometimes called the **noise reduction**, is obviously an indication of the effectiveness of the enclosure walls in preventing transmission of airborne sound from the machine to the receiver. However, as far as the receiver is concerned the more important difference is between the noise levels at the reception point before and after the installation of the enclosure around the machine. This difference is called the **insertion loss**. It may be estimated by first of all calculating the reverberant sound pressure level in the room produced by the machine, based on the sound power level of the machine and the room constant. The calculation is then repeated but using the sound power level of the enclosure. The two levels are then subtracted. This process, after certain approximations, leads to the following equation for the insertion loss (IL):

$$IL = SPL_{before} - SPL_{after}$$
$$= R - 10 \log_{10} S + 10 \log_{10} A$$

where SPL_{before} = sound pressure level in the room before enclosure fitted

SPL_{after} = sound pressure level in the room after enclosure fitted

S = total area of enclosure (four walls + roof)

A = area of absorption inside the enclosure

Note that in the insertion loss formula the area of absorption *inside* the enclosure appears explicitly. In the formula for noise reduction it is the absorption in the room *outside* the enclosure which is used — the absorption inside the enclosure does not appear explicitly, but it will of course affect the level inside the enclosure.

The enclosure walls also produce a direct sound field which will be important for enclosures situated outdoors, or in very large indoor spaces or for positions close to the enclosure. The estimation of the direct sound pressure level at a distance from the enclosure wall is rather similar to the estimation of noise from façades, discussed earlier. The radiating area to be used in calculations in this case is the area of the enclosure wall which faces the reception point, and not the total area of the enclosure.

Close-fitting enclosures

In order to save floor space and material costs many enclosures are built with the air gap between the machine and enclosure walls as small as possible. These close-fitting enclosures will not behave acoustically like a room since the source is radiating into a small volume and the sound inside the enclosure will not be as diffuse as in a larger enclosure. There will also be a close coupling between the source and the enclosure walls, and the stiffness of the air gap becomes important. The performance of close-fitting enclosures is difficult to predict but it is not as good as indicated by the above equations, which are really only valid for larger enclosures. The stiffness of a close-fitting enclosure is important, as well as its mass per unit area. The enclosure walls should be as stiff as possible.

Noise reduction and insertion loss

These are two different methods of assessing the performance of a noise-control device such as a silencer, a noise barrier or an acoustic enclosure.

The insertion loss is the difference in noise level at the receiving point before and after the installation of the noise-control treatment. It is the quantity which is most important to the receiver of the noise, or to the client who is paying for the noise-control treatment. The only difficulty with the measurement of insertion loss is that the measurements to be compared are taken at two different times, maybe months apart and it may be difficult to ensure that all other conditions affecting the test are repeatable.

The noise reduction is a measure of difference of levels 'across' or 'on opposite sides of' the noise-reduction treatment. The noise reduction of an enclosure, for example, may be defined as the difference in noise levels inside and outside the enclosure, and for a silencer in a duct system it is the difference in noise levels in the duct on either side of the silencer.

Unfortunately, the measured noise reduction may be different than the true insertion loss for a variety of reasons. In an enclosure, for example, the noise level inside the enclosure will almost certainly be increased by the build-up of reverberant sound inside the enclosure. Differences may also arise if the machine, before enclosure, was strongly directional since the reverberation inside the enclosure will mean that the sound radiated through the enclosure walls will probably be much less directional than that radiated from the machine surfaces. In a duct system the measured noise reduction does not take into account either the fact that the insertion of the silencer may cause a slight change in levels in the duct upstream, or that the length of the duct replaced by the silencer does itself produce some small attenuation of sound.

The above definitions of noise reduction and of insertion loss may not be universal, and confusion over terms does arise. The important principle, however, is clear: that any performance test parameters which purport to measure the effectiveness of a noise-control measure such as 'reduction' or 'attenuation' should be clearly defined, and the way in which the test relates to the real installation situation should be known.

Close shielding of noise-radiating surfaces — acoustic cladding or lagging

Where a complete enclosure is not feasible, noise from a particular location can be reduced by the use of a shield positioned very close to the noise-radiating surface. The shield consists of a layer of light resilient material attached to a second layer of heavier material, as illustrated in Fig. 8.14. The resilient layer may be foam or a fibrous material such as mineral wool. The outer layer should be as massive as possible and ideally be limp and well damped. Thin lead sheet is ideal, but other materials can be used, including damped sheet steel. Commercial 'sound barrier mats' are available which consist of a heavy but flexible layer of fabric or PVC which is loaded with particles of lead and lined with foam rubber.

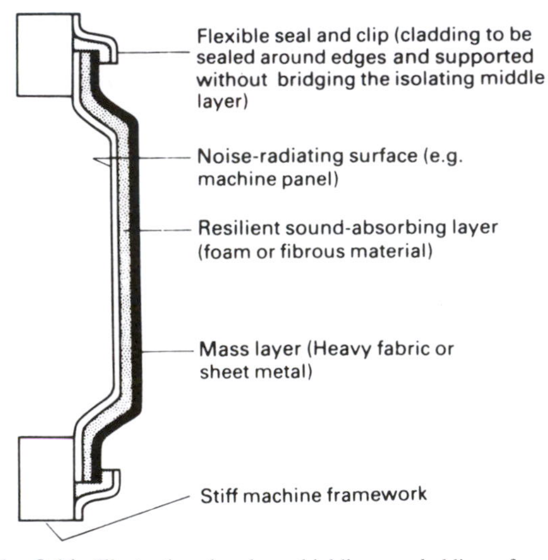

Fig. 8.14 Illustrating the close shielding or cladding of a noise-radiating surface

The shield acts in a complex way, incorporating in various degrees, at different frequencies the mechanisms of sound insulation and absorption, and vibration isolation and damping. The outer layer serves to reduce the radiation of sound by insulation and so it needs to be as heavy as possible. The foam or fibrous layer absorbs sound trapped between the outer layer and the radiating surface. In addition, the air in the fibrous or foam layer acts as a spring which at certain frequencies isolates the outer layer from the radiating surface. The presence of the shield may cause some damping of the vibration which is causing the noise radiation.

It is important that the shield is used with the resilient absorbing layer adjacent to the vibrating surface (and not the other way around) and that the method of fixing and supporting the shield does not allow contact between the shielded surface and the outer shielding layer. Any such bridging of the isolation between these two layers will reduce the effectiveness of the shield.

This type of shielding can be used for reducing noise from certain parts of a machine such as thin metal panels, and also for 'cladding', or 'lagging' pipes and ducts which are significant noise radiators. The advantage of a flexible shield is that it can easily be cut and shaped to curved surfaces. This type of noise-reduction treatment can be effective at high frequencies but gives only small reductions at low frequencies. In fact, increases in noise levels can arise at the frequency where there is a resonance of the mass of the heavy outer layer and the springiness of the air-filled foam or fibrous layer. To achieve bigger noise reductions at low frequencies, carefully designed cladding is required with a thick fibrous layer and very heavy outer skin. Alternatively, a double-skin system may be used with a second resilient layer and a second massive skin built on to the first layer of cladding.

8.24 The reduction of airborne sound and the uses of sound-absorbing materials

In principle this may be achieved in three ways:

1. Increase distance between noise source and receiver (when the unwanted sound is travelling direct from source to receiver).
2. Use insulation to restrict the unwanted sound from a particular locality using for example a partition wall, an enclosure or a screen or barrier. The sound energy is not absorbed, and appears elsewhere.
3. Use sound-absorbing materials to remove the sound by turning it (eventually) into heat.

The application of all three principles is discussed in this and in earlier chapters. However, sound-absorbing materials have such a wide variety of uses in noise control that it is thought worth while to list some of them as below.

(a) Control of reverberation time.
(b) Reduction of reverberant sound levels.
(c) When used with diffusers of sound, to reduce the undesirable effects of standing wave resonances in spaces.
(d) For lining the inside of ductwork, and in splitter and other types of resistive or absorptive silencers.

(e) For lining the inside of enclosures, and barriers (on the noise source side).
(f) For plugging small gaps and holes in insulation.
(g) For filling the cavity of double-skin partitions and in the reveals of double glazing, to reduce the build-up of sound in the cavity between skins.
(h) In conjunction with a massive layer as a cladding or close shield for pipes, ducts and panels.

8.25 Noise from ventilating and air-conditioning systems

The main source of noise is that generated aerodynamically by the motion of the fan blades through the air. This noise is transmitted via the duct system and enters the room at the grille or diffuser which terminates the ducts. Additional paths are transmission through the walls of the duct (called duct breakout) airborne noise direct from the fan itself, in the plant room and adjacent rooms, and to the outside of the building.

Structure-borne noise may also be generated by vibration of the fan casing and motor. This may be transmitted directly or via the duct to the building structure, or it may be radiated directly from the duct. To reduce these noise paths the fan and duct should be isolated from the building structure and the fan and duct connected by flexible (isolating) coupling.

Another source of noise is that generated aerodynamically by the flow of air through the duct system. This component of the noise (sometimes called regenerated noise) is very much dependent on the velocity of the air and the smoothness of the flow. It increases very rapidly as the flow velocity is increased and is also increased by obstacles to smooth flow and by changes in cross-section in the duct. Noise is also generated by the presence of bends, particularly if they are not radiused to smooth the flow, and by the flow of air through grilles and diffusers as it enters the room from the duct.

Figure 8.15 shows in schematic form the various components of a ventilation system, and Fig. 8.16 illustrates some of the various noise transmission paths. However, for many situations the most important source of noise in a ventilated room is that caused by transmission of fan noise into the room via the duct system.

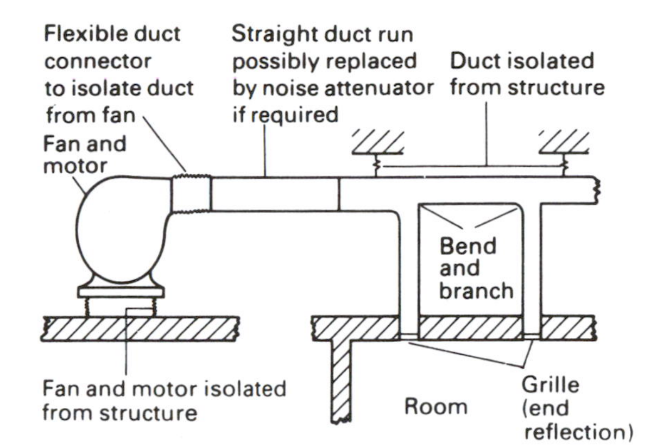

Fig. 8.15 Illustrating some of the features of a ventilation system

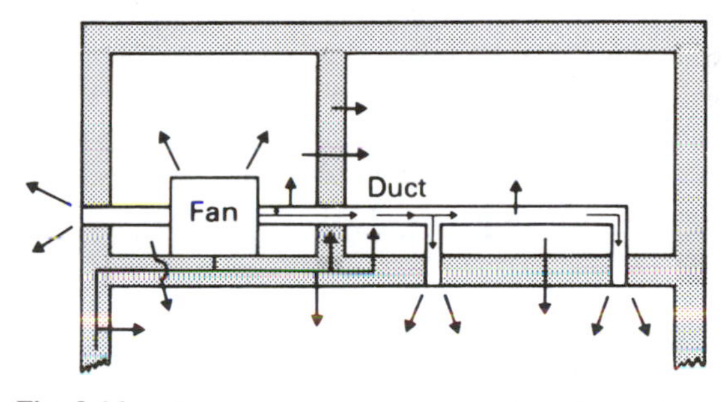

Fig. 8.16 Schematic diagram of fan and duct, illustrating some of the noise transmission paths in the building

Fan noise

The spectrum of fan noise consists of pure tones and a continuous spectrum. The frequencies of the pure tones are the fan blade passing frequency and its harmonics,

$$f_n = \frac{\text{RPM}}{60} \times N \times n$$

where RPM = fan rotation speed in revolutions per minute

N = number of fan blades

n = harmonic number (n = 1,2,3, . . .)

The continuous spectrum arises from turbulence created in the air in the wake of the moving blades. For most ventilating fans, correctly installed, the continuous spectrum noise is usually much more important than the pure tone radiation. Typical fan noise spectra are shown in Fig. 5.21. A knowledge of the spectral content of the noise as well as of the overall sound power is important when selecting the fan and in predicting noise levels from the fan–duct system.

The main steps in control of noise from a ventilation system which is already designed or built are:

1. Correct choice of fan.
2. Correct installation of fan.
3. Use of extra noise attenuation as required — usually in the form of a splitter silencer inserted in place of a length of straight duct, but also by lining ducts and bends with sound absorbing material.

In addition, of course, much can be done to minimize noise problems at the stage when the system is designed, by careful consideration of plant room location and positioning of duct runs with respect to noise-sensitive areas.

Choice of fan This will depend primarily of course on the ventilation requirements, i.e. on the volume airflow rate required and on the details of the air distribution system, and on cost. However, the acoustic output of the fan should be considered as well. Fan noise is very dependent on the tip speed of the fan blades, and larger, slower fans can be quieter than smaller fans moving the same quantity of air. The frequency spectrum of the fan is particularly important. Some high frequency noise will be removed by the 'natural attenuation' of the system (ducts, bends, etc.) and if necessary extra attenuation can be added in the form of a silencer. However, low frequency noise (63, 125 and 250 Hz octave bands) is much more difficult to remove. Therefore when choosing between two different fans of suitable duty, the one with the smaller sound output at the lower frequencies will be preferable from the noise point of view, even if the total sound powers of the two are the same.

Care should also be taken to ensure that the chosen fan will always be operating at or near its maximum aerodynamic efficiency, i.e. at the recommended static pressure for the volume flow rate required. Operating the fan well off its duty point will create extra noise.

It is necessary to know the sound power level which is produced by the fan in each octave band. This information should ideally be supplied by the fan manufacturer who should have obtained the data from noise measurements taken under special standard test conditions. The alternative (apart from arranging for such measurements) is to use one of the empirically based formulae which relates the sound power level of the fan to other of its properties such as the power of the fan motor, the static pressure developed and the volume flow rate of air delivered. An example of such a formula is given in Chapter 5. Correction factors for different types of fan allow the octave-band levels to be obtained from the overall level (see Fig. 5.21). These formulae will only give approximate levels and data from test measurements should always be used in preference, if available.

Installation of fan The sound power levels specified by the manufacturer will be the minimum noise level which it will emit, under the most favourable installation conditions. These occur when the flow of air into the fan is smooth and uniform. If as a result of poor installation design or practice the airflow is non-uniform or turbulent, then the fan will create a considerable amount of extra broadband noise. Such conditions can be caused by obstacles in ductwork which are close to and upstream of the fan, or if the fan is downstream of and too close to a bend. Obstacles and bends should be at least one fan diameter upstream of the fan to allow the airflow to smooth out before entering the fan. Air which has to 'turn sharp corners' into the fan will also cause extra noise, which may be reduced by using a tapered or a bell-mouth inlet to the fan. Obstacles such as struts, guide vanes and instrumentation probes which are close to the fan blade can also cause a very great increase in the pure tone component of the fan noise, as a result of the periodic chopping action on the air in the small gap between obstacle and blade.

Estimation of extra attenuation required

The procedure is as follows. The sound power level entering the air distribution system is known or may be estimated. From this figure must be subtracted the attenuation of sound energy produced by each component of the system. This then gives the sound power (expressed as a level, in decibels) entering the room at the grille. The total sound pressure level at any point in the room may then be calculated using the sound power level, the room constant and the distance of the grille from the receiver. This last part of the calculation proceeds as in Example 8.3. The calculated sound pressure level is then compared with the noise requirement of the room (e.g. an NC target) and if the estimated level is too great the difference give the extra attenuation which is required. This procedure is repeated for each octave band.

Attenuation is provided by the following components of the air distribution system (see Fig. 8.15): straight duct runs, bends, branches and duct terminations (end reflections).

Straight duct runs A smooth, rigid-walled duct would not provide any attenuation of sound striking it. However, sheet-metal ducts are not rigid and the sound causes the duct walls to vibrate and in doing so to remove energy from the sound waves in the duct. Some of this energy is re-radiated from the duct walls and some is dissipated by damping of the walls. The attenuation will depend on the mass per unit area of the duct walls, upon the frequency and the duct size. At low frequencies the stiffness of the duct walls will be more important than the mass and as a result low frequency attenuation will be different for round and circular cross-section ducts.

The attentuation of unlined sheet metal ducts is very small, ranging from 0·6 dB/metre for small rectangular ductwork at low frequencies to less than 0·1 dB/metre at high frequencies. The attenuation of ducts lined with acoustic absorption is discussed in Chapter 5 (see Example 5.8).

Bends The attenuation produced by a right-angled bend results mainly from the reflection of sound back towards the fan, thus reducing the amount of sound energy moving forward towards the room. It varies with the size of the duct and the frequency of the sound. For a square duct the bend attenuation is a maximum of some 7 or 8 dB at the frequency (octave band) for which the wavelength of sound in air is twice the duct width. At higher frequencies (an octave above) the attenuation drops to 3 or 4 dB and at lower frequencies it can fall to less than 1 dB.

If the bend contains turning vanes to help the air flow around the corner smoothly then the attenuation produced will be very much reduced (but so, of course, will the amount of noise regenerated by the bend). If the bend is lined with acoustically absorbing material then the attenuation will be greatly increased as shown in Table 5.5

Branches At a point where a duct splits into two branches it is assumed that the acoustic energy divides between the two branches in the same ratio as the division of airflow. Thus if the main duct divides into two equal branches the sound power level in each branch just below the junction is 3 dB less than in the main branch just above the junction. If the flow had divided into two parts down one branch and three parts down the other, then in the smaller branch the branch attenuation would be $10 \log_{10} (5/2) = 4$ dB, i.e. the sound power level in the smaller branch would be 4 dB less than in the main branch. In the other branch the attenuation would be 2·2 dB.

End reflection The sound near the end of the duct travelling towards the opening into the room 'notices' a change of acoustic impedance at the opening. This causes some of the sound energy to be reflected back down the duct rather than being transmitted into the room. This in effect provides an attenuation which depends on the wavelength of the sound and the dimensions of the opening. The attenuation is largest when the wavelength is much greater than the dimensions of the opening, i.e. for low frequencies and small openings. The attenuation is very small when the opening dimensions are greater than the wavelength, i.e. for large openings and high frequencies.

Example 8.10

Air enters a room via a single grille situated in one of the top corners of the room. A target level of NC 35 has to be achieved by the ventilation system which consists of a fan connected to a main duct of length 5 m. The airflow then splits into two equal parts, one of the branches of length 3 m serving the room in question via the single grille. There is one bend in the main duct prior to the division of the air flow, and one bend in each of the branches.

Calculate the extra attenuation required in the 250 Hz octave band to meet the NC 35 target at a point in the room 4 m from the grille, given the following information, all of which refers to the 250 Hz octave band.

Sound power level of fan = 88 dB

Attenuation of main duct = 0·3 dB/m

Attenuation of branch duct = 0·5 dB/m

Attenuation of bend in main duct = 6 dB

Attenuation of bend in branch duct = 3 dB

Room constant = 50 m^2

End reflection at grille = 4 dB

Step 1: Calculate attenuation provided by system

Main duct, 0·3 × 5	1·5 dB
Bend, main duct	6 dB
Branch (50% of total airflow)	3 dB
Branch duct, 0·5 × 3	1·5 dB
Bend, branch duct	3 dB
End reflection	4 dB
Total attenuation of system	19 dB

Therefore sound power level entering room

= 88 − 19
= 69 dB

Step 2: Calculate sound pressure level in room
Proceeding exactly as in Example 8.3,

$$\text{SPL} = L_W + 10 \log_{10} \left(\frac{Q}{4\pi r^2} + \frac{4}{R_c} \right)$$

Where $Q = 8$ (Table 8.1)

$L_W = 69$ dB

$r = 4$ m

$R_c = 50$ m^2

This gives SPL = 59·8 dB

or 60 dB to the nearest decibel.

Step 3: Compare with target and estimate extra attenuation required
From Fig. 6.2, the 250 Hz octave band level for NC 35 should be below 55 dB (approximately).

Therefore extra attenuation required = 60 − 55
= 5 dB

A similar calculation should be carried out for each octave band in turn, based on appropriate attenuation values for each band. The above example is of a very simple situation and is intended to illustrate the calculation method. A more accurate procedure also calculates the extra flow noise generated at bends, grilles and other obstacles. More specialized texts should be consulted for further details, and for data on attenuation by bends, straight duct runs and end reflections.

8.26 Silencers

There are two different types: dissipative (or absorptive) silencers and reactive silencers.

In absorptive silencers acoustic energy is converted to heat by the sound-absorbing processes which take place in the small interconnected air passages of fibrous or open-celled foam plastic materials. They are used to provide attenuation, of fan noise for example, over a broad band of frequencies. Because of the frequency characteristics of the absorbing materials they employ, this type of silencer is much more effective at medium and high than at low frequencies.

Reactive silencers work by providing an impedance mismatch to the sound waves, causing reflection back towards the source, and by using destructive interference to 'tune out' particular frequencies. The attenuation produced depends on the dimensions of the pipes and chambers of the silencer. They are usually used to reduce noise from pulsating gas flows such as the air inlet and exhaust systems of internal combustion engines. Reactive silencers can be very effective at reducing the amplitude of pure tones of fixed frequency, particularly if these are at low frequencies, where the absorptive type of silencer is ineffective. However, there can also be frequencies at which they allow sound to be transmitted with very little attenuation.

Many silencers use both reactive and absorptive mechanisms to achieve their effect.

Absorptive silencers

The simplest type of absorptive silencer is a duct with walls lined with sound-absorbing material. The attenuation produced, in dB per metre run of duct, depends on α, the sound absoprtion coefficient of the lining material, and the ratio of perimeter to cross-sectional area of the duct (P/S — see Chapter 5 and Example 5.8). In order to achieve maximum attenuation the shape of the duct should be designed to give the highest possible P/S ratio, which in effect means that for a given cross-sectional area of duct the sound is exposed to the greatest possible surface area of sound-absorbing material. It follows that the optimum shape for the cross-section of a rectangular duct is a long thin one. In commercial splitter silencers a rectangular duct is split into several such sections, each lined with sound-absorbing material (see Fig. 8.17).

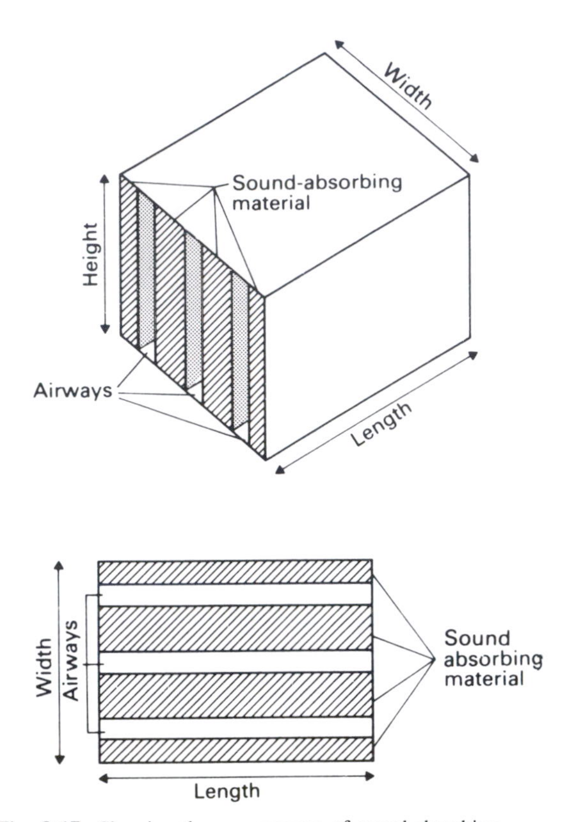

Fig. 8.17 Showing the arrangement of sound-absorbing materials in a splitter silencer with three airways

The attenuation produced by such a narrow lined section of duct is at greatest at medium frequencies, dropping off at the high and low frequency ends of the spectrum as shown in Fig. 8.18. The poor low frequency performance is simply because the sound absorption coefficient of the lining material is poor at low frequencies. The poor high frequency performance occurs because much of the high frequency sound energy tends to be 'beamed' down the centre of the duct airways, and is unaffected by the sound-absorbent duct linings. This occurs at frequencies where the sound wavelength is comparable with or smaller than the airway width. Therefore it can be seen that reducing the airway width will improve the silencer performance considerably (Fig. 8.18), particularly at high frequencies. The thickness of the sound-absorbing lining is also an important factor in determining the silencer performance; an increase in thickness producing improved attenuation, particularly at low frequencies (see Fig. 3.14).

Another very important factor which must be considered is the extra resistance to the flow of air which the silencer provides, which can be measured as a pressure drop across the silencer. Reducing the airway width too much will obviously increase this resistance to an unacceptable limit. Excessive restriction of airflow will also have an effect on

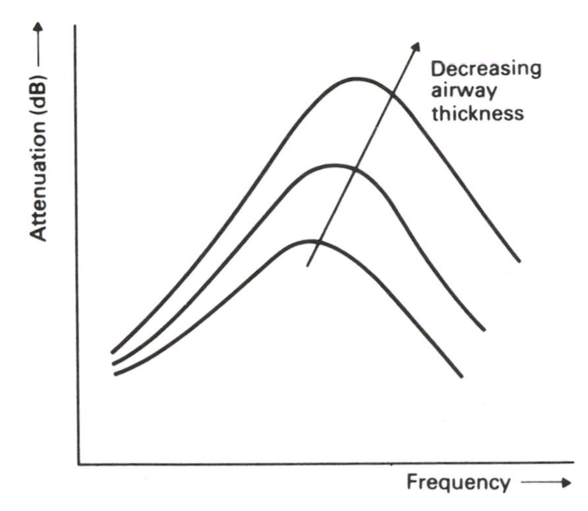

Fig. 8.18 Showing typical attenuation versus frequency characteristics of a splitter silencer section, illustrating the effect of changing airway thickness

another important silencer parameter, the noise generated by the flow of air through the silencer. Forcing the air through narrow airways will obviously cause an increase in flow velocity, and therefore in the amount of this self-generated noise. Thus there is a conflict between the requirements of good high frequency sound attenuation (i.e. narrow airways) and minimum flow resistance and silencer self-generated noise (requiring broad airways). Other factors which can affect the self-generated noise are changes of cross-section occurring within and at the ends of the silencer. It is also important that the sound-absorbent linings are kept as smooth as possible.

One way of obtaining the benefits of narrow airways without incurring the penalties of high flow resistance is to increase the number of airways, which increases the overall width of the silencer. Increasing the height has a similar effect. The alternative is to use a larger airway width, and increase the length of the silencer.

The acoustic performance of an absorptive silencer depends on four factors: sound absorption coefficient of the duct lining material, thickness of the absorbing lining, width of airway and length of the silencer. Silencer design is a compromise between acoustic performance, silencer size (height, width and length), acceptable flow resistance and material costs.

The position of the silencer in the duct system can be very important in determining its effectiveness in reducing the noise at the reception point. The optimum position can be governed by the possibility of noise breaking into the duct after (downstream as far as noise is concerned) the silencer, e.g. from noisy plant-rooms, or by possible break-out of noise from the duct before (upstream of) the silencer. The performance will depend on the flow conditions into the silencer section. Sound entering the silencer at random incidence, e.g. when the silencer is fitted close to a noise source such as a fan, will be attenuated more than sound flowing parallel to the duct walls. Positioning the silencer close to bends can cause increases in pressure drop and self-generated noise.

Other practical considerations are that the sound-absorbing materials used in the silencer should be able to withstand any adverse environmental conditions to which they might be subjected, such as oil or grease, mechanical erosion, high temperatures, and attack by insects, bacteria, fungi and vermin. The sound-absorbing materials may be covered with protective layers; for example by thin perforated sheet metal to protect against wear and tear, or by thin plastic films to protect against moisture or for reasons of hygiene.

Reactive silencers

A sudden increase in the cross-section of a pipe or duct produces a change in acoustic impedance which causes sound energy to be reflected back towards the noise source. The amount of attenuation produced depends on the ratio of the cross-sectional areas of the expanded and original sections of pipe. The expansion has the effect of a high pass acoustic filter, providing most attenuation at low frequencies when the wavelength of sound is much greater than the duct or pipe dimensions. A more gradual change of section as in the conical connector (Fig. 8.19(a)) works on the same principle. The attenuation produced is slightly less than for abrupt changes of section, but the gradual transformation allows a smoother airflow to be maintained. This type of device can be used to introduce a splitter silencer into a duct of different cross-sectional area.

A simple expansion chamber, as shown in Fig. 8.19(b), is the most basic form of reactive silencer. The attenuation it produces is a maximum for the frequency at which the length of the chamber is a quarter of the wavelength of sound in the gas in the chamber. At this frequency the chamber presents the greatest acoustic impedance mismatch to the sound in the inlet tube. The attenuation drops off away from this frequency and is a minimum when the chamber length equals half a wavelength. At this frequency there is resonant standing wave amplification of the sound waves in the chamber. Higher up the frequency scale there are other maxima (where chamber length is an odd number of quarter wavelengths) and minima (chamber length equal to an even number of quarter wavelengths) in the attenuation. The maximum attenuation at the quarter-wavelength condition depends on the expansion ratio.

The situation is complicated by the presence of the tailpipe. Resonances and antiresonances associated with the length of the tailpipe, and its relationship to the wavelength of the sound, produce more peaks and troughs in the attenuation spectrum of the silencer. Further expansion chambers can be used to provide more attenuation (Fig. 8.19(c)). These may be of the same length, or of different lengths in an attempt to fill in gaps in the attenuation provided by the first chamber. A further complication is provided by the presence of interconnecting pipes between chambers. There are yet more peaks and troughs in the silencer attenuation spectrum associated with the presence of these interconnecting pipes. The complete spectrum of the attenuation produced by a device such as that illustrated in Fig. 8.19(c) will therefore be very complicated, necessitating computer techniques for its prediction. It will depend on the lengths of the different expansion chambers and of the tail and interconnecting pipes and on the expansion ratio.

A reactive silencer of a different type can be designed to give high attenuation at a particular frequency. This type of silencer employs a resonant system which extracts sound energy from the duct at the resonance frequency in similar fashion to the Helmholtz or cavity resonator of Figs. 3.16 and 3.17. The resonant system may be a Helmholtz resonator in which a series of holes in the pipe wall communciate with a cavity surrounding the pipe. The resonator can be considered as a mass–spring system, in which the air in the holes acts as the mass and the volume of air in the cavity acts as the spring. Alternatively, a tuned length of closed-ended pipe acts as the resonant system (Fig. 8.19(d)). The advantage of these 'side-branch' types of resonator is that the silencer does not produce high flow resistance and back

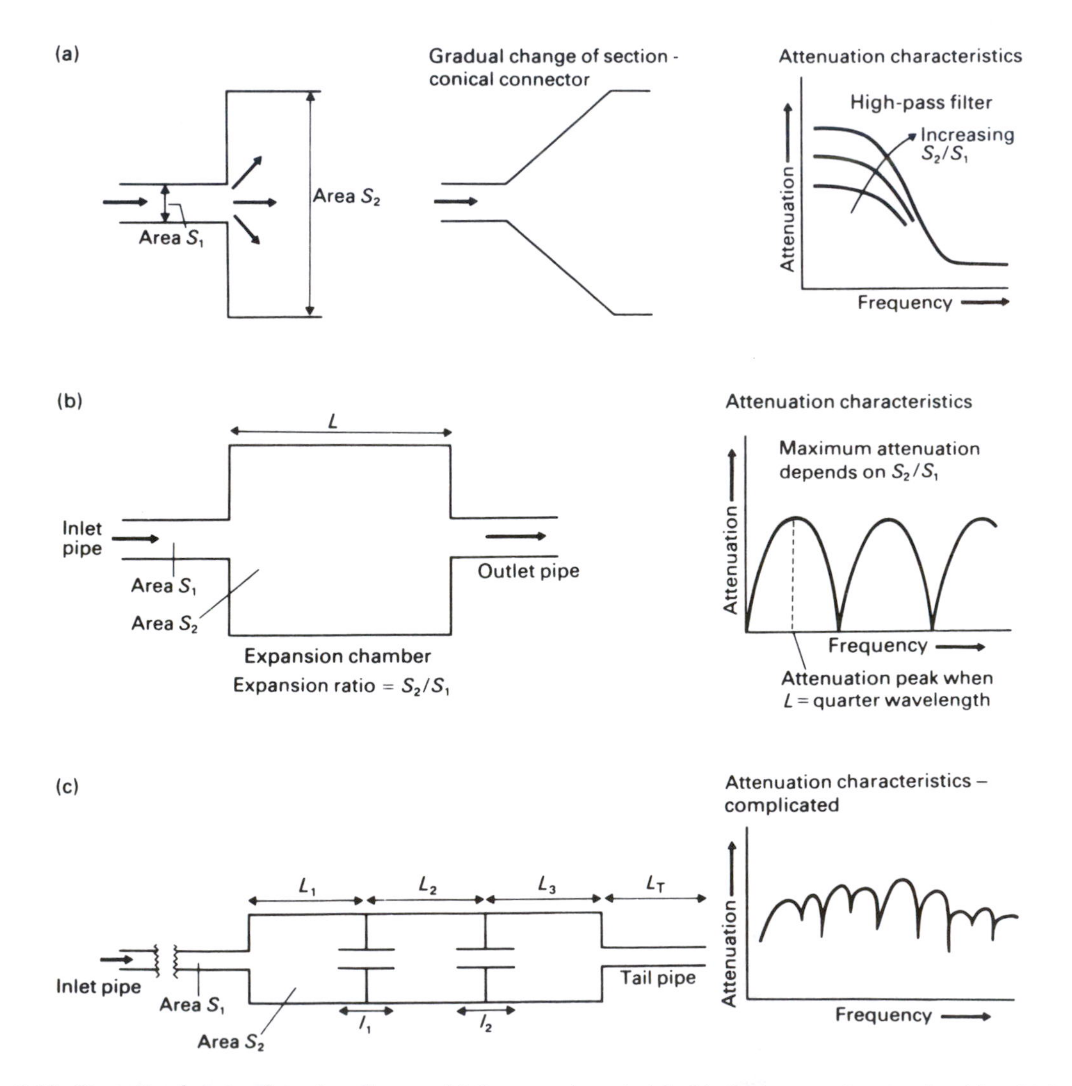

Fig. 8.19 Illustrating features of reactive silencers: (a) the expansion principle (b) simple expansion chamber (c) multiple expansion chamber with tail and interconnecting pipes; dimensions relevant to silencer performance are indicated

pressures which is sometimes the case with other types of reactive silencer. Another type of resonant device, this time 'in-line' with the main gas flow, is also shown in Fig. 8.19(d). This can also be compared to a Helmholtz resonator in which the air in the interconnecting pipe acts as the mass and the air in the two sections of the chamber acts as the spring. The resonant frequency depends on the length and cross-sectional area of the interconnecting pipe, and on the total volume of the chamber.

A plenum chamber in ventilating systems is an example of a device which incorporates dissipative and reactive sound-reducing mechanisms (Fig. 8.19(e)). Part of the attenuation, mainly at low frequencies, is produced by the expansion chamber principle, but the overall performance is greatly improved if the inside of the chamber is lined with sound-absorbing material and if baffles are used in the chamber to prevent sound travelling directly without reflection from the inlet to the outlet of the chamber.

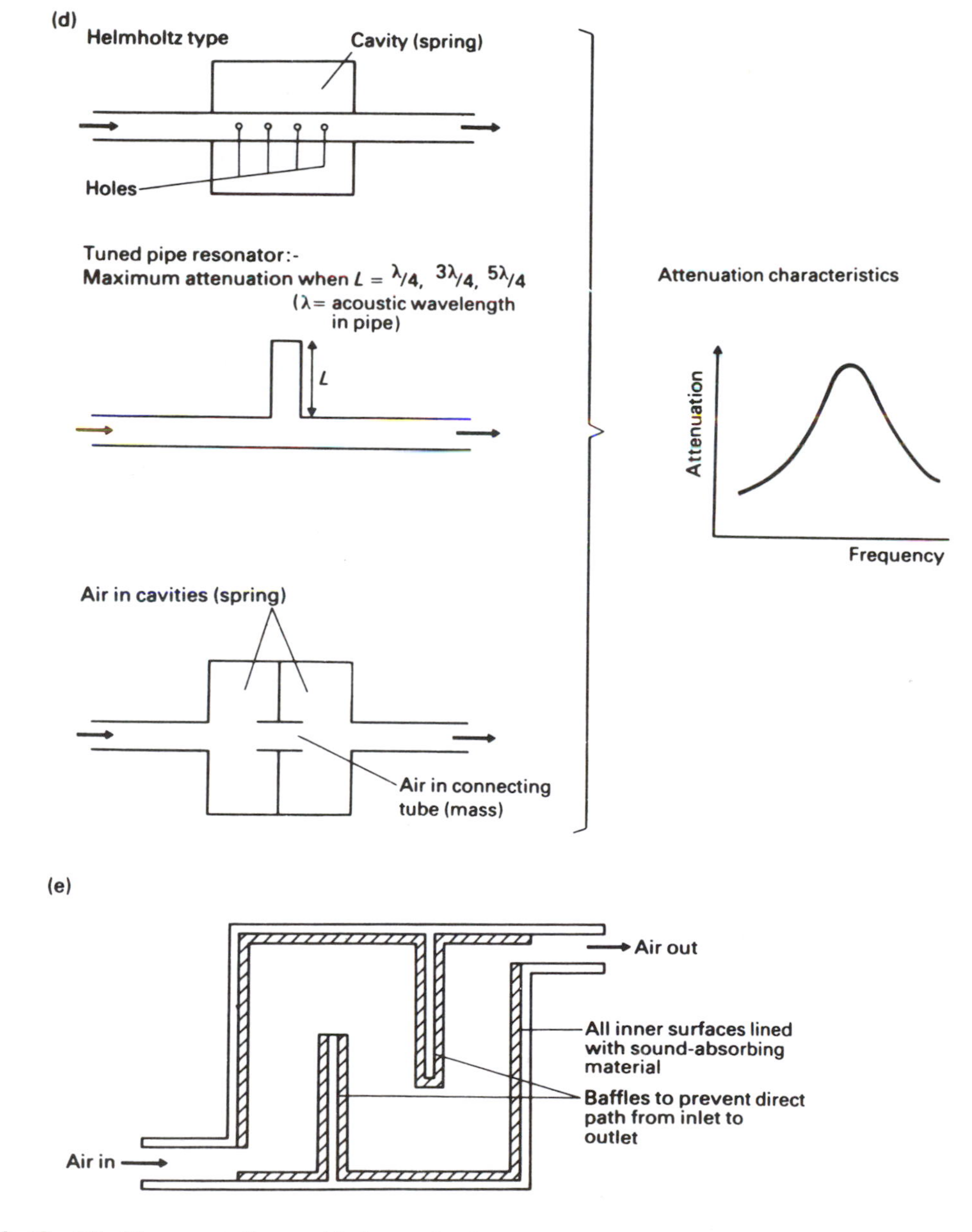

Fig. 8.19 (Cont'd) (d) resonant silencers (e) plenum chamber

8.27 Active noise control

Active noise-control systems apply the principle of destructive interference between waves to reduce noise. They use one or more loudspeakers to radiate sound which is equal in amplitude but opposite in phase to the noise waveform. The term **active** refers to the use of a source of energy, from the loudspeaker, in the noise reduction process, as compared to the so-called passive noise reduction measures, such as splitter silencers which rely on sound-absorbing materials to convert sound energy to heat.

Although the idea of cancelling the noise with antiphase signals (colloquially called antinoise) is simple in principle, it is difficult to carry out in practice. If significant noise reduction is to be obtained, the cancellation of the primary waveform (the noise) by the secondary waveform, produced by the loudspeaker, must be achieved to a high degree of accuracy. A 20 dB noise reduction, for example, requires a reduction of sound intensity to one-hundredth of its original value; this gives an indication of the degree of exactness with which the inverted secondary waveform must match the primary waveform it is designed to cancel.

The first active noise-control system was patented by Lueg in the 1930s. Practical difficulties meant that progress was slow until the 1970s and 1980s, when advances in electronics and signal processing techniques accelerated development.

When explaining the principles and methods involved in active noise control, it is convenient to refer to the cancellation of noise in a duct, a situation which has been the subject of much research. If the sound wavelength is more than twice the maximum duct cross-sectional dimension, the situation is simplified to a one-dimensional problem because only plane waves will travel along the duct. The elements of a simple active noise-control system are shown in Fig. 8.20. A microphone downstream of the noise source, perhaps a fan, is connected via the signal processor to a loudspeaker, called the secondary source, which is a further distance *L* downstream.

The condition for complete cancellation of noise to occur is that the secondary source must produce a waveform that is equal in amplitude but opposite in phase to the waveform received by the microphone a time L/c earlier, to allow for the travel time of sound upstream from microphone to loudspeaker. This assumes that there are no absorption losses in the duct, and it means the signal detected at the microphone must be delayed by an amount corresponding to the acoustic transit time minus the time taken to process the signal.

If the primary signal were a pure tone, in principle, cancellation could be achieved without any delay by fixing the distance between the microphone and loudspeaker to be half a wavelength. Even such a simple system would be sensitive to changes in temperature, which would affect the sound speed, and therefore the wavelength, or to changes in the phase response of the electronics and the transducers.

Another difficulty arises because the loudspeaker will radiate downstream towards the microphone and the fan as well as upstream towards the end of the duct. Therefore feedback from the loudspeaker will affect the signal at the microphone which is used to provide the cancellation waveform. Airflow in the duct will cause additional problems, especially if variable; it will affect the speed of sound in the duct and will possibly generate noise from turbulence.

It should by now be apparent that active noise control presents difficult signal processing problems, which will be further increased if the noise from the primary source is more complex, i.e. if it covers a wide frequency range and varies with time. Early attempts, in the 1970s and 1980s, at solving these problems used various configurations and placings of one or more loudspeakers and microphones in combination with various fixed signal processing systems. More recent research has successfully applied the techniques of adaptive digital filtering to produce active noise-control systems which can, within limits, adapt themselves to changes in operating conditions, such as changes in temperature, flow rate and characteristics of electronic components and transducers. One such system

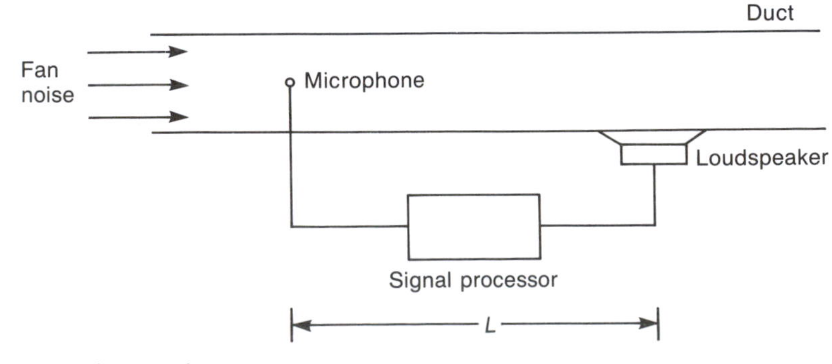

Fig. 8.20 Simple active noise-control system

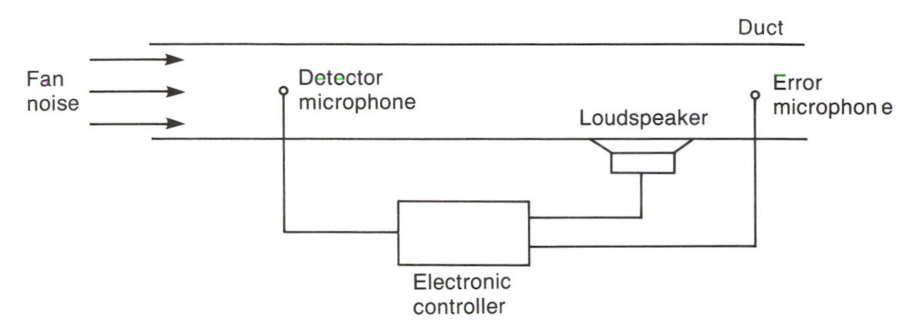

Fig. 8.21 Adaptive active noise-control system

(Fig. 8.21) involves the use of an error microphone downstream of the loudspeaker to detect imperfections in the cancelling process. The signal from the error microphone is fed back to the electronic controller, which adjusts the signal to the loudspeaker accordingly. A more detailed description of the operation of such systems is beyond the scope of this chapter, but more comprehensive treatments can be found in texts by Roberts and Fairhall, Beranek, and Nelson and Elliott (see Bibliography).

Active noise control has been successfully applied to a wide variety of noise sources, including industrial fans, air-conditioning systems in buildings, engines, generators, transformers, vehicle exhaust noise and noise from turbine exhaust stacks. The principle has been applied to the cancellation of noise in small enclosed spaces, such as aircraft cockpits and inside sports cars, as well as to hearing protectors and communication headsets. Noise reductions of up to 30 dB have been achieved at frequencies of up to 500 Hz. Some problems remain in developing loudspeaker and amplifier systems to provide the high levels of sound power that are sometimes required to cancel low frequency noise from some industrial sources, and to withstand long exposures to hostile industrial environments. Further developments in signal processing hardware and software will lead to improvements in the performance of future systems.

Active noise-control systems are most effective at low frequencies; the sound wavelengths are long, so the sound field changes only gradually with position. It is in the low frequency region that active noise-control methods are particularly advantageous, compared to passive systems, which are large and expensive at low frequencies. The use of active noise control can also lead to considerable savings in energy because the alternative passive silencers cause high flow resistance, which has to be overcome using increased fan power. Hybrid systems, which use active methods at low frequencies and passive methods at high frequencies, optimize the advantages of both methods of noise control.

Active vibration control Exactly the same principles and techniques used in active noise control can be applied to the reduction of vibration in machinery and structures. Vibration waveforms detected using accelerometers, or other vibration transducers, are fed in antiphase to one or more vibrators in an attempt to cancel the effect of the primary source of vibration.

8.28 Noise control at the receiver

As a last resort the solution to a noise-control problem may lie at this final link in the noise chain. There are a variety of possibilities, depending upon whether the problem is one of occupational or environmental noise exposure. These include:

1. Control of exposure time, i.e. time working on noisy processes.
2. Job rotation.
3. Provision of personal hearing protection (e.g. earmuffs or earplugs).
4. Provision of quiet working areas for time when not working on the noisiest processes.
5. Regular audiometric monitoring of the hearing levels of personnel.
6. Provision of double glazing and other extra noise insulation measures.
7. Provision of noise barriers, close to the receiver.
8. Provision of making noise (in offices, cafes, etc.).
9. Compensation
10. Relocation

Questions

(1) The sound pressure level close to a vibrating panel is 90 dB. The frequency of the panel vibration is 1000 Hz. Calculate the RMS displacement of the panel, assuming a radiation efficiency of 1·0.

(2) Calculate the sound pressure level expected close to

a panel vibrating at a frequency of 500 Hz. The RMS accleration of the panel is $0 \cdot 3$ m/s^2. Assume a radiation efficiency of $1 \cdot 0$.

(3) The sound pressure level from a noise source on hard reflecting ground outdoors is 82 dB at a distance of 3 m from the source, in a direction in which the directivity index is +3 dB. Calculate the sound power level of the source, and the sound pressure level at a distance of 15 m in a direction in which the directivity index is −2 dB.

(4) A canteen of dimensions 8 m × 4 m × 3·5 m high contains an extractor fan situated at ceiling height half way along one of the long walls. The sound power level of the fan is 48 dB in the 125 Hz octave band. Calculate the total sound pressure level in this band at a point 3 m from the fan. Also calculate the direct sound pressure level at this point. The reverberation time of the canteen has been measured and is 0·8 seconds in this octave band.

(5) An office is to be partitioned off from the rest of a busy typing pool, in which the noise levels in the 1000 Hz band can be as high as 76 dB. Calculate the sound reduction index required of the partition if the intrusive noise from the typing pool is to be kept below 40 dB in this band. The dimensions of the partition are to be 10 m × 4 m and it is estimated that the amount of absorption in the newly created office will be 25 m^2 in the 1000 Hz band. Explain the assumptions underlying your calculation.

(6) A diesel engine is situated in a plant room which contains one window of dimensions 3 m × 2 m. The walls and door of the plant room are massively constructed so that the window as 'weak link' is the major transmission path for noise to the outside. Calculate the sound pressure level in the 500 Hz octave band at a point 20 m from the window on a line perpendicular to the surface of the window, given the following information:

Measurements of the SPL from the machine, in the 500 Hz octave band, at 1 metre from the engine surface, at four points around the engine, under anechoic conditions, gave the following readings:

110 dB, 96 dB, 108 dB, 101 dB.

The dimensions of the plant room are 10 m × 6 m × 3 m high and the absorption coefficients of the inside surfaces, at 500 Hz are: walls $\alpha = 0 \cdot 02$, floor $\alpha = 0 \cdot 01$, ceiling α 0·4.

The sound reduction index of the window at 500 Hz is 30 dB.

(*Hint:* Use the four given measurements of the engine noise as the basis for calculating the sound power level of the engine (approximately), and assume that the sound inside the plant room will be completely reverberant.)

(7) Describe how the intensity at a given distance from a broad-band source is affected by the location of the source with respect to nearby surfaces.

An open window of dimensions 1 m by 2 m in the wall of a building has an average sound pressure level over its open area of 85 dB. Determine the effective sound power of the window opening acting as a noise source and the intensity level produced at distance of 100 m on the axis of the window. Discuss factors which might affect the accuracy of your calculation in the (a) 100 Hz, and (b) 1000 Hz octave band. (*IOA*)

(8) Describe the mechanism of sound transmission from a small source in free-field conditions. What is the effect of positioning such a source: (i) on the surface of a plane, (ii) in the edge enclosed by two planes, and (iii) in the corner where three planes intersect?

An electric buzzer generates a sound power level of 90 dB relative to 10^{-12} W. It is suspended in mid-air and radiates omnidirectionally. Calculate the sound pressure level at a distance of 3 m. What would the SPL be 3 m above the same buzzer laid on hard ground?

The same buzzer is now hung in a room of dimensions 5 m × 4 m × 3 m having an average surface absorption coefficient of 0·2. At what distance is the direct sound level equal to the reverberant sound level in this room and what is the sound pressure level at this distance from the buzzer? (*IOA*)

(9) Derive an expression for the sound pressure level at a distance r from a freely and uniformly radiating point source of known sound power level, situated at ground level. Indicate briefly why this expression may fail at small or large values of r for a real source under real propagation conditions. Define the terms 'directivity factor' and 'directivity index' of a source and show how directivity leads to a modification of the above mentioned expression.

The 500 Hz octave band directivity indices of a diesel engine at 90° and 50° to the longitudinal axis in the horizontal plane are known to be −5 dB and +7 dB respectively. The band SPL at 30 m at 90° is 85 dB. Calculate the band SPL at 50° at a distance of 90 m. (*IOA*)

(10) (a) Compare and contrast the use of sound power levels and sound pressure levels as means of specifying the noise output from machines.

(b) State four different methods for determining the sound power levels of machines and compare any two in terms of accuracy and equipment and facilities needed to carry out the tests.

(c) The sound pressure level at a distance of 2 m from a pump in an anechoic room is 82 dB. Calculate the level at the same distance from the pump when it is located on the floor in the corner of a plant room of dimensions 10 m × 6 m × 4 m. The average absorption coefficient of the surface of the plant room is 0·05. Assume that the pump behaves as a point source. *(IOA)*

(11) The following equation is often used to estimate the insertion loss of an enclosure used in a reverberant sound field

$IL = R - 10 \log(1/\alpha)$ dB

where R is the sound reduction index of the enclosure and α is the average sound absorption coefficient of the inside surfaces.

(a) Explain why the average absorption, α, is important in determining the transmission performance of the enclosure.

(b) Explain why this equation would be inaccurate if
  (i) the levels were measured close to the enclosure sides.
  (ii) the enclosure were placed near to a reflecting wall.

(c) An air compressor operates in a factory and generates a reverberant sound pressure level of 105 dB in the 1 kHz octave band.

As a noise control measure, it is planned to cover the compressor with an enclosure of 1·5 m × 2·5 m × 1·5 m high.

The enclosure is to be lined with 50 mm thick mineral fibre absorber having an absorption coefficient of 0·8 at 1 kHz. The floor of the enclosure is concrete having an absorption coefficient of 0·04 at 1 kHz.

Calculate the sound reduction index of the enclosure necessary to reduce the reverberant level to below 85 dB. *(IOA)*

(12) With reference to duct-borne noise in ventilation systems, discuss the contribution to sound attenuation made by

(i) Rectangular straight duct (sheet metal);
(ii) Circular straight duct (sheet metal);
(iii) Rectangular straight duct, internally lagged.
(iv) Elbows;
(v) Elbow with guide vanes; and
(vi) Elbow internally lagged.

Figure 8.22 shows an existing air supply system which incorporates a sound attenuator, but which nevertheless transmits too much noise to the rooms. Briefly suggest modifications to this low-velocity ventilation system which would lead to less plant room noise being transmitted to the rooms. *(IOA)*

(13) Discuss briefly the general principles relating the vibration of a surface and the associated radiation of sound. Explain the effects of variations of parameters such as vibration velocity of surface, the area of the surface and the radiation efficiency of the surface. If the vibration were in phase across the whole surface, such as in the case of a piston loudspeaker, what factors would determine the frequency of the transition from low to high radiation efficiency?

Briefly describe how the radiation efficiency of a vibrating surface varies with frequency.

A steel panel (5 m × 2 m) situated in the side of a ship (beneath the water-line) is found to be in heavy vibration in the 125 Hz 1/3 octave band due to structural excitation from a nearby pump. The spatially averaged 1/3 octave acceleration level is measured as 100 dB re 10^{-5} m s^{-2}, and the radiation efficiency of similar panels at this frequency has previously been found to be of the order of 0·003. Estimate the sound power radiation to the water (in W), and hence the expected underwater sound

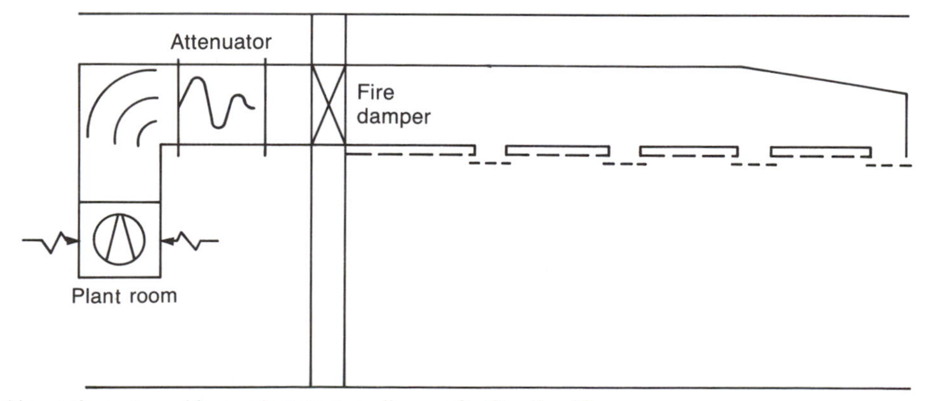

Fig. 8.22 Air supply system with sound attenuator: diagram for Question 12

pressure level at a distance of 50 m from the panel and express the result in terms of dB re 1 μPa.

$(\rho c)_{\text{water}} = 1050 \times 1500$ kg m^{-4} s^{-1}

(*IOA*)

(14) The noise from a new machine is specified in terms of L_W. Calculate L_W from measurements made at 500 Hz as follows:

(a) In a free field: with the machine placed on a reflective floor in an otherwise anechoic chamber, when an average L_p of 92 dB is measured over an enveloping hemisphere of 4 m radius.

(b) In a highly reverberant field: i.e. in a chamber which has a reverberation time of 6 seconds and a volume of 950 m^3, where the average L_p was measured as 104 dB.

Explain why one of the above measurement methods will give additional information about the noise source.

Describe briefly the principle of sound intensity measurement, and outline the practical equipment needed. Indicate how such measurements can be applied to the determination of L_W. Explain why this method can be carried out in a practical working environment, without the need for special acoustic chambers. (*IOA*)

(15) (i) Discuss the factors which affect the sound reduction index of a single-leaf partition, including how it varies with frequency. Explain the principles underlying the performance of double-skin partitions and the design measures necessary to achieve the maximum possible sound insulation.

(ii) Sound transmission through building façades can be predicted by:

$$L_2 = L_1 - R - 6$$

where L_2 = sound pressure level just outside the building,
L_1 = level inside
R = sound reduction index of the building façade.

Explain the basis of this formula and state any assumptions involved.

(iii) The sound pressure level inside a building is 98 dB in a certain octave band. Calculate the noise level in that octave band radiated from the façade of the building at a point which is 100 m from the centre of the façade in a direction at right angles to the façade.

The dimensions of the building façade are 5 m $\times$ 2·5 m and its sound reduction index in the required octave band is 25 dB. The ground between the façade and the reception point is soft and sound absorbing. Explain the steps in your calculation. (*IOA*)

(16) The occupants of a first-floor flat are complaining of the noise caused by the ventilation and extract system from the restaurant below. The restaurant kitchen is a flat-roofed extension built on to the restaurant. The extract fan unit is situated on the flat roof with the extract ductwork running across the roof and up the external wall of the flat (see Fig. 8.23). There is a window in the flat which overlooks the flat roof and the fan unit, and the ductwork passes close to this window and terminates at roof level.

Describe the various paths by which noise caused by the extract system arrives in the flat, and suggest possible noise-control remedies in each case.

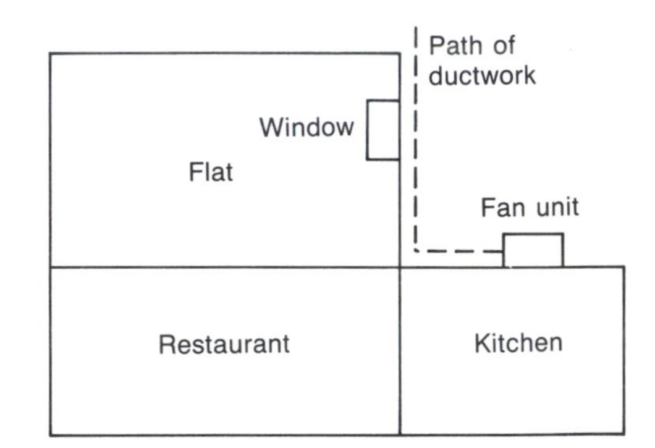

Fig. 8.23 Plan of restaurant and flat: diagram for Question 16

(17) One of the most frequent mistakes made by those who do not fully understand the principles of noise control is to confuse the use of *sound absorbing* and *sound insulating* materials. Carefully explain the difference between sound insulation and sound absorption.

(18) The manager of a factory which contains a large number of different sorts of machinery is aware that there are noise problems in the factory. He receives advertising literature from a firm producing sound-absorbing materials, showing examples of their use in reducing factory noise. He asks you to assess and advise on the suitability of this form of noise control for his particular factory. Describe how you would carry out the assessment. Say what information you would need to obtain, what measurements you would make, what calculations, if any, you would perform. What alternative noise control treatments might you recommend?

(19) A large area of disused gravel pits and surrounding waste ground is to be landscaped and developed as a leisure park with provision for car parking, bathing, picnicking, fishing, sailing, water-skiing and a nature reserve. One of the boundaries of the site is with the existing gravel extraction works and there are houses at various distances around the edge of the site. There is also the possibility of model aircraft, go-kart and clay pigeon shooting clubs being allocated the use of certain areas. Outline in general terms suggestions you would make, and conditions you would impose on the developers in order to minimize the noise nuisance to residents in the area.

(20) (a) Discuss the possible methods for reducing the noise from a vibrating machine panel. Explain why stiffening such panels is not always effective.

(b) What are the main mechanisms of noise generation by impacts. How may impact noise be reduced?

(c) Explain how the noise-control principles of absorption, insulation and isolation are demonstrated in the design of an effective acoustic enclosure. Discuss the advantages and disadvantages of using acoustic enclosures to reduce machinery noise.

(d) Distinguish between the techniques of sound absorption, sound insulation and vibration isolation used in noise control, with reference to the design of earmuffs, acoustic enclosures and double-leaf partitions.

(21) A large piece of rotating machinery is located on the first floor of a factory converted from an old mill, and is producing severe vibration throughout the building. Antivibration mounts are selected for the machine using an elementary design procedure which is based on a simple mass–spring model of isolation. Some time after the isolators have been fitted, vibration measurements show that the degree of isolation achieved is far less than expected.
Discuss possible reasons for this discrepancy (you may wish to refer both to practical points and to possible deficiencies in the model).

(22) A barrier is proposed to protect nearby houses from the noise from a regional airport, with some international airline traffic. Comment, in general terms, on the necessary size and likely effectiveness of such a barrier.

(23) Give arguments both for and against the use of enclosures as a solution to noise-control problems.

(24) A noise consultant giving a brief lecture to some engineers on the principles of noise reduction in industry uses the slogan: Slow and smooth. What does it mean?

(25) (a) Explain precisely what is meant by the damping of a flexible structure, distinguishing it clearly

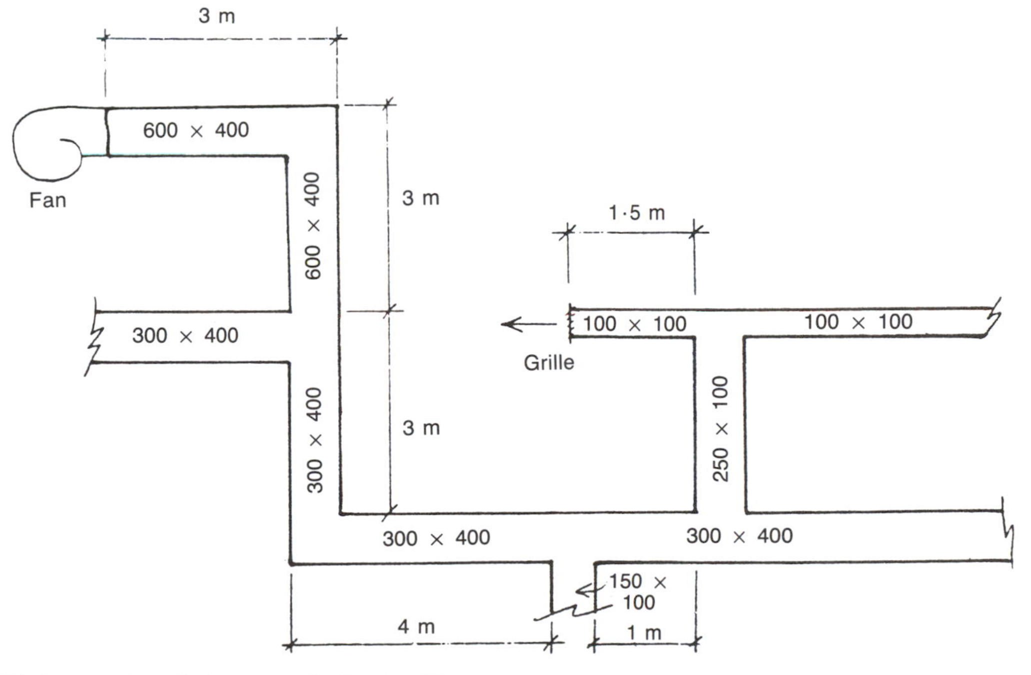

Fig. 8.24 Layout of ventilation system for Question 26

from its stiffness. Mention some of the mechanisms of damping which commonly contribute to damping of a structure.

(b) Under what conditions can the application of damping treatment to a structure, which is subject to continuous vibratory or impact forces, constitute an effective means of noise and vibration control?

(c) Describe the various forms of damping treatment available to the noise-control engineering, and discuss their practical limitations. (*IOA*)

(26) Work by Curle established the following relationship for sound power (in watts) generated aerodynamically at obstructions:

$$W \propto \rho v^6 c^{-3} L^2 f(Re)$$

where v = the air velocity
c = the velocity of sound
ρ = the air density
L = the characteristic dimension of the obstruction
$f(Re)$ = some function of the Reynolds Number.

What are the implications of this relationship in the design of air distribution systems?

The acoustic index run of a ventilation system is shown in Fig. 8.24. Estimate the sound power level spectrum emitted at the diffuser if the supply fan has the characteristics given. Include an allowance for tonal quality. Fan data and duct attenuation data are provided. (*IOA*)

Fan data

Centrifugal
Forward-curved blade impeller
No. of blades = 20
Speed = 3200 rev/min
Delivery volume = 4 m³/s
Fan static pressure = 1 kPa
L_W (fan) $= K + 10 \log_{10} Q + 20 \log_{10} P$

where Q is the fan volume flow rate (m³/s)
P is the fan static pressure (Pa)

Duct attenuation per metre run (dB/m) for a range of frequencies

Ducting dimensions (mm)	Frequency (Hz)						
	125	250	500	1000	2000	4000	8000
600 × 400	0·3	0·3	0·2	0·2	0·2	0·2	0·2
300 × 400	0·6	0·3	0·3	0·3	0·3	0·3	0·3
250 × 100	0·6	0·5	0·3	0·3	0·3	0·3	0·3
100 × 100	0·6	0·5	0·3	0·3	0·3	0·3	0·3

Elbow attenuation (dB) and end reflection loss (dB)

	Frequency (Hz)						
	125	250	500	1000	2000	4000	8000
Elbow attenuation							
600 mm × 400 mm	0	3	7	5	4	4	4
300 mm × 400 mm	0	1	6	7	4	4	4
End reflection loss	13	9	5	1·5	0	0	0

9 Instrumentation

9.1 Introduction

Accurate sound measurements depend upon the correct choice of the most suitable equipment for the particular measurement situation and upon the correct use of this equipment by the operator. This chapter describes some of the commonly used noise measuring instruments and explains the terms used to describe the performance and limitations of the equipment.

Any piece of equipment, whatever it measures, can be thought of as consisting of three parts (Fig. 9.1). First there is the transducer. This is the 'sensor' which converts the changes in the physical property to be measured into an electrical signal. A microphone is a sound pressure transducer, converting sound pressure signals into electrical signals. This electrical signal is then subjected to a variety of processes which condition the signal in the required manner. In sound level meters, for example, the various signal processes can include amplification, filtering, averaging and logging (i.e. producing a logarithmic value of the signal). The processed signal is then fed to the final stage of the instrument, where it is displayed as a deflection of a needle over the scale of a meter, or as a digital display. The displayed signal value is read and recorded manually. Alternatively the signal may be recorded automatically on a chart or a printer, or it may be delivered to another piece of equipment: a loudspeaker, a computer, an oscilloscope, even an alarm system such as a bell.

The correct choice of equipment can depend upon a number of factors including:

1. The type of noise and the way in which it is to be measured. Is it a steady noise for which the appropriate unit depends on an analysis of the frequency content of the noise? Is it a time-varying noise, such as traffic noise, which requires a statistical analysis over a period of time? Is it an impulsive noise or does it have pure tone components?
2. The accuracy required. Is it a precision measurement, involving controlled and standardized measurement conditions, or is it a preliminary survey under field conditions?
3. The conditions under which the instrument will be operating. Is it a quick field measurement, requiring a portable and battery-operated instrument, or is it a long-term survey, requiring permanent semi-automatic mains-operated equipment?

There is a very wide variety of equipment in use and, because of modern developments in electronics, particularly the introduction of digital techniques, rapid changes have recently occurred and are still taking place in sound measuring instrumentation. One such development, only briefly mentioned in this chapter, is the direct measurement of sound intensity.

Even the simplest item of equipment, the portable sound level meter, occurs in a bewildering variety of versions, and although they all have certain basic features in common, they differ in the detail of their performance, specification and method of use. All manufacturers provide a detailed handbook with their instruments, giving all this information, and a golden rule for anyone making sound measurements should be:

Always carefully read the instrument manufacturer's handbook before making measurements with an un-

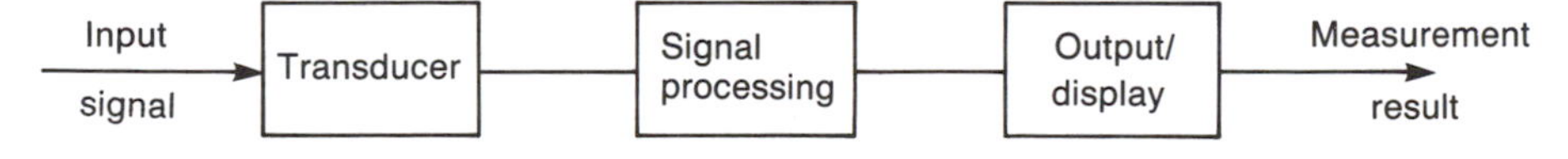

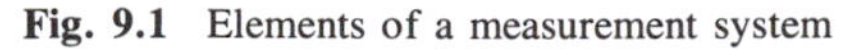
Fig. 9.1 Elements of a measurement system

familiar piece of equipment and keep it safely and readily available for reference thereafter. Always operate equipment in accordance with the manufacturer's instructions.

A section on basic electrical principles has been included in Appendix 5. It is intended for the benefit of readers with little or no background knowledge in this area and may help with comprehension of terms such as attenuator, impedance, capacitance, etc.

9.2 Sound pressure magnitudes: peak and RMS values

When the sound is a simple pure tone its magnitude may be represented by its amplitude, which in this case is also the same as its peak value. The average value of sound pressure over any period of more than a few cycles will be zero because the positive compression half-cycles will cancel the negative sound pressure half-cycles. For more complex waveforms, such as harmonic, transient or random noise, the expression of magnitude is not as simple, but the time-averaged value is still zero.

A commonly used expression of magnitude is the RMS value of the sound pressure, i.e. the square root of the mean (or average) of the square of the pressure. This gives a non-zero average, which is related to the average value of the square of the pressure, and so to the average sound intensity, and the energy content of the sound.

Other expressions of magnitude, which are particularly useful for impulsive sounds, are the peak, and peak-to-peak values.

For a sine wave signal there is a simple relationship between the peak and the RMS values:

$$\text{peak} = \sqrt{2} \times \text{RMS sound pressure}$$

There is no such simple relationship for more complex waveforms. The ratio of the peak to the RMS value is called the crest factor of the signal. For a sine wave the crest factor is therefore $\sqrt{2}$ or $1{\cdot}4$.

Peak and RMS values of signals are illustrated in Fig. 9.2.

9.3 The sound level meter

Figure 9.3 shows a simplified block diagram of a typical analogue sound level meter (i.e. with a meter and needle

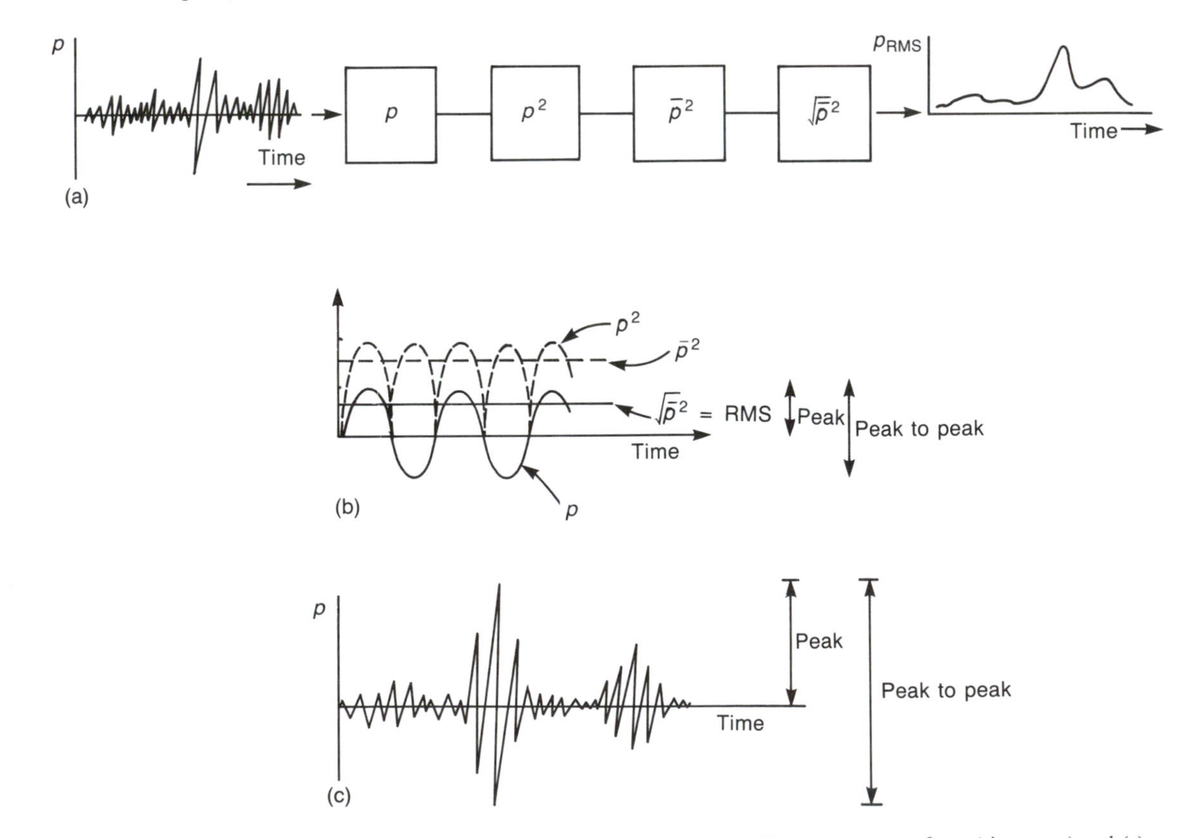

Fig. 9.2 Sound pressure signals, peak and RMS values: (a) processing stages (b) pure tone waveform (sine wave) and (c) a complex waveform

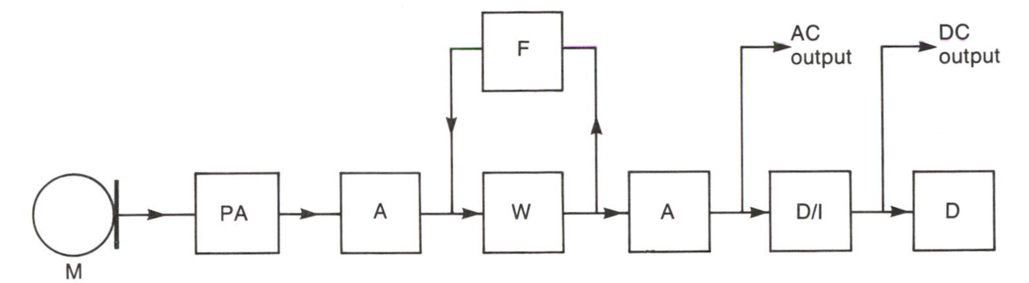

Fig. 9.3 Block diagram of an analogue sound level meter: (M) microphone (PA) preamplifier (A) amplifier and attenuator (D/I) detector/indicator circuits (D) meter or display (F) octave or one-third octave filter

display as opposed to a digital display) suitable for measuring fairly steady, non-impulsive noises. Equipment for measuring impulsive and time-varying noise will be considered later, as will digital equipment.

Microphone

The microphone faithfully converts the sound pressure signal into a corresponding electrical signal. If, for example, the sound is a 1000 Hz pure tone then the microphone produces a voltage which varies sinusoidally with a frequency of 1000 Hz. The amplitude of the voltage signal is proportional to the sound pressure amplitude. The microphone is the most important, and usually the most expensive, part of the whole instrument because any lack of faithfulness at this stage of the conversion can never be remedied by any subsequent electronics. The microphone is required, ideally, to match the performance of the human ear in the range of sound pressures and frequencies to which it can respond.

Amplifiers

The electrical signals produced by the microphone are very very small, often only a few microvolts (millionths of a volt). They have to be amplified several thousands of times before they can be processed by detector and indicator circuits then displayed on the meter. Amplification takes place in several stages but Fig. 9.3 shows it as two separate blocks, before and after the frequency weighting networks. The first amplification stage, the preamplifier, is particularly important. The microphone has an extremely high output impedance, which means it can deliver only very small currents, and its output falls when a load is applied. The preamplifier is designed to have an equally high input impedance so that it does not load the microphone significantly. In effect, the preamplifier acts as a buffer between the microphone and the rest of the equipment. Even the presence of a cable connecting the microphone to the preamplifier would load the microphone to an unacceptable degree, so the preamplifier is built into the microphone cartridge. The function of the preamplifier is not necessarily to magnify the signal — this is accomplished by the following stages — but to deliver a signal at a much lower output impedance, suitable for processing by the remaining circuitry in the sound level meter. The amount of amplification of the signal has to be adjusted according to the level of the signal produced by the microphone. This is achieved using a bank of signal attenuators which are controlled using the level range control switch of the instrument. In some instruments there is one range control for ease of use; others have two independently adjustable range controls which operate attenuators situated before and after the frequency weighting circuits. This arrangement minimizes the chance of the signal overloading the amplifiers and permits the widest dynamic range at high sound pressure levels.

Frequency weighting networks

Human hearing is not equally sensitive to all frequencies in the audible range. The weighting networks were originally designed to modify or weight the signal according to its frequency content in a way which attempts to simulate the human hearing frequency response. In this way the final decibel figure indicated on the meter should relate more closely to the average human response to noise than a simple indication of its RMS sound pressure and energy content. The weighting networks are based on the shape of equal-loudness contours derived from research investigations into the way in which the sensation produced by a sound, called its loudness, depends upon frequency. Loudness varies with the level or intensity of sound as well as its frequency, so three weightings A, B and C were originally proposed for different levels of sound. The B and C scales are now rarely used, and many sound level meters will only have the A scale. Nowadays there are more accurate, but more complex, methods, so the networks are no longer regarded as giving an accurate indication of loudness. But they have gained widespread acceptance, the A scale in particular, as the most convenient way of measuring almost all types of noise. The values of the weightings are fixed by international standards and are given in Chapter 1. An additional network, the D weighting, is based on human response to aircraft noise; it is intended for use in aircraft noise measurement. The A weighting is highest at low frequencies, i.e. it discriminates most against low frequencies, especially frequencies below 500 Hz; this is where human hearing is least sensitive. The A weighting is lowest in the range 1000–4000 Hz, where our hearing is most sensitive, and it starts to increase at higher frequencies, where hearing sensitivity drops off.

Weighted measurements, dB(A) for example, are often referred to as **sound levels** to distinguish them from unweighted measurements of the **sound pressure levels**. Not all sound level meters can measure the unweighted level, but where this is so, the measurement is often called the linear level, meaning on this occasion the 'unweighted' level, and is designated dB(LIN). The unweighted value is used only occasionally for measurement purposes but is a very useful facility in a sound level meter because it allows a faithful unweighted sample of a noise to be captured using a tape recorder, for a variety of subsequent analyses, or for replay through a loudspeaker.

As an alternative to the weighting networks, banks of octave or one-third octave filters may be attached to the instrument, allowing a frequency analysis of the noise to be performed. In most sound level meters these various possibilities are alternative options, i.e. one can have an A-weighted level or an octave band analysis, but some sound level meters do allow the A weighting to be followed by octave band analysis i.e. an A-weighted octave band analysis.

The detector and indicator circuits

The detector and indicator circuits have three main functions:

1. They provide an RMS value of the signal.
2. They average the RMS signal over an appropriate averaging time.
3. They produce a logarithmic value of the RMS signal so that the meter will read in decibels.

The signal entering these circuits, although modified by the weighting networks, is still an alternating waveform which changes rapidly from positive to negative, maybe several thousand times a second (depending on the frequency) in response to the compressions and rarefactions in the sound wave. The meter needle would not be able to follow these very rapid fluctuations, and in any case the average value of the signal would be zero. The signal has to be rectified to produce an RMS value, which is indicated by the meter. Even though the noise appears fairly steady to the human ear, the waveform of most complex sounds contains a random fluctuating component, from moment to moment, so the RMS value also fluctuates continuously. These fluctuations are smoothed out by averaging circuits. There are two different averaging modes available on most sound level meters, fast and slow. Usually when switched to fast mode, the meter needle will fluctuate over a range of 1–2 dB, even when measuring an apparently steady noise such as from a fan, vacuum cleaner or electric drill. These fluctuations may be evened out, visually, and an average value taken. The slow mode is intended for use when fluctuations of the meter needle in the fast mode are more than about 4 dB, making it difficult to judge an average value. Switching to slow mode reduces and slows down the fluctuations, making it much easier to read the meter. When measuring steady noises, the sound level reading, averaged over a few seconds, should be the same for both modes, fast and slow. The fast mode is more suitable when it is required to follow fairly rapid changes in the noise level, e.g. if it is required to measure the maximum noise level produced by a car or lorry as it passes by.

But when it is required to measure very sharply varying impulsive sounds, such as thumps, bangs and clatters produced by sudden impacts, or noise from explosions or gunfire, then neither the fast nor slow mode is suitable or adequate. These noise measurements require a sound level meter with detector and indicator circuits capable of dealing with impulsive sounds. Alternatively the output from an ordinary (non-impulsive) sound level meter may be used to make a tape recording of the impulsive sound for subsequent analysis with a more suitable instrument. The output to the tape recorder is taken before the signal reaches the detector–indicator circuits and is not affected by their impulse limitations.

The meter

The meter scale is graduated in decibels and has been calibrated to read directly in decibels relative to 20 μPa. On many older-style sound level meters only a limited decibel range (maybe 20 dB) is displayed on the meter scale, and in order to cover the entire measurement range of the instrument, it is necessary to operate a range selector switch, which controls the attenuators in the amplifier circuits.

Sound level meter controls

The sound level meter described above has the following control and selection switches:

On/off switch
Battery check switch: checks that the battery voltage is adequate to operate the instrument according to its specification
Frequency weighting selection switch: selects A, B, C or D weighting; selects LIN octave or one-third octave filters, or dB(LIN) as appropriate
Time weighting selection switch: selects either fast mode or slow mode
Level range selector switch

9.4 Additional features on sound level meters

Display

Developments in electronics have produced sound level meters with a wide variety of displays. Improvements in RMS circuits have made available meters with a greatly extended indicator range of up to 50 dB on the meter scale as compared with 20 dB on the older types. This makes the meter much easier to use because there is much less range switching involved in measuring over the complete noise level range. Some meters even have automatic range changing.

Meters may also be fitted with hold facilities, which 'freeze' the meter needle at its highest position during a noise measurement and makes it much easier to measure maximum readings. The meter needle is released by a reset button after every measurement.

Digital displays are updated after a certain refresh time, usually one second. In addition to the instantaneous sound pressure level at the moment of refresh, a number of other measures are often available, for example, maximum and minimum values within the last second, and the maximum and minimum values since the start of measurement.

Since a digital instrument will usually have integrating, statistical and peak measuring facilities, described later, a large number of measurement parameters may be available from a single measurement. The user must be sure of exactly what is being measured; this requires appropriate reference to the instrument handbook.

Outputs

The sound level meter may be fitted with output sockets to allow the signal to be taken to other instruments for different types of display, analysis or recording. The output may be taken from different stages of the signal processing chain.

1. An a.c. output taken after the weighting network but before the detector and indicator circuits. The signal at this stage, although weighted, still contains the sound pressure waveform information, and may be fed into a tape recorder, headphones or loudspeakers, or onto an oscilloscope for visual examination.
2. A d.c. output taken after the signal has been through the RMS circuits. This signal is suitable for feeding into a level recorder or other form of chart recorder to enable a permanent paper record to be made of the noise level and its variations. The signal may be linear or logarithmic depending on the location of the output point in the detector–indicator circuitry.
3. In a digital meter it may be possible to take the digitized signal to a printer for display, or to a computer for storage and further analysis.

Overload indicators

Overload indicators are flashing-light indicators fitted to some sound level meters, particularly those with impulse measuring facilities. They flash when the amplifier circuits are being overloaded by the signal, thus indicating that a change in the attenuator settings is needed to ensure an accurate measurement. In some meters there are two indicators, one for the amplifiers before the weighting networks and one for the amplifier stages after the weighting networks. Overload indicators are particularly useful in detecting sharp transient overloads in a signal, overloads which sometimes occur even though the meter needle is not fully deflected.

Accessories

The microphone may be separated from the body of the sound level meter by a flexible microphone extension bar. This may be helpful when it is required to reduce the effects of reflection from the body of the meter, the observer or any other nearby surface, or when measuring at inaccessible positions. The microphone may be located even more remotely from the meter by using a microphone extension cable and by mounting the microphone and meter separately on tripods.

Windshields are available which fit over the microphone and produce only minimal changes of sound level reading, but which protect the microphone from false readings caused by the passage of wind over the microphone. They are effective up to certain wind speeds, quoted by the manufacturer. It is good practice always to use windshields because they are always required for outdoor measurements in case of sudden gusts of wind, even on a calm day, and are often required indoors where there may be considerable air movement caused by natural draughts, forced ventilation or the cooling fans present in many items of machinery. The penalty for not using a windshield may be an invalid measurement; the windshield also provides some protection to the microphone against accidental knocks.

Calibration

An acoustic calibrator is the most important accessory of all; it is usually supplied by the manufacturer for each type of sound level meter. Acoustic calibrators are small, portable, battery-operated devices which, when located precisely over the microphone capsule, generate an accurately known sound level at the microphone. Calibrators are used to check that instruments are working properly and also to detect any small day-to-day changes in sensitivity. Any such changes may be corrected using a sensitivity adjustment screw. Sensitivity changes may arise from changes in the environment, e.g. temperature, or from changes in the circuit components, the battery or the microphone itself. The sound level meter should be calibrated at the beginning and at the end of each measurement, and at regular intervals throughout a measurement survey lasting several hours or more.

The acoustic calibrator is designed so that its output should remain accurately constant provided the battery voltage is adequate, but as with all sound measuring instrumentation, it should be checked and calibrated itself either by the manufacturer or by a calibration agency at regular intervals.

Sound level meters are also provided with an alternative, 'internal' calibration facility. It consists of a very stable signal generator that produces a signal of constant magnitude; this signal can be switched into the sound level meter circuit instead of the microphone signal. Based on the known sensitivity of the microphone, the expected indication on the display device (e.g. the meter) can be checked, and any deviation corrected if necessary. This method of calibration cannot detect or remedy any changes in the microphone sensitivity, so acoustic calibration is preferred whenever possible. Calibration is discussed in greater detail at the end of the chapter.

Alternative inputs

Some sound level meters allow signals from external devices to be fed into the input amplifiers of the meter instead of the microphone signal. Examples are signals from accelerometers and other types of vibration transducer or from a tape recording, allowing perhaps a signal recorded from a fairly basic type of sound level meter to be processed and measured by equipment with more sophisticated analysis facilities.

9.5 Performance specification for sound level meters

British Standard 5969:1981 (and International Standard IEC 651:1979) describes the specification of sound level meter performance. This is based on four characteristics of sound level meters:

1. Directional characteristics
2. Frequency weighting characteristics
3. The characteristics of detector and indicator circuits
4. Sensitivity to various environments

Four types of sound level meter are designated: types 0, 1, 2 and 3. Type 0, the most accurate, is intended for laboratory use only. Type 1, slightly less accurate, is also intended for laboratory work plus the most accurate type of field work. Types 2 and 3 are intended for field measurements only. The specifications in the standard are the same for each type of meter; it is the permitted deviations which vary. The permitted deviations are greatest for type 3 meters, the least accurate, and smallest for type 0 meters, the most accurate.

Ideally a sound level meter should respond equally to sound arriving at any angle of incidence, but this does not happen in practice because there are diffraction effects; the deviation from ideal increases with frequency. Maximum allowed changes in sensitivity within angles of $\pm 30°$ and $\pm 90°$ of the head on position (0° angle of incidence) are specified in the standard for different types of sound level meter and for different frequency ranges. For example, in the frequency range 1000–2000 Hz the allowed changes in sensitivity within a cone of $\pm 30°$ range from 0·5 dB for type 0 meters to ± 4 dB for type 3 meters. For directions within $\pm 90°$, the corresponding figures are 1·5 dB for type 0 and 10 dB for type 3 meters.

The frequency characteristics of the various weighting networks (A, B, C and D) are given in Chapter 1. The allowed tolerance on these values is the same for all the networks, for each type of instrument. For type 0 meters the tolerance is $\pm 0·7$ dB over most of the frequency range, increasing to ± 2 dB at the extreme high and low ends of the range. For type 3 meters the tolerances are ± 2 dB for the central frequencies, rising to ± 6 dB at the ends of the range.

The standard does not cover linear or unweighted measurements. This means that different sound level meters may operate over different frequency ranges, and tolerate different degrees of departure from an ideal flat frequency response. For this reason the use of C weighting, the standard frequency weighting which is closest to linear, is increasingly being specified in standard measurement procedures, and is being included in sound level meters as an alternative to dB(A).

The time averaging in the sound level meter is performed by exponential averaging circuits based on RC (resistor–capacitor) networks. The time constant for the fast mode circuit is one-eighth of a second; it is one second for the slow mode network. The performance of these circuits may be specified in terms of how quickly the meter needle rises when a sudden burst of pure tone signal is received. As an example of this specification, the meter should rise to within 1 dB of its maximum (i.e. continuous signal) value within 200 ms (one-fifth of a second) when switched to fast mode and it should rise to within 0·6 dB of maximum in 2·0 s when switched to slow mode.

The fast and slow circuits are not suitable for detecting very sharply changing, impulsive signals. A measure of the 'impulsiveness' of a signal is its crest factor, the ratio of the peak value to the RMS value of the signal.

Question

What is the crest factor of a continous sine wave signal? (Refer back to the beginning of the chapter if necessary for the relationship between peak and RMS values of a sine wave.)

[The answer is 1·414]

Sound level meters with special impulse circuits are needed for crest factors greater than 3.

Ideally the sound level meter should be completely unaffected by environmental changes other than the sound level. The standard specifies tests to check the effect on the sound level meter of variations in atmospheric pressure, mechanical vibration, magnetic and electrostatic fields, temperature and humidity.

It is impossible to give a precise overall figure for the accuracy of each type of sound level meter. This is because the accuracy depends on a range of factors, including the frequency content of the sound being measured and its direction relative to the microphone. A given sound level meter will therefore give a more accurate reading when measuring a beam of a 'middle frequency' sound approaching the microphone at 0° incidence than when measuring very high frequency sound approaching at large incidence angles. But here is a rough guide to the accuracy of meters of different types when used under typically normal conditions:

±0·7 dB for type 0 meters
±1 dB for type 1 meters
±2 dB for type 2 meters
±3 dB for type 3 meters

9.6 The measurement of impulsive sounds

The averaging time of the fast mode is one eighth of a second, but impulsive sounds such as explosions and gunfire have a much shorter duration, usually a few tens of milliseconds. It is obvious that fast is not a suitable mode for measuring such noises, but how should they be measured? Which is the most important characteristic of such noises? Is it the peak value of the sound pressure? Is it the duration of the impulse? Or is it some combination of both, such as the sound energy content of the impulse? Much research has been and continues to be undertaken into the way in which the human ear responds to sounds of very short duration.

It is known that bursts of sound (e.g. 1000 Hz pure tone bursts) with the same amplitude but different durations sound equally loud provided that the durations are greater than a few tenths of a second. But if the bursts are of very short duration, the loudness of the sound lessens as the length of the burst is reduced, even though the amplitude remains constant.

Based on the findings of some early research into human response to impulsive sounds, a standard impulse averaging circuit has been introduced into some sound level meters. This circuit follows the sharp increase of the sound pressure with an averaging time of 35 ms. The maximum value of the signal averaged in this way is detected and allowed to decay, with a much slower time constant of 1·5 s, thus allowing sufficient time for the sound level to be read on the meter.

All three averaging modes, slow (S), fast (F) and impulse (I) should give the same reading for steady, continuous sounds. For a single sound of short duration the impulse mode will give a higher reading on the meter than the fast mode, which is higher than for the slow mode. Some impulse sound level meters also have a detector which indicates the true peak value of the sound pressure waveform. For convenience, both impulse and peak modes may also be equipped with a hold facility, to enable the maximum meter reading to be 'frozen'. An alternative method of measuring the peak value is from a calibrated display of the impulse waveform on an oscilloscope screen.

For a pure tone signal the peak value will be 3 dB higher than the impulse-weighted value. For non-impulsive machinery noise (drills, fans, motors, etc.) this difference increases to about 10 dB, but for impulsive noises (punch press noise, gunfire, etc.) the peak value may be 25–30 dB or more above that measured using the impulse weighting.

Care should be taken to avoid confusion between the true peak sound pressure level, which represents the peak value of the unweighted signal waveform before it is processed by the weighting and RMS averaging circuits, and the various maximum levels available on certain digital sound meters. These are the maximum values of the frequency-weighted RMS signal which has been averaged with either fast or slow time weightings (see Fig. 9.2).

Accurate measurement of the peak value obviously requires a measuring circuit with a very quick response time. BS 5969 requires that measurement of peak sound pressure levels are carried out by a detector circuit with an averaging time of less than 50 μs, which is very much shorter than any of the other three (F, S and I). The accuracy of the peak value also depends on the bandwidth of the signal because sharp, short duration impulses contain a wide range of frequency components. For this reason it is important that the peak measurement is not carried out on an A-weighted signal. However, some meters have C weighting instead of linear, which is not defined by any standard. Because very high sound pressure levels from very loud impulsive sounds, such as gunfire or explosives, can cause permanent hearing damage, even over very short durations, the Noise at Work Regulations require measurement of the peak value of impulse noise.

The response of the human ear to sounds is an extremely complex subject and our knowledge of it changes as research continues into its various aspects. It is emphasized in BS 5969 that inclusion of the impulse time weighting characteristic does not imply any precise relationship between the subjective response to an impulsive sound and its impulse-weighted level, and a warning is given that the impulse reading cannot be reliably used for determining the loudness of impulses or for assessing the risk of their producing hearing impairment. This warning applies also to the other standard frequency and time weightings (A, B, C, D, F, S) which cannot be accurately related to subjective noise assessments. The main reason for including specifications for all these weighting networks in the standard is to ensure that sound measuring equipment always conforms to closely defined conditions, so that results obtained by the use of such equipment are always reproducible within stated tolerances.

9.7 Integrating sound level meters

Sound level meters which are able to measure the equivalent continuous sound level, L_{eq}, of a time-varying noise are called integrating meters. The name arises from the mathematical definition of L_{eq}, which involves the integration of the square of the varying sound pressure with respect to time, over the measurement period. Integrating meters are used for measuring fluctuating, intermittent and impulsive noises as well as being convenient for relatively steady noises where the sound level fluctuations are too great to allow accurate sound level readings to be taken, even using the slow time weighting. Some meters have the facility to measure sound exposure levels, particularly useful for measurement of noise from single events such as aircraft flyover, trains passing by or bursts of process noise.

L_{eq} is almost always measured as an A-weighted value, but it is possible to use other frequency weightings or to have linear or octave band L_{eq} values.

The performance requirements of integrating sound level meters are specified by BS 6698:1986 (IEC 804:1985). The standard is consistent with the requirements of BS 5969, for non-integrating meters, but it also specifies criteria and test methods for:

— integrating and averaging characteristics
— indicator characteristics
— overload sensing and indicating characteristics

Measurement of L_{eq} may be required over periods ranging from a few seconds to several hours. The measurement period may be under the control of the operator, using a reset button, or through the use of preset measurement intervals such as 1 min, 5 min, 15 min, 1 h. Some instruments may be programmed to start measurements at predetermined times, or when noise levels reach certain predetermined trigger levels. The meter may be fitted with an indicator of elapsed time since the start of integration. A pause button enables a burst of unwanted noise to be excluded from the measurement, for example, to exclude noise from an aircraft during measurement of traffic noise.

Because the noise level may vary over a wide range during the measurement of L_{eq}, integrating sound level meters are required to have a wide measurement range. They are fitted with overload indicators to warn if the sound pressure has exceeded the measurement range during the integration period.

9.8 Analogue and digital signal processing

The voltage signal produced by a microphone is called an analogue signal because it is exactly similar or analogous in its waveform to the sound pressure waveform. An analogue signal varies continuously both in size and in time, whereas a digital quantity can only vary in discrete steps, rather like a small quantity of money which can only increase or decrease in steps of one penny.

In an analogue instrument all the signal processing (amplifying, averaging, filtering, etc.) is performed by circuits which leave the processed signal still in its analogue form. In a digital sound level meter some of the processing is performed on a digital version of the signal. At the heart of any digital instrument is the analogue-to-digital converter (ADC). The ADC samples the continuous analogue signal at regular time intervals and digitizes it, i.e. divides it up into a number of discrete intervals or steps (Fig. 9.4). The digitized signal now consists of a series of numbers representing the value of the signal at successive time intervals. These numbers are stored in binary instead of decimal, i.e. to base 2 instead of to base 10.

Two very important properties of the ADC are the number of steps into which the signal is digitized and the rate at which the continuous signal is sampled. Both properties determine the degree of which the digitized signal

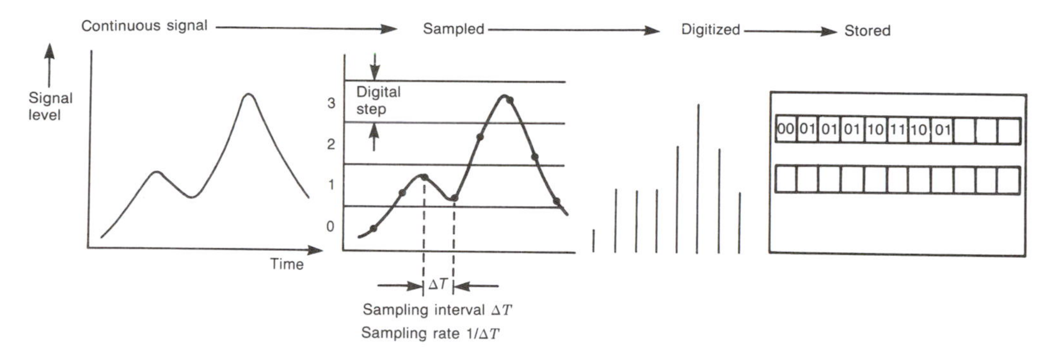

Fig. 9.4 The effect of an analogue-to-digital converter

faithfully resembles the original analogue form.

The number of digitization steps is related to the number of bits in the ADC. A bit is one piece of digital information, i.e. one digit in a binary number. An 8-bit ADC, for example, divides the signal up into $2^8 = 256$ different levels. The ADC is designed to work with analogue signals within a certain voltage range, e.g. ±2 V, and the circuits prior to the ADC perform the important task of ensuring that the signal is always amplified by the correct amount to fill this range, thus ensuring the maximum possible resolution of the digitized signal.

The dynamic range of a digitized signal from an N-bit ADC is from 1 to 2^-, i.e. from the lowest to the highest possible value. Expressed in decibels this is 20 log 2^-, or approximately 6 dB per bit. The earliest ADCs were 8-bit devices with a dynamic range of 48 dB, which was a severe limitation for use with wide range signals. Modern devices operate with 12 bits or more, i.e. a dynamic range of at least 72 dB.

Sometimes the sampled signal contains frequency components, e.g. if the signal is a sound waveform. Then the rate at which the ADC samples the signal determines the upper frequency limit of signals that may be faithfully processed by the converter. An important result of signal processing theory, called **Shannon's sampling theorem**, requires that a minimum of two samples per Hertz is needed adequately to represent the highest frequency component in the signal. A sample rate of at least 40 000 samples per second would be needed adequately to represent a signal with frequency components of up to 20 000 Hz. It is important that this condition is met otherwise false frequencies, called **aliases**, start to appear in the analysis. These arise at submultiples, e.g. one-half of those highest frequencies in the signal which are not adequately sampled because the sampling rate is too low. In order to reduce the possibility of aliasing a low pass filter called an **anti-aliasing filter**, is used to limit the upper frequency content of the signal entering the ADC to less than one-half of the sampling rate. The upper frequency limit of the signal, half the sampling rate, is called the **Nyquist frequency**.

There are many mathematically based signal processes, such as time averaging of the signal and statistical analysis, which may be performed much more easily with digital information than with analogue signals. Therefore parameters such as L_{eq}, SEL and L_N (e.g. L_{10} and L_{90}) which describe variability of sound level may be determined by digital processing of the digitized RMS signal. In this case a sample rate of a few tens of samples per second will be adequate to represent variations in the signal level. Frequency analysis of signals may be accomplished by either analogue or digital techniques. There are two distinct digital methods called fast Fourier transfer (FFT) analysis, and digital filtering. Both methods require the sound waveform to be sampled at a much higher sampling rate, i.e. up to 40 000 samples per second for the full audio range.

An advantage of digital circuitry in instrumentation is that it avoids limitations in analogue signal processing arising from the very small random electrical fluctuations which occur in all electronic components and in transducers. These fluctuations are called electronic noise; they are mixed in with the signal and set a lower limit, the noise level, to any meaningful value of the signal. Note that the term 'noise' in this context has nothing to do with acoustic noise, but simply means 'non-signal'. Each piece of analogue signal processing involves more electronic components, so it adds more noise to the signal. An advantage of digital processing is that, once the signal has been digitized, all the subsequent processes are effectively arithmetic or counting processes; they do not create extra noise, so there is no further loss or degradation of the information in the signal. However, any detail in the signal which is lost in the original digitizing process can never be replaced later, so the performance of the instrument depends very much on the quality of the ADC.

Modern sound measurement instrumentation contains a mixture of analogue and digital signal processing. A simplified block diagram of typical digital sound level

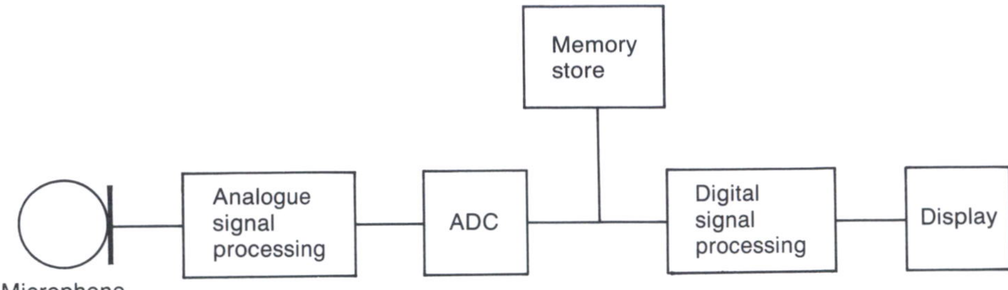

Fig. 9.5 Simplified block diagram of a digital sound level meter (mid 1980s version). Analogue signal processing includes preamplifier, amplifier stages, frequency weighting network or filters, RMS detector with time weighting. Digital sound processing includes control of sample interval, calculation of L_{eq} and L_N values

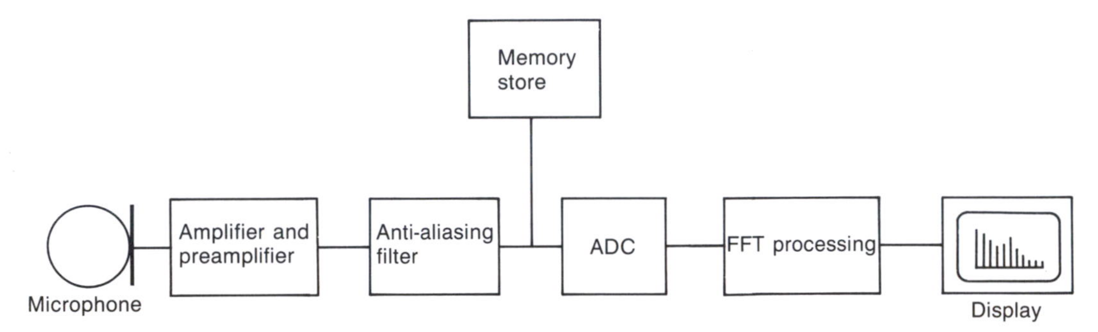

Fig. 9.6 Simplified block diagram of an FFT analyser

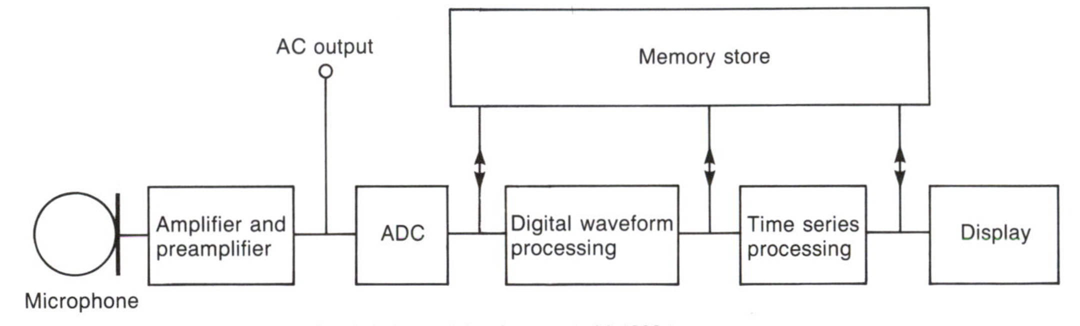

Fig. 9.7 Simplified block diagram of a digital sound level meter (mid 1990s)

meters developed in the early 1980s is shown in Fig. 9.5. The first part is similar to an analogue instrument; it contains the preamplifier, amplifiers, frequency weighting networks and the detector and indicator circuits which produce the RMS signal. The ADC is located after the RMS rectifier, and from then onwards all processes involve computations based on the digitized signal. The digitized signal is stored in the memory, thus allowing different types of analysis to be performed and a variety of different measurement parameters (e.g. L_{eq}, L_{10}, L_{90}, maximum and minimum period levels) to be stored in the memory and selected for readout via a digital display. Fig. 9.6 shows a simplified block diagram of an FFT analyser, in which the waveform signal is digitized immediately after it has passed through the anti-aliasing filter. The FFT algorithm is a set of numerical processes performed on the digitized waveform signal; it produces a narrowband line spectrum of the waveform at the moment of capture displayed on a screen.

Later developments in technology, in the early 1990s have allowed features of these two approaches to be combined. In the latest digital sound level meters (Fig. 9.7) the only analogue processes which are retained are the preamplifier and amplifier stages. Both frequency weighting and RMS averaging are performed digitally. The digitized waveform may also be frequency analysed, using digital filters, so that the variation in frequency spectrum with time may be monitored in real time, and time variation measures, L_{10}, L_{eq}, etc., are also available in octave or one-third octave bands, in real time.

FFT analysis and digital filtering are discussed again in the section on frequency analysis.

Information in digital form may be transferred to other equipment; a printer produces a permanent record of data analyses and a computer allows further analysis and storage of data. The analogue signal may be reconstructed from the digital information using a digital-to-analogue (DAC) converter.

9.9 Microphones

Before describing the different types of microphone, their main characteristics will be discussed. Frequency response and dynamic range, for example, are important properties not only of microphones but also of many other kinds of instrumentation.

Sensitivity

Sensitivity is the ratio of the output voltage produced by the signal to the sound pressure striking the microphone. It is measured in volts per pascal (V/Pa) or in dB relative to a sensitivity of 1 V/Pa. (Fig. 9.8). Ideally the microphone sensitivity should be as high as possible.

Example 9.1

Microphone A has a sensitivity of 50 mV/Pa and microphone B has a sensitivity of 12·5 mV/Pa. Compare these two sensitivities on a decibel scale, relative to each other and relative to a reference sensitivity of 1 V/Pa.

Two different sensitivities, S_1 and S_2 may be compared on a decibel scale by treating them as any other signal (pressure, acceleration, voltage, etc.) using the following formula to calculate N, the number of decibels:

$$N = 20 \log (S_2/S_1)$$
$$= 20 \log (50/12{\cdot}5)$$
$$= 12$$

Thus microphone A is 12 dB more sensitive than B, that is, if both microphones were subjected to exactly the same sound pressure then the electrical signal from A would be 12 dB higher than from B.

The sensitivity of microphone A, relative to 1 V/Pa (i.e. relative to 1000 mV/Pa) is:

$$N = 20 \log (50/1000)$$
$$= -26$$

The sensitivity of microphone A is therefore −26 dB relative to 1 V/Pa. A similar calculation for microphone B gives −38 dB. Thus the sensitivity of microphone A is 26 dB below the reference value, and microphone B is a further 12 dB lower.

Example 9.2

Microphone C has a sensitivity of −48 dB relative to 1 V/Pa. Express this in mV/Pa and calculate the magnitude of the microphone signal, in mV, produced by a sound pressure level of 74 dB.

From $N = 20 \log (S_2/S_1)$
it follows that $(S_2/S_1) = 10\ N/20$
$= 10^{-48/20}$
$= 0{\cdot}004$

S_1 is the reference sensitivity of 1000 mV/Pa

$\therefore\ S = 0{\cdot}0004 \times 1000$

$\therefore$ sensitivity of microphone C = 4 mV/Pa

A sound pressure level of 74 dB relative to 20 Pa corresponds to a sound pressure of:

$$20 \times 10^{-6} \times 10^{74/20} = 0{\cdot}1 \text{ Pa}$$

output voltage from microphone = 4 mV/Pa × 0·1 Pa
= 0·4 mV

The sensitivities of microphones A, B and C in these two examples are typical values for inch, half-inch and quarter-inch diameter condenser microphones, respectively.

Output impedance

Output impedance determines the way in which the microphone signal has to be handled by the amplifier circuits following it. It is an unfortunate fact of life that the microphones used in sound level meters, especially condenser microphones, have an extremely high output impedance. This means that a special preamplifier circuit has to be built and located next to the microphone as part of the microphone assembly. Without this preamplifier to buffer it from the rest of the sound level meter, the loading effect of the amplifier circuits would severely attenuate the microphone signal.

Linearity

The microphone output signal must be linear. This means that the sensitivity must remain constant throughout the measured range of sound pressures, or in other words, the output voltage signal of the microphone must be directly proportional to the sound pressure. This is necessary if the microphone is to give a faithful reproduction of the sound pressure waveform (Fig. 9.8).

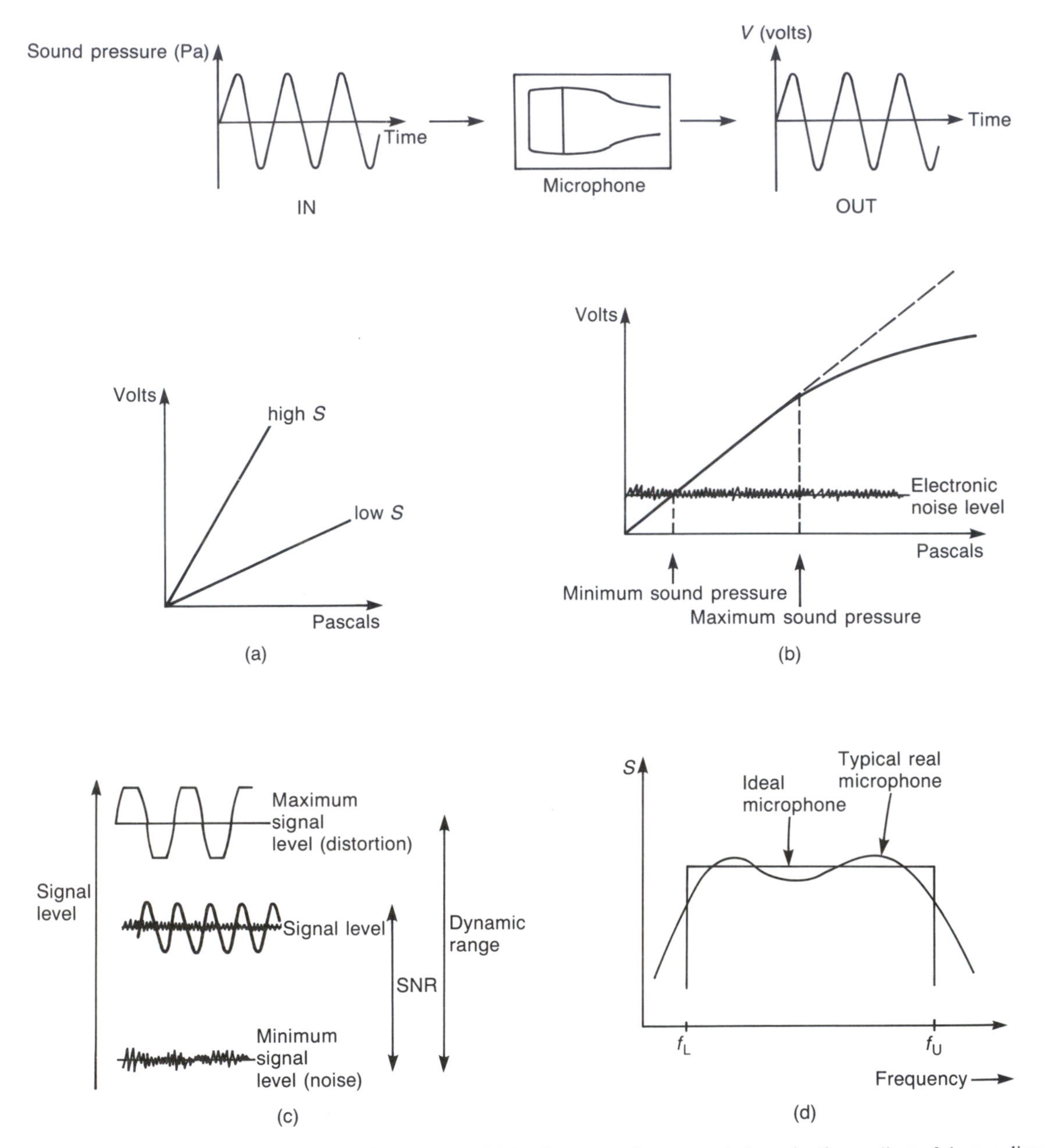

Fig. 9.8 Important properties of a microphone: (a) sensitivity S = output/input = volts/pascals, the gradient of the two lines on the graph; (b) linearity; (c) signal-to-noise ratio (SNR) and dynamic range; (d) frequency response, f_U = upper cut-off frequency and f_L = lower cut-off frequency

Dynamic range

Dynamic range is the measurement range of the microphone — the range of sound pressures over which the microphone operates and faithfully reproduces the sound pressure waveform. Measurement is limited at low levels by the electronic noise of the microphone and its preamplifier. Electronic noise produces a random output even when there is no sound pressure acting on the microphone. Measurement is limited at high levels because the microphone eventually becomes non-linear and produces a distorted waveform. At even higher levels the microphone may become damaged. The dynamic range is the range between these two extremes; it is often expressed in decibels. A more precise definition requires specification of the signal-to-noise ratio at one extreme, and the permitted percentage of distortion at the other extreme. The dynamic range of the microphone and preamplifier assembly usually determines the measurement range of the whole sound level meter.

The dynamic range is an important property of all instruments, e.g. tape recorders, amplifiers, rectifier circuits and accelerometers. In all cases there is an upper and lower limit for the signal, and the device will function according to its specification only if the signal lies within these limits.

Signal-to-noise ratio

Dynamic range is a property of the instrument, e.g. the microphone. A closely related term is the signal-to-noise ratio, which is a property of the signal. This again may be expressed in decibels. The two terms are closely related because the maximum signal-to-noise ratio is determined by the dynamic range of the instrument, as shown in Fig. 9.8.

Frequency response

Ideally the microphone should respond equally to all frequencies, i.e. it should have the same sensitivity to all frequencies. Together with linearity, this requirement is essential if the microphone is to give a faithful reproduction of complex sound waveforms, which will contain a range of frequencies. The frequency response of a microphone is represented by a graph of sensitivity versus frequency, and ideally this should be horizontal, i.e. a 'flat' frequency response, over the range of frequencies it is required to measure. In practice the frequency response of a good microphone will be approximately flat between upper and lower frequency limits, but will drop off at upper and lower frequency limits (Fig. 9.8).

Like dynamic range, the frequency response is an important property of amplifiers, loudspeakers, accelerometers, tape recorders and many other items of instrumentation. In all cases it is usual to represent the frequency response on a graph, for example, a graph of amplifier gain versus frequency, or a plot of the tape recorder ratio, replayed signal/recorded signal, versus frequency.

Some other important characteristics of microphones

Ideally the microphone sensitivity should remain absolutely constant with the passage of time, unaffected by all other changes in the environment (e.g. temperature, humidity, pressure, electric and magnetic fields, vibration) except sound pressure. And as far as possible, the microphone should be able to withstand, within limits, the effects of such changes in the environment, without suffering damage.

The microphone should be as rugged as possible, so it can withstand day-to-day handling without damage. Although some microphones are more rugged than others, *all microphones are extremely delicate and require the most careful handling and storage in order to avoid damage.* Another important property of a microphone is its cost.

The condenser microphone

Condenser microphones are very widely used in sound level meters and are essential if the highest possible degree of accuracy is required.

A diagram of a condenser microphone is shown in Fig. 9.9. Sound waves strike the flexible diaphragm, which forms one of the plates of a parallel-plate condenser. A rigid metal backing-plate forms the other plate, and there is a very small air gap between the two. A small hole behind the diaphragm ensures that the static atmospheric pressure is the same on either side of the diaphragm. The sound pressure fluctuations vibrate the diaphragm, causing variations in the thickness of the air gap between the plates. These in turn cause changes in the capacitance of the condenser. The condenser is incorporated into an electric circuit with a steady d.c. voltage, called the polarizing voltage, across its plates. The changes of capacitance produced by the sound are converted into changes in the voltage across the condenser plates. These electrical voltage changes form the microphone output signal, which is fed into the preamplifier, then to the rest of the sound level meter circuits.

The vibrating diaphragm of the microphone behaves rather like a mass–spring system, with its own natural or resonance frequency. Such a system only has a flat frequency response in the range of frequencies well below the natural frequency. This means that, in order to have a flat region extending across the whole audio frequency range (20–20 000 Hz), it is necessary for the microphone diaphragm to have a very high natural frequency, requiring a diaphragm which is simultaneously very stiff and very lightweight. This is achieved by making the diaphragm of a very thin sheet of a specially developed metal alloy, and stretching it like a drumskin to achieve the required stiffness.

The design of the microphone involves a compromise between its various requirements. The sensitivity depends on the stiffness of the diaphragm, the capacitance of the condenser and the magnitude of the polarizing voltage. Increasing the diameter of the diaphragm would have the advantage of increasing the sensitivity, increasing the capacitance and decreasing the output impedance, but it

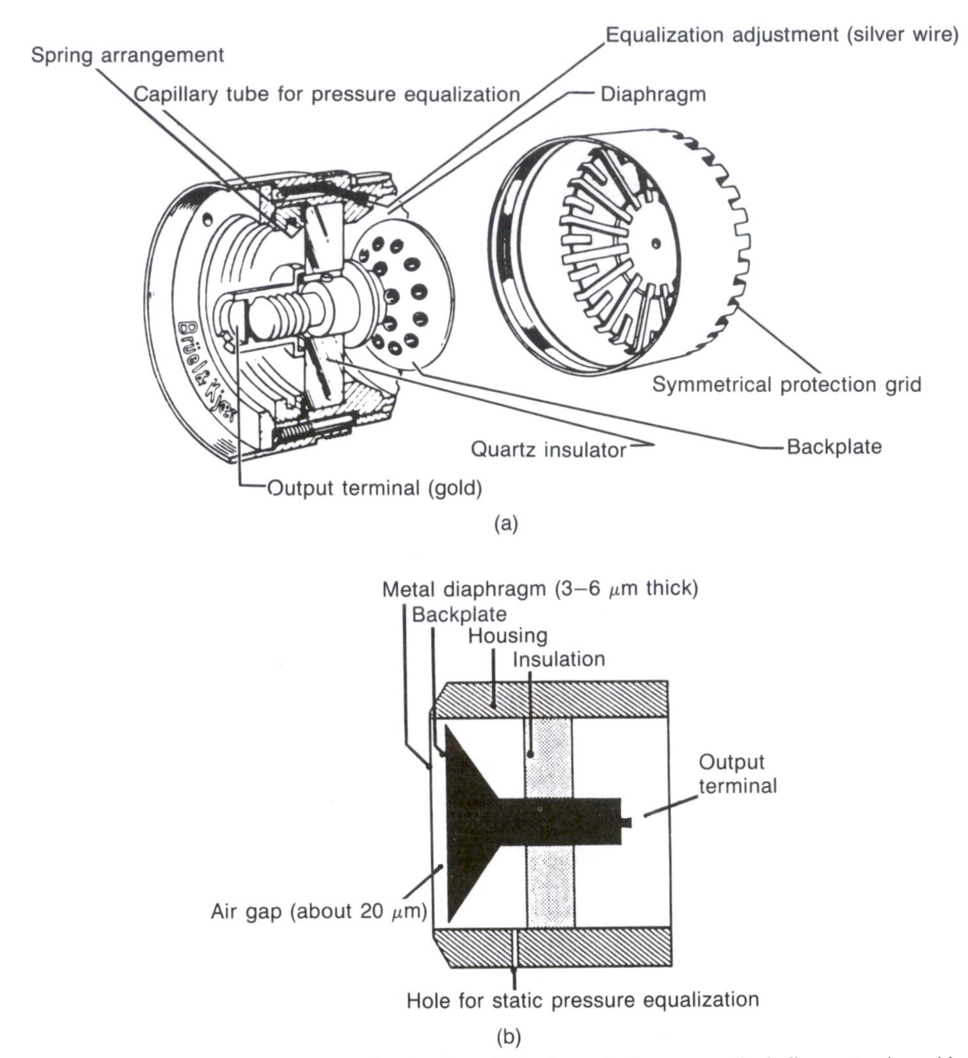

Fig. 9.9 The condenser microphone: (a) construction details and (b) schematic diagram, not including protective grid courtesy Bruel & Kjaer)

would also have the disadvantage of decreasing the high frequency performance and of making the microphone more directional in its response. Increasing the stiffness of the diaphragm will increase the natural frequency but will decrease the sensitivity.

The condenser microphone is generally considered to have a superior performance (stability, dynamic range, frequency response, sensitivity, etc.) compared to other types of microphone used for measurement purposes. But one disadvantage is its vulnerability to malfunction and damage in humid environments. Inside the microphone there is a fairly high potential difference (up to 200 V) across a very small air gap, and the problem arises from the possibility of condensation occurring within the air gap, leading to an electric discharge across the plates through the moist, conducting air. A discharge causes loss of signal, but more important, can lead to permanent damage to the microphone diaphragm. For this reason, unless special precautions are taken, measurements should never be taken in the rain or in other very humid situations. If malfunction is suspected when operating under such conditions, the sound level meter should be switched off immediately, to remove the polarizing voltage from the condenser.

It is possible to reduce these humidity problems by incorporating special modifications to the basic microphone design. They include a heating resistor in the preamplifier circuit, a thin quartz (insulating) coating to the diaphragm and a change to the pressure equalization arrangements so

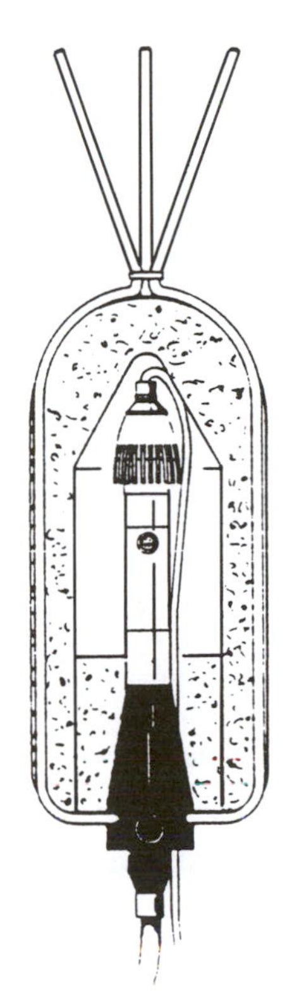

Fig. 9.10 Outdoor microphone arrangement: half-inch (12 mm) microphone and preamplifier fitted with rain cover and dehumidifier, and mounted inside a windscreen with bird spikes (Courtesy Bruel & Kjaer)

that all the air entering the microphone is dried by a silica gel dehumidifier. In addition it is possible to provide the microphone with a waterproof rainshield (Fig. 9.10) and, despite their basic vulnerability to high humidities, condenser microphones have been used successfully for permanent outdoor all-weather noise monitoring, near to airports, for many years.

The electret microphone

The electret microphone is a modified form of the condenser microphone (Fig. 9.11). The diagram consists of a polymer film coated on one surface with a thin metal film; the other surface is in contact with a metal backing-plate. These two metallic surfaces form the plates of the condenser; the polymer film in the gap takes the place of the air in the condenser microphone. The polymer is a specially developed material that is permanently polarized. Electric charges permanently reside on the two plates of the condenser—positive charge on one plate, negative charge on the other. This has the advantage that there is no need for the polarizing voltage required in the condenser microphone. The electret microphone is not vulnerable to the humidity problems of the condenser microphone because there is no air gap and no polarizing voltage. They are more rugged than condenser microphones and they have a higher capacitance because they use a polymer dielectric.

Electret microphones were developed in the early 1960s and since then have been widely used in hearing-aids and domestic cassette recorders. Their technical performance has been improved considerably over the years and they are now used for measurement purposes. Many sound level meters are designed to be used with either condenser microphones or electret (prepolarized) microphones.

The piezoelectric microphone

Piezoelectric microphones contain a thin slab of a piezoelectric material. They are also known as crystal microphones or ceramic microphones. The piezoelectric effect is discussed in Chapter 7. A diagram of a piezoelectric microphone is shown in Fig. 9.12. The sound waves strike a thin, flexible metal diaphragm. The vibration of the diaphragm is transmitted to the piezoelectric element, producing alternating stresses which in turn produce the alternating voltage.

The piezoelectric microphone is much cheaper than the other two types of microphone, but it has a greater capacitance and therefore a lower output impedance. It is rugged and does not require a polarizing voltage, but it is not as stable as the other two types. This is because the performance of the piezoelectric element can change with age. The frequency response of the piezoelectric microphone is limited and not as flat as the other two types. It is used in some of the cheaper sound level indicators.

The behaviour and performance of microphones in sound fields at high frequencies

At very low frequencies, where the sound wavelength is very much greater than the size of the microphone, the microphone can be considered as having almost point size; the microphone does not noticeably interfere with the sound wave. This is no longer true at higher frequencies; diffraction, scattering and reflection from the surface of the microphone become significant, and the presence of the microphone actually disturbs and changes the sound field it is intended to measure. These changes become important at frequencies where the wavelength of the sound becomes similar in magnitude to the dimensions of the microphone. These changes have nothing to do with the microphone's type or principles of operation; they occur simply because the microphone acts as an obstacle in the path of the sound waves.

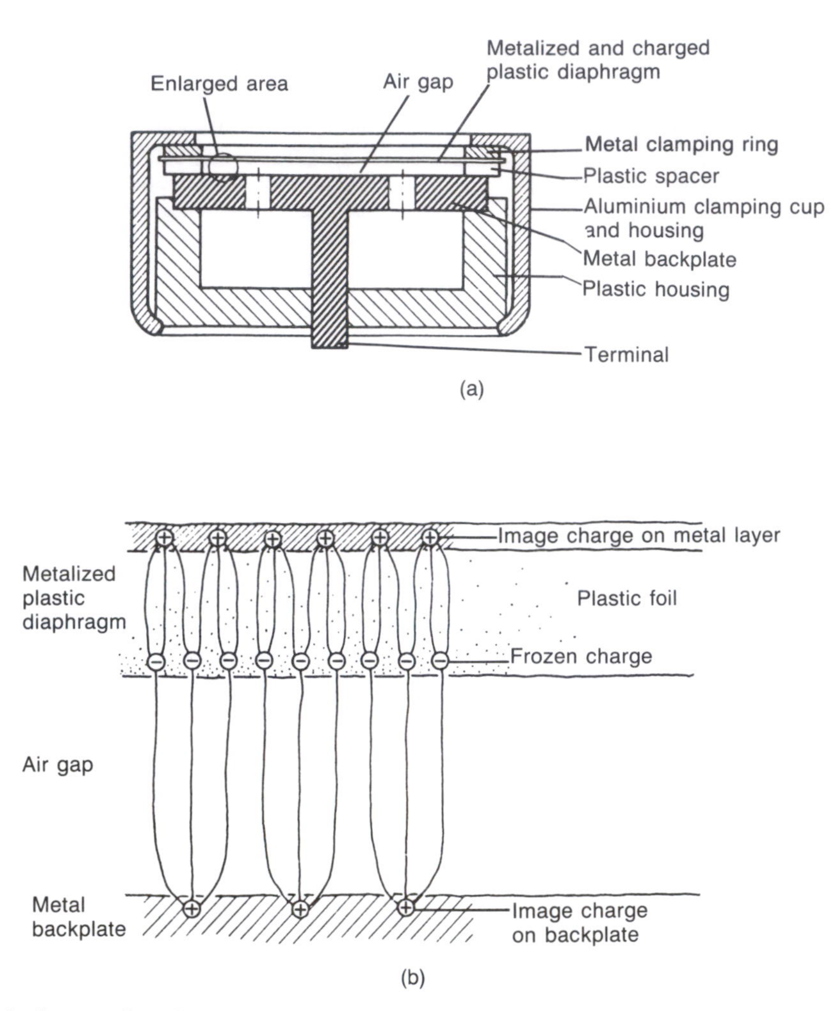

Fig. 9.11 Schematic diagram of an electret (prepolarized) condenser microphone: (a) construction and (b) enlargement to illustrate charge distribution (Courtesy Bruel & Kjaer)

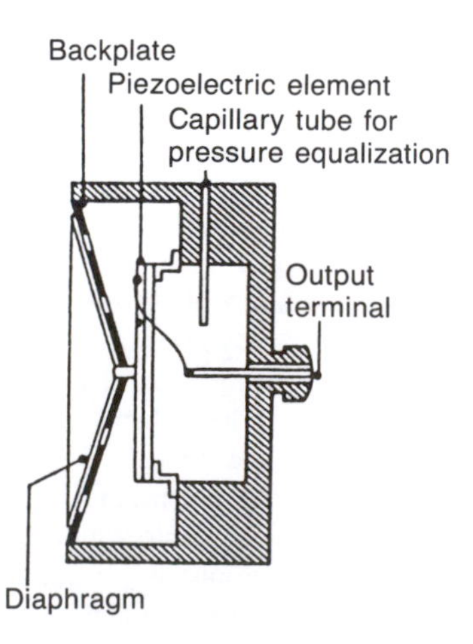

Fig. 9.12 Schematic diagram of a piezoelectric microphone

Question

At which frequency will the sound wavelength be equal to the diameter of a one-inch microphone?

$$f = \frac{c}{\lambda} = \frac{330 \text{ m/s}}{25{\cdot}4 \times 10^{-3} \text{ m}} = 12\,992 \text{ Hz}$$

i.e. ~ 13 kHz

Directionality of microphones

At low frequencies, sound waves spread evenly over the microphone and the sound pressure will be the same all over its surface. Under these circumstances the microphone will respond equally to sound approaching from all directions. As the frequency increases and the wavelength decreases, the distribution of sound over the microphone

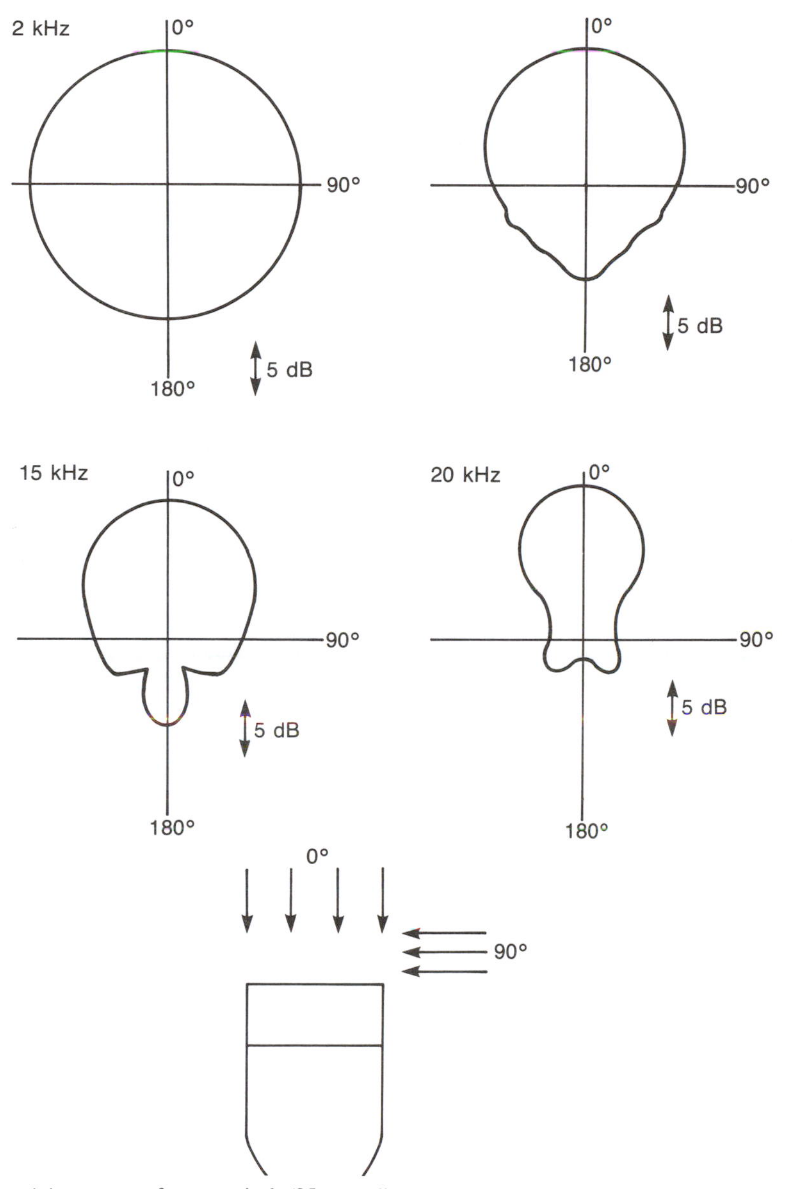

Fig. 9.13 Typical directivity patterns for a one-inch (25 mm) diameter microphone

becomes less uniform, so the microphone becomes more directional. Figure 9.13 shows the directionality pattern of a one-inch microphone at different frequencies. At low frequencies the pattern is approximately circular, indicating equal sensitivity to sound from all directions, but as the frequency increases, the region of maximum sensitivity becomes more and more confined to sound striking the microphone at small angles of incidence. At these high frequencies the orientation of the microphone relative to the direction of the sound becomes an important measurement detail.

Frequency response

The frequency response of a microphone depends on details of its internal construction, which determine the natural frequency of the diaphragm. But besides construction, the high frequency behaviour of all microphones is determined by the size of the diaphragm compared to the wavelength. This is easily understood by considering a microphone placed in a plane-wave sound field where the sound wavelength equals the microphone diameter (Fig. 9.14). The microphone is positioned so that sound waves approach

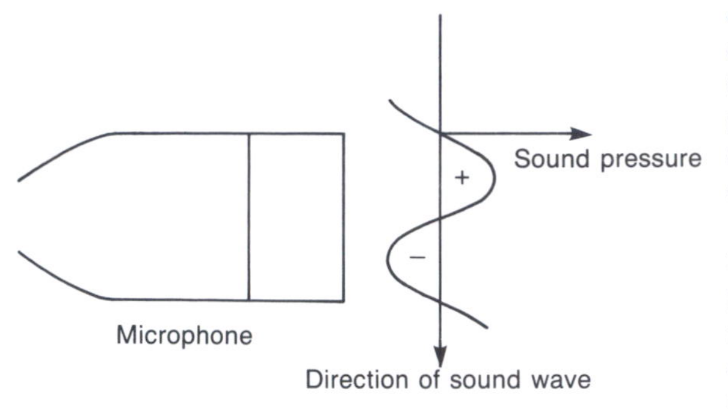

Fig. 9.14 The effect of high frequency sound on microphone response

at a 90° angle of incidence, i.e. so that the sound waves graze over the microphone surface. Half of the diaphragm is subjected to a positive sound pressure (a compression) while the other half is being subjected to a negative sound pressure (a rarefaction); the average pressure is equal to zero. The situation changes as the angle of incidence changes, until at 0° the sound pressure is in phase over the whole area of the diaphragm.

The general situation is that the sensitivity of the microphone drops off at high frequencies, where the sound wavelength becomes similar in magnitude to the microphone diameter. And at high frequencies the sensitivity depends on the angle of incidence of the sound. It follows that smaller microphones have a better high frequency response than larger microphones, although they tend to be less sensitive. At a given frequency, smaller microphones are also less directional than larger microphones.

Free-field, pressure and random incidence responses

Because the microphone acts as an obstacle to the very sound field it is to measure, the sound pressure at the microphone diaphragm will be different from the sound pressure at that point when the microphone is absent. The difference is known as *the free-field correction factor* and depends on the angle of incidence, the frequency of the sound and the microphone diameter. It may be calculated using diffraction theory. Manufacturers of microphones usually designate three different types of microphone for use in different types of sound field: free-field response, pressure response and random incidence response (Fig. 9.15).

Free-field microphones are intended to give a reading which indicates the sound pressure level at the measurement position in the absence of the microphone. This is achieved by including the free-field correction factor in the calibration of the microphone. Free-field microphones are usually designed for use at 0° incidence and are the correct type to use for measuring direct sound which comes predominantly from a particular direction; the microphone should be pointed in that direction towards the source of sound.

Pressure response microphones do not include the free-field correction in their calibration; they indicate the sound pressure actually present at the microphone diaphragm, including a component caused by scattering and diffraction due to the presence of the microphone itself. This type of microphone is the most suitable for measuring sound in confined spaced, e.g. when using acoustic couplers or when calibrating audiometers, and for measuring the sound pressures acting on surfaces, e.g. at the wall of a duct, when the microphone would be flush-mounted in the wall. When they are used for measuring free-field sound, pressure response microphones are intended to be oriented so that the sound strikes the microphone at a 90° angle of incidence.

Random incidence response microphones are calibrated and intended for use in situations where the sound is diffuse and strikes the microphone randomly from all directions, e.g. in a reverberant sound field indoors. Free-field microphones may also be adapted to measure diffuse sound by fitting a diffusing device called a random incidence corrector to the microphone.

Figure 9.16 shows the differences between the frequency responses of the three microphone types. They only become appreciable at high frequencies, being approximately 1 dB at 2000 Hz, 3 dB at 5000 Hz, and 10 dB at 10 000 Hz for a one-inch diameter microphone.

Conclusion

It is important when measuring sounds with a significant high frequency content to choose the correct type of microphone for the type of sound field to be measured, and to orientate the microphone correctly with respect to the direction of the sound. The manufacturer of the sound level meter usually gives advice concerning the correct choice and use of microphone in the instrument handbook.

9.10 Frequency analysis of acoustic signals

Most noise signals are extremely complex and have a continuous frequency spectrum, i.e. all frequencies are present to some extent. Filters are electronic circuits which act like an open gate to certain frequency components in the signal, allowing them to be transmitted, but are closed to other frequencies. They may consist simply of passive circuit components (resistors, capacitors and inductors) which do not require any electrical energy, or they may

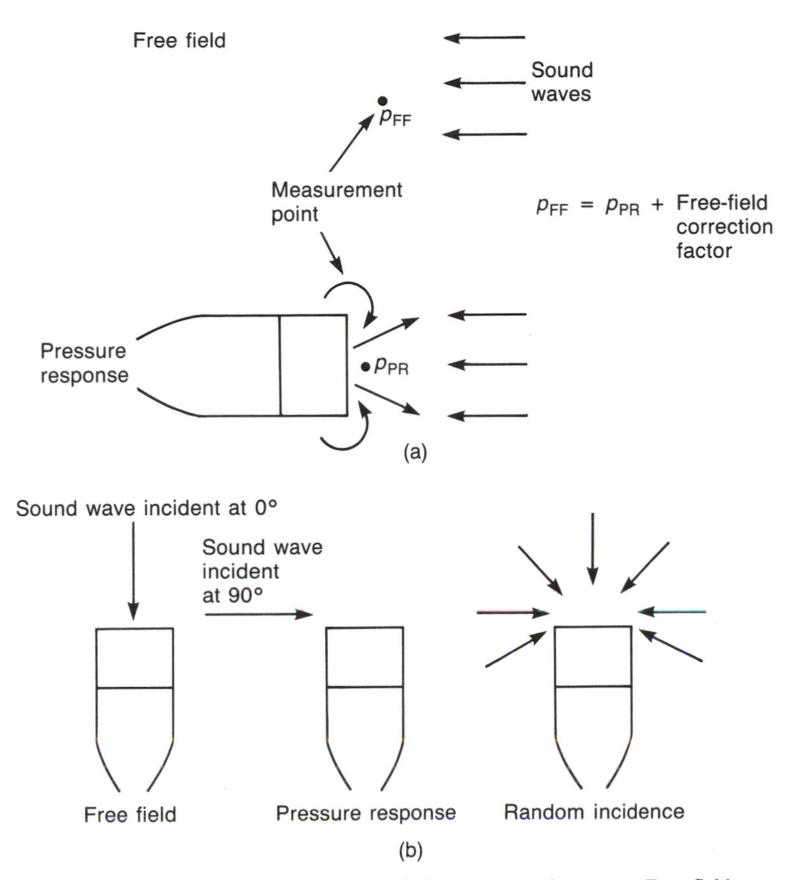

Fig. 9.15 Illustrating free-field, pressure response and random incidence microphones: (a) Free-field measurements indicate the sound pressure as though the microphone were not present; pressure response measurements indicate the sound pressure actually present at the microphone diaphragm. (b) Free-field microphones include the free-field correction in their calibration and should be used at 0° incidence. Pressure response microphones should be used at 90° incidence and random incidence microphones should be used in reverberant sound fields

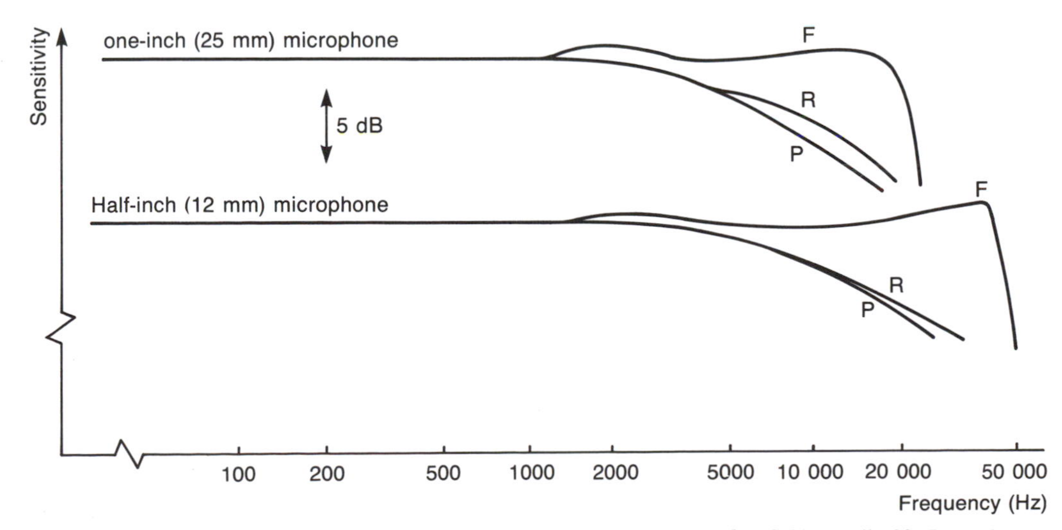

Fig. 9.16 Typical frequency response characteristics of condenser microphones: (F) free field, usually 0° (R) random incidence (P) pressure response

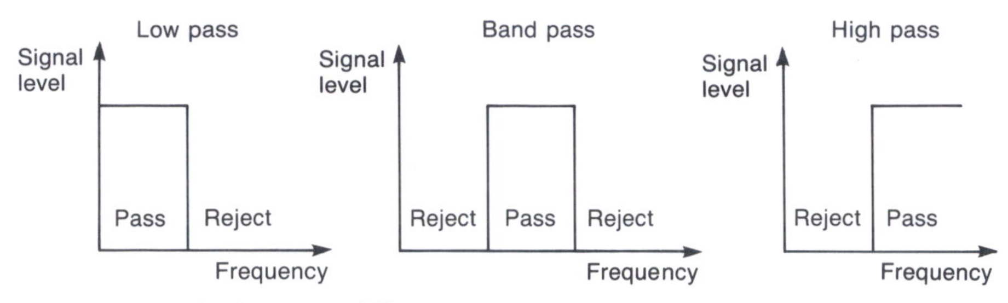

Fig. 9.17 Frequency response for three types of filter

be active, containing transistors and other semiconductors which require a battery or power supply of their own. They may be analogue in their operation or they may employ digital techniques.

Filters may be low pass, high pass or band pass (see Fig. 9.17). A frequency weighting network, e.g. the A weighting, is a special kind of band pass filter. High and low pass filters are sometimes built into instruments to exclude unwanted signals, outside the frequency range of interest. For example, a vibration meter may be fitted with a high pass filter which cuts out all signals lower than say 1 Hz to eliminate the possibility of spurious signals caused by temperature fluctuations. A low pass filter may also be used with a cut-off frequency of say 1000 Hz to eliminate unwanted high frequency signals caused by accelerometer resonances.

Band pass filters are used to analyse the signal, i.e. to measure the sound energy in each frequency band. A band pass filter is characterized by the midband frequency and by the width of the band. The most common are the octave and one-third octave band filters. An octave is a range extending from one frequency to double that frequency. Ten contiguous octave bands cover the entire audio range. The lowest band is centred on 31·5 Hz and the highest on 16 kHz, but for most work the eight bands from 63 Hz to 8000 Hz are sufficient. The centre frequencies of the bands (see Fig. 9.18) and the bandwidths are laid down in international standards. An octave band spectrum gives a good broad indication of the frequency content of a noise; it is useful in assessing possible noise control measures because acoustic properties of materials, such as absorption coefficient and sound reduction index, are usually given in octave bands. Octave band analysis is also required in many of the standard noise measurement methods (e.g. noise criteria and speech interference levels). One-third octave filters give a more detailed picture of the frequency composition of the noise. There are three one-third octaves within each octave band. They are also specified in some standard methods of noise measurement (e.g. aircraft noise measurement and building acoustics).

For an even more detailed investigation, narrowband filters are available. They may be used to measure accurately the frequency of a sharp peak in the spectrum and thus enable its cause to be diagnosed, perhaps to relate

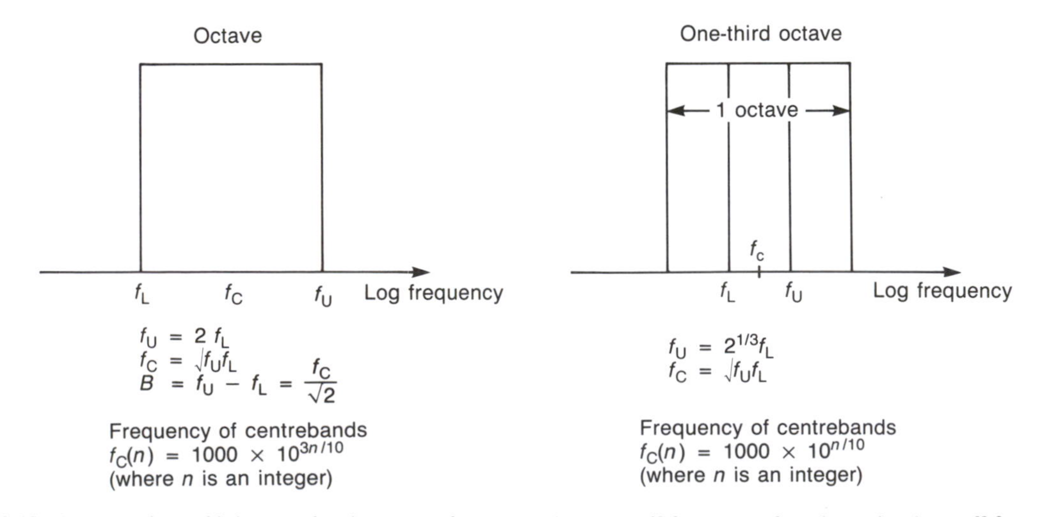

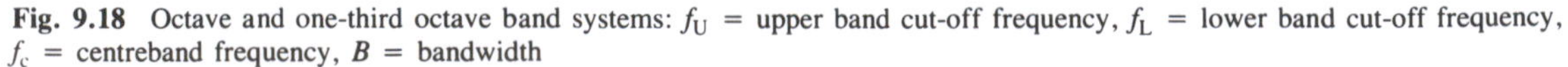
Fig. 9.18 Octave and one-third octave band systems: f_U = upper band cut-off frequency, f_L = lower band cut-off frequency, f_c = centreband frequency, B = bandwidth

the sound frequency to the meshing frequency of a particular gear, to the blade passing frequency of a fan or to the resonance frequency of a vibrating panel.

Narrowband filters may have a continuously variable centreband frequency. The bandwidth may be specified on a constant bandwidth basis (e.g. 10 Hz, 3 Hz or 1 Hz), which does not change across the frequency range, or it may be a constant percentage bandwidth (e.g. 10%, 3% or 1% of the centreband frequency), in which case the actual bandwidth increases as the centre frequency increases. One-third octave filters have a bandwidth of approximately 23% of the centreband frequency.

When carrying out narrowband analysis of signals, it is necessary to take into account the effect of the **time–bandwidth product theorem** of signal processing, according to which the averaging time (T/s) required to achieve a certain degree of measurement accuracy depends upon the bandwidth of the signal (B/Hz):

$$\epsilon = \frac{1}{2\sqrt{BT}}$$

where ϵ is the ratio of the standard deviation to the mean of the measured value. Thus narrowband frequency analysis requires longer averaging times than would be necessary for wider bandwidths.

The ideal band pass filter has the square-shaped characteristic shown in Fig. 9.17, i.e. the attenuation is constant, ideally zero, across the band, the cut-off is perfectly sharp at the ends of the band and rejection of signal outside the band is total. In practice filters are not perfect and have a characteristic shape which only approximates to the ideal (see Fig. 9.19).

Banks of octave (or one-third octave) filters are available for attachment to many types of sound level meter. If the noise is constant or steady, the analysis involves switching the signal through each filter in turn and reading the band level from the meter. This may be done manually and the band levels tabulated and plotted graphically. This process may be made easier using modern digital equipment which automatically switches through the filter range, stores the data in memory for later retrieval and controls the production of calibrated plots of the frequency spectra from printers or chart recorders. It is common practice to allocate equal space along the frequency axis to each octave (or one-third octave) band. This really amounts to using a log frequency scale. A linear frequency scale may sometimes be used for narrowband analysis to show even spacing between successive harmonic components.

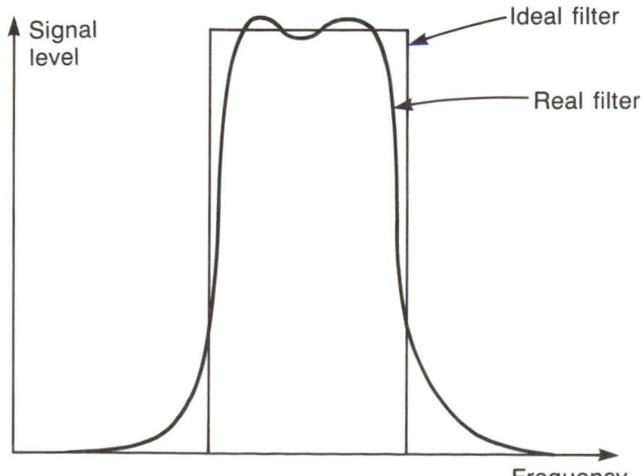

Fig. 9.19 Frequency response for real and ideal filters

9.11 FFT analysis

Fast Fourier transform (FFT) analysers use digital signal processing techniques to produce very rapid narrowband frequency analysis of acoustic signals.

The method has its origin in the work of the French mathematician J.B. Fourier (1768–1830) who showed that any complex periodic signal, with repetition time T seconds, can be broken down or analysed into a series of component times having frequencies which are multiples (harmonics) of the repetition rate $1/T$, called the fundamental frequency (Fig. 9.20). Conversely, complex harmonic signals may be built up or synthesized from combinations of harmonically related single frequencies.

In theory the Fourier series for a harmonic waveform may be infinite, i.e. it may consist of an infinite number of harmonics, but usually it is strongly convergent, so the amplitude of the higher harmonics reduces rapidly with increasing harmonic number. The amplitude and phase (compared to the fundamental) of each frequency component may be obtained from a formula derived by Fourier, which involves mathematical integration over one complete cycle of the signal, i.e. over period T. The Fourier series for simple periodic signals is covered in many mathematical textbooks. The result for a square wave signal is illustrated in Fig. 9.21; the signal is broken down into a series of odd-numbered harmonics (1, 3, 5, etc.).

Figures 9.20 and 9.21 show that Fourier's theorem allows periodic signals to be represented by a line spectrum of frequencies. The frequency spacing between the lines is $1/T$; if the period T of the signal increases then $1/T$ decreases and the line spacing decreases, so the lines move closer together.

The extension of Fourier's theorem to non-repetitive signals involves the extension of the period T to infinity, and the evaluation of an integral over this time, called the Fourier transform, which produces a continuous frequency spectrum, i.e. a spectrum in which an infinite number of lines are separated by infinitesimally small frequency

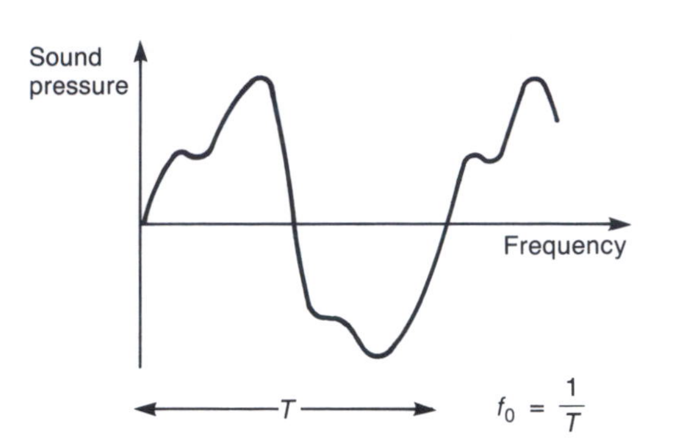

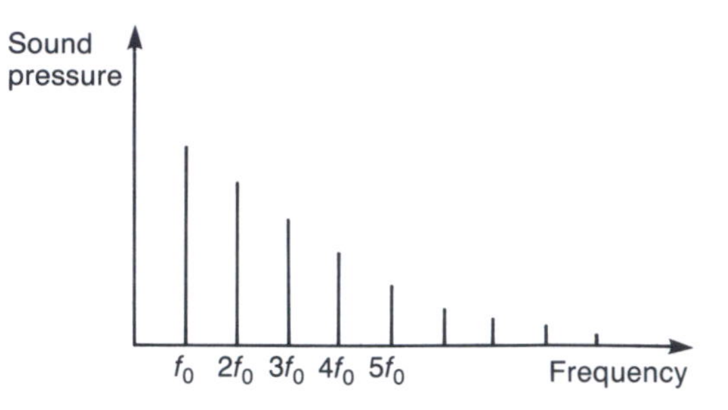

Fig. 9.20 Fourier series or line spectrum for a periodic waveform

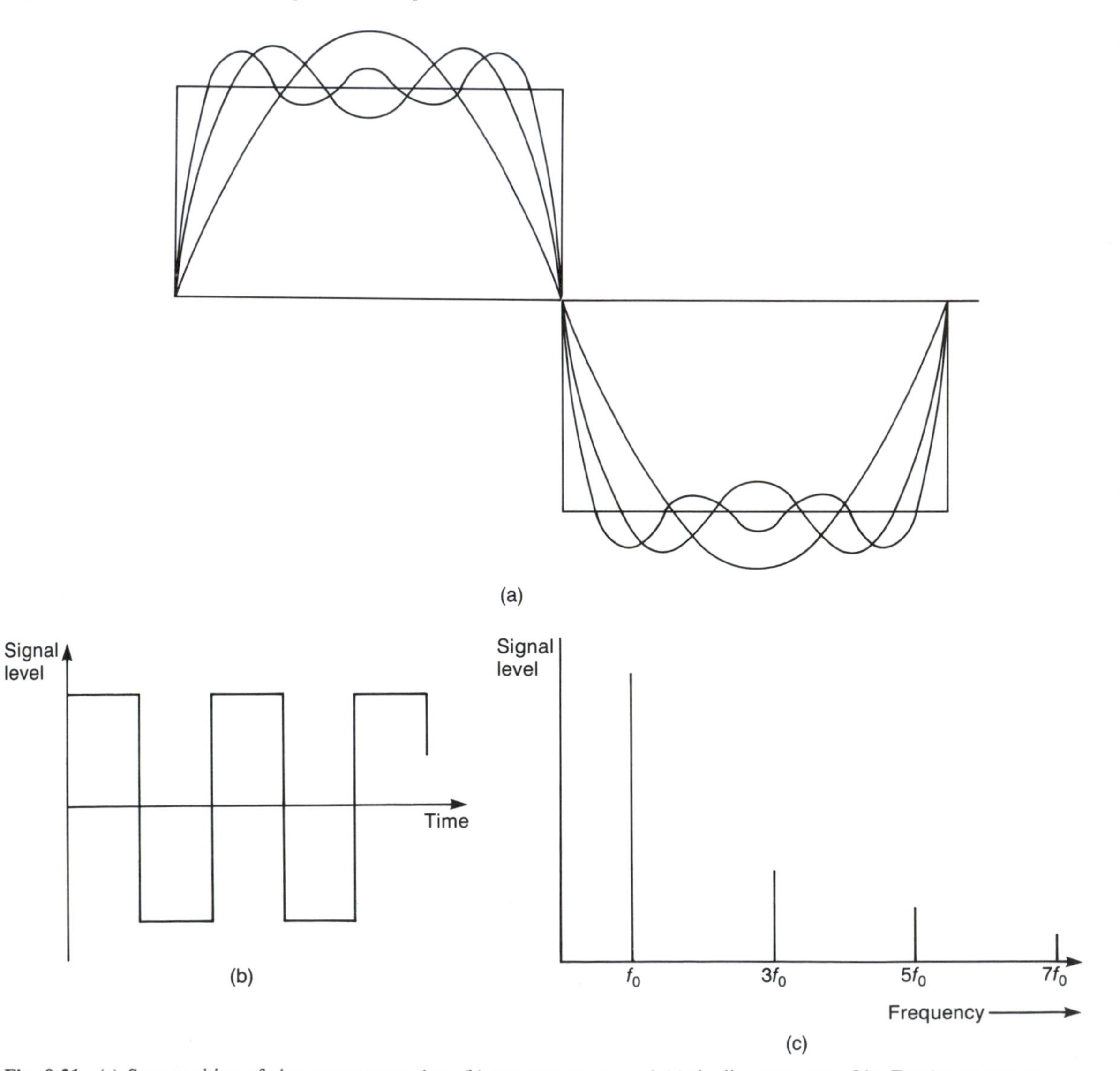

Fig. 9.21 (a) Superposition of sine waves to produce (b) a square wave, and (c) the line spectrum of its Fourier components

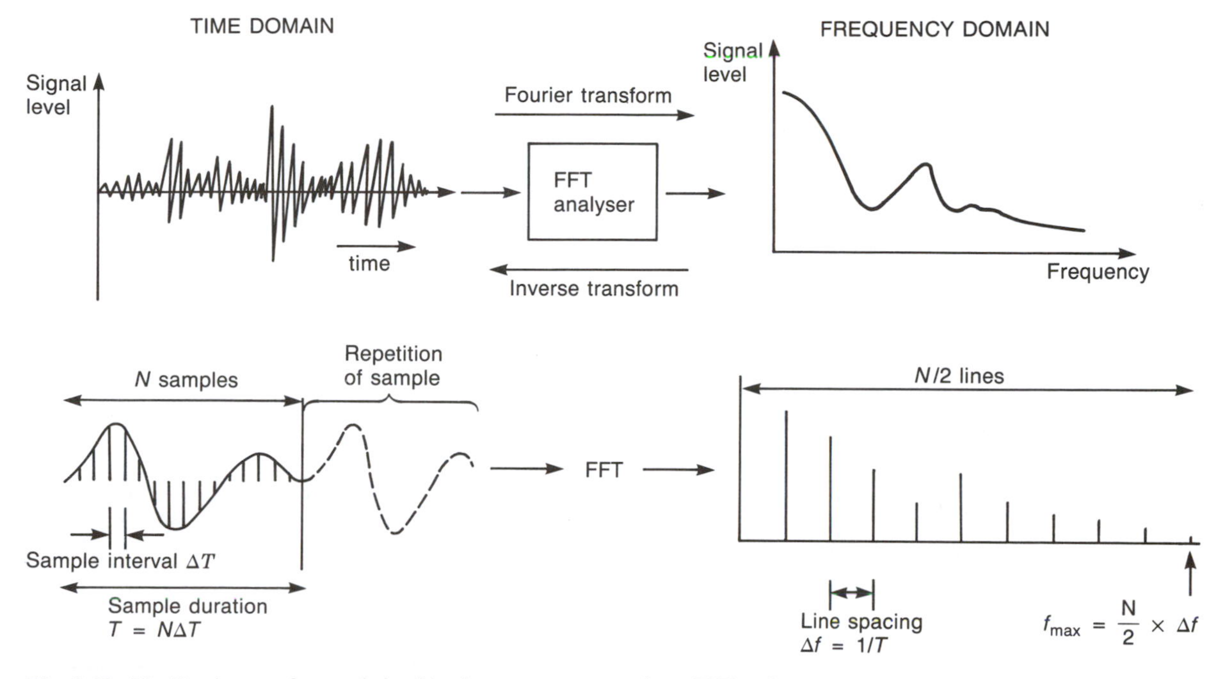

Fig. 9.22 The Fourier transform: relationships between parameters in an FFT analyser

increments. The Fourier transform may be thought of as a mathematical process which converts a description of the signal in the time domain, i.e. a waveform, into a description of the signal in the frequency domain, i.e. a frequency spectrum. Conversely, the inverse Fourier transform converts a frequency spectrum into its corresponding waveform (Fig. 9.22).

Although the Fourier transform of some simple signals may be evaluated exactly, using techniques of mathematical analysis, most complex signals require the use of numerical methods of integration which involve the performance of very large numbers of calculations. In the 1960s Cooley and Tukey, in the United States, devised a method for vastly reducing the calculation time for evaluating the Fourier transform; FFT analysis derives from their algorithm.

In an FFT analyser a sample of the signal is digitized, and the FFT calculations performed on the digitized sample to produce a line spectrum. The fundamental frequency in this spectrum is $1/T$, where T is the total duration of the sample in seconds; the frequency spacing between the lines, $\Delta f = 1/T$. Thus, in effect, the FFT produces the spectrum which would result from an indefinite repetition of the sample.

An FFT performed on a sample of N points produces a spectrum of $N/2$ frequency lines, each defined by an amplitude and a phase. The time interval between each digital sample is T/N, so the sampling rate is N/T. The maximum frequency component in the line spectrum, f_{max}, is $N/2 \times 1/T$, which is the Nyquist frequency relating to the requirements of Shannon's sampling theorem, described earlier.

In summary these variables and their relationships are (Fig. 9.22):

number of samples	$= N$
interval between successive samples	$= \Delta T$ seconds
total duration of sample	$= T = N\Delta T$ seconds
sampling rate	$= N/T = 1/\Delta T$ samples per second
line spacing, Δf	$= 1/T$ hertz
number of frequency lines	$= N/2$
maximum frequency, f_{max}	$= N/2 \times 1/T$ hertz

The value of N is usually fixed for a particular type of FFT analyser. Any one of the three variables f_{max}, Δf and T may then be selected over a range of values; this selection then fixes the value of the other two variables. The sampling rate is automatically adjusted to be at least twice the value of f_{max}, and an anti-aliasing filter having an appropriate cut-off frequency is automatically selected.

An example may illustrate the interrelationship between these variables and their selection ranges. A particular FFT analyser operates on 1024 data sample points (note that N is usually an integral power of 2). In theory this should produce a 512-line spectrum, but in practice only 400 lines are used; this allows for the fact that the anti-aliasing filter cannot have a perfectly sharp, cut-off characteristic. The frequency range, f_{max}, may be selected within the range 20–20 000 Hz in steps of 2, 5, 10 (i.e. 20 Hz, 50 Hz, 100 Hz, 200 Hz . . .). On the highest range the 400 lines are spread over a frequency range of 20 000 Hz, so that the line spacing, Δf, is 20 000/400 = 50 Hz. This means that the total duration of the waveform sample, during which 1024 values are collected, is 1/50 second or 20 ms, and the sampling rate is therefore 1024 × 50 = 51 200 samples per second, which is, as required, more than twice the value of f_{max}. At the other extreme, if an f_{max} value of 20 Hz is selected, this gives a much higher frequency resolution with $\Delta f = 0{\cdot}005$ Hz, but with a much higher total sample duration of 20 s. The sampling rate in this case is 51 samples per second.

The fact that the Fourier transform is carried out on a finite number of data samples rather than on a continuous waveform has some inevitable consequences for the accuracy of the results of FFT analysis. The possibility of aliasing has already been discussed. Another difficulty arises because random signals usually begin and end on different non-zero values. The repetition of these samples in the FFT process will create steps or discontinuities, which will cause false values in the frequency spectrum. The difficulty will not be present in transient signals, where the capture of the sample can be arranged to completely include all of the transient, so the sample starts and ends with zero values. It may be partially overcome for non-transient signals by using a time window which weights or shapes the sample in a way which smooths out the discontinuities at the start and end of the sample, but which has minimum effect in between.

To improve the statistical reliability of the results, FFT analysers usually have the facility to compute and display the average spectrum derived from analysis of several samples of the signal. Some analysers also have the facility for zoom-FFT which enables a limited portion of the frequency range to be analysed with a greater degree of frequency resolution, i.e. with reduced line spacing.

It is possible to generate one-third octave and octave band values from an FFT line spectrum, and some analysers offer this facility, but care should be taken at low frequencies when accuracy will be reduced if only a few lines are contributing to the band value.

Dual-channel FFT analysers are able to sample two signals simultaneously and compute various joint functions of the two input signals such as correlation, coherence and transfer functions. The two signals could be the vibration from a particular machine, or part of a machine, and the noise signal from a nearby microphone; or they could be vibration (or noise) from two different parts of a structure.

The main use of FFT analysis is for performing very rapid frequency analysis of signals which have prominent tonal and harmonic characteristics, such as the noise or vibration from gearboxes and turbines. It enables noise sources to be identified from a detailed examination of their frequency spectrum. All vibrating machines, including motors, pumps and fans, produce a vibration spectrum which contains components at the rotation speed and its harmonics. The level of these components gives a good indication of wear and onset of faulty operation, so FFT analysis is widely used in condition monitoring of machinery.

This has been a simplified and non-mathematical introduction to FFT analysis. More advanced texts should be consulted for a more detailed and rigorous treatment.

9.12 Digital filters

In a digital filter a sequence of digital operations involving addition, multiplication and time delay are performed on the digitized signal. The effect of these operations on the digital signal is intended to be equivalent to the effect of an analogue filter on the corresponding analogue signal.

There are a number of important differences between digital filtering and FFT analysis. Digital filtering is a continuous process, in which a continuous input of signal to the filter results in a continuous output of filtered signal, exactly as for an analogue filter. In contrast, FFT analysers operate on discrete blocks of data; each sample block is captured then analysed while the next block is being captured, and so on. FFT analysers produce a line spectrum, whereas the digital filtering process is usually used for octave and one-third octave analysis. Digital filters may be designed to produce band pass filters of any desired bandwidth and shape, including those that meet the requirements of British and international standards. The centre frequency of the band is determined by the signal sampling rate, and the filter shape by the details of the sequence of filtering operations. Changes in the configuration of digital filters may therefore be accomplished very rapidly by changes to software, whereas changes in physical electronic components would be required for the equivalent analogue filter.

9.13 Real-time frequency analysers

The level and frequency content of the noise from a machine may change slowly, over days and weeks, and these changes may be adequately followed using a simple sound level meter fitted with a manually operated set of octave band filters. In contrast, it takes only a few seconds for changes to occur in noise level and frequency content as an aircraft flies over. Changes during human speech or during a handclap take place over even shorter time periods; they require the use of a real-time analyser, which gives a very quick, almost instantaneous analysis and display of frequency spectra. This enables rapid changes in the signal to be observed as they happen.

Analogue real-time analysers use a band of analogue filters, in parallel, so that the signal is divided and passes through all the filters at the same time. More modern devices use digital filters which can operate so quickly that a complete band spectrum may be obtained rapidly enough to follow any changes as they occur. The spectrum, which may be octave, one-third octave or narrowband, is usually displayed on a screen. There are facilities to compute and display various noise indices, to store the analyses, and to produce hard copies of the spectra. And it is often possible to build up a three-dimensional picture (level, frequency and time) called a waterfall plot; this shows how the frequency content of the signal changes with time.

One definition says that real-time frequency analysis includes all of the signals in all of the frequency bands at all times, displaying the results on a continuously updated screen.

According to this definition, FFT analysers are not always capable of operating in real time. The time taken for carrying out the FFT analysis on a block of data points will be constant for any particular analyser, independent of the choice of frequency range, f_{max} (see earlier discussion). For a 400-line spectrum this is typically 0·045 s. But the time taken to collect the data sample will depend on the choice of f_{max}, which determines the sampling rate. Consider the values in the earlier discussion: for the lowest value of f_{max}, f_{max} = 20 Hz, the time taken to collect 1024 data points at a sampling rate of 51 samples per second results in a sampling time of 20 s. By comparison, the analysis time of 0·045 s is negligible, and the analyser operates in real time. At the highest frequency range, when f_{max} is 20 000 Hz, the sampling rate is 51 200 samples per second, and the sample is collected in 0·02 s, which is shorter than the analysis time. Under these circumstances the analyser cannot possibly display analysis of all the signal, so it does not operate in real time. On this basis such an analyser would operate in real time only for frequencies up to 5000 Hz.

9.14 Graphical records of noise and vibration data

A wide variety of printers, plotters and recorders are available to produce permanent graphical records of the measurement and analysis of signals from sound level meters, tape recorders and similar instrumentation. They include digitally controlled devices which can provide hard copies of displays from video screens, as well as more traditional level and chart recorders.

The types of graphical plot which may be useful include the following:

1. Waveform plots: graphs of instantaneous sound pressure or vibration level against time, taken from the screen of an oscilloscope or FFT analyser or directly from the microphone or vibration transducer, before RMS rectification, are particularly useful for investigating peak values and decay rates of transient signals.
2. Time histories of varying noise levels: graphs of dB(A) against time, together with frequency spectra, are the most common method of presenting noise data. Different types of noise have different patterns of noise variation; road traffic noise and train noise show good contrast. During the investigation of environmental noise complaints, it is often possible to identify different noise sources from their temporal pattern and to determine when sources are switched on and off from an investigation of the time history, as well as measuring noise levels, durations and background noise. In the workplace level versus time plots (e.g. from data-logging dosemeters) are useful in identifying and controlling noise exposure of employees. On a much shorter time-scale, this type of display is used in the measurement of reverberation time of rooms. On a much longer time-scale, graphs of period values (e.g. hourly L_{eq} or L_{10}) are useful in illustrating variations in noise level over hours, days, weeks or longer periods.
3. Frequency spectra of steady noises: graphs of sound pressure level versus frequency and three-dimensional (waterfall) plots show how the spectrum varies with time in some situations.
4. Statistical analyses of fluctuating noises: probability histograms show the percentage of total duration at various noise levels.
5. Graphs of various parameters calculated from measurement data, such as sound absorption coefficient, or standardized level difference, are plotted against frequency.

In a level recorder the pen moves vertically up and down the paper, indicating the level of the signal, and the paper moves to produce the horizontal axis, which is usually time or frequency. In other recorders the pen moves in both vertical and horizontal directions. In either case an important property of the instrument is the response time of the pen movement system, which must be fast enough to follow changes in the fluctuating signal, but not so fast as to be unstable. The recorder may have its own signal processing facilities, e.g. for RMS averaging or for producing logarithmic (decibel) versions of the signal, or it may be that a fully processed signal is delivered to the recorder.

In all cases it is important that the charts and graphs are accurately calibrated; axes should be labelled and annotated with all relevant details about the noise.

9.15 Tape recorders

A chart recorder gives a permanent record of the analysis of a noise, but a tape recording allows a permanent record of the noise signal itself to be stored, on magnetic tape. Tape recordings of noise may be made for a variety of reasons:

1. They save time carrying out lengthy analysis procedures at the noise measurement site.
2. They enable a variety of different analyses to be made, including analyses which require equipment not available at the measurement site.
3. They capture a rare or unrepeatable noise event.
4. They enable the sound to be listened to for subjective assessments and comparisons.

The tape recorder must have a sufficient dynamic range and an adequate frequency response to enable it to store and reproduce the signal faithfully. It must be possible to adjust the level of the signal being recorded on the tape, and the recorder must have a signal level meter. These facilities are necessary to ensure that the signal is large enough to fully use the dynamic range of the tape recorder so that the signal-to-noise ratio is as big as possible, but not so large that overload of the recording system occurs, producing distortion (or clipping) of the signal. The danger of overload is greatly increased when the noise contains an impulsive component because the recording level meter may indicate an average level, which is well below the peak level. The operator must learn to make allowance for this difference. Tape recorders with automatic level control and noise reduction circuits are not suitable for recording noise for measurement and analysis unless these facilities can be switched off.

Most tape recorders have a variety of different tape speeds. The upper frequency limit of the recorder depends on the tape speed and is highest for the maximum tape speed. The choice of the appropriate tape speed for a recording may therefore be a compromise between frequency response and economy in the use of tape, since the highest tape speed gives the lowest recording time per tape. It is possible by recording at one speed and replaying at another to transform the frequencies in the signal, for example, a recorded signal of 10 000 Hz may be replayed at half the recording speed to produce a signal of 5000 Hz.

The noise signal may be processed by the recorder in a variety of ways before being recorded onto the tape. The direct record mode is the simplest and most common, in which it is the audio frequency signal itself which being recorded directly onto the tape, although it is frequency weighted to allow for the fact that the frequency characteristics of the magnetic record and replay processes are not constant with frequency. However, the weighting of the recorded signal is 'equalized' or 'undone' by an opposite (or inverse) weighting of the replay signal, to produce a total frequency response which is reasonably flat over the operating frequency range. The direct record made is quite capable of responding faithfully to signals of up to 20 000 Hz, but because of the nature of the magnetic recording process, tape recorders of this type have a low frequency limit, which varies for different machines but is roughly in the region of 20 Hz. If it is necessary to record and measure frequencies below 20 Hz then a different mode of recording must be used in which the audio frequency noise signal is used to frequency modulate a much higher carrier frequency. The signal is demodulated during the replay process. Frequency modulation (FM) tape recorders are capable of responding equally to signals of frequencies down to 0 Hz. FM tape recorders are often essential for recording very low frequency, infrasonic signals and for many vibration signals.

The performance of the tape recorder, whether direct record or FM, depends on the type and quality of the tape, so tape of a high quality and of a type suitable for use with the particular tape recorder should always be used. The tape should be replayed on the same machine as used to make the recording.

The use of multiple-track tape recorders allows simultaneous recordings of two or more signals, e.g. noise at two different locations, noise and vibration from one machine or vibration in three directions (*x*, *y* and *z*). Comparisons between the different signals can help in the location of noise sources and the identification of transmission paths.

It is very important that all tape recordings are calibrated by the inclusion of a signal of known level (e.g. from an acoustic calibrator attached to the microphone) on the tape. The details of all relevant instrument settings (such as the attenuator range controls on the sound level meter) and any changes to these settings during the recording must be carefully noted. Other details relevant to the measurement situation (e.g. date, time, microphone position and details of noise source) must also be noted. Ideally all these details should be noted both on the tape itself, i.e. via the spoken voice, and also on documentation attached to the tape.

9.16 DAT recorders

Since the late 1980s digital audio tape cassette recorders (DATs) have been increasingly used for noise and vibration measurement. These instruments have built-in ADCs and the acoustic signal is recorded on the tape in coded digital form. They have good frequency response over the audio range and can be adapted to operate down to 0 Hz; they have a very high dynamic range, about 80 dB. DACs enable the signal to be replayed through headphones or loudspeakers.

The very high dynamic range means that it is possible to record the entire range of environmental noise likely to be encountered in most situations without the need to adjust recording levels or sound level meter ranges. This allows environmental health officers and noise consultants to use DATs in the investigation of noise complaints. Often the noise being complained of occurs only intermittently, or during night-time periods when it is difficult for the investigator to be present to witness or measure the noise. A combination of DAT recorder and sound level meter, secured in a tamper-proof box, is set up, calibrated and left in the complainant's home so that he or she can switch on the instrument when the offending noise occurs. The tape recording can be used by the investigator in a number of ways. It can help to verify the claims of the complainant about the noise, and it can be used in negotiations with the author of the noise to help secure its reduction.

9.17 Environmental noise analysers

The level of many environmental noises varies widely with time, and statistically based measurements have to be taken over periods of up to 24 hours, or even longer. Examples are road traffic noise, train and aircraft noise, noise from building sites, factories and commercial premises, general community noise and noise from entertainment events.

Microprocessor-controlled instruments with a wide dynamic range have been developed which can be left to monitor the signal from a microphone automatically over long periods of time. A variety of noise indices, such as L_{eq}, L_{10}, L_{50} and L_{90}, may be selected, measured then indicated by digital display or by printout on paper tape. The value of these indices may be measured and indicated automatically at regular specified time intervals, e.g. every hour, or their cumulative value over the total period (e.g. 18 hours) may be measured.

Some of the early signal processing in these instruments (e.g. A weighting and RMS averaging) is performed by analogue circuits, but at the heart of such instruments is an analogue-to-digital converter which samples and digitizes the signal, a memory which stores the sample for subsequent analysis and a microprocessor which directs and organizes the sampling process, the calculation, and the display of the selected noise indices from the data in the store.

These instruments give tremendous flexibility to the operator, who has control over sample rate and duration of sampling (within the available memory capacity), selection of indices to be measured and the time intervals over which measurements are to be made. The adaptability of microprocessor control sometimes allows selection of a different range of indices; they are plugged in using electronic circuit boards containing different signal processing routines.

In some situations, particularly at night-time, noise levels are usually very low, punctuated by occasional bursts which may be causing a disturbance and which may be of interest. The analyser may be programmed to sample and analyse only noise events over a certain threshold, and at the end of the monitoring period, to list them with their onset time, duration and values of the selected noise indices. It is even possible for a tape recorder to be switched on automatically to record each event (then switched off afterwards) so that the events can be listened to and identified later.

9.18 Noise dosemeters

Noise dosemeters (also called personal sound exposure meters) are small portable instruments designed to be worn by a worker throughout the day in order to measure his/her personal noise exposure. A small microphone is attached to the collar (the hat or helmet if worn) about 2 cm from the ear. The microphone is attached by a flexible lead to the small battery-operated instrument which is carried in a pocket.

The instrument is a specially developed form of integrating sound level meter. The microphone signal is A-weighted then squared; this is because the sound intensity is proportional to the square of the pressure. The noise energy received by the microphone also depends on the exposure time; the instrument integrates the square of the pressure and the exposure time to produce a true measure of A-weighted sound energy. This is displayed as a count, i.e. a number, which may be converted to a personal daily noise exposure level, $L_{EP,d}$, in dB(A) using a calculation chart. Equivalent to an eight-hour L_{eq}, this value is used to evaluate the risk of hearing damage to the worker. In some instruments the result is expressed in terms of 'percentage noise dose', where 100% represents an $L_{EP,d}$ of 90 dB(A), which is the second action level of the Noise at Work Regulations (1989).

Simple dosemeters are useful for determining whether or not a noise exposure has exceeded the action levels of the Noise at Work Regulations (1989), but they will not give any information about the pattern of noise exposure throughout the measurement period. More complex data-logging dosemeters are available which can give a statistical analysis of the noise levels and can be downloaded onto computers to give a detailed profile showing how the noise level has varied throughout the measurement period. This will enable periods of exposure to high noise levels to be identified and linked to particular workplace activities or locations. Remedial action can then be taken to reduce the noise exposure in the future.

Dosemeters are usually classed as type 2 instruments according to BS 6698, even if the highest quality microphones are used. They cannot achieve the accuracy required of type 1 instruments because of uncertainties in the measurement associated with the diffraction and scattering of sound close to the microphone by the head and body of the wearer. The measurement error may be of the order of ± 2 dB.

9.19 An introduction to the principles of the direct measurement of acoustic intensity

The intensity of an acoustic wave is defined as the rate at which acoustic energy passes through unit area perpendicular to the direction of propagation of the wave. It is measured in W/m^2.

The instantaneous intensity (I) at a point in a wave can be shown to be the product of the acoustic pressure (p) and the acoustic particle velocity (u) at that instant:

$$I = pu$$

To find the average intensity of the wave at that point it is necessary to find the 'time average' of this product because both p and u vary from moment to moment throughout the cycle of the sound wave, so:

$$\bar{I} = \overline{pu}$$

where the bar symbol ($^{-}$) denotes 'time averaged'.

It was stated in Chapter 1 that the acoustic intensity is proportional to the square of the sound pressure ($I \propto p^2$) and that for plane waves $I = p^2/\rho c$. Sound pressure may be easily measured using a microphone, and this relationship is the basis of indirect measurements of acoustic intensity. There is, however, a very important distinction between the intensity and the square of the pressure to which it is so closely related. Since intensity describes a **rate of flow** (of energy), it is a quantity which has a direction, like force, momentum or velocity. It is called a vector quantity. The quantity square of the pressure has no direction attached to it. It is a scalar quantity like mass or volume. Therefore indirect measurements of acoustic intensity based on sound pressure measurements may give the correct magnitude of the intensity but can never indicate the direction of flow of acoustic energy.

Returning to the equation $\bar{I} = \overline{pu}$, the direct measurement of acoustic intensity at a point in a sound field requires two transducers at that point, one to measure p (i.e. a microphone) and another to measure u. The signals from these two transducers must then be multiplied together and time averaged to produce an analogue signal which directly represents the acoustic intensity.

Unfortunately it is not easy to measure acoustic particle velocity directly. There is, however, a relationship between p and u (not involving the unknown I) which can help. The correct statement of this relationship (called the equation of motion) involves the use of the notation and symbols of differential calculus, but the general idea may be obtained by considering a small box of air in the path of a plane wave striking one face of the box (of area A) at right angles (Fig. 9.23).

The length of the box is Δx; the symbol Δ indicates a small increment in the direction x in which the wave is travelling. The pressure and particle velocity at the front face of the box (facing the approaching sound wave) are p and u, respectively; at the opposite face they have undergone very small changes and are $p + \Delta p$ and $u + \Delta u$, respectively.

The required relationship is obtained by applying Newton's second law to the motion of the air in the box caused by the passage of the sound wave:

$$\text{force} = \text{mass} \times \text{acceleration}$$

Remember that:

$$\text{force} = \text{pressure} \times \text{area}$$

So the net force on the box, to the right is:

$$pA - (p + \Delta p)A = -A\,\Delta p$$

The mass of air in the box is:

$$\text{density} \times \text{volume} = \rho A \Delta x$$

where ρ = density of the air

Now recall the formula for acceleration:

$$\text{acceleration} = \frac{\text{change in velocity}}{\text{time taken}}$$

$$\text{acceleration} = \frac{\Delta u}{\Delta t}$$

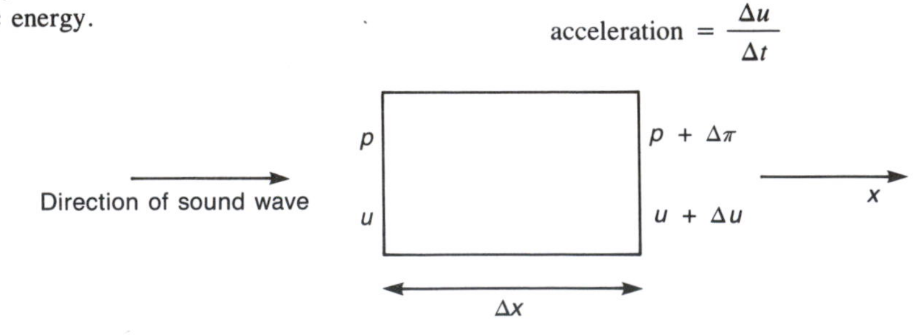

Fig. 9.23 Incremental changes in acoustic pressure and particle velocity over a small distance

where Δt = small interval of time in which the particle velocity has changed from u to $u + \Delta u$

Combining everything together into the force equation:

$$-A\Delta p + \rho A \Delta x \cdot \frac{\Delta u}{\Delta t}$$

A cancels leaving:

$$-\Delta p = \rho \Delta x \cdot \frac{\Delta u}{\Delta t}$$

And on rearranging:

$$\rho \frac{\Delta u}{\Delta t} = -\frac{\Delta p}{\Delta x}$$

This is the required relationship which shows that **the rate of change of particle velocity (with time) is related via the density of the medium to the rate of change of acoustic pressure with distance, i.e. to the sound pressure gradient**. Therefore, if the sound pressure gradient can be measured, the rate of change of particle velocity can be found, and by integrating it with respect to time, the particle velocity itself can be obtained.

The sound pressure gradient in a sound field may be measured, approximately at least, by having two microphones close together in the sound field. If the two sound pressures are p_1 and p_2 and the distance between the microphones is Δx then:

$$\text{sound pressure gradient} = \frac{\Delta p}{\Delta x} = \frac{p_1 - p_2}{\Delta x}$$

The two microphone arrangement is the basis of the sound intensity meter. The sound pressure used for the $\bar{I} = \overline{pu}$ calculation is the average of the two signals, i.e. it is given by:

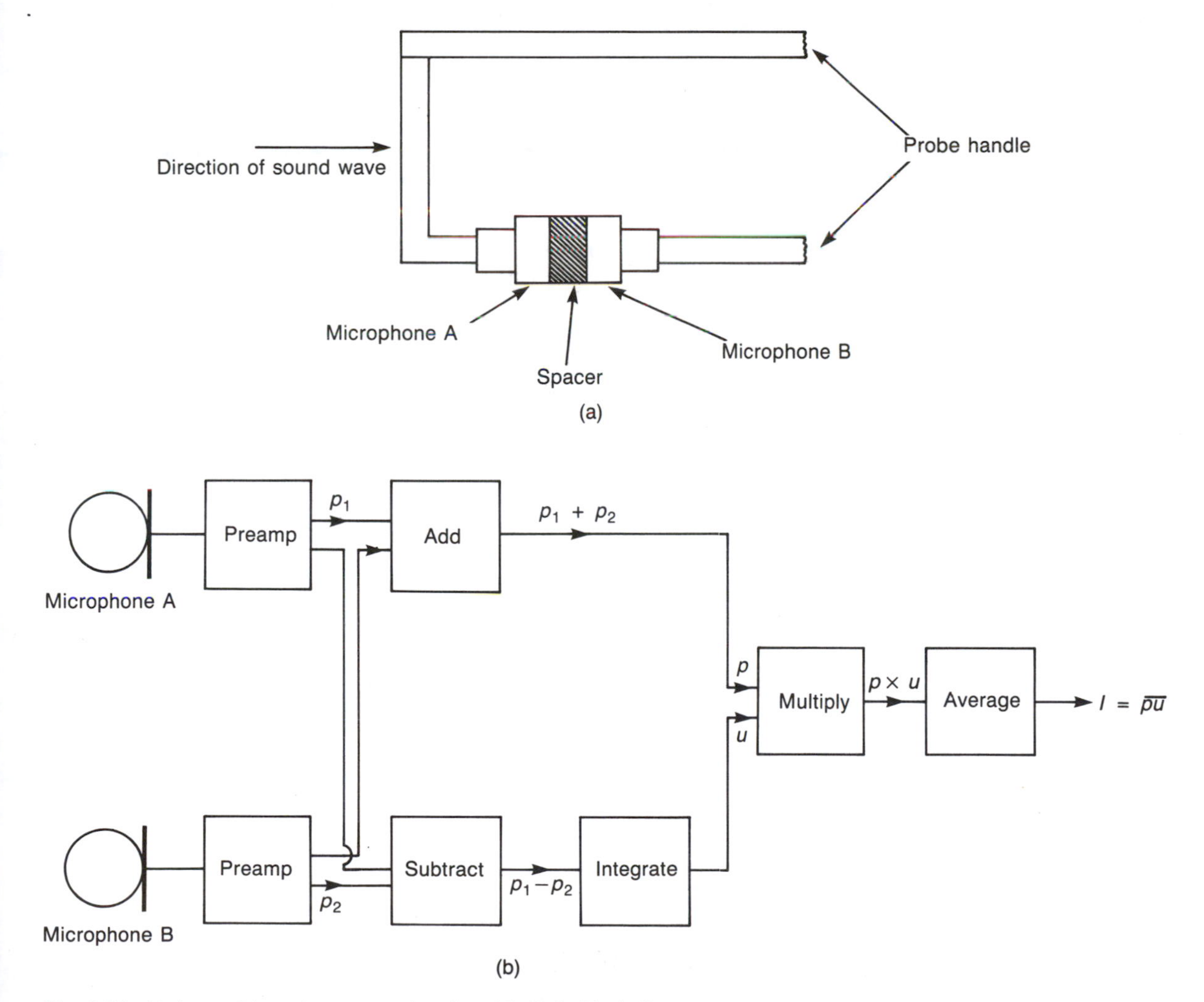

Fig. 9.24 (a) A sound intensity meter and probe with (b) its block diagram

$$p = \frac{p_1 + p_2}{2}$$

The sound intensity may therefore be measured by processing the two signals and combining them in the following way. The difference between the two signals ($p_1 - p_2$) is obtained then integrated with respect to time to produce a signal which is proportional to u. The sum ($p_1 + p_2$) of the two signals is obtained to produce a signal which is proportional to p. The signals proportional to p and u are then multiplied together and time averaged to produce a signal which may be calibrated to represent the average acoustic intensity. A block diagram of a sound intensity meter is shown in Fig. 9.24.

The vector nature of acoustic intensity means that the two-microphone intensity probe is directional in its response, giving the maximum signal when the axis of the probe (i.e. the line joining the two microphones) is lined up in the direction of the acoustic energy flow. It is this ability to identify the direction of flow of acoustic energy, as well as measuring its magnitude, which is the basis of many of the applications of the sound intensity meter. One of these applications is in the measurement of the sound power emitted by noise sources. Using an acoustic intensity meter it is possible to measure the sound power level of a noisy machine *in situ*, in the presence of background noise and without the need for a specialist acoustic environment, i.e. an anechoic or reverberant room. This is described in Chapter 8. Other applications include the location and identification of noise sources and the detection of leaks and weak-link areas in the transmission of sound through partitions and panels.

9.20 Calibration

Although calibration has already been mentioned earlier in the chapter, it is sufficiently important to deserve a section of its own in any chapter on instrumentation.

The principle behind the use of the acoustic calibrator is that **the whole measurement system from the transducer right through to the display device should be calibrated** and not just individual parts of the system. If the measuring system ends in a level recorder, the level of the calibrator signal applied to the microphone should be marked on the level recorder paper. When tape recordings are made, it is imperative that a sample of the calibration signal is recorded on the tape. It is always important to note the settings of all relevant instrument controls, e.g. the positions of gain control switches; adjustments need to be made to the calibration if the settings are subsequently changed.

Many acoustic calibrators operate at only one level and at one frequency — a common type generates 94 dB at 1000 Hz at the microphone. Confidence that the sound level meter also reads correctly at other levels, say 84 dB and 74 dB, is based on the accuracy and reliability of the attenuators; the above case produces attenuations of 10 dB and 20 dB respectively. The reliability of measurements at other frequencies is based on the frequency response of the whole instrument, especially the microphone. Some calibrators allow calibration over a range of different levels and frequencies.

Automatic calibration

Fully automated noise measuring stations operating continuously over long time periods have to be calibrated in a different way. It is obviously impossible to automate the process of fitting an acoustic calibrator over the microphone. A fine grid, called an electrostatic actuator (Fig. 9.25), is permanently fitted over the diaphragm of the microphone. Calibration is achieved by remotely applying a known high voltage to the actuator. The resulting electrostatic force between diaphragm and actuator deflects the diaphragm in a similar way to the pressure from a sound wave. The signal level produced by the actuator voltage is displayed and may be compared to that expected from the known microphone sensitivity.

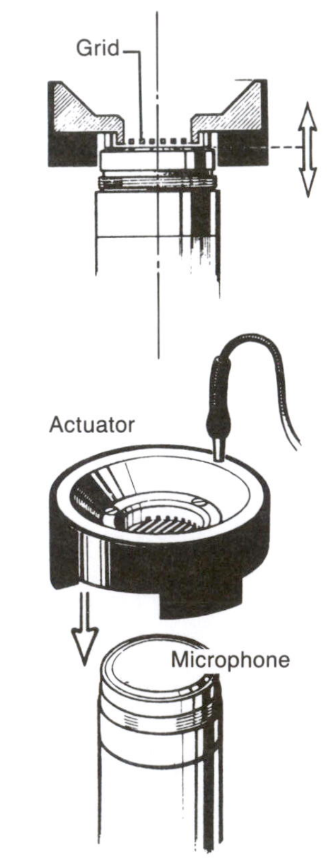

Fig. 9.25 Electrostatic actuator (Courtesy Bruel & Kjaer)

The principle of total calibration of the entire measurement system applies equally to vibration measuring equipment; equivalent to an acoustic calibrator is a vibrating table. The vibration transducer is mounted on the vibrating table, which then produces an accurately known vibration level.

Internal calibration

Sound level meters and vibration meters do contain their own internal calibration systems. They consist of voltage signals of accurately known amplitude, generated within the instrument, which can be applied to the input amplifiers instead of the transducer signal. Alternatively a calibrated signal from an external signal generator may be used. In either case this form of calibration can serve only to check the signal processing and display systems; it cannot detect changes in sensitivity of the transducer.

Traceability and verification

In addition to the day-to-day checking of sound level meters using acoustic calibrators, it is good measurement practice and a requirement of some measurement standards, including BS 4142, to ensure that both sound level meters and calibrators are periodically calibrated by a laboratory belong to the National Measurement and Accreditation Service (NAMAS). In the laboratory the equipment is calibrated against more accurate equipment, which itself has been calibrated against even more accurate equipment, and so on, thus ensuring **traceability** of calibration to a national or international standard.

It is also desirable to verify that the sound level meter is still operating in accordance with manufacturer's specifications and the requirements of standards such as BS 5969 and BS 6698. Full **verification** against these standards, called pattern evaluation or type testing, is only required of the manufacturer before a new instrument is introduced onto the market, and the procedures would be too time-consuming and expensive to be applied as a subsequent periodic verification test of the instrument in use. Furthermore, these standards, although giving performance specifications, do not indicate how they should be tested. Therefore BS 5780 has been issued to describe procedures for the periodic verification of sound level meters. BS 5969 and BS 6698 are under review, and the new versions will include procedures for periodic verification; eventually BS 5780 will no longer be needed, and will be withdrawn.

Laboratory calibration

An acoustic calibrator is a **transfer standard** which has itself been calibrated using a microphone of accurately known sensitivity. The calibrator is then used to transfer the calibration of other uncalibrated microphones, but with a slightly lesser degree of accuracy. Somewhere along the line a microphone must be calibrated by other means, without using a transfer standard. There are three methods available:

1. Electrostatic actuator method
2. Comparison method (i.e. direct comparison with other microphones)
3. Reciprocity method

The electrostatic actuator method has already been described. It is also useful for measuring the frequency response of microphones and sound level meters in the laboratory as well as for remote calibration.

The comparison method is simple in principle. The sound pressure level is measured at a certain point using a calibrated microphone. This is then replaced by the uncalibrated microphone, which is thus experiencing the same sound pressure. The indicated levels are compared. Although simple in principle, an anechoic room is usually needed in practice to achieve sufficient control of the acoustic environment and to ensure a high degree of accuracy.

The reciprocity technique is an absolute method, i.e. it does not rely on the availability of an already calibrated microphone. The method, which will not be described here, relies on the relationships between the performance of the device when operated as a microphone and when operated in reverse as a loudspeaker.

9.21 Noise measurement: the effective use of equipment

It is entirely fitting that this chapter should end with a section on how equipment should be used in order to carry out measurements which are appropriate, representative, reliable and accurate. There is no instrument which is foolproof, and the adage about computers — rubbish in, rubbish out — is applicable. Results of noise measurements taken with even the most sophisticated, accurate and expensive equipment may be useless if the operator does not fully understand how to use the instrument, and does not have sufficient knowledge of acoustic principles to appreciate its limitations or how to make the necessary decisions about what, how, where and when to measure.

Measurement aims

The starting point is to decide on the main aims of the noise measurement because they will influence further decisions. As an example, consider the possible reasons for needing to measure noise levels produced by a piece of machinery such as a compressor:

To compare with noise levels from other, alternative equipment.

To compare with manufacturers' noise specifications.

To compare with company's own (i.e. customer's) noise specification.

To assess the need for, or effect of, machine maintenance.

To assess need for noise reduction requirements and to quantify them.

To diagnose noise sources and mechanisms, and to specify control methods.

To assess the effectiveness of noise control measures.

To measure the sound power level of the machine.

To assess the noise exposure level of the machine operator.

To use as a basis for predicting noise levels at other distances.

To use as part of a noise complaint investigation.

To use as part of a planning application investigation.

To use as part of an environmental impact assessment.

Some of these different aims will involve very different measurement procedures, for example, choice of measurement position, type of sound level meter and measured noise parameter.

Choices and checks

Choices and checks, factors and influences plus information to be noted may include:

Checks

- battery condition
- calibration
- use of windshield
- background noise levels

Choices

— type of sound level meter
— type of microphone (free-field, pressure response, random incidence)
— position of microphone relative to source or receiver
— height of microphone above ground
— orientation of microphone (frontal or grazing incidence)
— noise parameter(s) to be measured
— measurement time(s) and duration(s)

Factors and influences

— other noise sources
— nearby sound reflecting, absorbing or shielding surfaces
— room characteristics (indoors), e.g. size, surface finishes
— sound propagation conditions (outdoors), e.g. wind, weather, ground type, topography
— type of sound field, e.g. near-field, far-field, direct or reverberant sound, presence of standing waves

Information

— all the above information about checks, choices, factors and influences
— manufacturers' type numbers and serial numbers of all equipment
— results of measurements and relevant instrumentation settings
— subjective impressions of the noise, e.g. steady, intermittent, impulsive, tonal
— maps and plans showing measurement positions, source and receiver positions, with distances

Measurement standards and codes of practice

In many cases these measurement decisions will be based on the operator's knowledge and experience, but there are some situations in which conditions are specified by standards or codes, which ensure uniformity of good practice.

Examples of some standardized procedures include:

— Measurement method given in Department of Environment publication *Calculation of Road Traffic Noise*
— Measurement method given in Department of Environment (draft) publication *Calculation of Train Noise*
— Aircraft noise, ISO 3891:1978, Procedure for describing aircraft noise heard on the ground
— Commercial and industrial noise, BS 4142:1990
— Construction site noise, BS 5228:1984
— General environmental noise, BS 7445:1991
— Workplace noise, Noise at Work Regulations Guidance Notes 3 to 8, 1990

Measurement reports

Measurement reports should contain relevant information in sufficient detail to enable a repeat of the measurements to be carried out, should this be necessary, under, as far as possible, identical conditions. The sort of information to be reported is listed above. It can conveniently be grouped as follows:

- Basic information such as dates, times, locations and name(s) of persons carrying out the measurement
- Information about the source(s) of noise
- Information about the instrumentation and the noise measurement data

- Information about the acoustic environment in which the measurements were taken

Many of the British Standards on noise measurements end with sections on the information to be recorded and reported, and they give detailed guidance in these specific measurement situations.

Where a report is lengthy or detailed, it may be helpful to the reader if it is split into separate sections, such as:

- aims of the measurements
- measurement procedures
- results of measurements
- calculations and analysis of measurement data
- discussion of results
- conclusions
- summary or abstract

Accuracy of measurements

Every measured value has associated with it a degree of uncertainty, or error. The word **error** is used here in a specialist, technical sense, not in its everyday sense; it does not mean a mistake. A statement of the experimental error in a measured quantity is also a statement of its accuracy, e.g. reverberation time = 1·6 ± 0·2 s. Accuracy is not the same as precision, which is a statement about the smallest detectable change in a quantity. Thus a measurement of a noise level of 86·2 dB(A) from a machine implies that the measuring instrument was capable of a precision of 0·1 dB, but this does not necessarily mean that measurement is accurate to the nearest 0·1 d(B), since random fluctuations, or errors might mean that a repeat measurement gave a value one or two decibels higher or lower.

Errors may be of two types, random or systematic. Random errors will show up as fluctuations in the measured value when the measurement is repeated; systematic errors will not. Systematic errors, once detected, may either be allowed for or eliminated, as in the case of a sound calibrator or sound level meter which is systematically reading 0·3 dB too low.

Errors in the measured value of a noise level, perhaps from a machine, may arise from a number of sources:

- due to the limitation of the accuracy of the sound level meter, and variability in environmental conditions affecting it
- due to variability in the way the operator takes the measurement
- due to variability in the source of noise
- due to variability in environmental conditions affecting noise propagation
- due to variability caused by the statistical nature of the quantity being measured, e.g. traffic noise
- due to limitations in the accuracy of theoretical (or other) assumptions underpinning the measurement (e.g. in the measurement of sound power level or sound intensity level)

Careful measurement procedures can often minimize errors, but if they are of the random type they will never be eliminated completely. The magnitude of random errors is usually estimated from the scatter in results. A common method is to estimate and quote the standard deviation of a group of measurements from the mean value.

Example 9.3

The sound pressure levels measured in eight different microphone positions in a room during a sound insulation test are given below. Calculate the mean and standard deviation values.

measured levels: 92 dB, 95 dB, 91 dB, 96 dB, 90 dB, 92 dB, 95 dB, 93 dB

mean level = (92 + 95 + 91 + 96 + 90 + 92 + 95 + 93)/8 = 93 dB

deviation from mean: −1, +2, −2, +3, −3, −1, +2, 0

deviation squared: 1, 4, 4, 9, 9, 1, 4, 0

mean square deviation = (1 + 4 + 4 + 9 + 9 + 1 + 4 + 0)/8 = 4

standard deviation = root mean square deviation = 4 = 2 dB

∴ sound pressure level in room = 93 ± 2 dB

If the number of results is large enough, and their scatter about the mean is random, the usual normal or Gaussian statistics can be applied to interpret the standard deviation in terms of probabilities and confidence limits, e.g. there will be a 68% probability that any one result will lie within one standard deviation of the mean, and a 95% chance that it lies within two standard deviations of the mean.

When the results are sound pressure levels, in decibels, the question sometimes arises as to whether the mean and standard deviation should be based on the decibel values, as in the above example, or on the squares of the sound pressure underlying them. Beyond the scope of this chapter, the question is discussed in the book by Yang and Ellison (*Machinery Noise Measurement*) together with other aspects of errors in noise measurements.

The measurement uncertainty, or error, in measured values determined using standard methods are usually given, in terms of a standard deviation in the appropriate British or International Standard. BS 4196 describes three different grades of measurement of sound power level — precision, engineering and survey — with three different levels of accuracy.

Any more detailed discussion of accuracy and measurement errors is beyond the scope of this chapter. The reader should at least be aware that every measured quantity has associated with it a degree of uncertainty, or error, which specifies its accuracy. A knowledge of the likely degree of accuracy of measurements is important when assessing the significance of their contribution to the overall result or conclusion. Even when a formal statement of uncertainty is not required, the statement of the result should be consistent with its likely accuracy. Thus it may be appropriate to quote a measured sound level only to the nearest decibel if it is known to vary over a range of a few decibels, even though the sound level meter indicates to the nearest 0·1 dB. This is a requirement of BS 4142 in particular. The quotation of results to the nearest 0·1 dB may sometimes be justified, when the measurements are very accurate or as the results of intermediate steps in a calculation, where rounding errors may accumulate. But the statement of a result to more than one decimal place, arising perhaps from a calculation, can never be justified and should be avoided.

Questions

(1) Draw a block diagram of a simple sound level meter and describe the function and characteristics of the various parts. Describe the special features of (a) impulsive sound level meters; (b) integrating sound level meters and (c) digital sound level meters.

(2) Describe the principle of operation of the following types of microphones and compare and contrast their properties: (a) condenser microphone; (b) electret microphone and (c) piezoelectric microphone.

(3) Explain the difference between analogue and digital processing of signals. Discuss the impact of digital techniques on sound measuring instrumentation.

(4) Explain the meaning of the following terms used in connection with the performance of microphones: sensitivity, linearity, stability, output impedance, frequency response and dynamic range.

(5) How may acoustic equipment be calibrated so that confidence may be placed in the reliability of measurements?

(6) Explain the principles of acoustic intensity measurement. What are the advantages of the direct measurement of intensity over the indirect method, based on sound pressure measurements? Give some applications of direct intensity measurement.

(7) Explain why it is sometimes necessary to make tape recordings of sound and vibration signals. Discuss the requirements of a suitable tape recorder.

(8) State, with reasons, the equipment you would choose for the measurement and analysis of noise in the following situations, giving any special requirements:
  (a) To assess the noise received by a factory worker who is continually moving from place to place; the noise level in the factory varies at different positions and throughout the day.
  (b) To monitor the noise inside the bedroom of a dwelling from a nearby factory, over a 24-hour period; you are particularly interested in measuring and identifying bursts of noise from the factory which occasionally occur in the middle of the night.
  (c) To continuously monitor noise outdoors near to an airport over a six-month period.
  (d) To measure and identify the source of an annoying whine from a piece of factory machinery.
  (e) To measure the noise from a clay pigeon shoot; L_{AE}, peak sound pressure and maximum octave band levels are required.

(9) Explain why the sensitivity of a microphone at high frequencies depends on the size of the microphone and the angle of incidence of the sound. Explain free-field response, pressure response and random incidence response.

(10) Explain the meaning and significance of the following terms used in connection with sound measurement: frequency weighting network, crest factor, averaging time, real-time analysis and filter bandwidth.

(11) (a) Define the equivalent continuous sound level.
  (b) List what sounds can be evaluated using the equivalent continuous A-weighted sound level, and mention one example in which its use is not appropriate, giving brief reasons.
  (c) Use a labelled sketch to describe the basic elements of an integrating sound level meter.
  (d) Briefly describe the basic characteristic of a digital frequency analyser based on the Fast Fourier Transform (FFT), and mention one application.

  Assuming that the sampling frequency is $2f_u$ samples per second, where f_u is the upper limit of the frequency range of interest, and the FFT execution time for a 1024 sample FFT is 0·2 s, determine whether the analyser can be used for real time analysis for the frequency range from 0 to 4000 Hz. *[IOA 1992]*

(12) (a) Describe the construction and action of an air condenser instrumentation microphone. Discuss the advantages and disadvantages of this type of microphone.
  (b) Describe the frequency response of a piezoelectric accelerometer.

When a piezoelectric accelerometer is calibrated, the output voltage is 96 mV rms for a sinusoidal input of 9·81 ms^{-2} at 80 Hz. The accelerometer is then placed on an object vibrating sinusoidally at 50 Hz. The output voltage is now 54 mV rms. Calculate the peak velocity. (*IOA 1991*)

(13) Describe with the aid of a block diagram, the operation of a simple analogue sound level meter fitted with octave band filters. Explain the difference between AC and DC output signals and indicate where they would occur on the block diagram. Explain which part of the system limits the ability of the meter to measure impulsive sounds accurately. (*IOA 1990*)

(14) For a sinusoidal wave, show that the crest factor = 1·414 or approximately 3 dB and the form factor = 1·11 or approximately 1 dB.

Discuss the importance of crest factor on measurements made when using either a sound level meter or a level recorder. (*IOA 1990*)

(15) Write brief notes on the following, with reference to the design and use of sound level meters.

(a) IEC 651 Type 0, 1, 2, and 3 instruments.
(b) Fast, Slow and Impulse response.
(c) A, B, C and D weighting networks.

Describe how you would use a graphic level recorder in conjunction with a sound level meter to obtain a chart recording of sound levels.

A manufacturer quotes the dynamic range of a sound level meter as 38 to 140 dB when used with a microphone of sensitivity 50 mV Pa^{-1}. Estimate the internal electrical noise generated in the instrument as an equivalent input voltage. What would be the limits of the dynamic range of the instrument when fitted with a 1/4″ microphone of sensitivity 4 mV Pa^{-1}? (*IOA 1989*)

(16) There is a proposed industrial development at the edge of an existing industrial zone. To the north of the site is an existing housing estate, to the west a main road, to the east public open space and the rest of the industrial zone lies to the south.

You are instructed to determine the existing background noise levels at the dwelling nearest the the site. What instrumentation would you use and what measurements would you take for this purpose taking all relevant factors into account?

Why may the L_{Aeq} value be determined to the same degree of accuracy as the background level but in a shorter period of time? (*IOA 1988*)

(17) Outline the construction and action of an analogue sound level meter.

A sound level meter is to be used as a vibration detector with an accelerometer connected to its input amplifier. Calibration of the accelerometer at 10 m s^{-2} gave a meter reading of 92 dB (linear). When the accelerometer was attached to a vibrating surface, the reading was 75 dB (linear). Determine the acceleration of the surface in dB re 10^{-6} m s^{-2}. If the vibration frequency is 30 Hz, calculate the displacement of the surface.

(18) Describe how a noise which varies with time may be measured, including both analogue and digital techniques and discuss the problems of assessing this type of noise. Include reference to industrial, traffic and construction site noise, the appropriate British Standards and investigations of community response to time varying noise. (*IOA 1987*)

(19) What precautions should be made when measuring on the ground the noise of aircraft in flight?

(20) (i) Draw a block diagram of a noise dosemeter and explain how it works.

(ii) An industrial process produces a fairly steady noise of equivalent level 84 dB(A) with a superimposed impulsive noise of level 130 dB(A) and duration 5 ms occurring regularly once every minute. What is

(*a*) the equivalent level of the impulsive sound?
(*b*) the total L_{eq}?

(iii) A plant worker wearing a noise dosemeter calibrated to read 100% for a noise dose of 90 dB(A) for 8 hours receives a dose of 100%. If he receives an increase of dose of 50% during a period of 30 minutes spent in a particularly noisy area of the plant, what is the L_{eq} of this noise and what is the maximum permissible L_{eq} for the remaining 7·5 hours of the operator's working day? (*IOA 1987*)

(21) Discuss the main principles of operation and constructional details of the capacitor microphone. Account for the differences between the *free-field* and *pressure* responses for this type of microphone and indicate how these responses can be used to advantage.

Explain what design modifications are necessary for capacitor microphone systems used for external, unattended noise monitoring purposes.

If open circuit output voltages of 7·94 μV and 25·2 V were recorded corresponding to the internal noise level and the 3% distortion limit respectively for a microphone (sensitivity −38 dB re 1 V per N/m^2) estimate the dynamic range for the microphone (*IOA 1986*)

(22) (*a*) A microphone amplifier is calibrated to read dB re 20 μPa when used with a microphone of sensitivity 50 mv/Pa.

(i) Calculate the rms acceleration if the meter reads 78 dB when the microphone is replaced by an accelerometer of sensitivity 2 mV/m s^{-2}.

(ii) If the vibration frequency is known to be 50 Hz calculate the corresponding rms velocity and displacement.

(*b*) A one-third octave filter has a centre frequency of 1 kHz. What are its upper and lower frequency limits?

Describe briefly with the aid of sketches, how real band pass filter amplitude transfer functions (attenuation curves) differ from ideal ones. (*IOA 1985*)

(23) Discuss qualitatively, with the help of a suitable block diagram, the operation of a noise dosemeter and explain how the readings from such a meter can be interpreted in terms of the equivalent continuous sound level, L_{eq}. (*IOA 1983*)

(24) Give an explanation of the following terms:

(i) The signal to noise ratio of a tape recorder.
(ii) The Fast and Slow response of a sound-level meter.
(iii) The averaging time of a level recorder.
(iv) The bandwidth of a spectrum analyser.

Describe how the 'A' weighting response of a sound-level meter could be tested using a signal generator.

Discuss the significance of the deviations from the standard dB(A) measurements on the assessment of noise nuisance from an industrial complex at high, medium and low audio-frequencies. (*IOA 1982*)

(25) Given an accurately calibrated sound-level meter, describe the procedure you would adopt when making a noise survey of a workshop.

Include in your description reference to:

(*a*) Various sound fields.
(*b*) Areas near noisy machines.
(*c*) Assessment of the effects on personnel-criteria to be considered.
(*d*) Fluctuation of noise levels — movement of personnel.

Indicate how you might determine whether the presence of impulsive noise is of practical importance.

Define the term 'crest factor' and explain the importance of 'capturing' the complete noise event.

Show by calculation that an impulsive sound of 3 s duration, at a level of 50 dB above the prevailing background noise level, would raise the 8 hour L_{eq} by about 10 dB. (*IOA 1982*)

(26) (i) Explain why the frequency response of a microphone is directional at higher frequencies and distinguish between pressure response and free-field response.

(ii) Describe the characteristics of octave band filters suitable for the analysis of noise.

(iii) A sound level meter correctly registers 124 dB SPL when undergoing a calibration check.

If the microphone sensitivity is -26 dB re 1 V/Pa, what is the input voltage to the meter? (reference pressure = 20 μPa). The microphone is then replaced by an accelerometer, with a sensitivity of 2 mV/m s^{-2}, attached to a vibrating surface. If the reading is 85 dB, what is the r.m.s. acceleration of the surface? (*IOA 1981*)

(27) Explain the terms *frequency response* and *dynamic range* when applied to a sound-level meter. Discuss the factors which are likely to affect the frequency response of the meter.

What determines the lowest sound level that can be measured accurately by a sound-level meter?

By means of a block diagram, explain the functioning of a digital sound-level meter. Why does such a meter have a mode in which the maximum sound level in a measuring period is stored in the display?

Comment upon the statement: 'It is not the inherent performance of the sound-level meter but the method of application that is the dominant factor in determining the quality of measured and reported data'. (*IOA 1979*)

10 The law relating to noise

This chapter is a general guide to noise and the law, so many topics have been dealt with in outline only. Some have been dealt with more fully, e.g. the Environmental Protection Act 1990, the Noise and Statutory Nuisance Act 1993 and the Control of Pollution Act 1974 because of their direct applicability to noise control. A bibliography has been provided at the end of the book and this should aid further study. In order to help the reader with no legal knowledge, the following outline on English law may be useful.

10.1 English law

The law found in the following pages relates only to the law laid down in England and Wales. Within the political state of the United Kingdom of Great Britain and Northern Ireland and its dependencies, there are a number of different legal systems with different histories. Scotland, Northern Ireland, the Isle of Man and the Channel Islands all have their own. Often, however, Acts of Parliament legislate for all the United Kingdom. But, because each system has its own methods of putting law into practice, sometimes separate acts have to be passed to take account of these differences.

Categories of English law

There are a number of categories of law within the English legal system. Firstly, law can be divided into two types, public and private. Public law is for the benefit of society in general and is designed to prevent the breakdown of law and order. For this reason, any such rules, if broken, are punished by agents of the State. The obvious example is criminal law. If someone murders or steals, he is first apprehended by the police and then brought before the criminal courts and prosecuted by the Crown Prosecution Service. He or she can be punished in many different ways such as imprisonment and fines, disqualification in the case of driving offences, and community service for certain examples of antisocial behaviour such as vandalism. In today's sophisticated society many types of bad behaviour are dealt with by the criminal law, even if this does not appear to be wholly appropriate at first. For example, driving offences are dealt with in the criminal courts. Breaches of the Environmental Protection Act 1990, the Noise and Statutory Nuisance Act 1993, the Control of Pollution Act 1974, the Health and Safety at Work, etc., Act 1974 and local government bye laws, are also dealt with by the criminal law. In the first three examples the law is enforced by environmental health officers (EHOs); the fourth example is enforced by EHOs and the Health and Safety Inspectorate instead of the Crown Prosecution Service. However, serious breaches of safety legislation will be dealt with in the Crown Court and be prosecuted by the Crown Prosecution Service.

Private or civil law is for the benefit of individuals and thus only protects individual interests. The person who has suffered from noise-induced illness caused by an employer's disregard for the Health and Safety at Work Act and the Noise at Work Regulations 1989, cares little for the prosecution of the employer. He or she would prefer to recover compensation in the form of damages by suing the employer in the civil courts for the tort of negligence or the tort of breach of statutory duty (see pages 232 & 237). Private law covers many different types of law. For us, the most important is the law of tort. A tort is a civil wrong which is generally not bad enough to be treated as a crime. Examples include negligence, nuisance, trespass to land, person or goods, libel and slander. Some torts may also be dealt with as crimes in the criminal courts, e.g. assault and battery, which are types of trespass to the person. This is why one may hear of people taking private actions when the police have failed to prosecute for some reason. Other less relevant forms of private law include contract, family law, company law, land law and succession on death.

The courts

As there are different forms of law, so there are appropriate courts which are part of a hierarchical system, with the most important court at the top of the ladder, the House of Lords and many different magistrates' courts dealing with the least serious criminal matters at the bottom. Figure 10.1 shows a simplified version of the court system.

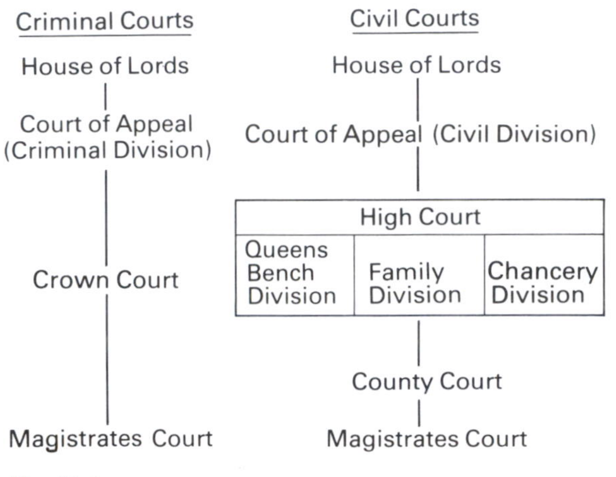

Fig. 10.1

The courts most relevant to us are the magistrates' court, which deals with minor criminal offences under the Environmental Protection Act and Control of Pollution Act and the Health and Safety at Work Act. The Crown Court deals with more serious offences under those Acts.

If someone has committed a tort, then they can be sued in either the Queen's Bench Division of the High Court or the county court. Generally, claims for damages for more than £50 000 or injunctions will be dealt with in the Queen's Bench Division and for cases under £25 000 in the county court. Cases that fall between the limits may go either way, depending on certain factors such as the financial substance of the action, whether matters of public interest are involved, whether the facts, legal issues, remedies or procedures involved are particularly complex and whether transfer is likely to result in a speedier trial. Those factors could also be used to recommend transfer of an action worth less than £25 000 to the High Court and one worth more than £50 000 to the county court.

The sources of English law

Like a river, English law is created from a number of sources. The oldest source has been that of *customary law*, where people have exercised some right for centuries and eventually it has been recognized in a court of law, e.g. a right of way. Nowadays, only local customs become law and then only if they satisfy certain criteria.

More important to us is law created by *judicial precedent*. Here the judges in court cases apply existing principles of law to new situations as and when they arise. Their decisions on points of law are called precedents; if they are made or confirmed by courts high up the ladder, they bind all lower courts. Lower courts must obey the ruling on that point unless the later case can be distinguished on the facts. One of the advantages of the precedent system is that it only makes law for situations which have actually arisen, whereas the next source, *legislation*, legislates for situations which may never arise.

Legislation is law created by a body specially set up for that purpose. In England and Wales this is the Houses of Parliament. The law is primarily found in Acts and if directly enforceable, i.e. all the rules relating to a particular aspect are contained in the Act, then this is known as *direct legislation*. The Environmental Protection Act, Noise and Statutory Nuisance Act and Control of Pollution Act are good examples, as all the basic law is found within the Acts themselves. If the Act merely enables somebody else to make the rules, such as a government minister in consultation with others, then the Act is known as an *enabling Act* and the law made thereunder, as *delegated, indirect* or *subordinate* legislation. Such laws are usually called *rules*, *regulations*, orders-in-council or *bye laws*. Thus, the Health and Safety at Work Act is an enabling Act in relation to the Noise at Work Regulations 1989, and a local bye law forbidding the sounding of motor-car horns outside a hospital would be made under the Local Government Act 1972.

A more recent source of law is now found in the *European Union* and further details may be found later in the chapter.

Common Law and Equity

In English law we also talk of *Common Law* and *Equity*. The common law is that law that has been and is still being created by custom and precedent (see above). It was originally described as common by the Norman kings following the conquest of 1066. In their desire to unite the differing peoples of England under one king, visiting judges were sent around on circuit throughout the kingdom, trying cases using, wherever possible, the king's law and gradually, by absorbing and eliminating undesirable local customary law, a homogeneous law was applied, common to all the land. Thus, wherever one lived, in theory, they had the opportunity to use the common law and were able to avoid the corrupt local courts.

Unfortunately, because of power struggles between the Crown and the barons, the developing common law was dealt a serious blow by a number of procedural restrictions. It became very rigid and inflexible. The expressions 'going by the book', 'keeping to the letter of the law' or 'too much red tape' could all have been applied to the common law at that time. As a result, many people were unable to go to law and would have had to fall back on the local courts if a new development had not occurred. The king was regarded as the 'fountain of all justice' and as such it was possible to ask him directly for a remedy where no others existed. Travelling to see the king in the Middle Ages was not easy and sometimes it even involved going behind enemy lines abroad. Even so, many found this method of getting justice so successful that the sheer numbers overwhelmed successive kings. Gradually, this right to hear direct appeals from his subjects was passed by the king to the Lord Chancellor. He was the king's priest and confessor and gradually a new approach to the law and new remedies evolved. This new body of rules was called equity. The Lord Chancellor was more concerned with doing justice and achieving equality than going by the letter of the law. Eventually, he was given his own court, the Court of Chancery. Other judges were appointed and equity developed in a different way to the common law, with its own remedies such as injunctions. Equity offered *not* an alternative system but a supplement 'to stop the gaps' in the common law relating to private legal matters.

Equity did not use precedents (see above) but it did use maxims which even today are at the root of equitable decisions. An example is 'He who comes to equity must come with clean hands.' This means that a plaintiff asking for an equitable remedy or for relief in equity in some way, must have behaved fairly throughout the affair and not merely complied with his or her legal duties.

Another equitable maxim is 'Delay defeats equity.' You cannot expect the court to help you in equity if you delayed seeking that help as it may have adversely affected the opposition's position.

Today the two systems of rules are applied in all the courts, but if there is a conflict between a legal rule and an equitable rule, the equitable rule prevails.

10.2 An outline of the law relating to noise

Technological and scientific knowledge today indicates the undesirability of unwanted sound, i.e. noise. But this knowledge has been gained over a relatively short period of history, for although it was recognized long ago that a noisy workplace could cause deafness or discomfort, it was another matter to prove it. That is why the law relating to noise has grown up bit by bit, protecting different interests with different remedies.

This chapter is set out in a series of subsections. Sometimes more than one of the topics will be applicable, and wherever possible, this has been indicated. But treating the subject in only one chapter has required considerable brevity.

10.3 The eradication or reduction of noise or vibration at common law

The common law has treated noise and vibration to be within a category of behaviour known as *nuisance*. A nuisance may be so bad as to constitute a crime or it can be treated as a *tort*, i.e. a civil wrong.

Nuisance

There are three types of nuisance. A *public* nuisance is a crime created by the common law, i.e. it is considered to be so harmful that people breaking the law should be punished by the State for the good of the people. A *private* nuisance, on the other hand, is a tort. The person suffering may thus sue the wrongdoer, known as the tortfeasor, and may be awarded a civil remedy such as damages or an injunction. Statutory nuisances are nuisances created by Acts of Parliament and are dealt with in the criminal courts.

Public nuisance Public nuisance has been defined as 'an act or omission which materially affects the reasonable comfort and convenience of life of a class of Her Majesty's subjects' (*AG v. PYA Quarries*, 1957).

Because of its ancient beginnings, it has embraced different types of behaviour which one would not normally consider to be nuisances. Examples include obstructing the highway, keeping a brothel, polluting a public water supply, causing dust, noise and vibration by quarrying activities and causing dirt, noise and an unreasonable amount of traffic by holding a pop festival. Recently, a doctor was prosecuted for performing operations while suffering from an infectious form of hepatitis. In *R. v. Holme* (1984) an eccentric but sane man was found guilty of committing a public nuisance. Apart from threatening people, banging on car roofs and blocking the public highway, he also imitated an ape, provoked dogs into barking in the early hours of the morning, played one chord on the piano through the night, played a radio at top volume all day and night while it was hanging from a rope out of his window, kicked a dog up the street and assaulted people! There seems to be little in common with each example, other than the element of annoyance or inconvenience to the public. Inconvenience and annoyance are also common to private nuisance as we shall see later.

As public nuisance is technically a crime, it could be tried on indictment in the Crown Court like other serious crimes, e.g. murder. But it is more usual for the Attorney-General, who is the principal law officer of the Crown and head of the Bar, to start the action in the High Court on behalf of a sufficient portion of the public. This is called a relator action. An application has to be made to the Attorney-General to which he may or may not agree. Similarly, a local authority may take action in the same way on behalf of a community within its area under S222 of the Local Government Act 1972. S222 states

> (1) Where a local authority consider it expedient for the promotion or protection of the interests of the inhabitants of their area — (a) they may prosecute or defend or appear in any legal proceedings and, in the case of civil proceedings, may institute them in their own name, and (b) they may, in their own name, make representations in the interests of the inhabitants at any public inquiry held by or on behalf of any Minister or public body under any enactment.

How many constitutes a sufficient portion of the public for it to be dealt with as a public nuisance depends on the circumstances of each case. In *R. v. Lloyd* (1802) the occupants of only three barristers' chambers in Cliffords Inn in London affected by noise was held not to be a sufficient number for a public nuisance. In *AG v. PYA Quarries* (1957) Denning LJ referring to quarrying activities affecting a wide area stated that 'a public nuisance is a nuisance which is so widespread in its range or so indiscriminate in its effect that it would not be reasonable to expect one person to take proceedings on his own . . . but that it should be taken on the responsibility of the community at large'. Thus, a noise nuisance affecting a whole village or a number of streets could probably be dealt with as a public nuisance, whereas if only a few houses are affected then it seems not.

If a member of the public has been particularly affected by the commission of a public nuisance, provided the person can prove he or she has suffered damage over and above that suffered by others in the neighbourhood, then he or she may sue in tort. Such damage could take the form of personal injury, loss of business or physical damage to property. Such action was taken in *Halsey v. Esso Petroleum Co. Ltd.* (1961) (see below). The damage must have been a foreseeable consequence of the nuisance, e.g. loss of sleep due to vehicles being driven during the night, as in Halsey's case. The person suffering 'extra' damage does not have to have an interest in land nor can the person only sue in relation to the effect of the nuisance on his or her land, as in private nuisance, which is a tort designed to protect use of land only (but see page 214).

As public nuisances are essentially crimes, it is no defence to say there has been consent to the nuisance, as one can never consent to a crime. But one can always consent to commission of a tort (see page 215).

Finally there does not have to be any indirect invasion of private land, as there has to be in private nuisance.

Private nuisance The tort of private nuisance has been defined by Professor Winfield as an unlawful interference or annoyance which causes damage to an occupier or owner in respect of his or her use and enjoyment of his or her land, or of certain rights over or in connection with land. These 'certain rights' are easements — legal rights attaching to land in respect of another's land over which the owner of the easement has, for example, the right to light, water or a right of way.

Private nuisance therefore protects only interests in *land*, not personal interests as in negligence (see below). Thus, damages for personal injuries cannot be recovered under this tort (cf. public nuisance).

Behaviour amounting to a private nuisance may take the form of noise, preventing certain uses of the property such as for sleeping or use of the garden. Vibrations of course can actually physically harm buildings. Nuisance can also take other forms, such as that created by smoke, water, gas, smells, fumes, roots undermining walls and other unneighbourly behaviour. The courts agree that other types of nuisance may occur in the future.

The behaviour amounting to nuisance may be intentional, negligent or perhaps even unintentional and not negligent.

Criteria for suing in nuisance Unlike negligence there is only one essential element in order to sue for nuisance. But many other factors have to be taken into account when deciding whether a nuisance has or has not been committed.

(i) Damage or harm must have been caused The essential element nuisance is like negligence. There must be some damage or harm suffered on the part of the plaintiff in relation to his or her land. This may take the form of physical damage to the property, such as cracks caused by vibrations. If there is physical damage, the court looks no further at whether damage has been caused and is less likely to investigate its triviality.

If, on the other hand, the damage is only an interference with the use of land or its enjoyment, then it must not be trivial. The interference must be substantial. In *Andreae v. Selfridge* (1938) it was said by the Master of the Rolls that the loss of only one night's sleep through excessive noise was not trivial and no injury to health need be proved.

The damage that results in these situations is that the plaintiff is unable to use a room or a garden in the normal way because of the nuisance. This could be quantified by a valuation in respect of the reduction in value of the house. Imagine buying a house, part of which you could not use because of the noise or vibration from next door. You would expect to receive a reduction in price. In the case of the hotel, the proprietors could by reference to their bookings and accounts show the loss in business caused by the noise.

(ii) The behaviour must be unusual, excessive or unreasonable The courts take the view that there should be 'give and take' between neighbours. So an occasional

noisy party or do-it-yourself session would not be actionable. Weekly parties or continuous drilling would almost inevitably amount to nuisance. Normal, reasonable and moderate use of property cannot be actionable. In *Halsey v. Esso Petroleum Co. Ltd.* (1961) there was an excess of noise, smell and acid smuts caused by the defendant's business. In *Andreae v. Selfridge & Co. Ltd* (1938) demolition contractors caused an unreasonable amount of dust and noise which affected the plaintiff's hotel trade to its detriment. In *Dunton v. Dover District Council* (1977) the noise from a playground adjacent to a private hotel was found to be excessive.

In *Toft v. McDowell* (1993) however, the court found that *ordinary* use of premises did, partly in this case, give the right to successfully sue in nuisance where the parties, living in converted flats, shared the responsibility of maintaining an adjoining floor/ceiling, which experts showed contained no form of sound insulation. This decision is quite unusual and may relate only to the particular facts involved.

(iii) The behaviour must have gone on for some time Generally, in relation to noise nuisance, isolated acts are not actionable. But they may give rise to another type of tort, e.g. an explosion could be caused by negligence.

(iv) The nuisance must be caused by another person on neighbouring property and not on the plaintiff's own premises This is partially self-explanatory. If the nuisance was actually on the plaintiff's land he or she should obviously take action to stop the antisocial behaviour as he or she is in charge of that land. The land does not have to be adjacent to the plaintiff's land but obviously it will not be far away for a noise nuisance.

It seems that the nuisance could be committed on a highway and not necessarily on the defendant's own land. In *AG v. Gastonia Coaches* (1977) a coach business operated from a house and yard in a residential area. The business expanded and the defendants were unable to park all their vehicles on their own premises. They began to park their coaches on the roadway outside and they also carried out maintenance and repairs. The coach drivers also parked their own cars there. The local residents complained of the noise and fumes from the vehicles and also about the obstruction of the road. In a relator action they claimed for both public and private nuisances and asked for damages and an injunction. The court held that only the obstruction was a *public nuisance*. The neighbour living next to the premises established that the obstruction to access to his own home by parking on the highway and the noise and fumes from the vehicles using the highway did amount to a private nuisance. However, the noise of the maintenance did not amount to a private nuisance.

(v) Character of the neighbourhood Thesiger LJ is often quoted from his judgement in *Sturges v. Bridgeman* (1879) in which he said 'what would be a nuisance in Belgrave Square would not necessarily be so in Bermondsey.' This is not strictly true, however, as one cannot create nuisances merely because one is in an industrial or urban area. In *Rushmer v. Polsue and Alfieri* (1907), for example, the defendants moved some new printing equipment into their works in Fleet Street, then a very noisy area. The noise from the equipment was over and above what would have been normal in that area. Thus they were liable in nuisance. Nevertheless, it is obvious that one cannot expect the quietness of a rural community in the centre of a city. If the area is in a noise abatement zone (see below) then this should obviously influence court decisions in actions for private nuisance. Unfortunately, at present, the small numbers of noise abatement zones that have been set up under the Control of Pollution Act 1974 (the relevant part has not been repealed by the Environmental Protection Act) means that in the majority of cases the court continues to listen to expert evidence and actual evidence of the noise itself in order to decide whether the noise is excessive for that area.

(vi) Abnormally sensitive plaintiffs The law can only protect the average person with average sensitivity to noise. If the plaintiff is unusually sensitive, he or she will not be able to succeed in an action, unless the noise is over and above what would be acceptable to the average person. For example, in an unreported case in 1977 the ringing of church bells did not constitute a legal nuisance to the neighbours. Their objections indicated that they were unusually sensitive. But if a nuisance is established, the plaintiff can recover damages for interference with a special use or in relation to his or her sensitivity.

(vii) The defendant's conduct Cases show that behaviour amounting to a nuisance may be intentional, negligent or occasionally unintentional and even non-negligent. The unthinking noise producer is probably most common, but after being given notice of the nuisance by the sufferer, those people who continue to commit noise nuisance are probably intentional or negligent. Malice need not be shown to succeed in nuisance, unlike some other torts, such as malicious prosecution, for which actual malice must be proved. But if it can be shown that the defendant intended to annoy people, this will better support the plaintiff's case. In *Hollywood Silver Fox Farm Ltd. v. Emmett* (1936) the defendant encouraged his son to shoot his gun with the intention of interfering with the breeding of foxes on the plaintiff's farm. In *Christie v. Davey* (1893) the defendant's neighbour, annoyed by the music lessons being given by the plaintiff, banged trays and made other noises with the intention to annoy. In *Fraser v. Booth* (1948) a neighbour got his son to let off fireworks in order to discourage the plaintiff's homing pigeons from returning to their loft.

(viii) The behaviour must be indirect in nature Nuisance by its very nature is an indirect tort; the bad behaviour is on the defendant's own land or possibly on the highway (see above). If it were direct, e.g. throwing things on to

the plaintiff's land, this would amount to trespass to land, and a different basis for action would have to be used.

(ix) Could the nuisance have been prevented easily? If it could have been prevented for a relatively small cost, the nuisance should have been stopped and the court will take this into account. Switching off record-players costs nothing. This does not mean to say that the court will not find that a nuisance exists if the cost of its eradication is huge.

What sort of plaintiff may sue? As the tort only protects interests in land, it has been thought until recently that the plaintiff must prove that he or she has a right to enjoy the land itself or an easement over that land. Thus, he or she must normally be in possession of the land, unless the damage is physical and affects the property in some permanent way, in which case persons out of possession may sue. Thus anyone in legal or equitable occupation of the land may sue, e.g. an owner-occupier, tenant in possession, licensees in possession and lessors (but the last only if their reversionary interest has been damaged, e.g. by cracks in the property). (Lessors or landlords lease or rent their property to lessees or tenants, who at the end of the allotted time must deliver up possession of the land back to the landlords. While they are out of possession, the landlords are said to have *reversionary* interests as the land will revert back to them.) Otherwise, as noise is of such a transient nature, people such as lessors, who are not in possession of the land, are generally unable to sue for nuisance.

Who can be sued?

(i) The creator of the nuisance In nuisance generally, the creator will always be liable, even if he or she gives up occupation of the land. This seems to be inappropriate in the case of noise, as noise is normally created by persons currently occupying land.

(ii) The occupier Obviously the occupier should be primarily liable as he or she is in control of the premises and can tell people what they must or mustn't do. The occupier will therefore be responsible for his or her own and his or her family's acts. By the principle of vicarious liability, an employer is also responsible for the torts committed by any employees during the course of their employment. Thus, a discotheque owner would be liable for the nuisance created by a disc jockey or band unless he or she expressly forbade the particular behaviour causing the nuisance. But even in these cases, the courts often take the view that the prohibition must be so strong as to put the act outside the scope of the servant's employment. A disc jockey playing music loudly would thus be doing what he is employed to do, but in a wrongful way, making his employer vicariously or indirectly liable in tort.

The occupier may also be liable for his or her independent contractors, i.e. those people with whom he or she has a contract, not a contract of service like an employee, but a contract for services such as window-cleaners, building contractors, etc. However, the occupier will be responsible only if he or she has a great deal of control over the independent contractor. Otherwise, the plaintiff would have to sue the independent contractor as the creator of the nuisance.

The occupier may also be liable for nuisances created by invited guests, such as young cousins addicted to loud music while sunbathing and persons who have licences to enter his or her property. They have no legal or equitable rights to the land, but they have been given a licence by the occupier to enter the land for a particular reason. Anglers may enter for fishing by right of a written fishing licence, or someone such as a neighbour may have been asked to feed the cat and water the plants while the occupier is on holiday. But in such cases, by attempting to abate the nuisance, the occupier may reduce or eliminate his or her liability.

Liability for the nuisance of a trespasser only occurs if the occupier has been negligent in discovering the existence of the nuisance.

An occupier who carries on or adopts a nuisance created by a former occupier will obviously be liable for nuisance, *Sedleigh Denfield v. O'Callaghan* (1940). The making of noise is a more active example of nuisance, whereby the noise-maker will be sued as the creator rather than the adopter even if, for example, he or she is merely taking over a noisy business (see prescription below).

However, in *Sampson v. Hodson-Pressinger* (1981) a new landlord of a building, previously converted into flats, was liable in nuisance because, at the time he was assigned the freehold reversion, he knew that a roof terrace allowed noise to penetrate to the flat below. He was effectively authorizing the nuisance (see below).

(iii) The person who authorizes the nuisance If a person lets land for a purpose which by its very nature entails the commission of a nuisance, then that person may be liable in tort. This applies even if it is only foreseeable that the nuisance *may* be committed by the occupiers. In *Tetley v. Chitty* (1986) a local authority gave planning permission to a go-kart club on a site leased to them by the council specifically for go-karting. The court held that the council was liable in nuisance as the noise was an ordinary, natural and necessary consequence of the letting and as such the council had given express or implied consent. But in *Smith v. Scott* (1972) a local authority let property to a known 'problem' family who, it could have been foreseen, would create nuisances by causing damage and noise. By including in the tenancy agreement a clause prohibiting such behaviour, the authority was held not to have authorised

the nuisance.

Any landlord who retains the power to inspect premises let by him or her without a prohibition clause may be open to action even though he or she has not in fact authorized the nuisance.

Defences

(i) Prescription Generally, as a defence to a nuisance action, one may gain the legal right to continue to commit a nuisance by prescription. However, in relation to noise nuisances this seems to be less likely as the behaviour amounting to a nuisance should be capable of existing as an easement. Easements are legal rights attaching to land, such as rights of light, way and water, and it is open to conjecture whether a prescriptive right can be obtained in relation to noise, especially in the light of the Environmental Protection Act 1990. A right to commit a public nuisance which is a crime can never be obtained by prescription.

For a prescriptive right to be gained, the owner of the land must prove that he or she has been committing the nuisance for 20 years at least. He or she must not have committed the nuisance with the other person's permission, nor must it have been done secretly (usually impossible in any event in relation to noise), nor must force have been used to exercise it (*Nec vi, nec clam, nec precario* — without force, secrecy or permission). If the neighbour disputes the nuisance then this, too, would negate the prescriptive right.

The behaviour must have amounted to a *nuisance* for the period of 20 years. In *Sturges v. Bridgeman* (1879) a confectioner operated at the rear of property abutting onto a garden belonging to a doctor in Wimpole Street. The doctor moved his consulting rooms from the main house to the bottom of the garden, where the noise from the confectioner became a nuisance and the doctor sued. The confectioner, although able to prove that he had operated his business for over twenty years in the same manner, was not able to show that it amounted to a *nuisance* for those years, so the doctor was successful in his action. In other words, the noise only became a nuisance when the doctor moved.

(ii) 'Coming to the nuisance' What happens if someone knows that a nuisance is being committed in relation to property he or she is proposing to occupy? It is sometimes argued by the defence that where the behaviour amounting to a nuisance was being committed before the plaintiffs arrived and now continues, the plaintiffs cannot then successfully pursue a claim in nuisance. This is *no* defence. But there have been cases where the court in such situations, when deciding whether or not to grant an injunction, has weighed up the general interest of the public, who may have been benefitting from the nuisance behaviour, as opposed to the person coming to the nuisance in full knowledge of its existence; the court has refused to grant the injunction. (The court, however, could not refuse to grant damages once the tort has been proved, see page 216). *Miller v. Jackson* (1977) involved the nuisance of cricket balls from a village cricket green landing in an abutting garden. Although a nuisance was proved to exist, the injunction requested was refused because the plaintiffs came to the nuisance and indeed were probably attracted by the closeness of the cricket ground. A similar decision was reached in a county court case in 1994 when it was decided that no nuisance had been committed in the first place, even though cricket balls had landed on a number of occasions in the plaintiff's garden.

(iii) Consent It is usually a good defence if a plaintiff consents to the nuisance being committed. Even if not a defence, it may still have the effect of defeating a plea for an injunction. There have been unreported cases where builders, for example, have informed local people of the likelihood of building noise and occasionally made payments to them or even paid for holidays while the worst of the work has been carried out. Consent could be given gratuitously or could be paid for. But before refusing to grant a remedy, the courts would look carefully at the nature of the consent and could take the view that the plaintiffs may not have realized to what extent of noise they had consented. Thus, to be successful as a defence, one would probably have to prove that there was consent and secondly that the plaintiff knew exactly to what he or she was consenting. If there was a written contract between the parties, this would make it more difficult for the plaintiff's case, especially if the plaintiff in accepting payment specifically stated that he or she gave up his or her right to sue for nuisance.

(iv) Mitigating the nuisance In law it is always the duty of a plaintiff to mitigate his or her loss. This means that, whatever the damage suffered, he or she has tried to reduce it accordingly. But only reasonable steps need be taken, and if such action requires spending money, the defendant would not be able to say that the plaintiff had not mitigated his or her loss.

(v) Statutory authority If an Act of Parliament or a relevant rule or regulation imposes a *duty* on someone or somebody to do something which inevitably will cause a nuisance, the authority will be a defence, provided that no negligence was attached to the action. It will be up to the defendant to prove that he or she took care to avoid negligence, even if the Act expressly made the defendant liable for nuisances.

There are slightly different rules in relation to *powers*. Duties *must* be carried out; powers *may* be carried out. Once again there must be no negligence on the part of the defendant, but this time the statute must either expressly exclude liability for nuisance or it must make no mention of the matter. In *Allen v. Gulf Oil Refinery Ltd.* (1981) powers given by the Gulf Oil Refinery Act 1965 permitted the building of an oil refinery. Local villagers complained about noxious odours, vibrations and excessive noise. The House of Lords held that the section of the Act relied on gave the company immunity from an action for nuisance which might be the inevitable result of building such a refinery.

Remedies

(i) At common law damages On proving his or her case, the plaintiff is entitled to compensation (damages) for actual damage caused to his or her land, such as depreciation or loss of business. Entitlement is as of right because it is a *legal* remedy, and even if the plaintiff has behaved inequitably (see below), the court cannot refuse damages if he or she satisfactorily proved the case and there are no adequate defences. The court will take into account and offset the loss caused by reasonable behaviour, when the job which caused the nuisance was necessary. For example, in *Andreae v. Selfridge & Co. Ltd* (1938) there would in any event have been loss of custom due to work, but not as much as was actually caused.

(ii) In equity. (a) Injunction An injunction is a court order which orders someone to do or not to do some specific thing. In relation to noise nuisance it usually orders the defendant to stop the work or behaviour which amounts to a nuisance. Such injunctions are called *prohibitory* or *restrictive* injunctions. They may be awarded before the case actually comes to court for trial (*interlocutory* or *interim*); if they are awarded at the trial they are called *perpetual*. However, this does not mean that the injunction necessarily carries on in perpetuity.

Because an injunction is an equitable remedy, it is *discretionary* and will only be awarded if damages are inadequate on their own. If there is no likelihood of a repeat of the nuisance, an injunction would not be the appropriate remedy.

In using the court's discretion, the equitable maxims 'He who seeks equity must do equity' and 'Delay defeats equity' will be taken into account by the court. The first means that the plaintiff must have behaved equitably or fairly if he or she is going to be availed of an equitable remedy, e.g. if he or she had told a neighbour that he or she didn't mind the senior citizens' tap-dancing class practising every week next door, he or she may be refused an injunction. (But if he or she proves a nuisance exists, he or she can still get damages by right.)

The second maxim means that if one requires an equitable remedy the plaintiff must not delay; he or she must go to court as soon as possible.

A request for an injunction often accompanies a request for damages. In *Kennaway v. Thompson and Another* (1980) the Court of Appeal reversed the decision of the trial judge, thus allowing an injunction where previously only damages had been granted for past and future nuisance. The nuisance caused by a motor-boat club was restrained only as to behaviour, which did amount to a nuisance. Thus, the club members could still enjoy their water sports but not at the expense of the plaintiff. The court provided an agenda for a season of events with suitable gaps in between. Furthermore, noise limits were set for boats not taking part in national or international events, and no more than six boats were to be used at any one time. The injunction may be permanent or it may request modification of behaviour, e.g. operations to be carried out during specified hours only. An injunction may also be granted but suspended in order to give the defendant the opportunity of modifying his behaviour. See also *Miller v. Jackson* (1977).

Injunctions may be granted but suspended giving the defendant time to modify his or her behaviour.

Quia timet injunctions may be granted even though the nuisance is not actual, only threatened. But in order to obtain such an injunction, one would have to show that there was imminent danger of irreparable damage (in nuisance, damage to land). One cannot really imagine a situation in relation to pure noise nuisance which would give such a result. However, the prospect of an unauthorized pop festival, perhaps resulting in many different types of nuisance, could possibly give rise to a *quia timet* injunction.

Historically, injunctions were only granted by the Court of Chancery (the court applying equity) and the court had no power to grant damages, as this was a legal remedy only available in the common law courts. However, the Chancery Amendment Act 1858 (Lord Cairns Act) changed the law and permitted the award of damages in lieu of, or in addition to, an injunction, e.g. where only an injunction had been asked for. The remedy is now found in S50 Supreme Court Act 1981. (Damages under this heading could be granted even in lieu of a *quia timet* injunction, but this would be extremely unusual as it would allow the award of damages, even though there had been no tort committed and thus no possibility of damages being awarded at common law!)

The leading case, *Shelfer v. City of London Electric Lighting Co.* (1895), involved nuisance caused by noise and vibration. It stated that damages in lieu of injunctions would only be awarded if (i) there was only a small injury; (ii) the injury could be estimated in money terms; (iii) the injury could be adequately compensated by a small amount of damages, and (iv) the award of an injunction would be oppressive. But each case must be looked at individually and in the light of the equitable maxims.

If someone carries on with acts in breach of terms of the injunction, they will be in contempt of court and could be fined or sent to prison. Anyone knowingly aiding them in breach could also be dealt with for contempt, even if not party to the original action. If a company or other type of corporation is in breach, its property may be sequestrated and its officers committed for contempt.

(b) Non-judicial action: abatement of nuisance Self-help is not usually advisable. The person suffering from the nuisance, after giving the other party notice, enters the land of the person committing the nuisance and does whatever necessary to abate it. He or she must not do anything which causes unnecessary damage and if there are alternative methods of abatement, then he or she must choose one that causes least harm, e.g. a person infuriated with the noise of a neighbour's defective burglar alarm entering the property and cutting wires. However, in so doing he or she is committing the torts of trespass to land and property. Fortunately, abatement is a defence to trespass. But this type of approach is still not advisable because one could possibly make oneself liable for negligence if there were a subsequent burglary. The burglary might have been prevented had the defective alarm been noticed and mended by keyholders. If someone does exercise his or her right to abate, he or she cannot then sue for damages in nuisance. (There are now better remedies under legislation see below.)

10.4 Statute law

Environmental Protection Act 1990
Noise and Statutory Nuisance Act 1993

The common law is undoubtedly adequate in awarding compensation or injunctions when a nuisance has been committed. But the civil procedures in the county court, or High Court if an injunction is sought, are quite complex and relatively slow, requiring in nearly every case the services of a solicitor. If the case is uncertain, a barrister may be asked to give 'counsel's opinion' on its merits; if the case actually goes to the High Court, a barrister or an advocate solicitor must be briefed to act as advocate. All this adds to the expense and time involved.

In the main, common law actions can only be made *after* the event, and it has been left to Parliament and local authorities to lay down rules preventing noise or restricting it. This legislation is cheaper to enforce, relying on the criminal court structure. Noise control under legislation, whether under the following legislation or under byelaws made by local authorities, is also quicker and easier, and does not necessarily require the services of lawyers.

The Control of Pollution Act 1974 (the 1974 Act) made great strides in noise control and part of this Act remains in force in relation to noise abatement zones (see pages 230–1).

The Environmental Protection Act 1990, which for the most part supersedes the 1974 Act, specifically allows recourse to the civil courts for relief if the Act falls short, as in *Hammersmith London Borough Council v. Magnum Automated Forecourts* (1978) and *London Borough of Lewisham v. Saunders* (unreported).

The main legislation regarding noise control is now the Environmental Protection Act 1990 (the 1990 Act or EPA) Part III sections 79–82 as amended and supplemented by the Noise and Statutory Nuisance Act 1993 (the 1993 Act). Part III of the 1990 Act is concerned with statutory nuisances in general as amended by the 1993 Act.

S79 defines certain matters which constitute 'statutory nuisances'. For this book, however, I will only deal with those relating to noise. The subsections of importance to us are therefore:

S79(1)(a) Any premises in such a state as to be prejudicial to health or a nuisance.
S79(1)(f) Any animal kept in such a place or manner as to be prejudicial to health or a nuisance.
S79(1)(g) Noise emitted from premises so as to be prejudicial to health or a nuisance. (Note by S79(7) noise includes vibration.)
S79(1)(ga) Noise that is prejudicial to health or a nuisance and is emitted from or caused by a vehicle, machinery or equipment in a street. (This was introduced by the 1993 Act.)

Furthermore, under S79 '. . . it shall be the duty of every local authority to cause its area to be inspected from time to time to detect any statutory nuisances which ought to be dealt with under S80 below or sections 80 and 80A below and, where a complaint of a statutory nuisance is made to it by a person living within its area, to take such steps as are reasonably practicable to investigate the complaint.

S79(1)(a) is a re-enactment of S91(1)(a) of the Public Health Act 1936 and the similarities between the new Act and the old Public Health Acts leads one to believe that much of the old case law will continue to be of relevance in the future.

Thus, in each case a statutory nuisance is one which is either prejudicial to health **or** a nuisance. As far as noise control in general is concerned, this is a big improvement because the previous legislation was only concerned with nuisances.

S79(7) specifically defines 'prejudicial to health' as 'injurious, or likely to cause injury to health'.

The word *nuisance*, however, is not defined. It is and has been implicit in noise legislation that the word *nuisance* is to be read in terms of both public and private nuisance, and no understanding of the present legislation is possible without this; the reader is referred to pages 211–17.

The new approach under EPA is therefore a great improvement on the previous legislation. Now, in applying the EPA to a situation, the local authority's environmental health officer (as enforcing agent) has two approaches in each statutory nuisance situation.

Before looking at each specific statutory nuisance, it should be noted that the word *premises* as defined by S79(7) includes land and any vessel other than one powered by steam reciprocating machinery S79(12). Thus, premises do not have to be indoors, like a house or factory, but will include gardens, fields, etc.

Furthermore, by S79(7), the definition clause has been extended by the 1993 Act to define the word *equipment* to include a musical instrument. Also the word *street* means 'a highway and any other road, footway, square or court that is for the time being open to the public'.

S79(1)(a) Any premises in such a state as to be prejudicial to health or a nuisance

This section is not intended to cover *noise-making* activities. Instead it is concerned with the passive state of the premises which by their very nature are either prejudicial to health or causing a nuisance. In *London Borough of Southwark v. Ince and Another* (1989), a case brought under a similar section of the Public Health Act 1936, the local authority was taken to court as owners of premises which were inadequately insulated against noise from adjacent railway lines and roads. The noise was making the occupants ill, and the court held that premises in such a state as to admit noise could be prejudicial to health.

S79(1)(f) Any animal kept in such a place or manner as to be prejudicial to health or a nuisance

The barking of dogs or the sound of other animals, such as cockerels, has long been a ground for neighbourhood noise disputes. The problem with the 1974 Act's approach was that one could only deal with the situation if the barking, etc., amounted to a nuisance in common law terms. Also prosecutions under the Public Health Act 1936 S92(1)b, an identical section to S79(1)(f), have cast doubt on whether one could apply it to animals who were *only* noisy. See *Galer v. Morrissey* (1955), where it was held not to apply, and *Coventry City Council v. Cartwright* (1975), where doubt was cast on this viewpoint. It remains to be seen whether action can be taken, under this section and/or under S79(1)(g), if it can be proved that the noise is *only* prejudicial to health, e.g. if the noise is only occasional. Probably, if there is only noise involved, subsection (g) may be more appropriate. (Remember that the keeping of animals can also be dealt with under other legislation such as Town and Country Planning, see pages 244–7, and local byelaws, see page 258.) If cruelty is involved, the police and RSPCA can also deal with this under the Protection of Animals Act 1911.

S79(1)(g) Noise emitted from premises so as to be prejudicial to health or a nuisance

This has replaced and extended the previous law found in Part III of the Control of Pollution Act 1974. It should be emphasized that this subsection allows action to be taken whether the noise is or is likely to be harmful to health, or if it is a nuisance. This subsection does not apply to premises occupied on behalf of the Crown for Navy, Army or Air Force purposes, for the Department of Defence or for the purposes of a visiting force as defined by the Visiting Forces Act 1952 S79(2). Furthermore, no action can be taken under these provisions in relation to noise caused by aircraft in general, except for model aircraft, which are subject to a code of practice (see page 232).

Under this section the noise must be emitted from *premises*. In *Tower Hamlets London Borough Council v. Manzoni & Walder* (1984) the word *premises* was held not to cover noise created in streets or public places. Here demonstrators used megaphones to protest against the sale of animals at a street market. This continued for a number of weeks and the council served a noise nuisance notice under S58 of the 1974 Act. On appeal to the Crown Court against the notice, the court held that the section should be read in the light of the whole of Part III of the 1974 Act and S58 should therefore be construed to apply to noise emanating from premises.

As a result of the limitations of this subsection, the 1993 Act has introduced the new subsection S79(1)(ga).

S79(1)(ga) Noise that is prejudicial to health or a nuisance and is emitted from or caused by a vehicle, machinery or equipment in a street

This subsection does not apply to noise made by (a) traffic; (b) any Navy, Army or Air Force of the Crown or by a visiting force as defined previously or (c) by a political demonstration or a demonstration supporting or opposing a cause or campaign (S79(6A)).

Guidance to the new subsection is being prepared by the Department of the Environment. It appears to the writer that car radios, cars being worked on or revved up, outdoor do-it-yourself work on the roadside and car alarms will all fall within the scope of this subsection. Defective exhausts will probably come within its scope, too. It will have to be seen how this section will be interpreted by the courts. On the whole, the new subsection seems to have been greeted with enthusiasm by most EHOs. Some doubts have been expressed about the problems caused by underfunding of local authorities, but this is a perennial condition and one which is not going to go away. The statutory duty, which is set out below, cannot be avoided and the new 1993 Act has not changed this part of the legislation.

The local authority's duty S79(1) continues after defining the different matters that constitute 'statutory nuisances' by stating:

> and it shall be the duty of every local authority to cause its area to be inspected from time to time to detect any statutory nuisances which ought to be dealt with under S80 below and, where a complaint of a statutory nuisance is made to it by a person living within its area, to take such steps as are reasonably practicable to investigate the complaint.

Duty In law if a *duty* is imposed on someone or somebody then that duty must be carried out. It is not a mere *power* which may or may not be carried out. This does indeed mean that local authorities are in somewhat of a dilemma in this seemingly endless recession. They must carry out this duty but may not have sufficient money to fund it. What could happen if they do not make their investigation? Schedule 3 para. 4 of the 1990 Act specifically deals with this duty. If the Secretary of State is satisfied that the local authority has failed, in any respect, to discharge this function, he or she may make an order declaring the authority to be in default. Such an order may direct the defaulting authority to perform the function specified in the order and may specify the manner of its performance and the time(s) when it should be performed. Should the authority fail to comply with the order, the Secretary of State, instead of enforcing the order by mandamus, may make an order transferring to himself or herself the function of that authority. (Mandamus means 'we command'; it is a High Court order used to order the performance of a public duty, often by a local authority.) Any expenses incurred by the Secretary of State must be reimbursed by the defaulting authority. The Secretary of State can vary or revoke any such order. It is argued that, as the Act has specifically enacted these procedures, it is unlikely any other remedies for maladministration are available.

Local authorities *must* therefore inspect for statutory nuisances. At the time of writing it appears that some local authorities are now contracting out their enforcement authority.

Complaints made to the local authority This reflects what has usually always happened in practice. If anyone living in the local authority's area complains to the local authority of a statutory nuisance, the local authority must take 'such steps as are reasonably practicable to investigate the complaint.' It is much more likely that claims for maladministration in this respect would arise where the local authority ignores the complaint.

Following the investigation

S80 abatement notices S80(1) states that where a local authority is satisfied that a statutory nuisance *exists*, or is likely to *occur* or *recur*, in the area of the authority, the local authority shall serve a notice ('an abatement notice') imposing all or any of the following requirements:

(a) Requiring the abatement of the nuisance or prohibiting or restricting its occurrence or recurrence.
(b) Requiring the execution of such works and the taking of such other steps as may be necessary for any of those purposes.

And the notice shall specify the time or times within which the requirements of the notice are to be complied with.

This section imposes a *duty* on the authority. It *must* serve an abatement notice if a statutory nuisance:

- exists
- is likely to occur
- is likely to recur

Compare this with a civil action for nuisance, where the sufferer can only sue when there has been nuisance behaviour on more than one occasion (see page 212). Normally, civil action can never be used for behaviour which may occur in the future.

The abatement notice must impose certain requirements. It can require:

1. Abatement of the nuisance, e.g. asking someone to reduce the sound from their television.
2. Prohibition of a nuisance likely to occur or recur, e.g. a forthcoming pay party that will probably cause a nuisance.
3. Restriction of such nuisances, e.g. imposing days and times during which the nuisance behaviour must be stopped.
4. Execution of such works and pursuance of such other steps as may be necessary for those purposes.

(Mnemonic, PEAR: prohibition, execution, abatement and restriction)

The notice must contain the time or times within which the requirements must be complied with. Normally this should be a reasonable time, but in cases under S58 of the 1974 Act it has been held that no time need be specified; indeed in certain cases, a very short time would be sufficient, depending on the facts. In *Strathclyde Regional Council v. Tudhope* (1983) the City of Glasgow Council served an S58 notice on the Strathclyde Authority in respect of noisy roadworking operations, requiring that all pneumatic drills should be fitted with effective exhaust silencers and dampened tool-bits. No time was stated for compliance; the court held that the notice should come into effect at midnight following the date of service and that this was not unreasonable. However, to avoid appeals being made in relation to the reasonableness of time limits imposed by abatement notices, it is much more sensible to state a reasonable time at the outset. If someone is being asked to build something or modify machinery, tasks which require expert advice, a longer time should be given.

Should the abatement notice contain a prohibition on the occurrence or recurrence of the nuisance then this prohibition continues indefinitely. In *Wellingborough Borough Council v. Gordon* (1991) the council had served a prohibition notice in 1985 banning the recurrence of a noise nuisance. The recipient subsequently held a birthday party to which all his neighbours were invited and to which there were no complaints. The council took the recipient to the magistrates' court for breach of the notice. The magistrates' court agreed with the recipient that this was an isolated incident and this was a reasonable excuse, especially since there had been no complaints. This was overturned on appeal. No appeal had been made against the notice in the first place and the reasonable excuse defence should not have been allowed by the magistrates. The loud music at the party was exactly the sort of noise the notice had been designed to prevent in the first place.

A recent House of Lords decision has settled a vexed question once and for all. In *Aitken v. South Hams District Council* (1994), even though it could be argued that S58 of the Control of Pollution Act 1974 seemed to have been repealed completely, without any provision for the continued validity of the notices, the Lords held that the notice did, in law, continue to be effective. They reversed the decision taken in the divisional court. The case concerned the appellant who appealed against a conviction for contravening an S58 notice served in November 1983, during the period between August and October 1991, by allowing his dogs to bark so as to cause a nuisance. The legislation involved both the Environmental Protection Act 1990 and S16(1) of the Interpretation Act 1978.

Who must the notice be served upon? S80(2) states that subject to section 80A(1), below, the abatement notice must be served as follows:

(a) On the person responsible for the nuisance.
(b) On the owner of the premises if the nuisance arises from any defect of a structural character (as in the Ince case, see page 218).
(c) Where the person responsible for the nuisance cannot be found or the nuisance has not yet occurred, on the owner or occupier of the premises.

The 'person responsible' is defined in S79(7) as the person to whose **act**, **default** or **sufferance** the nuisance is attributable. This shows that the person may be responsible in a positive way, by actually doing the act, or in a negative way, by not doing something he or she should be doing or by allowing a state of affairs to continue.

Trespassers who are by definition in occupation, even though they are not the legal owners of the land, may be served with notices if they create the nuisance or as occupiers.

'Where the person responsible . . . cannot be found' may cover situations where the local authority would only be able to ascertain the person truly responsible for the nuisance after extensive and costly research, and the court should be told what and how extensive the enquiries were.

The 1993 Act has inserted a new section, S80(A), which applies to statutory nuisances under S79(1)(ga). In such cases, where the nuisance (a) has *not* yet occurred or (b) arises from noise emitted from or caused by an unattended vehicle or unattended machinery or equipment, the abatement notice shall be served in accordance with subsection (2) S80(A)(1).

SS80(A)(2) states that the notice shall be served:

(a) Where the person responsible for the vehicle, machinery or equipment *can* be found, on *that* person.
(b) Where the person cannot be found or where the local authority determines that the paragraph should apply, by fixing the notice to the vehicle, machinery or equipment.

S80(A)(3) states that where

(a) an abatement notice is served in accordance with subsection 2(b) above by virtue of a determination of the local authority, and
(b) the person responsible for the vehicle, etc., *can* be found and served with a copy of the notice within an hour of the notice being fixed to the vehicle, etc., a copy of the notice shall be served on that person accordingly.

This would cover the situation where someone parks his or her car with the radio playing while visiting a shop. The notice is attached to the windscreen before the driver returns within the hour. A copy of the notice is then served on the driver. If anyone without authority removes the notice, it is a criminal offence liable on summary conviction S3(7) of the 1993 Act.

By S80(A)(4) if an abatement notice is served under subsection 2(b), i.e. the local authority determines that the person responsible can't be found, the notice affixed to the vehicle, etc., must state that the time specified in the notice for compliance with requirements will be extended by a further specified time if a subsequent copy notice is served under subsection (3).

Thus S80 gives the local authority wide-ranging powers. Not only may it give notice to stop or reduce existing noise, but it may also restrict the emission of noise to certain parts of the day or week. It can prohibit noise nuisance which has not yet occurred or nuisance that may happen again, and it can order works to be carried out. The owner or occupier of premises may be held responsible. In *Cooke v. Adatia* (1989) the court, in dealing with an S58 Control of Pollution Act notice, stated that the notice should clearly and separately specify what is required under subsections 1(a) and (b). This should now be kept in mind when dealing with S80 of the Environmental Protection Act.

Execution of works The local authority no longer needs to specify what works must be carried out to comply with the notice. Now the authority must only 'require' the execution of such works and the taking of such steps as may be necessary to fulfil the terms of the abatement notice. These are vaguer words which allow the authority to show that work must be done without having to specify what exactly has to be done. It has happened in the past that the specified works have been carried out yet the nuisance remained.

It should be noted that by S81(3) if an abatement notice has not been complied with, the local authority may abate the nuisance itself and do whatever may be necessary in execution of the notice. This is so even if the authority does or does not decide to take proceedings in court for an offence under S80(4), see below.

Furthermore, by S81(4) any expenses reasonably incurred by a local authority under S81(3) may be recovered from the person who caused the nuisance, and if that person is the owner of the premises, from *any* person who is for the time being the owner. Presumably this means it is possible to claim against people besides the person who was an owner of the property at the time the notice was served, thus avoiding a hiatus if the property is sold while the proceedings are carried on. Under this subsection the court may also apportion the expenses between persons whose acts or default have caused the nuisance, in such a manner as it thinks fair and reasonable.

Coresponsibility for nuisances What happens if a number of people have contributed to a statutory nuisance but individually their behaviour would not necessarily amount to a nuisance. S81(1) clearly states that S80 will apply to each person in such situations. This has been added to by the 1993 Act. An additional S81(1A) states that in the case of an S79(1)(ga) statutory nuisance where more than one person is responsible (whether or not their individual behaviour would amount to a nuisance), S80(2)(a) shall apply with the substitution of 'any one of the persons' for 'the person'.

Nuisance created outside the local authority's area By the very nature of noise nuisances, it could happen that the point of immission, where the noise nuisance actually has its effect, is different from the point of emission, where it has been created. Indeed, the two sites could occur across local authority borders. But S81(2) allows that in such situations the local authority may still take action under S80 as if the act or default was wholly in its area, but any appeal made against the notice must be heard by the magistrates' court having jurisdiction where the act or default is alleged to have taken place. Liaison between the authorities to obviate reliance on this section is obviously desirable, so that reciprocal arrangements are made whereby the authority from whose area the nuisance is emitted takes action.

Appeals By S80(3) the person served with the notice may appeal against the notice to a magistrates' court within the period of 21 days beginning with the date on which he or she was *served* with the notice.

Under Schedule 3 to the Act the appeal is made *by way of complaint for an order* and the Magistrates' Court Act 1980 applies (Schedule 1(2)).

Any appeal from the magistrates' court in relation to an appeal under S80(3) will be made to the Crown Court at the instance of **any** party to the proceedings in which the decision was given (Schedule 1(3)).

The Secretary of State has made regulations in relation to such appeals. These are the Statutory Nuisance (Appeals) Regulations 1990 made under this Act and the Control of Pollution Act 1974.

The following are grounds for appeal under S80:

(a) That the abatement notice is not justified by the terms of S80.
(b) That there has been some informality, defect or error in or in connection with the abatement notice.
(c) That the authority has refused unreasonably to accept compliance with alternative requirements, or that the requirements of the abatement notice are otherwise unreasonable in character or extent, or are unnecessary.
(d) That the time, or where more than one time is specified, any of the times, within which the requirements of the abatement notice are to be complied with is not reasonably sufficient for the purpose.
(e) Where the noise to which the notice relates is a nuisance falling within categories a, f, g (or ga) of the Act and arises on industrial, trade or business premises, and that the best practicable means have been used for counteracting the effect of the noise.
(f) That in the case of category g (or ga) the requirements imposed by the abatement notice by virtue of S80(1)(a) are more onerous than the requirements for the time being in force, in relation to the noise to which the notice relates, of:

(i) Any notice served under S60 or S66 of the Control of Pollution Act
(ii) Any consent given under S61 or S65 of the Control of Pollution Act
(iii) Any determination made under S67 of the Control of Pollution Act

(g) That the abatement notice *should* have been served on some person instead of the appellant, being:
(i) The person responsible for the noise
(ii) In the case of a nuisance arising from any defect of a structural character, the owner of the premises
(iii) In the case where the person responsible for the nuisance cannot be found or the nuisance has not yet occurred, the owner or occupier of the premises

(h) That the abatement notice *might* lawfully have been served on some person instead of the appellant, being:
(i) In the case where the appellant is the *owner* of the premises, the *occupier* of the premises
(ii) In the case where the appellant is the *occupier* of the premises, the *owner* of the premises
And that it would have been equitable for it to have been so served.

(i) That the abatement notice might lawfully have been served on some person in addition to the appellant, being:
(i) A person also responsible for the nuisance
(ii) A person who is also an owner of the premises
(iii) A person who is also an occupier of the premises
And that it would have been equitable for it to have been so served.

If the court is satisfied that there was an informality, defect or error in the abatement notice but that it was not material, then the court shall dismiss the appeal (Reg. 2(3)).

If the grounds for appeal include one specified in para. 2(h) or (i), the appellant must serve a copy of the notice of appeal on any other person referred to, or any person having an estate or interest in the premises in question.

On hearing the appeal, the court may quash the notice, vary it in favour of the appellant or dismiss the appeal.

In *Johnsons News of London v. Ealing London Borough Council* (1989) it was held that the time for assessing the noise nuisance, when a notice has been served, is at the time the appeal is heard, not at the time the notice is served.

Regulation 3 specifies two situations in which a notice will be suspended pending appeal:

(a) When the notice requires expenditure on works before the appeal has been heard.
(b) In the case of category g (or ga), when the notice to which the noise relates is caused in the course of the performance of a legal duty, e.g. by a statutory undertaker such as British Gas.

However, the notice shall not be suspended pending appeal in the above situations if:

1. The noise is injurious to health.

OR

2. The noise is of such limited duration that suspension would render the notice of no practical effect, or the expenditure incurred on the works before the hearing would not be disproportionate to the public benefit.

In such cases the notice must include a statement that it shall not be suspended pending appeal. Such a statement was included in the leading case *Hammersmith London Borough Council v. Magnum Automated Forecourts Ltd.* (1978).

Powers of entry Schedule 3 para. 2 of the Act provides that any person authorized by a local authority may, on production of his or her authority (if required), enter any premises at any reasonable time:

(a) In order to ascertain whether or not a statutory nuisance exists.

OR

(b) For the purpose of taking any action, or executing any work, authorized or required by Part III of the Act.

Para. 3(2) further provides that in the case of wholly or mainly residential premises, normally 24 hours notice of the intended entry under the previous subparagraph must be given to the occupier, unless there is an emergency.

Should it be shown to the satisfaction of a justice of the peace in a sworn information in writing

(a) that admission to any premises has been refused, or that refusal is apprehended, or the premises are unoccupied, or the occupier is temporarily absent or that the case is one of emergency or that an application for admission would defeat the object of the entry

AND

(b) that there is reasonable ground for entry into the premises for the purpose for which entry is required THEN the justice may by warrant under his hand authorise the local authority by any authorised person to enter the premises, if necessary by force. (Para. 2(3))

When entering the premises under any of the above circumstances, the authorised person may:

(a) take other people with him and such equipment as may be necessary
(b) carry out such inspections, measurements and tests as he considers necessary for that purpose. (Para. 2(4))

After such authorized entries, the person must then leave the premises *as effectively secured against trespassers as he or she found them*.

> Any warrant issued under sub-para 3 shall continue in force until the purpose for which the entry is required, is satisfied. (Para. 2(6))

> An emergency is to be interpreted as a situation where the authorised person has reasonable cause to believe that there is immediate danger to life or health and that immediate entry is necessary to verify the facts or to ascertain their cause and to effect a remedy. (Para. 2(7))

The 1993 Act has extended the powers in Schedule 3 to provide any person authorized by a local authority with powers to:

(a) Enter or open a vehicle, machinery or equipment, if necessary by force. OR
(b) Remove a vehicle, machinery or equipment from a street to a secure place.

Both powers are for the purpose of taking any action, or executing any work, authorized by or required under Part III in relation to a statutory nuisance within S79(1)(ga).

Once again, on leaving the unattended vehicle, etc., it must be left properly secured as effectively as it was found according to S4(2) of the 1993 Act. Otherwise it should be immobilized or removed from the street (S4(3)). No more damage should be committed than necessary (S4(4)). Before any of the above action is taken, the police should be notified; they should also be notified if the vehicle, etc., is removed and should be informed of its new location.

Offences relating to entry Para. 3(1) of the third Schedule to the Act provides that if a person *wilfully obstructs* someone exercising the above powers he or she shall be liable on summary conviction (in the magistrates' court) to a fine not exceeding level 3 on the standard scale.

Trade secrets Someone carrying out the powers conferred under para. 2 must be careful that he or she does not disclose a trade secret discovered while exercising those powers, otherwise he or she will be liable on summary conviction to a fine not exceeding level 5 on the standard scale. This is so unless the disclosure was made in the performance of his or her duty, or with the consent of the person having the right to disclose. (Para. 3(2))

Breach of notice By S80(4), as amended, if a recipient of an abatement notice contravenes any requirement contained in the notice without *reasonable excuse*, he or she commits an offence against this part of the Act. Summary proceedings will then proceed in the magistrates' court.

If the offence is committed on industrial, trade or business premises, the guilty person will be liable to a fine on summary conviction not exceeding £20 000, according to S81(6). But if the offence is committed elsewhere, the fine will be not more than level 5 on the standard scale, with a further fine equal to one-tenth of that level for each day the offence continues.

In *Cooke v. Adatia* (1989) it was held that, in proving there has indeed been a breach of the notice, it is not necessary to show that a particular occupier of premises has suffered from the nuisance (cf. an action for private nuisance). Thus, if it is impossible for one reason or another to produce admissible evidence from a particular occupier, then noise meter evidence or other evidence can be produced by the environmental health officer of the noise. There is no specified type of evidence that need be produced under the Act.

Defences

(i) Reasonable excuse Firstly, it is a defence if the recipient had 'reasonable excuse' to breach the abatement notice (S80(4)).

This will depend on the facts of the situation but would probably cover emergency situations.

In *Wellingborough Borough Council v. Gordon* (1990) (see page 220) reasonable excuse was not held to be a proper defence in those circumstances, although the recipient's behaviour could be pleaded in mitigation. In *Lambert (A) Flat Management v. Lomas* (1981) the 'reasonable excuse' defence was said to have been provided in the 1974 Act (Control of Pollution Act) only as a defence to a criminal charge; it was not to provide an opportunity to challenge the notice, which should properly have been done by appealing against its issue. (Read the grounds for appeal against notice; reasonable excuse does not figure among them.)

(ii) Best practicable means If the noise was caused in the course of a trade or business, then it is a defence to prove that the 'best practicable means' have been used to counteract the effects of the nuisance (S80(7)).

This defence is available in the case of all the possible noise-related statutory nuisance situations, S79(1)(a), (f), and (g) (ga), but only if the premises or vehicle, machinery or equipment are used for trade or business.

S79(9) gives a statutory definition of what is meant by best practicable means. The word *practicable* means reasonably practicable having regard, among other things, to local conditions and circumstances, to the current state of technical knowledge and to the financial implications. Although a court may have sympathy with a factory, vital to a community, it will not necessarily acquit the owner because of financial implications. In *Wivenhoe Port v. Colchester Borough Council* (1985), a case concerned with dust nuisance, the court held that the defence was not maintained by proving that the business would require extra money to operate or even would become unprofitable. The *means to be employed* include the design, installation, maintenance and manner and periods of operation of plant and machinery, and the design, construction and maintenance of buildings and structures.

The test, according to the section, is to apply only so far as is compatible with any duty imposed by law. Thus, if there are legal duties imposed at common law in nuisance, in negligence or in statute in the same situation, this test must be compatible with those duties. Finally the test is to apply only so far as compatible with safety and safe working conditions, and with the exigencies of any emergency or unforeseeable circumstances. If there is a code of practice under S71 of the Control of Pollution Act (noise minimization), this must be regarded when applying the best practicable means defence.

(iii) Noise nuisance special defences The above defences apply on the whole to all types of statutory nuisances. There are additional special defences relating to S79(1)(g) noise nuisances, introduced by the 1974 Act, and to S79(1)(ga), introduced by the 1993 Act.

By S80(9)(a) it will be a defence if the alleged offence under S80(4) was covered by a notice served under S60 of the 1974 Act, or consent had been given under S61 or S65 of the 1974 Act. These sections relate to noise on construction sites and noise level registers (see below).

There will also be a good defence if an S66 Control of Pollution Act noise reduction notice was in force at the time of the alleged offence and the noise was not in contravention of that limit (S80(9)(b)).

It will be a defence if a noise reduction notice had been made in relation to a particular day or time and the alleged offence was committed outside the period for which the limit was laid down (S80(9)(b)).

Finally, if an S67 Control of Pollution Act noise level (new buildings liable to an abatement *order*, not an abatement notice) had been fixed and the noise of the alleged offence did not exceed that level, then this would be a good defence (SS9(c)).

In the last three paragraphs, the defences apply whether or not the relevant notice was subject to appeal at the time when the offence was alleged to have been committed.

The right to take High Court action, S81(5) Section 81(5) allows the local authority to take proceedings in the High Court, if it feels that proceedings for an offence under S80(4) would be an inadequate remedy. The intention would be to secure the abatement, prohibition or restriction of the nuisance, even though the local authority has suffered no damage from the nuisance.

This may be instead of, or as is possibly more usual, following failure of summary proceedings in the magistrates' court.

This section can only be relied on if the nuisance complained of is a statutory nuisance. If the nuisance is only a private nuisance, the local authority or the person suffering only have the right to proceed in the High Court if they have suffered interference with their own land. If the nuisance is a public nuisance, the local authority would have to use S222 of the Local Government Act 1972 (see page 212).

Compare the use of S222 in respect of breach of a construction abatement notice under S60 of the 1974 Act, where no statutory right was given within the Act itself, unlike S81(5) above (see the Bovis case below).

An example of a situation where this section was used is the leading case of *Hammersmith London Borough Council v. Magnum Automated Forecourts Ltd.* (1978) which sorted out the confusion over jurisdiction when using the equivalent section under the old legislation.

In 1976, the defendant company erected a new building in a quiet street and used it for a 'taxi care centre'. This provided 24-hour service for taxis, selling fuel, washing facilities and vending-machines. People living nearby complained about the noise, particularly in the early hours of the morning. On being satisfied that a noise amounting to a nuisance existed, the local authority served notice under the then S58(1), requiring Magnum Ltd. to stop all activities on the premises between 11:00 PM and 7:00 AM. In the notice the authority had stated that the expenditure necessary to comply with the notice was not disproportionate to the public benefit. Thus, the notice was not suspended pending any appeal. The defendants continued to operate as before. They also appealed against the notice. The local authority therefore decided to apply to the High Court for an injunction. When the appeal came before the magistrates' court, the magistrates said that the High Court must now deal with the matter. When the case was heard in the High Court, it decided that the magistrates must determine the case. Before the magistrates' appeal was heard, the appeal from the High Court was heard by the Court of Appeal. It decided that the present situation was adequately covered by S58(8) and that recourse to the High Court for a civil remedy was thus available. The notice had specifically stated, as required, that it would not be suspended pending an appeal. When the nuisance continued, in contravention of the notice, action had to be taken, and if the authority was of the opinion that proceedings under S58(4) would be inadequate, there would be no alternative but to seek an injunction. The Court of Appeal thus granted an interlocutory injunction (i.e. pending the outcome of the action) pending determination of the matter in the magistrates' court.

If the local authority felt at the outset that there would

be no response at all to an abatement notice, it could proceed immediately under S81(5) in the High Court and the whole matter would be dealt with in the High Court, without any reference to the magistrates' court.

If such proceedings are brought under S81(5), in relation to a statutory noise nuisance S79(1)(g) or (ga) only, then it will be a good defence to prove that the noise was authorized by a notice under S60 or a consent under S61 (construction sites) of the Control of Pollution Act.

A person aggrieved's rights, S82 In recessionary times the local authority's ability to take action under S80 can often be reduced through lack of money. Night-time activities, such as pay parties, cannot be properly monitored if there are no environmental health officers on duty. Irregular nuisances are also extremely difficult to monitor. So, with all the will in the world, it may be necessary to turn away a person suffering from a noise nuisance because the local authority is unable, for whatever reason, to take action. Fortunately, S82 is available to help in such situations. (Remember that this section can be used *against* a local authority, especially in respect of category (a) statutory nuisances.)

This section is a considerable improvement on its old equivalent, S59 of the 1974 Act, which only gave *occupiers of property* rights to complain to a magistrates' court. Furthermore, the 1993 Act has greatly extended this section by covering category (ga) cases, too.

Now a magistrate's court may act on a complaint made by *any person* that is *aggrieved by the existence of a statutory nuisance*. Who is a person aggrieved? It appears from cases that the aggrieved person should not be 'a mere busybody who is interfering in things that do not concern him' as per Lord Denning in *Att.-Gen. (Gambia) v. N'Jie* (1961). There should be some connection between the aggrieved person and the statutory nuisance complained of. Thus, if someone's health is being affected or someone is suffering from a nuisance directly, he or she should be able to make a complaint. Similarly, if his or her family is being affected, he or she will be able to complain.

The magistrates may, if satisfied that the alleged nuisance exists or that, although abated, it is likely to recur on the same premises (or in the case of a category (ga) nuisance, the same street), make an order for either or both of the following purposes:

(a) Requiring the defendant to abate the nuisance, within a time specified in the order, and to execute any works necessary for that purpose.
(b) Prohibiting a recurrence of the nuisance and requiring the defendant, within a time specified in the order, to execute any works necessary to prevent the recurrence.

The court may also impose a fine not exceeding level 5 on the standard scale.

This section does not allow action to be taken if there is only a likelihood of a statutory nuisance occurring. One must either *exist* or there must be a likelihood of its recurrence (cf. the local authority's duties).

The date on which it must be decided whether a nuisance exists or, if abated, is likely to recur is the date of the hearing of the complaint.

By S82(3) if the magistrates are satisfied that the alleged nuisance exists and in their opinion renders the premises unfit for human habitation, they may, by order under subsection (2), prohibit the use of the premises for human habitation until the premises are to the satisfaction of the court rendered fit.

The proceedings should be brought against the same people as in S80 and the coresponsibility section also applies (S82(4) and (5)).

The local authority's rights are different to those of the aggrieved person, in that magistrates' court action would only occur following the service of an abatement notice by the authority. However, before instituting proceedings under S82, the aggrieved person must now give notice in writing of his or her intention to bring proceedings, and the notice must specify the matter complained of. In relation to category g noise nuisances, at least three days notice must be given. For all other categories, a minimum of 21 days notice is necessary.

The order must be unambiguous and need not necessarily specify what works must be done to abate the nuisance, unless it would otherwise be unclear what is required. Compare a simple noise nuisance situation with a category a situation, which could require the soundproofing of walls and floors; the order must then be served as soon as possible on the defendant.

If after receiving the court order, the defendant, without reasonable excuse, contravenes any requirement or prohibition imposed by that order, he or she will be guilty of an offence and liable to summary conviction in the magistrates' courts to a fine not exceeding level 5 on the standard scale together with a further fine of an amount equal to one-tenth for each day the offence continues after the conviction (S82(8)).

Defences The 'best practicable means' defence is available in relation to category a, f and g (and ga) statutory nuisance, but only if the nuisance is committed on industrial, trade or business premises. It is not available if the nuisance is such as to render the premises unfit for human habitation (S82(9) and (10)).

Local authority works By S82(11), if someone is convicted of an offence under S82(8), the magistrates, *after* giving the local authority the opportunity of being heard on the subject, may direct the authority to do anything which the convicted person was required to do by the order to which the conviction relates.

Also by S82(13), if the potential defendants can't be found, then after listening to representations by the local authority, the magistrates may direct the authority to do anything which the court would have ordered those people to do.

Compensation, S82(12) Section 82(12) gives the complainant an opportunity to be compensated by the defendant. If it is proved that the nuisance existed at the date of the complaint, whether or not it is shown that the nuisance still exists or is likely to recur, the courts shall order the defendant(s) to pay the complainant a reasonably sufficient sum to compensate him or her for any expenses properly incurred by him or her in the proceedings. Expenses are not the same as damages in a civil case, where the plaintiff would be compensated for the interference with the use of his or her land. Nevertheless, it is an improvement on the previous law, although there was a similar provision under the Public Health Act 1936 for non-noise statutory nuisances.

10.5 Construction sites

We now return to the Control of Pollution Act 1974 which, as I explained at the beginning, was not completely repealed by the Environmental Protection Act 1990.

One of the most difficult areas involving noise control relates to construction sites, which by virtue of the type of work involved, often create noise nuisance. The Act lays down special rules peculiar to construction site work. Such work is defined by S60(1) as:

(a) The erection, construction, alteration, repair or maintenance of buildings, structures or roads.
(b) Breaking up, opening or boring under any road or adjacent land in connection with the construction, inspection, maintenance or removal of works.
(c) Demolition or dredging work.
(d) Any work of engineering construction.

Section 60 notices

If it appears to a local authority that any of the above work is to be carried out or is being carried out, it may serve a notice imposing requirements as to the way the work is to be carried out and, if necessary, publish the notice of the requirements in such a way as the authority thinks appropriate.

Such a notice may specify:

(a) Plant or machinery which is or isn't to be used
(b) Times of operation
(c) Levels of noise for:
  (i) emission from the premises
  (ii) emission from any particular part of the premises during specified hours

It may also provide for change in circumstances (S60(3)).

According to S60(4)(a), the local authority must have regard of any codes of practice issued under this part of the Act, i.e. the Control of Noise (Code of Practice for Construction Sites) Order 1975, which adopted the British Standards Institution Code of Practice BS 5228 (revoked and reformulated in 1984 and 1987 in the Control of Noise (Codes of Practice for Construction and Open Sites) Orders 1984/92 and 1987/1730).

They must also ensure that the 'best practicable means' are employed to minimize noise S60(4)(b).

Under S60(4)(c) the local authority must, before specifying a particular method, plant or machinery, have regard to the desirability of specifying other methods, plant or machinery which would be almost as effective in minimizing noise and more acceptable to builders, in the interests of the recipients of notices. It seems from this subsection that it is important to look at the work from the viewpoint of the builder, with noise reduction achievable using acceptable alternative equipment.

The local authority must have regard to the need to protect persons from the effects of noise in the locality in which the premises are situated (S60(4)(d)). In view of the complaints that the local authority receives in relation to construction work, one could argue that this is a most important consideration.

The notice must be served on the person who is going to carry out the works and on such persons as have control over the operations, as the local authority thinks fit (S60(5)).

Under S60(6) the notice *may* specify the time within which the notice must be complied with, and it may require works to be executed, or the taking of steps as may be necessary or specified, to comply with the notice.

By S60(7) appeals may be made to the magistrates' court within 21 days and the Control of Noise (Appeals) Regulations 1975 apply. The following are the grounds of appeal:

(a) That the notice is not justified by the terms of S60.
(b) That there has been some informality, defect or error in, or in connection with the notice.
(c) That the authority has refused unreasonably to accept compliance with alternative requirements, or that the requirements of the notice are otherwise unreasonable in character or extent, or are unnecessary.
(d) That the time — or where more than one time is specified, any of the times — is not reasonably sufficient for the purpose.
(e) That the notice should have been served on some person instead of the appellant, being a person who is carrying

out the works or is going to carry out the works, or a person who is responsible for, or has control over, the carrying out of the works.

(f) That the notice might lawfully have been served on some person in addition to the appellant, being a person who is carrying out the works or is going to carry out the works, or a person who is responsible for, or has control over, the carrying out of the works. And that it would have been equitable for it to have been so served.

(g) That the authority has not had regard to some or all of the provisions of S60(4).

If the person served with a notice fails to do anything required within the specified time and without reasonable excuse, he or she commits an automatic offence S60(8). But if consent has been given under S61, it will be a defence.

In *Walter Lilly & Co. v. Westminster City Council* (1994) building contractors were served with S60 notices; they were also asked under S93 of the Act to give details of the work to be carried out and the proposed time for completion. Noise nuisances were committed over a weekend period but they were in relation to a second contract, not specified in the original works to which the S60 notice related (nor contemplated originally when the S93 details had been supplied). The council then took magistrates' court action against the builders. The contractors appealed to the divisional court, stating that the notice should only relate to the works being carried out at the time of the notice. The builders could have appealed against the notice in the first instance, but the problem works, the second contract, had not been contemplated at that time. The court upheld the contractors' appeal. Thus, it appears from this case that an S60 notice can only be used in relation to the work being carried out at the time it is served. A statement providing for changes in circumstances is only meant in relation to the works in hand or contemplated at that time. It seems as if this decision could cause problems in the future, although the law has been logically applied.

It should be noted that, like proceedings under S82 of the Environmental Protection Act 1990, if recipients of notices give the appearance that they will continue to ignore the notice, recourse can still be made to the High Court. The Control of Pollution Act 1974 does not specifically state this, so either S222 of the Local Government Act must be used or there must be an unlawful interference with the local authority's land to enable them to invoke private nuisance. Indeed, *City of London Corporation v. Bovis Construction Limited* (1988) particularly looked at situations where proceedings under S60 of the Control of Pollution Act 1974 had commenced but an injunction was being sought. In that case, the court said, injunctions were to be used with caution. Only if the recipient was obviously going to ignore the notice and there was behaviour over and above a breach of the criminal law should an injunction be sought. In that case S222 was used to instigate the course of action.

Prior consent on construction sites, S61 amended by 1990 Act Section 61 gives those people responsible for construction work, and the local authorities, an opportunity to settle any problems relating to the potential noise before the work starts. Indeed, Department of the Environment Circular 2/76 suggested that there should be some sort of 'early warning system' by the local planning authority, giving attention to work with potentially serious noise problems. Furthermore, advice is given in the BSI Code of Practice 5228 that the noise requirements of a local authority should be ascertained before the tender documents are sent out, so that they can be incorporated into those documents. This is only right as one of the main reasons why builders ignore S60 notices is that the financial implications of fixed damage clauses in the building contract make it imperative that they finish work on time. To restrict their working hours to the notice requirements reduces the time available to finish the work before incurring fixed damages (the erroneously named penalty clauses). Had the local authority told the employer *before* the work was contracted out that they would expect work to be carried out only between certain hours and without using certain equipment, then the tender documents could have alerted the potential contractors to the problem in advance and allowed them to make a sensible judgement as to the amount of time needed to complete the work.

Application may be made before the work begins and consent may be given to the applicant. The parliamentary draftpersons probably envisaged that the contractor would make such an application. It has been suggested, however, that the consulting engineer would be a more suitable person and that the consent should be sought before the tender is made, in order to take account of any requirements made by the local authority.

By S61, if building regulation approval is required, the application must be made at the same time or later than the request for approval. The application must contain the following particulars:

(a) The works and the method by which they are to be carried out.

(b) Noise minimization steps (S61(3)).

According to S60(4), (5) and (6), if the local authority considers that sufficient information has been given and that, if the works were carried out in accordance with the application, it would **not** serve a notice under the preceding section in respect of those works, the authority shall grant its consent within 28 days and must not serve a notice under S60, but in so doing may:

(a) Attach conditions.
(b) Limit consent where there is a change in circumstances.
(c) Limit duration of consent.

If there is a contravention of any of these items, an offence will have been committed (S60(5)). The consent may be published if the authority thinks fit (S60(6)). Where the local authority does not give its consent or gives its consent but subject to conditions, the applicant may appeal to a magistrates' court within 21 days (S60(7)). Should proceedings be brought under S60(8), it would be a good defence to prove that the alleged contravention amounted to the carrying out of the works in accordance with consent given under this section.

Any consent given does not itself constitute a defence to proceedings under S82 of the 1990 Act (an aggrieved person's right to take proceedings, S61(9)).

Where consent had been obtained by someone other than the site worker, e.g. an employer, an architect or a consulting engineer, they must bring the consent to the notice of the site worker, otherwise the applicant will be guilty of an offence (S61(10)).

10.6 Noise in streets, S62 amended by 1990 Act and Noise and Statutory Nuisance Act 1993

Section 62 is concerned not with general noise in streets, but specifically with the use of loudspeakers.

By S62 a street is defined for this section as a highway and any other road, footway, square or court which is for the time being open to the public. A highway has at common law been defined as a right of way over which the public have the right to pass and repass, thus a street probably does not have to be metalled.

Thus, under S62(1) a loudspeaker in a street shall not be operated:

(a) Between 9:00 PM and 8:00 AM the following morning for any purpose (S62(1)(a)).
(b) At *any* other time, for the purpose of *advertising* any entertainment, trade or business (S62(1)(b)).

This section has been amended by the 1993 Act (SS7, 8), and Schedule 2 S62(1A) and (1B) has been inserted after S62(1). This states that 'Subject to subsection (1B) of this section the Secretary of State may by order amend the times specified in subsection (1)(a) of this section.' Furthermore, by S62(1B) any order under the new subsection shall *not* amend the times so as to *permit* the operation of a loudspeaker in a street at any time between the hours of 9:00 PM and 8:00 AM.

Any person contravening the above is guilty of an offence.

Subsection (B) would obviously affect the trade of traditional travelling food vendors, so S62(3) states that it shall not apply to the operation of a loudspeaker between 12:00 PM and 7:00 PM of the same day, provided *all* the following apply:

(a) The speaker is fixed to a vehicle being used to sell perishable foods (for humans).
(b) The speaker is operated solely for informing members of the public that the commodity is for sale from the vehicle.
(c) The speaker is operated so as not to give reasonable cause for annoyance to people in the vicinity.

Note: The Control of Noise (Code of Practice on Noise from Ice-Cream Van Chimes, Etc.) 1981 made under S71 is also appropriate here (see page 232).

Exceptions to the general prohibition under S62

Obviously there have to be exceptions. By S62(2) the general prohibition does not apply to:

(a) Operation of loudspeakers by the police, fire brigade, ambulances, National Rivers Authority, water undertakers, sewerage undertakers exercising their functions, or a local authority within its area. (Mnemonic, SWAN FLAP).
(b) Operation of loudspeakers by persons for communicating with someone on a vessel in order to direct its movement or the movement of another vessel.
(c) Operation of loudspeakers if they are part of a public telephone system.
(d) Operation of loudspeakers if they are in or fixed to a vehicle and they are operated either for the entertainment of the driver or passengers (such as a radio or television), or they are horns or other devices for warning traffic **and** they are operated so as not to cause reasonable annoyance to persons in the vicinity.
(e) Operation of loudspeakers not on a highway, when in connection with a transport undertaking used by the public, such as British Rail or a bus company, when making announcements to passengers or employees.
(f) Operation of loudspeakers by travelling showmen on land being used for a fair.
(g) Operation of loudspeakers in cases of emergency.

The 1993 Act has inserted S3(A), which states that the prohibition section of the Act (banning loudspeakers between 9:00 PM and 8:00 AM) shall not apply to loudspeakers operating in accordance with a consent granted by a local authority under Schedule 2 of the 1993 Act.

Schedule 2 consents By S8(1) of the 1993 Act, a Local Authority *may* (not shall) resolve that Schedule 2 is to apply to its area. Such a resolution must be published for two consecutive weeks in a local newspaper in circulation in that area. The notice must state, in addition to the fact that the resolution has been passed, the general effect of Schedule 2 and the procedure to apply for a consent under the schedule.

Schedule 2 para. 1 states that any person may apply to the local authority for consent to operate a loudspeaker in contravention of S62(1) of the 1974 Act in its area. Para. 2 states that a consent shall not be given to the operation of a loudspeaker in connection with any election or for the purpose of advertising any entertainment, trade or business. The consent may be granted subject to such conditions as the local authority considers appropriate.

Procedure The application must be made in writing giving sufficient information as may be reasonably required by the local authority. From the date of receipt, the application must be considered and the decision notified within 21 days. If consent is granted subject to conditions, the conditions must be specified. The authority may publish details in a local newspaper.

Burglar alarms

Section 9 and Schedule 3 of the 1993 Act have introduced new rules relating to alarms. By S9 a local authority *may* make a resolution that Schedule 3 applies to its area *after consultation* with the chief officer of police (S9(1)).

If a local authority makes such a resolution, then Schedule 3 (apart from para. 4) shall come into force on a specified date called the first appointed day (at least four months after the resolution), then para. 4 shall come into force in its area on the second appointed day (at least nine months after the first appointed day); paras. 2 and 3 cease to have any effect. A notice must be published in a local newspaper for two consecutive weeks ending at least three months before the first appointed day.

New alarms By para. 1, if a *person* installs an audible intruder alarm on or in any premises, he or she must ensure that the alarm complies with any prescribed requirements and that the local authority is notified within 48 hours of the installation. Such requirements will be contained in regulations to be made by the Secretary of State. Contravention of either compliance or notification amounts to a summary offence punishable by fine. This would normally apply to installers of burglar alarms who should be well apprised of the legal requirements.

Operation of alarms before or after second appointed day By para. 2 an *occupier* of premises in which or on which an alarm is installed on or after the first appointed day, or by para. 4, on or after the second appointed day, shall *not* permit the alarm to be operated *unless* para. 5 is satisfied. Failure to comply with para. 5 requirements amounts to a summary offence punishable by fine.

Also under para. 3 a person who *becomes* an occupier on or after the first appointed day similarly must not allow the alarm to be operated or face the same consequences.

Para. 5 requirements for operating alarms There are three requirements:

1. The alarm must comply with any prescribed requirements made by regulation of the Secretary of State.
2. The police must be notified in writing of the names, addresses and telephone numbers of the current keyholders.
3. The local authority must have been informed of the address of the police station to which notification has been given.

Entry to premises Para. 6 gives the much needed statutory authority to turn off the alarm. If the alarm is operating audibly more than one hour after it was activated and the noise is such as to give people nearby *reasonable cause for annoyance*, a generally or specially authorized local authority officer may enter the premises to turn off the alarm. No force may be used under this paragraph.

But under para. 7, the officer may apply to a justice of the peace for a warrant to enter the premise by force if need be. The JP must be satisfied:

(a) That the alarm has been operating for more than one hour after it was activated.
(b) That the audible operation of the alarm is such as to give persons living or working nearby reasonable cause for annoyance.
(c) Where notification of details of the keyholders has been given, that the officer has taken steps to obtain access with their assistance.
(d) That the officer has been unable to gain access without using force.

By para. 7(2), before applying for a warrant, the officer shall leave a notice at the premises stating that the alarm is operating so as to cause annoyance to those close by and that an application is being made to a JP for a warrant to enter and turn off the alarm.

The officer must enter the premises accompanied by a police constable when using this paragraph's authority.

The warrant shall continue in force until the alarm has been turned off and the officer has complied with para. 10 (see below).

The officer may take with him anyone, such as a

locksmith, and such equipment as may be necessary (para. 8).

Anyone entering premises by virtue of para. 6, 7 or 8 must not cause more damage or disturbance than necessary. (Should they do so, they would be liable in tort for negligence or trespass to land or property.)

By para. 10 an officer using para. 6 or 7 to enter premises which are unoccupied, or from which the occupier is temporarily absent, shall:

(a) After the alarm has been turned off, reset it if reasonably practicable.
(b) Leave a notice giving details of action taken.
(c) Leave the premises, so far as is reasonably practicable, as effectually secured against trespassers as he or she found them.

The local authority may recover expenses connected with breaches of paras. 2, 3 and 4, or in relation to entering property (para. 11).

Finally, providing anything done by the officers or anyone authorized is done in good faith for the purposes of this schedule, they are protected from liability except against those powers of the district auditor and court under the Local Government Finance Act 1982.

In London only, under the London Local Authorities (Miscellaneous Provisions) Act 1991, local authorities already have the right to prosecute owners or occupiers of buildings whose audible intruder alarms contravene similar requirements to those under the 1993 Act. But there is a slight difference between the two; the London Act allows action to be taken if the alarm causes only annoyance. The 1993 Act's 'reasonable cause for annoyance' is probably more onerous.

10.7 Noise abatement zones S63

By the remaining part of S57 of the Control of Pollution Act 1974 a local authority is under a statutory duty to cause its area to be inspected from time to time to decide how to exercise its powers concerning noise abatement zones. There is no need, however, for the local authority to make an inspection before making an order. *Morganite Special Carbons v. Secretary of State for the Environment* (1980).

By S63 a local authority may, by order, designate all or any part of its area a noise abatement zone. Department of the Environment Circular 2/76 explains the technical aspects of noise abatement zones. Their purpose is to prevent a deterioration in environmental noise levels and, wherever practicable, to achieve reduction in noise levels.

The order must specify the classes of premises to which it applies (S63(2)). Examples include public utility installations, such as waterworks, power stations, gasworks, launderettes, and entertainment halls. The control of noise that will be achieved is limited to controlling noise from such individual premises and not by laying down a standard maximum noise level for the zone. The classes of premises specifically exclude domestic premises, as it was felt that the other methods of controlling noise, under the then Control of Pollution Act 1974 and now S80 and S81 of the Environmental Protection Act 1990, would be more appropriate. The order designating the zone is called a Noise Abatement Order. Do *not* confuse it with the Noise Abatement *Notice* under S80.

The order may subsequently be revoked or varied by another order S63(3).

Register of noise levels, S64

Where a local authority has designated its area or any part of it a noise abatement zone, it must measure the level of noise emanating from the premises falling within the particular category noted in the order, e.g. commercial premises (S64(1)). Which category is noted will depend on whether it is considered to be a problem in that area.

The measurements shall then be recorded in a noise level register to be kept by the local authority (S64(2)). A copy of the record shall be served upon the owner and occupier of the affected premises and he or she may appeal to the Secretary of State within 28 days, who may then give such directions as he or she thinks fit (S64(3) and (5)). The Control of Noise (Appeals) Regulations 1975 are the appropriate regulations for such an appeal. Otherwise, once the levels have been recorded, the record cannot be properly challenged, except as to whether it was properly served. The register shall be open for inspection and copies may be obtained (S64(7)). The Secretary of State, by virtue of S64(8), has made regulations giving the methods by which the measurements are to be made, the Control of Noise (Measurement and Registers) Regulations 1976; the measuring device should comply with BSS 4197/1967.

Exceeding the registered level, S65 Once a noise level has been determined, it must not be exceeded, unless written consent has been given previously (S65(1)). Such prior consent must have been entered on the Register S65(2). If the level or the higher permitted level has been exceeded, then an offence will have been committed (S65(5)). The magistrates' court convicting a person of an offence may, if satisfied that the offence is likely to recur, make an order requiring the execution of any works necessary to prevent it continuing or recurring. Failure to carry out the works is also an offence (S65(6)). The local authority may be heard on the matter and may be empowered to do any such work and recover the cost from the offender (S65(7) and S69).

The consent mentioned above must be applied for if the owner wishes to be able to emit noise above the registered level. Unless a decision is made in writing within two months, it will be deemed refused by the local authority unless a further period is agreed in writing (S65(3)). The failed applicant may appeal to the Secretary of State (S65(4)). The Control of Noise (Appeals) Regulations 1975 apply. Such consents may be made subject to such conditions as necessary, e.g. the amount by which the noise levels may be increased or the periods during which the level may be increased (S65(2)). All these particulars must be recorded on the noise level register. The consent will contain a statement to the effect that the consent does not of itself constitute any ground of defence against proceedings instituted under S82 (aggrieved person's rights) of the 1990 Act S65(8). It may, however, be a defence to an offence under S80 (notice from local authority).

Reduction of noise levels, S66 If it appears to the local authority that the level of noise emanating from any premises under a Noise Abatement Order is not acceptable, having regard to the purposes for which the order was made, and that reduction is practicable at reasonable cost and would serve a public benefit, then the authority may serve a notice on the person responsible (S66(1)). Such a notice is known as a Noise Reduction Notice. Do *not* confuse it with the Noise *Abatement* Notice under S80.

According to S66(2), the notice shall require:

(a) Reduction of noise to a specified level.
(b) Prohibition of an increase without consent.
(c) Such steps to be taken so as to achieve those purposes.

Such steps and the abatement of the noise must be taken within a period of not less than six months from the date of service (S66(3)). Contravention of the requirements amounts to an offence (S66(8)). In such proceedings, if the offence was committed in the course of a trade or business, the best practicable means defence can be used (see page 223).

The notice may specify times, dates and different noise levels (S66(4)). S66 Noise Reduction Notices *override* the consent given under S65 (S66(5)). The local authority must record the details of the noise reduction notice in the noise level register (S66(6)).

The recipient has three months in which to appeal to a magistrates' court (S66(7)) and the 1975 Control of Noise (Appeals) Regulations apply. The following are the grounds for an appeal:

(a) That the notice is not justified by the terms of S66.
(b) That there has been some informality, defect or error in, or in connection with, the notice.
(c) That the authority has refused unreasonably to accept compliance with alternative requirements, or that the requirements of the notice are otherwise unreasonable in character or extent, or are unnecessary.
(d) That the time — or where more than one time is specified, any of the times — within which the requirements of the notice are to be complied with is not reasonably sufficient for the purpose.
(e) Where the noise to which the notice relates is noise caused in the course of a trade or business, that the best practicable means have been used for preventing, the noise or for counteracting the effect of the noise.
(f) That the noise should have been served on some person instead of the appellant, being the person responsible for the noise.
(g) That the notice might lawfully have been served on some person in addition to the appellant, being a person also responsible for the noise, and that it would have been equitable for it to have been so served.

New buildings, S67 Where a building is being constructed within a noise abatement zone or works are being done to a building, so that they fall within a class specified within the Noise Abatement Order, the local authority may determine the level of noise which will be acceptable as that emanating from the premises (S67(1)). This does not apply to private houses. The level must be placed on the noise level register (S66(2)). This must be done on the authority's own initiative or in response to a request by the owner or occupier of the premises, or even a developer who is in the process of acquiring the premises. The applicant or the recipient of the notice stating the determination of the level has a right of appeal within three months to the Secretary of State (S66(3)). The Control of Noise (Appeals) Regulations 1975 apply. The usual two-month refusal period applies where an application has been made to determine the acceptable noise level (S66(4)).

If a Noise Abatement Order comes into force and later any premises become premises to which the order would apply, as a result of construction work (from scratch) or works carried out on the building, but no noise level has been determined, then S66 applies but the recipient of a determination of noise level has six months in which to appeal (S66(5)).

Noise from plant and machinery, S68

Regulations may be made under section 68 in relation to the reduction of noise caused by plant and machinery by the use of devices or arrangements. Also, regulations may be made to limit noise from plant or machinery used on construction work or by machinery operating in a factory, as defined by the Factories Act 1961. Standards, specifications, methods of testing and descriptions not forming part of the regulations may be used. Before these regulations are made, the Secretary of State must consult with relevant persons and bodies representing producers and users in order to prevent rules being made which are impracticable or involve unreasonable expense. Contravention means the commission of an offence, but the best practicable means defence may be used.

Codes of practice, S71

In order to provide guidance on methods of minimizing noise, the Secretary of State may, by order, issue codes of practice or approve codes of practice made by other bodies. The relevant Orders to date are the Control of Noise (Code of Practice for Construction and Open Sites) Order 1984 and 1987, Control of Noise (Code of Practice on Noise from Ice-Cream Van Chimes, Etc.) Order 1981, Control of Noise (Code of Practice on Noise from Audible Intruder Alarms) Order 1981, Control of Noise (Code of Practice on Noise from Model Aircraft) Order 1981. Codes of practice have been planned to cover such matters as audible bird scarers, powerboats, clay pigeon shooting and off-road motor-cycle sport.

Thus it can be seen that the Acts provide a comprehensive treatment of most noise problems; enforcement is achieved at local authority level. Many of the problems with the legislation are caused by lack of funding, so that local authorities are unable to monitor noise nuisances properly.

10.8 Noise and personal health and welfare

As we have already seen, the tort of nuisance, the Environmental Protection Act, the Noise and Statutory Nuisance Act, etc., try to provide a quieter society and, in so doing, indirectly reduce the effects of noise on health. However, only in exceptional circumstances will compensation be awarded for illnesses or deafness caused by noise. Until fairly recently there was not even statutory guidance on reducing the effects of noise in the workplace.

The law deals with personal health and welfare in a piecemeal way, through the tort of negligence, social security legislation, the Health and Safety at Work, Etc., Act 1974 and the Noise at Work Regulations 1989. Compensation may be awarded in some cases; state pensions may be given in others. Wrongdoers, under health and safety legislation, unmindful of others' safety, may be dealt with in the criminal courts.

Negligence

The tort of negligence is a tort which protects personal interests and not just interests in land (cf. nuisance). If someone causes damage to person or property by being negligent, he or she may be sued for damages. (For criteria as to whether the case should be started in the county court or Queen's Bench Division of the High Court see page 210.) Most damage caused by noise will be to health, but vibrations or an explosion could damage property.

Requirements In order to sue successfully in negligence, the plaintiff has to prove that *all* the following elements exist:

1. That a duty of care was owed in law by the defendant to the plaintiff.
2. That this duty has been breached.
3. That the actual damage was caused as a direct result of the breach.

The duty of care In law there can be no action taken unless there are laws giving people rights and duties. Often there are statutory rights and duties contained in Acts or statutory instruments. The problem with common law rights and duties is that one often has to test whether they existed *after* the event. We cannot predict what is going to happen to us in the future. Fortunately, in one respect involving employer/employee situations, the existence of this elusive duty of care is predetermined. If the person suffering from noise-induced illness is an employee and he or she wishes to sue his or her employer, then it has long been recognized that in common law a duty of care is automatically owed by employers to their employees while they are still at work or affected by work.

In other situations the matter is not so simple and has been made more complex by a series of legal decisions in recent years. The history of negligence as a tort in its own right, not dependent on mere precedents, began with the famous case of *Donoghue v. Stevenson* (1932). This case was important in that the Law Lords put forward a formula for determining whether or not a duty of care existed in a given case. This was called the *neighbour principle* and stated 'You must take reasonable care to avoid acts or omissions which you can reasonably foresee would be likely to injure your neighbour.' Who then, in law, is my neighbour? The answer seems to be 'persons who are so closely and directly affected by my acts, that I ought reasonably to have them in contemplation as being so affected, when I am directing my mind to the acts or omissions which are called in question' as per Lord Atkin. Over the years this formula was considered too wide in its application, but many cases have been decided on this basis and would still provide a precedent in 'same facts' situations. An updated version was provided in 1977 with *Anns v. London Borough of Merton*. Lord Wilberforce said that whether a duty of care exists can be approached in two stages:

> First, one has to ask whether as between the alleged wrongdoer and the person who has suffered damage,

> there is sufficient relationship of proximity or neighbourhood such that in the reasonable contemplation of the former, carelessness on his part may be likely to cause damage to the latter, in which case a prima facie duty of care arises.
> Secondly, if the first question is answered affirmatively, it is necessary to consider whether there are any considerations which ought to negative or to reduce or limit the scope of the duty or the class of person to whom it is owed or the damages to which a breach of it may give rise.

Over the years both formulae were considered to be too wide. In 1990 in *Caparo Industries Plc v. Dickman*, a case which has been confirmed by subsequent cases, a formula was arrived at incorporating facets of the other two decisions reducing the scope of the duty of care. (*Anns v. Merton LBC* was also overruled on its particular facts in another decision.) The Lords decided that three requirements were necessary:

(a) There must be reasonable foreseeability of the relevant loss.
(b) It must be just and reasonable that a duty should exist.
(c) There must exist a sufficient relationship of proximity between defendant and plaintiff.

This test is more stringent than the earlier ones. Let us apply it to a hypothetical situation. A lady lives next door to a factory operating noisy machinery. She now suffers deafness and tinnitus. No medical explanation can be given other than it has been caused by the noisy machinery. Is she owed a duty of care? Is there reasonable foreseeability of the relevant loss? Could that factory owner have reasonably imagined that a neighbour could be affected by that work and suffer from deafness, especially if he has provided his workers with ear defenders in accordance with the Noise at Work Regulations? Secondly, is it just and reasonable that she, a neighbour, should be owed such a duty? Probably, yes. Thirdly, is there a sufficient relationship of proximity between the parties? In all three cases one could argue in favour of a duty being owed.

Breach of the duty Having established that a duty of care was owed in the circumstances, the plaintiff must then prove to the court that there was a breach of that duty, i.e. the defendant was negligent in fact and law.

This is done in two stages. First, taking the circumstances of the case, the court decides on the standard of care, i.e. what should have been done by the defendant. Second, the defendant's actual behaviour is measured against the established standard. If it falls short, the defendant is negligent; if it matches or is above the standard, the defendant has committed no breach.

(i) The standard of care The standard of care is not of superman nor that of a fool. The court 'fixes' the standard by examining evidence which would show what the *average reasonable* person should have done when placed in the circumstances of the case. Experts may be called to give evidence of trade practice. So, the average factory owner has to meet the standard of the average factory owner doing that particular sort of work. The same factory owner driving his or her car home at night has to meet the standard of care of an average driver. If he or she decides to repair his television set, the standard of care is that of the average layperson. It makes no difference whether he or she is extremely intelligent, habitually lazy or unusually careless because this introduces a further subjective element, which has been rejected by the courts.

Furthermore, the knowledge which one is supposed to possess depends on the current state of knowledge. So, at present, we are probably doing many things which in years to come will prove to have been dangerous. Once knowledge becomes available to society, any failure to use it appropriately could amount to negligence, even if individual defendants are ignorant of that knowledge.

(i) Other factors which may raise or lower the basic standard of care (a) Is the risk reasonably likely? The courts are not going to be impressed with a defendant who failed to avoid an obvious risk. Once again, the reasonable foresight test is used. Thus failure to see that ear defenders are being worn in noisy factories or on building sites will amount to negligence because of the extreme likelihood of impairing one's hearing.

Even if the risk is not likely but is reasonably foreseeable, it should be taken into account when determining the standard of care. This follows dicta in the *Wagon Mound No. 2* (1967), a Privy Council shipping case which involved an unusual chain of events. For example, in *Haley v. London Electricity Board* (1965) the electricity board's employees had dug a hole in a pavement and, although they erected adequate barriers to prevent a sighted person from falling down the hole, they were inadequate for a blind person. Although the risk of harm being caused by their action was slight, it was reasonably foreseeable that a blind person could have fallen down the hole.

If the risk is a 'fantastic possibility' then probably there would be no need to do more than take the usual precautions.

If you are dealing with a known hazardous situation, such as working in a noisy industry or with noisy equipment, the courts take the view that you must take extra care of those people affected by it.

(b) Are the plaintiffs at particular risk? Having decided that certain categories of people are owed a duty of care, if it then appears they would be at particular risk because of age, infirmity or other special reason, such as being

women of child-bearing age, then the standard of care is much higher than normal. So schoolteachers owe a high standard of care towards their pupils. Employers of handicapped people must see that such people are safe, e.g. they must make sure that emergency warning alarms are easily seen by deaf workers, as they cannot hear normal sirens. Apprentices or anyone doing a new job are at particular risk until they understand what is required of them. They should be adequately supervised; warnings and advice should be given and seen to be understood. If they do not speak English, translators and notices in their own language should be provided. This need is increasing with increased mobility of workers in the European Union, where sub- and co-contractors may be from other countries, and it exists in areas where there are many immigrants who may not necessarily speak English.

In *Paris v. Stepney Borough Council* (1951) the employing council was held to be negligent in failing to give goggles to a one-eyed workman, although it would not have been negligent (then) had he been fully sighted. This seems an odd decision, but the court specifically mentioned the high degree of risk to a partially sighted person. Obviously, failure to provide earmuffs in situations where they are required by the Noise at Work Regulations 1989 would amount to negligence (and be grounds for suing for another tort, breach of statutory duty, see page 237). By analogy with the Paris case, it is possible that failure to provide earmuffs for a partially deaf person, working in relatively noisy circumstances, would amount to negligence, but not if he or she were of good hearing.

(c) Could the risk of injury or harm have been eradicated by a small outlay? No court will be impressed with the defendant arguing that he or she could only have avoided the risk of harm at great expense (unless coupled with the argument that the risk was a fantastic possibility). But, if the harm could have been avoided at small cost and the defendant failed to take avoiding action, the courts are more likely to find that the standard of care was higher than normal and there would be a case of negligence.

(d) Is the defendant an expert? If I fall down a pothole, smashing my leg so the only way to free it is by amputation, then the standard of care to be shown by an unqualified rescuer is that of the average rescuer in an emergency. But if my leg is amputated in hospital, the surgeon is expected to show such skill and experience as is appropriate to his or her position, qualifications and experience. Thus, in the first case the standard is quite low, and in the second extremely high. In *McCafferty v. Metropolitan Police District Receiver* (1977) the plaintiff, a ballistics expert, sued his employers for damage to his hearing. The employers were distinguished scientists who knew more of the effect of such work than the ballistics expert (see below).

Amateurs doing work which could be better done by experts should show reasonable care. This is because their standard of care would be that of the average amateur doing a particular job. In determining such a standard, the courts assume that all people have a certain degree of common sense and knowledge.

(e) Are there regulations, codes of practice, etc., to follow? If there are statutory regulations laying down conditions to be observed (such as the Noise at Work Regulations 1989, see page 240), they should be complied with; then the prime cause of action will be for breach of statutory duty (see page 237). If it is impossible to sue in this way, due to lack of one of the essential preconditions, the plaintiff can sue in negligence and plead that the regulations established a particular standard of care which was not met. But the plaintiff will usually sue for both negligence and breach of statutory duty as it would be impossible to add claims at a later date if one claim failed for some reason. (If there has been a breach of the regulations, there will probably be a criminal prosecution first; this will provide good evidence for the subsequent civil action.)

Any statutory codes of practice are not themselves absolute legal requirements, but they will usually establish a set standard of care.

If there are house or company rules or regulations, they will indicate a certain level of care, which may or may not be what is acceptable in the circumstances. For example, some company rules may be far too stringent; their breach would perhaps be grounds for dismissal under a contract of employment but would not amount to negligence.

(ii) Circumstances which may lower the required standard of care (a) Was there an emergency? Whether or not an emergency lowers the standard of care depends on the circumstances of each case. If, for example, there is a fire, rules should already exist so as to avoid worsening the situation. If these rules are not followed, there may be a case of negligence. But, if there are no rules and the defendant does his or her best under the circumstances, then he or she will have acted as the average reasonable person; there can be no negligence. Ear defenders may have been removed during an emergency in order to follow instructions; the resulting damage to hearing may not be actionable.

(b) Is the plaintiff an expert? If the defendant is working with a plaintiff who is an expert, this may reduce the standard of care, so as to negative any claim of negligence against the defendant, provided that the expert knew what was being done. But if negligence is proved, the defendant could claim that the plaintiff was contributorily negligent or even that there was consent (see page 236).

(c) What did the defendant do? Having determined the standard of care which should have been exercised in the

given situation, the plaintiff must then prove that the defendant did not conform with that standard. If car drivers should normally be sober, then a defendant causing an accident while inebriated will be found to be negligent.

Normally the burden of proving the breach will fall on the shoulders of the plaintiff. It is not for the defendant to show that he or she was not negligent. Evidence must be produced to satisfy the court that, on a balance of probabilities, the defendant was negligent in the circumstances. Consequently, all evidence must be properly produced and preserved. The type of evidence will depend on the situation; it could include photographs, pieces of equipment or machines, medical evidence, witness accounts, expert evidence, accident reports, letters, contracts, drawings and plans. It is up to the plaintiff to present sufficient evidence and sufficiently persuasive evidence so the court will agree that the defendant was negligent.

A conviction in criminal proceedings will be good evidence in subsequent civil actions.

If, from the outset, the evidence is overwhelmingly in favour of the plaintiff, it is highly unlikely that the case would ever actually go to a court hearing. The lawyers would then settle out of court. But sometimes, even though the defendants have admitted their liability, they still dispute the amount of damages, known as the quantum. This is especially true of personal injuries or fatalities because the courts have a highly complex approach to arriving at a final global amount. Such cases will go to court, but merely for the judge to decide on the amount of damages to be awarded; no evidence need be produced.

Damage must result from the breach

(i) Damage must be caused First, there must be damage caused. If there is no damage, there can be no action, even if one can prove the first two elements. This is because negligence is *not* actionable *per se*, like nuisance, but unlike trespass, where no damage need be proved. Damage may be caused to people — injuries — or to buildings. In noise cases the injuries may be occupational deafness or tinnitus; in vibration cases the injuries may be white finger or carpal tunnel syndrome, caused by vibrating tools. Physical damage to buildings could also be caused by noise and vibration.

(ii) The damage must not be too remote from the negligent act in fact If there is damage, it must not be too far removed from the negligent act. There has to be a causal link between the initial damage and the resultant damage. If no audiometric tests are undertaken at the commencement of a worker's employment, how can he or she later prove that any occupational deafness is as a result of his or her noisy work conditions? Conversely, if the employer is trying to defend an action, having performed no tests at the outset, how can he or she claim that the worker was already deaf before being taken on? It is obviously in the interests of truth that tests are undertaken in noisy industries. Medical evidence will have to be produced in these cases, but is not always conclusive.

(iii) The damage must not be too remote in law This is the reasonable foresight approach again. To limit the liability for all the direct *factual* consequences of negligence, the courts now recognize that one should only be liable for those consequences which could have been *reasonably foreseen* at the time of the negligence. The test was approved by the Privy Council in *Wagon Mound No. 1* (1961). Thus, it is reasonably foreseeable that excess noise can cause a number of symptoms. If, on the other hand, noise also causes conditions that are currently unknown to society, the employer could not be sued in the future because the damage would not have been reasonably foreseen at the time of the alleged negligent acts. Once a hazard becomes known, one must guard against that hazard.

It is never a defence in law to claim that you did not realize what you were doing was legally wrong. Ignorance is no defence. Thus, employers in business have a particularly difficult task in taking note of all modern developments affecting their work.

Defences The defendant in court does not sit back and cross his or her fingers, hoping the plaintiff will fail to prove a case. He or she will attempt to raise appropriate defences which could cancel out any negligence.

The burden of proving a defence obviously falls on the defendant, and he or she must produce evidence to support this defence.

Necessity It is occasionally possible that the defendant may claim that the tort was committed necessarily to prevent greater harm occurring. However, most of the cases involving necessity as a defence are concerned with other torts, where the act was deliberate.

Volenti non fit inuria *Volenti non fit inuria* is consent, literally 'there is no injury to a willing person.' This is an absolute defence and, if proved, cancels out the negligence, thus defeating the plaintiff's claim. It arises where the plaintiff:

- knew that the defendant was being negligent
- knew of the risk
- continued to cooperate with the defendant
- suffered injury of the type one would expect to result from such negligence

Nowadays the courts are very reluctant to find a case of consent existing in all but the most open-and-shut cases. Consequently, its application adheres to strict guidelines.

(i) The plaintiff must know that the defendant is being negligent There can be no valid consent if the plaintiff did not know there was any negligence. Thus, the defendant must prove there was either express or implied consent *to the negligence*. Express consent will be either verbal, (and therefore witnessed in order to stand up in court) or written, which may be witnessed. But under the Unfair Contract Terms Act 1977, liability for what amounts to negligence that causes death or personal injury cannot be excluded *at all*, by contract or notice, in the course of a *business*. Consent may be implied where the plaintiff continues to cooperate despite the negligence.

Consent concerns the negligence, not merely the hazards of a job. People who do dangerous work may be paid danger money and they consent to run the risk of the ordinary hazards of the work (and possibly sign an agreement to that effect), but they are *not* consenting to actual negligence. So if an employee arrived at work to find the safety equipment vandalized and the employer unwilling to do anything about the situation, this unwillingness will amount to negligence. Furthermore, if the employee carries on working, it could be argued that he or she is willing to run the risk of the employer's negligence.

(ii) The plaintiff must have known of the risks which could be caused by the negligence Someone who consents to run the risk of a negligent act will not lose his or her action for negligence if he or she was injured in some other way.

(iii) Consent must be freely given The defence will fail if the worker is under duress from his or her employer to carry on work regardless of the negligence.

Contributory negligence Contributory negligence is not a complete defence, but it acknowledges that, although negligence may have been proved against the defendant, the plaintiff may at the same time, by his or her own fault, have contributed to his or her own injuries or loss, e.g. by failing to wear earmuffs.

The Law Reform (Contributory Negligence) Act 1945 introduced this defence and by S1(1) 'the damages recoverable . . . shall be reduced to such an extent as the court thinks fit and equitable, having regard to the claimants' share in the responsibility for the damage'.

The wording of the Act allows the defendant to plead this without having to prove *negligence* on the part of the plaintiff. All he or she needs to do is to demonstrate the plaintiff was at fault in some way which contributed to the damage. The plaintiff should have reasonably foreseen that his or her behaviour would contribute to his or her injury, otherwise the defence will fail.

As a result, the court works out the damages in the usual way then reduces them proportionately, taking account of the blameworthiness of the plaintiff, e.g. that by failing to wear his or her earmuffs the plaintiff is 50% to blame and therefore gets only 50% damages.

Obviously, it is in the worker's own interests to do all he or she can to prevent greater personal injury by complying with all safety requirements, such as wearing safety clothing.

Limitation Act 1980 It is unfair that plaintiffs do not take advantage of their right to sue within a reasonable period. The Act lays down various periods in which the action must be brought. For most torts this is a period of six years. The exceptions are actions brought for damages for negligence, nuisance or breach of duty, when the damages claimed include damages for personal injuries and when the period is three years. The limitation period begins to run from the date on which the cause of action accrued. This is difficult to calculate in relation to noise-induced illnesses, as such illnesses usually start imperceptibly and progressively get worse. Deafness caused by an explosion would be easier to deal with.

By S2(A) and (B) of the Limitation Act 1980, the time now runs from the date of the accrual of the cause of action or from the date of the plaintiff's knowledge. The knowledge is that the injury is significant, that the injury is due to the alleged negligence of the defendant and the identity of the defendant. Furthermore, the court has the right to override the time limits. These points were raised in *McCafferty v. Metropolitan Police District Receiver* (1977). The plaintiff was a senior experimental officer in charge of a ballistics section of a police laboratory. Although he possessed considerable knowledge of ballistics, he had no scientific or medical qualifications. Part of his job involved shooting rounds of ammunition. From 1965 this took place in a room 22 feet by 6 feet which had no sound absorption. On the advice of his superiors, who were distinguished scientists, the plaintiff protected his hearing by putting cotton wool in his ears. By 1967 he was suffering from tinnitus. On consulting a specialist, he was told to get earmuffs immediately. Earmuffs were supplied by his employers, but evidence was produced at the trial that they would only be effective in a room which had special acoustic protection. He continued to use the earmuffs and noticed no deterioration, although his tinnitus continued. In 1973 he underwent hearing tests, which showed that he was suffering from severe acoustic trauma and, on advice, he quit his job. In 1974 he started proceedings to claim damages for negligence resulting in damage to his hearing by 1967 and the premature termination of his job, caused by the defendants' failure to protect his hearing.

The defendants pleaded contributory negligence, claiming that the plaintiff had been told to find out about precautions and, furthermore, that the claim was statute-barred by being outside the limitation period.

The Court of Appeal held that there had been no contributory negligence because, despite the delegation of responsibility regarding precautions, he was not the proper person to do this. He was an expert, but not in medical and acoustic matters. Also he had always complied with instructions given him.

As far as his claim about his hearing was concerned, the date of knowledge that the injury was significant, for the purposes of S2(A), was in 1968, and by then he realized the gravity of the matter. Thus, the action was prima facie statute-barred. Nevertheless, the court could exercise its discretion and override the three-year period. They took into account such matters as the reluctance of the plaintiff to start an action in order to keep good relations with his employers and his obvious interest in his work.

Vicarious liability If a worker is negligent and causes injury to someone, it is often the case that his or her employer will be sued either with the worker or instead. This is because the employer is vicariously or indirectly liable for the torts of his or her servant, committed during the course of that servant's employment, even though that servant is obviously working badly. If the tort was committed while the worker was on a 'frolic' of his or her own, i.e. doing something totally outside his or her duties, the employer will not be responsible. Whether the worker is doing the act in the course of his or her employment is a question of fact in each case.

Employers' insurance By the Employers' Liability (Compulsory Insurance) Act 1969, every employer in Great Britain, with certain exceptions, must insure and maintain proper insurance against liability for bodily injury and disease sustained by his or her employees arising out of their employment. It has the effect that in most cases of negligence caused by employees, or by the employer, the employer's insurance company will make payment directly to the sufferer. Only where the offer is too low, or where the insurers refuse to make any payment, will the sufferer need to go to court.

Breach of statutory duty

Certain Acts of Parliament and regulations impose duties on employers and employees to do or not to do specific things, e.g. the Noise at Work Regulations 1989.

A breach of such duties is primarily punished by criminal proceedings. Whether the relevant Act or regulations allow someone to be able to sue in *civil law* initially depends on what was intended by the criminal Act. Thus there is no right to sue for breach of one of the general duties under the Health and Safety at Work, Etc., Act 1974 because this is specifically stated in the Act (S47). Following a prosecution under the Act, an action would merely have to be brought in ordinary negligence. But, if there has been a breach of a specific duty in a set of regulations, such as the Noise at Work Regulations, action could be brought for the tort of breach of statutory duty.

What must be proved to be able to sue? There are five things which must be proved:

1. That the regulation or Act imposes a duty to do something.
2. That the particular regulation has been broken.
3. That the plaintiff is within the class which the regulation was designed to protect, e.g. a worker in a shipyard.
4. That the Act or regulation conferred a right to sue on the plaintiff.
5. That the plaintiff is suffering from an injury which the regulation was designed to protect against, even if the injury did not occur in the precise way anticipated.

There are similar defences to those of negligence and some may be specified in the Act or the regulations themselves.

State compensation for accidents at work or diseases or conditions contracted during the course of employment

The law governing this area is found principally under the Social Security Contributions and Benefits Act 1992 Part V, the Social Security Administration Act 1992 and the Social Security (Consequential Provisions) Act 1992.

In order to claim for a benefit as a result of an accident, the accident must have occurred:

- at work or in the course of the employee's employment
- as a result of no one's fault
- as a result of another's misconduct or act of God

And the employee must not have caused the accident himself or herself, either directly or indirectly.

Prescribed diseases The insured must be shown to be suffering from one of the prescribed diseases or conditions set out in the Social Security (Industrial Injuries) (Prescribed Diseases) Regulations 1985. There are four categories. Occupational deafness and white finger come within category A in relation to conditions due to physical agents, and are reproduced here in full. By the time this book is published, carpal tunnel syndrome, caused by using hand-held vibrating tools, will have been added to the list.

A10 Sensorineural hearing loss amounting to at least 50 dB in each ear, being the average of hearing losses at 1, 2 and 3 kHz frequencies and being due in the case of at least one ear to occupational noise (occupational deafness). The occupation which must have caused the condition are:

(a) the use of or work wholly or mainly in the immediate vicinity of pneumatic percussive tools or high speed grinding tools, in the cleaning, dressing or finishing of cast metal or of ingots, billets or blooms; or
(b) the use of or work wholly or mainly in the immediate vicinity of pneumatic percussive tools on metal in the shipbuilding or ship repairing industries; or
(c) the use of or work wholly or mainly in the immediate vicinity of pneumatic percussive tools on metal or for drilling rock in quarries or underground or in mining coal; or
(d) work wholly or mainly in the immediate vicinity of dropforging plant (including plant for drop-stamping or drop hammering) or forging press plant engaged in the shaping of metal; or
(e) work wholly or mainly in rooms or sheds where there are machines engaged in weaving manmade or natural (including mineral) fibres or in the bulking up of fibres in textile manufacturing; or
(f) the use of or work wholly or mainly in the immediate vicinity of machines engaged in cutting, shaping or cleaning metal nails; or
(g) the use of or work wholly or mainly in the immediate vicinity of plasma spray guns engaged in the deposition of metal; or
(h) the use of or work wholly or mainly in the immediate vicinity of any of the following machines engaged in the working of wood or material composed partly of wood, that is to say multi-cutter moulding machines, planing machines, automatic or semi-automatic lathes, multiple cross-cut machines, automatic shaping machines, double-end tenoning machines, vertical spindle moulding machines (including high speed routing machines), edge banding machines, band-sawing machines with a blade width of not less than 75 millimetres, and circular sawing machines, in the operation of which the blade is moved toward the material being cut; or
(i) the use of chain-saws in forestry.

A11 Vibration white finger, i.e. episodic blanching occurring throughout the year affecting the middle or proximal phalanges, or in the case of the thumb, the proximal phalanx, of

(a) in the case of a person with five fingers (including a thumb) on one hand, any three of those fingers; or
(b) in the case of a person with only four such fingers, any two of those fingers; or
(c) in the case of a person with less than four such fingers, any one of those fingers or, as the case may be, the one remaining finger.

The occupations that must have caused this condition are:

(a) The use of hand-held chain-saws in forestry; or
(b) the use of hand-held rotary tools in grinding or in the sanding or polishing of metal, or the holding of material being ground or metal being sanded or polished by rotary tools; or
(c) the use of percussive metalworking tools, or the holding of metal being worked upon by percussive tools, in riveting, caulking, chipping, hammering, fettling or swaging; or
(d) the use of hand-held powered percussive drills or hand-held powered percussive hammers in mining, quarrying, demolition or on roads or footpaths, including road construction; or
(e) the holding of material being worked upon by pounding machines in shoe manufacture.

Unfortunately, for occupational deafness, the claimant must have worked in any of the occupations listed in A10 for a period or periods or not less than 10 years.

Disablement benefit Disablement benefit takes the form of a pension or gratuity paid after a period of 90 days, excluding Sundays. The period begins at the time of the accident or the onset of the industrial disease or condition. The money is paid where there remains a loss of physical or mental faculty when entitlement to other benefits has ceased. It is based on medically assessing the person's loss of faculty as a result of his or her incapacity as compared with another person of the same age and sex. Besides this pension, if the person is severely disabled or has particular difficulties, he or she may also claim other benefits, e.g. constant attendance allowance, but this is unlikely in relation to occupational deafness. It is also unlikely that occupational deafness would give rise to 100% disablement but, if it did, the 1996 benefit would be £95.30 per week if aged over 18.

Claims are made to the local office of the Department of Social Security; after filling in the appropriate claim form (there is a special form for occupational deafness), he or she is then referred to an otologist, who will provide a report to a special medical board. These boards are appointed under the Social Security Act to determine whether the claimant actually suffers from occupational deafness, or any other prescribed disease, and to what extent his or her hearing has been affected.

The decision is first given by the insurance officer. But if the claimant is dissatisfied, he or she has the right to go to the local medical tribunal. Appeals go from there to the National Insurance Commissioner and may be made by the claimant, an employees' association such as a trade union, or by the insurance officer. The commissioner's decision is final.

Health and Safety at Work, Etc., Act 1974

Although the majority of accidents occur in the home, unfortunately, it would be impossible to monitor domestic safety conditions. Legislation has therefore concentrated on trying to control accidents at work.

Legislation was originally introduced in a piecemeal way, initially covering only the most dangerous of occupations and later covering the most common — factories, offices and shops. This piecemeal approach to legislation has now been tied together by the Health and Safety at Work, Etc., Act 1974, which covers all work situations. Under its umbrella come all the old Acts and regulations, which remain in force until they are gradually replaced by new regulations under the 1974 Act.

The 1974 Act does *not* set out to enable compensation to be recovered from wrongdoers, although compensation may be awarded if there is a breach of a *specific* duty under the Act or regulations made thereunder (S47). In relation to noise, they would be the Noise at Work Regulations 1989 (see page 240). The Act's prime responsibility, therefore, is to impose *criminal* liability on those breaching the Act or regulations made under it.

Reasons for enactment The 1974 Act came into effect in 1975 and was designed to provide, for the first time, health, safety and welfare protection for *all* people at work (except domestic servants), *and* for people *affected* by the dangers of work.

The Act thus provides a framework on which complex technical and scientific regulations can be formulated to cover all work processes, but which will be administered and enforced in a standardized way.

The Health and Safety Commission The Health and Safety Commission has corporate status; it was set up by the 1974 Act with the purpose of achieving better health and safety by undertaking research and training, formulating policy, providing advice and information and acting in close liaison with appropriate government departments. Its present address is Bernards House, 1 Chepstow Place, Westbourne Grove, London W2 4TF. The Commission may also issue codes of practice; they do not actually form part of the legislation or general law, but they do provide good evidence in court if any recommended practice has not been followed.

The Commission may also approve existing codes of practice issued by other specialist bodies or government departments. It must have the consent of the Secretary of State before approving a code and it must first consult appropriate bodies.

The Commission may also hold informal investigations and formal inquiries into accidents or situations when it thinks necessary.

The Health and Safety Executive Before the 1974 Act there were a number of different inspectorates having similar responsibilities but whose authority stemmed from their controlling Acts, e.g. the factory inspector was empowered by the Factories Act 1961. All the inspectorates have now been brought under the control of the executive, which is the health and safety enforcement agency.

Some of the inspectors' powers may be transferred to other enforcement agencies. For example, activities involving catering, offices and the sale or storage of goods in shops, warehouses and launderettes are dealt with by local authority environmental health departments. They have exactly the same powers (under the 1974 Act) as other health and safety inspectors.

In brief, under S2 of the 1974 Act an employer must ensure, so far as is reasonably practicable, the health, safety and welfare of his or her employees. This duty is expanded by S2(2), which specifies five areas of responsibility:

1. The employer must provide and maintain plant and systems of work which are, as far as is reasonably practicable, safe and without risks to health.
2. He or she must ensure the safety and absence of risks to health in the use, storage, handling and transport of substances and articles.
3. He or she must provide sufficient information, instruction, training and supervision in order to ensure the health and safety of his or her employees.
4. He or she must maintain the place of work in a condition which is safe and without risk to health and ensure safe access and egress to it.
5. He or she must provide a safe working environment with adequate welfare facilities.

Furthermore, by S3 every employer and every self-employed person is under a duty to ensure so far as is reasonably practicable that persons *not* in his or her employment are not exposed to risks to their health or safety by the conduct of the business. Thus passers-by, visitors and neighbours are owed a duty under S3.

The term *reasonably practicable*, which qualifies many of the duties, means that the employers are not required to go to unreasonable lengths of trouble and expense to eliminate a small risk of accident.

S6 imposes duties on manufacturers, designers, importers and suppliers of articles for use at work or in fairground equipment. They must ensure that the article is so designed that it will be safe and without risks to health at all times during their use at work. As a result of the Noise at Work Regulations 1989 (see page 240), this now includes 'a duty to ensure that, where any such article as is referred to therein is likely to cause any employee to be exposed to the first action level . . . or above or to the peak action level . . . adequate information is provided concerning the noise likely to be generated by that article'. Also, by S6, arrangements must be made for testing and examining the articles or equipment and the employees should be provided with such information as is necessary to secure their health when using it.

S7 imposes two duties on every employee at work:

(a) To take reasonable care for the health and safety of himself or herself and of other persons who may be affected by his or her acts or omissions at work.
(b) As regards any duty or requirement imposed on his or her employer, to cooperate with that employer so far as is necessary to enable the duty or requirement to be performed or complied with.

Failure to abide by these duties amounts to a criminal offence.

By S8 a person who wilfully or recklessly interferes or misuses anything provided in pursuance of a relevant statutory provision, such as safety apparatus, is guilty of an offence.

In order to supplement the 1974 Act, regulations have been enacted, such as the Noise at Work Regulations 1989 (see below). Codes of practice may also be issued or approved by the commission. Once approved they may then be used in evidence in court to show that someone has infringed the Act.

Powers of the inspectorate The inspectors have wide-ranging investigatory powers; they can enter premises, examine and take photographs, take samples and measurements, dismantle machinery, take possession of anything relevant and require anyone to give information. If an inspector finds that there has been a breach of an Act or regulations, he or she may take the following enforcement steps. By S21 the inspector can issue an improvement notice requiring the situation to be remedied within a specified period. The inspector can issue a prohibition notice if there is a risk of serious personal injury, ordering the work to cease immediately or to be suspended for a specific period. Appeals against both notices may be made to an industrial tribunal. The inspector can also seize, render harmless or destroy dangerous articles. Finally, the inspector can prosecute either in the magistrates' court for less serious offences or in the Crown Court for serious breaches. Fines and imprisonment can be ordered. Anyone can be prosecuted, including corporations and companies.

An offence under the 1974 Act would provide good evidence in any civil action taken by an employee for compensation, but it does not in itself allow for damages to be paid. Action could, however, be taken for breach of statutory duty if there are relevant grounds.

Noise at Work Regulations 1989

The long-awaited Noise at Work Regulations came into force on 1 January 1990. They were made under the Health and Safety at Work, Etc., Act 1974 and implemented in the United Kingdom the contents of EC Directive 86/188/EEC *The protection of workers from noise*, making redundant the Department of Employment's Code of Practice for Reducing the Exposure of Employment Persons to Noise.

Under Regulations 2 and 3, the provisions apply to *all employees, trainees and self-employed* in workplaces, excluding those on aircraft and shipping and, by analogy with the Act, all those at home.

Regulation 2 is a definition clause:

- *Daily personal noise exposure (dpne)* means the level of dpne of an employee ascertained in accordance with Part 1 of the schedule to the regulations, but taking no account of the effect of any personal ear protector used. This is calculated by reference to the formula, $L_{EP,d}$.
- *Exposed* means exposed while at work.
- *First action level* means a dpne of 85 dB(A).
- *Peak action level* means a level of peak sound pressure of 200 Pa. This is relevant where using guns or cartridge-operated tools in an otherwise quiet working situation.
- *Second action level* means a dpne of 90 dB(A). This has been criticized as too high.

By Regulation 2(2), an employer includes reference to self-employed persons and to their workers, as well as to himself.

The employer's general duty By Regulation 6, every employer shall reduce the risk of noise damage to the hearing of employees to the lowest level *reasonably practical*.

By Regulation 4, employers must make adequate arrangements to assess the exposure of workers to noise at or above the first action level or the peak action level. This must be made by a *competent person* who can advise on action to be taken, including taking specialist advice. He or she must identify those who will be exposed and provide the employer with such information as will help him to comply with his duties under Regulations 8, 9 and 11.

Assessments must be reviewed if there has been a significant change in the work or a suspicion that the assessment is no longer valid.

By Regulation 5 adequate assessment records must be kept.

If the dpne is likely to be at or above the second action level or the peak action level, employers must ensure that exposure is reduced to the lowest level reasonably practicable *other* than by providing ear protectors (Reg. 7).

Provision of ear protections, Reg. 8 Regulation 8(1) requires that where employees are likely to be exposed to the first action level or above and where dpne is likely to be less than 90 dB(A), then suitable and efficient personal ear protectors must be available to those who request them.

But by Regulation 8(2) ear defenders *must* be provided where there is exposure at or above the second or peak action level. When properly worn, such defenders must reasonably be expected to reduce risk of hearing damage to below that caused by unprotected exposure at those levels.

By Regulation 8(3) personal ear protectors provided under this regulation must comply with any legislation in Great Britain that implements any provision on design or manufacture with respect to health and safety in any relevant EC directive listed in Schedule 1 to the Personal Protective Equipment at Work Regulations 1992 which is applicable to ear protectors.

Ear protection zones, Reg. 9 Every employer must demarcate as an ear protection zone (EPZ) those parts of his or her premises where there is exposure to the second action level or above, or to the peak action level or above. Ear protection zones are indicated by a sign (see BS 5378) as specified in the Safety Signs Regulations 1980. The sign should indicate that the area is an EPZ and that defenders must be worn.

Maintenance of equipment, Reg. 10(1) Employers must ensure, so far as is reasonably practicable, that any items provided by them are properly used. Equipment should be maintained properly; it should be in good working order and in good repair.

Employees' duties, Reg. 10(2) Every employee must, so far as is reasonably practicable, make full use of anything the employer is required to supply and should report any defects to the employer.

Information, Reg. 11 All employers must supply employees who are likely to be exposed to the first action level or the peak action level with adequate information, instruction and training on:

(a) The risk of hearing damage caused by such exposure.
(b) Actions to minimize risk.
(c) What must be done to obtain personal ear protection under Regulation 8.
(d) Employees' duties.

Duties of designers, makers, importers and suppliers of articles for use at work or at fairgrounds, Reg. 12 (noise generated by machines) This regulation extends the duties under S6 of the Health and Safety at Work, Etc., Act 1974 and requires such people to supply adequate health and safety information with such articles, where the articles are likely to cause exposure at or above the first action level or the peak action level. Thus machinery manufacturers, suppliers, etc., need to provide noise data with all machines likely to cause exposure above 85 dB(A), first action level, or 200 Pa peak action level.

Exemptions Regulation 13 permits the Health and Safety Executive to grant exemptions from:

(a) Regulation 7 requirements where $L_{EP,d}$ dpne averaged over a *week* is below 90 dB(A) *and* there are arrangements to ensure that this is not exceeded.
(b) The obligation to use ear protectors at all, where it would be detrimental to overall health and safety, i.e. where a greater catastrophe could happen if the ear protectors were used.
(c) An employer's duty to provide protectors to the level specified in Regulation 8(2) where not reasonably practicable, e.g. where noise levels are so high that *no* protector can meet normal attenuation requirements.

The EC directive required member states to provide facilities for workers to have their hearing checked. This was not incorporated into the 1989 regulations as it was felt that National Health Service facilities were sufficient. It is extremely difficult to prove that someone is deaf as a result of his or her work, or vice versa, if no audiometric tests have been undertaken at the outset. Nor can their deteriorating hearing be checked.

By Regulation 14, the Secretary of State for defence, in the interest of national security, may use a certificate to exempt from any requirement a member of Her Majesty's Forces or visiting forces, a member attached to headquarters or a member attached to an organization.

10.7 The Land Compensation Act 1973: compensation for noise caused by public works

There are many Acts empowering government departments and local authorities to compulsorily purchase property and give compensation, where property is adversely affected by public works, but the Land Compensation Act extends these rights. Part I of the Act confers a right to compensation for depreciation in the value of land caused by public works. Part II, as amended by the Highways Act 1980 and the Local Government, Planning and Land Act 1980, confers on public authorities power to acquire land and to do works to reduce the harmful effects of public work on their surroundings.

Part I: the right to compensation

By S1, certain specified landowners in certain specified circumstances may be paid compensation in respect of the depreciation in value of their land as a result of certain physical factors, including noise, vibration, smell, fumes, smoke and artificial lighting, caused by the use of a new or altered *highway*, *aerodrome* or *other public works* provided in exercise of statutory powers.

By S9(7), altering or changing the use of public works does not cover the intensification of use. The compensation will be paid by the appropriate highway authority or else by the person or body managing the works. The source of the physical factors must be situated on or in the public works, or in relation to aerodromes, it must be caused by the use of the aerodrome as such or by aircraft departing or arriving.

In the case of public works, according to S1(6), the authorities must have statutory immunity against actions for nuisance, otherwise compensation will not be awarded, e.g. immunity under the Civil Aviation Act 1982 (see page 248). If they do not have immunity then the appropriate course of action for a person suffering is to sue in tort for nuisance. In *Vickers v. Dover District Council* (1993) the council exercised its power under the Road Traffic Regulation Act 1984 to provide a car park. There was nothing in the Act giving immunity from action, and the plaintiff could and should proceed in nuisance. The Lands Tribunal said that there was no right to compensation under the Land Compensation Act.

By S2, where the claim relates to a dwelling-house, the claimant must own the legal fee simple or must have a tenancy of at least three years to run and be in occupation of the property.

Claims By S3 the claim is made by serving the responsible authority with a notice containing *inter alia* particulars of the land, the interest held, the public works to which the claim relates and the amount of compensation sought. No claim shall be made before the expiration of twelve months from the relevant date. The authority may enter the land and survey it for valuation purposes. If there are any disputes, they must be determined by the Lands Tribunal.

Part II: mitigation of injurious effect of public works

By S20(1) the Secretary of State may make regulations requiring authorities responsible for public works to insulate buildings against noise from construction or use of public works, or to make grants towards the cost of such insulation. The regulations are the Noise Insulation Regulations 1975/1763 as amended by the Noise Insulation (Amendment) Regulations 1988/2000.

These regulations apply to buildings affected by the construction and use of new highways and additional carriageways to existing highways, subject to certain time limits, and impose duties or give powers to insulate against noise or give grants to do such works. Buildings which are eligible under the regulations are dwellings or other buildings used for residential purposes not being more than 300 metres from the nearest point of the carriageway. Certain types of buildings are specifically excluded.

By Regulation 3(1), the appropriate highway authority is under a duty to carry out or make a grant in respect of the cost of carrying out insulation work on an eligible building, where the *use* of a relevant highway causes or is expected to cause noise at a level not less than the 'specified level', i.e. L_{10} (18-hour) of 68 dB(A). For the purpose of this regulation, the use of the highway will cause noise at a level not less than the specified level if the 'relevant noise level', i.e. the level of noise expressed as a level of L_{10} (18-hour) 1 metre in front of the most exposed of any windows and doors in a façade of a building caused by the traffic using or expected to be using the highway is greater by at least 1 dB(A) than the 'prevailing noise level' (measured in much the same way as the relevant noise level but before the works for the construction of the road began) and is not less than the specified level. Additionally, the noise caused or expected to be caused by traffic using or expected to be using that highway makes an effective contribution to the relevant noise level of at least 1 dB(A).

In the above case, the highway or additional carriageway must have been first open to the public after 16 October 1972.

The noise levels must be calculated in accordance with memoranda published by HMSO: Calculation of Road Traffic Noise 1975 and 1988.

Acquisition of land in connection with highways S246 of the Highways Act 1980 empowers a highway authority to acquire, compulsorily or by agreement, land for the purpose of reducing any adverse effect caused by any highway, either existing or proposed.

By S26 of the Land Compensation Act, an authority may acquire land by agreement in order to mitigate any adverse effect which the public works will have on the surrounding area. Such land could be seriously affected by the construction, alteration or use of the public works. By S27 of the same Act, an authority having S26 powers to acquire land may carry out works to reduce the injurious effect on land it already owns or the land it has acquired under the previous section. Such work may include planting trees, shrubs or plants, or laying out the area as grassland.

Thus the Land Compensation Act provides a number of remedies for people who are affected by works which in the opinion of the state are important to the rest of the community.

10.8 Insulation requirements under the Building Regulations

One way of reducing sound at source is to provide sufficient insulation when constructing buildings. The Building Regulations 1991 made under the Building Act 1984 provide a framework of control of building standards supplemented by guidance documents, called approved documents, on different aspects of construction work. The approved documents provide the detail that used to be found in the pre-1985 regulations. These documents are approved or issued by the Secretary of State. In relation to noise, the relevant document is Approved Document E: Airborne and Impact Sound.

The 1991 regulations are less regulatory than the previous regulations; failure to work to Building Regulation standards does not necessarily mean there has been a breach. Failure to comply with the guidance documents does not in itself render a person liable to civil or criminal proceedings, but it would provide good evidence of a general failure to abide by the regulations (S7 of the 1984 Act). If the builder has followed the guidance in the approved document, it will be evidence tending to show that he or she has complied with the regulations, and vice versa. If he or she chooses other methods of building, the onus falls on the builder to show that he or she has satisfied the requirements of the regulations. The builder can do this by showing that he or she has followed an appropriate British Standard, or an equivalent national technical specification of any member state of the European Union; he or she has used a product bearing an EC mark in accordance with Construction Products Directive 89/106/EEC; he or she has followed an appropriate technical specification as defined by the Directive or by the British Board of Agrément Certificate.

The purpose of the provisions is to control noise from *other* parts of the building, not to control noise *entering* through external walls.

The regulations apply to any 'building work' or 'material change in use' of a building. Regulation 3 defines *building work* to include *inter alia* the erection or extension of a building and the material alteration of a building. The rest of the definition (Reg. 5) applies to sanitary fittings, bathroom fittings, sewers, etc. A material change covers changes to dwellings where not previously so used; similarly, flats, hotels or institutions, public buildings (Reg. 6). Where there is a material change in use to a dwelling, the requirements of resistance to sound are required.

By Regulation 7, 'any building work shall be carried out with proper materials and in a workmanlike manner'.

By Regulation 8, the requirements of the regulations do not require anything to be done, except for the purpose of securing reasonable standards of health and safety for persons in or about the building.

Part E of Schedule 1 to the Building Regulations 1991 refers to airborne sound through walls, airborne sound through floors and impact sound on floors.

> **Airborne sound through walls E1.** A wall which:
>
> (a) separates a dwelling from another building or from another dwelling, or
> (b) separates a habitable room or kitchen within a dwelling from another part of the same building which is not used exclusively with the dwelling
>
> shall resist the transmission of airborne sound.
>
> **Airborne sound through floors and stairs E2.** A floor or stair which separates a dwelling from another dwelling, or from another part of the same building which is not used exclusively as part of the dwelling, shall resist the transmission of airborne sound. [This seems to be a stronger requirement than before. Would you agree?]
>
> **Impact sound on floors and stairs E3.** A floor or a stair above a dwelling which separates it from another dwelling, or from another part of the same building which is not used exclusively as part of the dwelling, shall resist the transmission of impact sound.

These are the three requirements in relation to sound. The approved document describes three ways to satisfy them. The first, and easiest, is to follow an example of a widely used construction method given in the document. The second is to adapt a form of construction similar to one that has been shown by field tests to comply with the requirements. The third is by testing a part of the construction in a specified type of acoustic chamber. For comparatively small builders, it would be far more sensible to use the first method because small builders are unlikely to have the resources or the expertise to test unorthodox constructions.

The approved document refers to two British Standards: BS 2750 *Methods of measurement of sound in buildings and of building elements*, Part 4:1980 Field measurements of airborne sound insulation between rooms and Part 7: 1980 Field measurements of impact sound insulation of floors; and BS 5821 *British Standard method for rating the sound insulation in buildings and building*, Part 1:1984 Method for rating the airborne sound insulation in building and of internal building and Part 2:1984 Method for rating the impact sound insulation. Other relevant standards are BS 1142:1989 *Specification for fibre boards* and BS 5628 *Code of practice for use of masonry*, Part 3:1985 Materials components, designs and workmanship.

The Secretary of State's view is that the requirements of E1, E2 and E3 are met if the relevant parts of the building are designed and built in such a way that noise from normal domestic activities in an adjoining dwelling or other building is kept down to a level that will not threaten the health of the occupants of the dwelling and will allow them to sleep, rest and engage in normal domestic activities in satisfactory conditions.

Work must be supervised either by the local authority's building control officers or an 'approved inspector', i.e. one approved by the Secretary of State. The builder must provide the local authority with a building notice (for small and minor works) or deposit full plans. The local authority can ask for more information. If full plans are deposited, they must be passed or rejected by the authority within five weeks. Provided the plans show that the work will be carried out in accordance with the Building Regulations, the local authority must approve them. During the course of the work, notification must be made to the local authority at various times. Firstly, at least 48 hours notice of commencement. Secondly, at least 24 hours before covering up of any excavation for a foundation or a damp-proof course or any concrete or other material laid over a site material. Thirdly, at least 24 hours notice before covering up drains and finally within seven days of completion of the work.

Breaching the regulations is an offence triable in the magistrates' court under S35 of the Building Act 1984. Each day the offence continues will incur a further fine after summary conviction. The local authority could also serve notice on the owner, requiring demolition or removal of the works, or order the owner to make such alterations as are necessary (S36(1)(a) and (b)). Similarly, if work is carried out and no plans were deposited (and they should have been) or the plans were rejected or the work was carried out otherwise than in accordance with any requirements made by the local authority, then the local authority can require the demolition, removal or alteration as before (S36(2)). The local authority may also do the works themselves and charge the owner (S36(3)). No S36 notice can be served after the expiration of twelve months from the date of completion of the works (S36(4)). By S36(6) there remains the right of the local authority, the Attorney-General or any other person to apply for an injunction for the removal or alteration of any work on the ground that it contravenes any regulation or provision of the Act. However, if plans were deposited and they were passed by the local authority, or notice of their rejection was not given in the relevant period and the work was executed in accordance with the plans, then the court, on granting an injunction, has power to order the local authority to pay compensation as the court thinks just and can join the local authority as party to the action, if it is not one already.

10.9 Town and country planning

Town and country planning can obviously play a large part in the control of noise. It may predetermine what should be located in a particular area by the use of structure and local plans, thus avoiding new problems. And it may refuse planning permission to build potentially noisy premises or lay down specifications or conditions, if planning permission is given.

The Department of the Environment (DOE) produced a circular in 1973, *Planning and Noise*, currently being updated, which gave advice to local authorities to achieve those aims, indicating sources of material and information and encouraging liaison between authorities affected by the same problems.

The relevant legislation is found in the Town and Country Planning Act 1990 as amended by the Planning and Compensation Act 1991. At the base of all planning control are three types of general plans, the *structure* plan, the *local* plan and the *unitary* plan. The structure plan is usually undertaken by the county councils and the local plan by the district planning authority. Unitary plans are prepared for former metropolitan counties by metropolitan district councils. The plans are based on a survey of the area, taking into account *inter alia* the principal physical and economic characteristics of the authority's area and effect of neighbouring areas; the population; communication and transport systems and traffic plus anything else of relevance. The plans lay down aims and objectives, policies and goals, but they do not have to be followed; they are merely guides for planners.

A structure plan consists of a written statement describing the planning authority's policy and general proposals. It is accompanied by general diagrams and illustrations as appropriate. There is now a specific duty to include 'measures for the improvement of the physical environment'. The plan should be relevant for 15 years, at least, and it should be kept under review. During its formative stages, adequate publicity must be made of the draft so that ordinary people and other relevant interests may make representations.

Planning permission

The Act can only control 'development' which is defined by S55(1) of the 1990 Act as 'the carrying out of building, engineering, mining or other operations in, on, over or under land, or the making of any material change in the use of any buildings or other land'. SS1A inserted by the 1991 Act now specifically includes in the definition of 'building operations' (a) demolition of buildings; (b) rebuilding; (c) structural alterations of or additions to buildings and (d) other operations normally undertaken by a person carrying on business as a builder. Subsection 1A has put an end to doubt in this area and has particular relevance to noise because demolition work is often just as noisy as straightforward building work.

Thus, any operation falling outside the definition will not require planning permission.

By S57(1) planning permission is required for the carrying out of any development of land.

If there is any doubt whether a project may require planning permission then, under S192 of the 1990 Act, a person may apply to the local planning authority as to whether the work to be carried out or the proposed change of use is lawful. If, on an application under this section, the local planning authority is supplied with such information as would satisfy it, the local planning authority shall issue a certificate to that effect and in any other case it shall refuse the application (S192(2)). If a certificate is issued, by S192(4) there is a conclusive presumption of lawfulness of any use or operations unless there is a material change before the specified use or work has started. Thus, this gives the local planning authority's seal of approval to work which would not require permission. If a certificate is refused then planning permission must be applied for.

There is no development if there is a change of use within the same use class, as specified in the Town and Country Planning (Use Classes) Order 1987 and amended by the Town and Country Planning (Use Classes) (Amendment) Order 1991. But a change from one use class to another use class does require permission. No classes in the order include use as a theatre, amusement arcade, centre or funfair; the sale of fuel for motor vehicles or the sale or display of motor vehicles; a scrapyard or a yard for storing or distributing minerals (like coal) or breaking motor vehicles. They are potentially nuisance-causing and would require specific planning permission.

Here are some examples of the sorts of premises included in the order. *Class A1* includes shops with certain specific exceptions, e.g. take-away hot food; even sale of cold food and drink off the premises will come into Class 3. *Class A2* is concerned with financial and professional services covering banks and building societies. *Class B1* covers offices (but not A2), including those connected with research and industrial processes, being those uses which could be carried on in any residential area without detriment to the amenity of that area by reason of noise, vibration, smell, fumes, smoke, soot, dust, ash or grit. *Class B2* is general industrial for industrial processes not falling within class B1. *Class B8* covers storage and distribution centres. *Class C1* covers hotels and hostels. *Class C3* covers houses used by single people or families or those with not more than six residents living together as a household. *Class D2* covers assembly and leisure arenas, concert halls, bingo halls, casinos, dance-halls, swimming-baths, skating-rinks, gymnasia or places for other indoor or outdoor sports that do not involve motorized vehicles or firearms.

Under another order, the Town and Country General Development Order 1988 as amended, para. 3 and Schedule 2 gives *automatic* planning permission for certain developments specified in 28 categories. In contrast to the Use Classes Order, which says changes within a use class are *not* development, works carried out under the General Development Order are definitely development, but authorized without specific planning permission under the order.

The order is extremely detailed, each category having strict limitations and exclusions. Obviously, the development proposed is of a relatively minor nature and such that, had it been necessary, planning permission would have been granted. This also prevents waste of time and money in the local planning authority. Any planning permission granted by virtue of this order is subject to any relevant limitations or conditions specified in Schedule 2. Application still has to be made to the relevant local planning authority.

Applying for planning permission Section 62 of the 1990 Act, the Town and Country Planning (Applications) Regulations 1988 and the General Development Order 1988 lay down the requirements for applying for planning permission. Applications are made on forms supplied by the local planning authority, to which all necessary information, site plans and drawings are attached. By S70 the local planning authority may grant permission, either unconditionally or subject to such conditions as it thinks fit, or it may refuse permission altogether. Model conditions are contained in Circular 10/73, currently being updated. Case law has shown that conditions must be relevant to planning and to the development to be permitted; they must also be reasonable. The government has also said in its circular that 'conditions should be necessary, precise and enforceable. The key test is whether planning permission would have to be refused if the condition were not imposed. If not, then such a condition needs special and precise justification.' Refer to additional guidance in DOE Circular 1/85 *The Use of Conditions in Planning Permissions*. There should be close liaison with the environmental health department in order to get expert advice on noise as well as to consider the imposition and monitoring of compliance with the conditions. Here are some examples of relevant conditions suggested in the draft DOE circular *Planning and Noise*:

1. Development shall not begin until a scheme for protecting the proposed dwellings from noise from the ______ has been submitted to and approved by the local planning authority; all works which form part of the scheme shall be completed before any of the permitted dwellings are occupied.

In relation to aerodromes examples are:

3. The total number of movements shall not exceed (   ) per annum

6. Except in an emergency, the runways shall not be used by (class of aircraft)

In relation to noise emitted from industrial or commercial buildings and sites:

11. Before the use commences, the building shall be insulated in accordance with a scheme agreed with the local planning authority in order to secure an acceptable level of noise emanating from the building.

16. No (specified machinery) shall be operated on the premises before (time in the morning) on weekdays and (time in the morning) on Saturdays nor after (time in the morning) on weekdays and (time in the evening) on Saturdays, nor at any time on Sundays or Bank Holidays.

In dealing with applications the local planning authority must have regard to the provisions of the development plans so far as they are material to the application and to any other material considerations.

Nowadays, one material consideration is the environmental impact assessment, made under the Town and Country Planning Assessment of Environment (Effects) Regulations 1988 and the Town and Country Planning General Development (Amendment) Order 1988. The enabling Act is not the 1990 Act (or earlier Town and Country legislation) but the European Communities Act 1972 as the regulations were made as a result of EEC Council Directive 85/337/EEC. Under these regulations, in certain situations before development has begun, an assessment must be made of the likely effects of the works on the environment. By Annex 1 of the directive in some cases there *must* be an assessment, e.g. for motorways, airports and long-distance railways. In other cases an assessment is required if the development is *likely* to have a significant environmental effect because of its size, nature or location. Guidance is found in DOE Circular 15/88.

In other cases there are many regulations on whether an assessment is necessary. The assessment could be voluntary, ordered by the local planning authority or ordered by the Secretary of State. If required, a statement must be prepared by the developer, who must identify those objects which may be affected by the works — humans, climate, animals, etc. — how the works affect them and the proposed methods of reducing or avoiding the harmful effects of the works. The local planning authority (if the appropriate body) will then prepare the assessment which must be taken into account before permitting the development.

Other material considerations include overlap with certain areas of statutory control, perhaps in relation to noise clashes with building regulation control and noise control under the Environmental Protection Act 1990.

An interesting decision was given in *Gillingham Borough Council v. Medway (Chatham) Dock Co. Ltd.* (1992). Here a local authority granted planning permission to develop part of the former Chatham Royal Dockyard into a commercial dockyard. Initially it was clear that the new development would be operating 24 hours a day. Between 300 and 400 lorries would use the local access roads, all residential streets. But after the yard began operating, complaints came from local residents and the local authority applied for an injunction under S222 of the Local Government Act 1972 on the basis that the noise and other discomfort amounted to a public nuisance (see page 212). The local authority was refused an injunction, so it appealed to the High Court. The High Court turned down the appeal principally, it said, because the nature of the locality had changed, a change authorized by the local authority when it granted planning permission in the first place. The usage of the neighbourhood was now characteristic of the environs of a commercial port. It could even be argued that changes in the development plan could have a similar effect on an area. It was in some respect analogous to the defence of statutory authority (see page 215). When granting the permission, great account had been taken of the financial advantages for the area, which had suffered greatly as a result of the closure of the dockyard. No environmental assessment had been undertaken (there was no need at the time). It may be that this decision could be used as a precedent in non-planning cases because it relates to the nature of the neighbourhood. Local planning authorities must therefore be even more wary of granting permissions in environmentally difficult situations and the average person must be more vigilant in voicing objections before permissions are granted.

Enforcing the law

By S171A, if no permission has been given and the developer has carried out the work or there has been a contravention of a condition, the local planning authority can take the following steps.

Firstly, it can issue an enforcement notice under S172. A copy must be served on the owner and the occupier of the land and any other person with an interest in the land. By S173 the enforcement notice must specify the breach and whether the breach is caused by unauthorized development or by failure to comply with a condition or limitation under S171A; it must also be clear to the recipient what this means.

The notice must also specify the steps which the authority requires to be taken or the activities which the authority requires to be stopped in order to achieve the remedying of the breach. Thus, according to S173(5), the enforcement notice may require:

(a) Alteration or removal of works.
(b) Carrying out of building or other operations.
(c) Any activity not to be carried on, except as limited by the terms of the notice.

When specifying the steps to be taken they must not be too vague, otherwise the courts may regard the notice as a nullity. Thus mere directions to 'install satisfactory soundproofing' is not sufficiently precise (*Metallic Protectives Ltd. v. Secretary of State for Environment*, 1976).

The recipient of the notice may appeal to the Secretary of State against it (S174).

If the notice is ignored, the local planning authority has the right to enter the land and carry out the required work at the expense of the developer (S178).

By S179, if at the end of the period for compliance with the enforcement notice, the required steps have not been carried out or work directed to cease has not ceased, then the *owner* of the land is in breach and is guilty of an offence. The local planning authority can then prosecute and, on finding the defendant guilty, will be liable on summary conviction to a fine not exceeding £20 000 and on conviction on indictment to an unlimited fine.

Where a condition has been imposed by S187A, the local planning authority may serve a breach of condition notice, requiring the steps to be taken. Ignoring the notice amounts to an offence.

Finally, the local planning authority also has the power to serve a stop notice, prohibiting work carried out in contravention of planning conditions, if it considers it expedient that a relevant activity should cease before the expiry of the period for compliance with an enforcement notice. The stop notice ceases to have effect once the enforcement notice has been quashed or withdrawn at an appeal or when the period for compliance has expired. There are no rights of appeal; failure to comply with the notice constitutes an offence as above.

S187B introduced by the 1991 Act has introduced a similar right to take High Court action, as we have already seen, under the Environmental Protection Act. Thus, if a local planning authority considers it 'necessary or expedient for any actual or apprehended breach of planning control' to be restrained by injunction, it may apply to the court for an injunction, whether or not it has exercised or is proposing to exercise any of its other powers under the Act.

10.10 Civil aviation

The problems of intensification in air travel are obvious in the field of noise control. People suffering the most are those living close to airports, such as Gatwick and Heathrow, especially when further runways are threatened. How are these problems being tackled? Because of the different and localized problems, special laws have been enacted. Many Acts have been passed and will be passed to try and diminish the effects of what is an essential part of a modern transport system.

The relevant legislation is mainly found in the Civil Aviation Act 1982 and all subordinate legislation made thereunder.

Statutory immunity

The early air industry was considered to require protection from litigation in case large payouts crippled this new mode of transport. As early as 1920 legislation was passed to prevent lawsuits.

Thus, in normal flying situations there is immunity from legal action. By S76(1) of the Civil Aviation Act 1982:

> No action shall lie in respect of trespass or . . . nuisance by reason only of the flight of an aircraft over any property at a height above the ground which having regard to wind, weather and all the circumstances of the case is reasonable, or the ordinary incidents of such flight, so long as the provisions of any Air Navigation Order and of any orders under S62 above have been duly complied with and that there has been no breach of S81.

S62 orders concern control of navigation in time of war or emergency and S81 concerns dangerous flying — a criminal offence.

In normal circumstances, even if a landowner is troubled by incessant noise from overflying aircraft, he or she will not be able to sue for noise nuisance.

Furthermore, by S77(1):

> An Air Navigation Order may provide for regulating the conditions under which noise and vibration may be caused by aircraft on aerodromes and may provide that ss(2) below shall apply to any aerodrome as respects which provisions as to noise and vibration caused by aircraft is so made.

S77(2) states:

> No action shall lie in respect of nuisance by reasons only of the noise and vibration caused by aircraft on an

> aerodrome to which this subsection applies by virtue of an Air Navigation Order as long as the provisions of any such Order are duly complied with.

Similarly, if our landowner finds that the nearby aerodrome is the source of noise nuisance while the aircraft are on the aerodrome, no action can be taken, providing they comply with any Air Navigation Orders (see below).

The 1989/2004 Air Navigation Order provides in article 83 of the order, that the Secretary of State may prescribe the conditions under which noise and vibrations *may be caused* by aircraft in virtually all types of aircraft and on virtually all types of aerodromes.

The Secretary of State has issued the Air Navigation (General) Regulations 1993. Regulation 13 states:

> For the purpose of Article 83 the conditions under which noise and vibration *may be caused* by aircraft (including military aircraft) on Government aerodromes, aerodromes owned or managed by the Civil Aviation Authority, licensed aerodromes or aerodromes at which the manufacture, repair or maintenance of aircraft is carried out by persons carrying on a business as manufacturers or repairers of aircraft shall be as follows:
>
> (a) the aircraft is taking off or landing; or
> (b) the aircraft is moving on the ground or water; or
> (c) the engines are being operated in the aircraft:
>   (i) for the purpose of ensuring their satisfactory performance; or
>   (ii) for the purpose of bringing them to a proper temperature in preparation for, or at the end of, a flight; or
>   (iii) for the purpose of ensuring that the instruments, accessories or other components of the aircraft are in a satisfactory condition.

In other words, Regulation 13 lays down the situations in which aircraft may make noise and vibrations without there being any legal comeback from people suffering as a result.

These sections therefore cut down the possibility of litigation in the majority of cases, but they do not rule out an action being brought, where the noise or vibration was being caused outside the protection of a section, e.g. by ignoring regulations, flying too low or where there is another cause of action in addition to nuisance (or trespass). Many groups over the years have criticized this special protection and have pointed out that in other countries, such as the United States, which has no similar immunity, civil aviation has continued to flourish. The Batho Report also considered that S76 is no longer appropriate. The report especially differentiated between commercial aircraft and private and leisure aircraft, the latter being less deserving of protection. The report indicates there are currently about 280 airfields in the United Kingdom with approximately 7000 aircraft registered in Britain alone, not counting the foreign-registered aircraft; it adds that most of the 7000 aircraft are used for private flying.

The protection afforded by this legislation does not extend to situations where persons or property are damaged by the aircraft itself or things falling from the aircraft, including people. In such cases the owner of the aircraft is strictly liable, i.e. without proof of negligence or intention, as it is obviously in the interests of public safety that bits don't fall off aeroplanes. Thus by S76(2):

> Where material loss or damage is caused to any person or property on land or water by or by a person in or an article, animal or person falling from, an aircraft while in flight, taking off or landing, then unless the loss or damage was caused or contributed to by the negligence of the person by whom it was suffered, damages in respect of the loss or damage shall be recoverable without proof of negligence or intention or other cause of action, as if the loss or damage had been caused by the wilful act, neglect or default of the owner of the aircraft.

Such loss or damage includes that caused by noise or vibration. Thus, the Act differentiates between mere nuisance and direct damage caused by aircraft.

Legal control of aircraft noise

Legal control of aircraft noise is achieved in a number of ways. Firstly, by the making of regulations under the Act to control noise and vibration caused in the operation of aircraft. Secondly, by phasing out noisy aircraft through aircraft certification. Thirdly, by providing compensation under the Land Compensation Act 1973 as amended (see page 242) for diminution in value of property, and in certain areas, by providing grants towards the insulation of buildings affected by aircraft noise. There exist other practical measures to reduce noise, but they are not the concern of the law.

Regulations By S78(1) the Secretary of State may by notice impose a duty on aircraft operators taking off or landing at a designated aerodrome to secure compliance with any requirements specified in the notice. At present Heathrow, Gatwick and Stansted have been designated. The requirements of the notice will be for the purpose of limiting or mitigating the effect of noise and vibration at landing and take-off. By S78(2) if such persons do not comply with the requirements, then after being given the opportunity of making representations to the Secretary of State, the manager of the aerodrome may be directed to withhold airport facilities from the operator and his or her servants. The manager is under a duty to comply. Such notices include requirements for minimum noise routes and night levels of noise on take-off.

By S78(3) the Secretary of State may also make an order to reduce the effects of noise and vibration by limiting the number of occasions on which aircraft may take off or land during certain periods. To that end the Secretary of State may specify the *maximum* number of take-offs or landings during these periods in relation to particular types of aircraft. Furthermore, the Secretary of State can determine which aircraft operators may take off and land within these periods and the number of occasions particular types of aircraft belonging to the operators may take off and land. It is the responsibility of the aerodrome manager to see that these rules are complied with.

In *R. v. Secretary of State for Transport ex. p. Richmond-upon-Thames London Borough Council* (1993) an application for judicial review of a decision of the Secretary of State was made by various local authorities. The Secretary under S78(3)(b) had issued a press announcement that he intended to introduce a new system of night-flying restrictions at Heathrow, Stansted and Gatwick. The numbers would no longer be *fixed*; they would *vary* according to the noise produced by the aircraft. The court granted the application. The section specifically states that the requirements are to state the maximum number of movements 'the linchpin of any order made' as per Laws J. Thus, the decision was not authorized by the section and was invalid.

If it appears to the Secretary of State that an aircraft is about to take off in contravention of limitations imposed, then without prejudicing the powers of the airport manager, any person with the Secretary's authority may detain the aircraft for such period as necessary and may enter on any land for that purpose (S78(5)(c) and (d)).

The Secretary of State by S78(6) may give the manager of the aerodrome directions in order to limit or mitigate the effect of noise and vibration and the manager must comply.

By S78(7) the Secretary of State may also, after consulting with the manager of the aerodrome, order the manager to provide, maintain and operate in a particular area, at a specified time, such equipment for measuring noise in the vicinity of the area as is specified. The manager must then make reports and permit any authorized person to inspect the equipment.

Failure to comply with any of the duties will make the person defaulting guilty of an offence liable to summary conviction, and continued non-compliance will lead to the commission of separate offences for each day on which the default occurs (S78(9)).

S63(2)(b) of the Airports Act 1986 authorizes the power to make byelaws for controlling the operations of aircraft within or directly above the airport for the purpose of limiting or mitigating the effect of noise, vibration and atmospheric pollution caused by aircraft using the airport.

S38 of the Civil Aviation Act 1982 fixes, by reference to noise factors, the charges that may be levied for using licensed aerodromes.

By S35 the Secretary of State may, by order, require the management of any specified aerodrome to provide adequate facilities for consultation regarding the management of aerodromes to the following parties where their interests may be affected: the users of the aerodrome, interested local authorities and any organizations representing the interests of persons concerned with the locality in which the aerodrome is situated. All national and regional airports have been designated in this way (as well as some general aviation aerodromes). In 1994 Heathrow, Gatwick and Stansted airports have taken over the responsibility for aircraft complaints from the Department of Transport.

By S6 there is a power enabling the Secretary of State to direct the Civil Aviation Authority to take action to prevent noise pollution caused by civil aviation, providing it already has powers to act. Following the completion of the M25, London's orbital motorway, S6 powers were used to prevent a temporary helicopter link between Heathrow and Gatwick from continuing in operation beyond its original permission.

Finally, by S5 the Secretary of State is empowered to require the Civil Aviation Authority to consider environmental matters when licensing or renewing a licence for an aerodrome. But this power appears never to have been exercised.

Noise certification Noise certification is a concept which lays down rules prescribing noise limits for aircraft calculated in relation to their maximum certificated weight. The authority for relevant legislation is found in S60 of the Act and EC Council Directive 89/629.

By S60 'Her Majesty the Queen may by Order in [the Privy] Council (called here an Air Navigation Order) make such provisions as authorised in ss3.'

> S60(3) . . . an Air Navigation Order may contain provisions . . .
>
> (r) for prohibiting aircraft from taking off or landing in the United Kingdom unless there are in force in respect of these aircraft such certificates of compliance with standards of noise as may be specified in the Order and except upon compliance with the conditions of those certificates and
>
> (s) for regulating or prohibiting the flight of aircraft over the United Kingdom at speeds in excess of Mach 1.

S60 also empowers the implementation into English law of the provisions of Annex 16 of the 1944 Chicago

Convention on International Civil Aviation. (Remember that international law found in conventions such as this, or accords, or treaties, does not become part of an internal legal system until ratified in some way.)

The relevant order is the Air Navigation (Noise Certification) Order 1990/1514.

The order applies (subject to the right of the Civil Aviation Authority, after consultation with the Secretary of State to make exceptions) to the following cases:

(a) every propeller driven aeroplane having a maximum total weight authorised of 9000 kg or less
(b) every aeroplane which is capable of sustaining level flight at a speed in excess of flight Mach 1.0, being an aeroplane in respect of which applicable standards are specified in Art. 6.9 of this order
(c) every microlight aeroplane
(d) every other subsonic aeroplane (of certain specifications)
(e) every helicopter (of certain specifications)

By this order no relevant aircraft may take off or land in the United Kingdom unless the Civil Aviation Authority, or other competent authority of the state where the aircraft has been registered, has issued a noise certificate and the conditions have been complied with. Military aircraft and visiting foreign military aircraft are exempt. The certificates in this country are issued by the Civil Aviation Authority, and the applicant must supply all necessary information and satisfy all the relevant tests to his or her aircraft. The certificate has no time limit but will end if the aircraft is modified in such a way as to affect its noise emission.

Council Directive 92/14EEC, issued in 1992 by the European Commission, was incorporated into English law by the Aeroplane Noise (Limitation on Operations of Aeroplane) Regulations 1993/1409. These regulations apply, with certain exceptions, to every civil subsonic jet aeroplane with a maximum take-off mass greater than or equal to 34 000 kg and with more than 19 passenger seats. In such cases no aircraft is permitted to take off or land within the United Kingdom unless it is carrying a noise certificate, as required by its country of registration, issued to certain specified standards.

Noise insulation Despite all attempts to reduce noise at source, it is impossible to eradicate the problem for those living close to airports. By S79 of the Act, the Secretary of State may make schemes requiring the person for the time being managing the aerodrome to make grants to certain people occupying certain types of buildings close to designated aerodromes towards the cost of insulating such buildings. Heathrow and Gatwick both have schemes but some other airports have schemes authorized by private Acts of Parliament.

The insulation required involves double glazing of windows and the installation of specified ventilation systems.

Compensation may be also awarded under the Land Compensation Act 1973 (see page 242).

10.11 Road traffic noise

Road traffic noise is created in a number of ways, from operating hooters, noisy exhausts, car radios, revving up, to mere high density of traffic. The problems can be dealt with in many ways, some of which we have already seen. Under the Land Compensation Act 1973, land can be purchased or compensation be given in certain circumstances; under the Noise Insulation Regulations 1975 as amended, works may be carried out or grants given; in some situations the Environmental Protection Act 1990 or Noise and Statutory Nuisances Act 1993 may be implemented and local byelaws may be of particular use.

Also regulations have been made under the Road Traffic Act 1972 and S41 Road Traffic Act 1988 to control the use and construction of certain motor vehicles, i.e. the Road Vehicles (Construction and Use) Regulations 1986 as amended. These regulations cover a number of different aspects of noise from manufacturing to use. Breach of any regulation is an offence.

Thus, a vehicle propelled by an internal combustion engine must be fitted with an exhaust system including a silencer and the exhaust gases from the engine must not escape into the atmosphere without first passing through the silencer (Reg. 54(1)). All exhaust systems and silencers must be maintained in good and efficient working order and must not be altered after manufacture so as to increase the noise made by the escape of exhaust gases (Reg. 54(2)).

Reg. 55 applies to any wheeled motor vehicle with at least three wheels, such as motor cars, first used after 1 October 1983, subject to certain exceptions such as motorcycles, tractors or road rollers. These vehicles must be constructed so that they comply with the requirements set out in the Table made under Reg. 55(3) and which comply with certain European Union Directives so that the sound level from the specified vehicle does not exceed the set limits as measured using the specified methods and apparatus. Regs. 56, 57 and 57A and 57B cover similar matters in relation to agricultural motor vehicles and industrial tractors, motor cycle construction, motor cycle exhaust systems, and motor cycle maintenance respectively. Reg. 58 applies to motor vehicles not covered by the preceding Regulations and which were first used after 1 April 1970. Reg. 59 excepts vehicles from the above requirements if they are proceeding to a place where noise emission levels are going to be measured in order to determine whether they comply with the Regulations or if they are going to be adjusted or modified so that they do comply with the Regulations.

Apart from the technical control of noise emission in the construction and use of motor vehicles, the Regulations also lay down rules restricting the driver in the way he or she drives. Reg. 97 lays down a general prohibition on noisy use. 'No motor vehicle shall be used on a road in such manner as to cause any excessive noise which could have been avoided by the exercise of reasonable care on the part of the driver.'

By Reg. 98 the driver of a vehicle must stop the engine when stationary unless he or she is stuck in traffic or it is necessary to keep the engine running in order to examine or repair it.

Reg. 99 deals with the problem of warning devices such as a car horn. Thus, no person (note not just the driver) must sound or cause to be sounded any audible warning instrument such as a horn, gong, bell or siren fitted or carried on the vehicle, if the vehicle is stationary **at any time** unless to warn of danger to another moving vehicle on or near the road. Nor must a person sound such devices while moving on restricted roads between 23.30 pm and 7.00 am the following day. Those provisions do not apply to reversing warning devices provided they are on certain specified types of large vehicles such as goods vehicles or buses. In addition, subject to certain exceptions, no-one must sound or cause to sound a gong, bell, siren or two-tone horn fitted to or carried on a vehicle whether it is stationary or not. The exceptions are for emergency services when it is necessary to warn other road users of the urgency of the situation or to warn them of the vehicle's presence; theft alarms (but this does not include two-tone horns); bus alarms to summon help for the bus personnel; an apparatus designed to inform the public that the vehicle is conveying goods for sale but only between 12.00 midday and 19.00 pm and subject to S62 Control of Pollution Act 1974 (see page 228).

Finally, town and country planning can also play an important part in reducing noise caused by traffic. This can be done by predetermining what should be located in a particular area by the use of the structure and local plans and thus preventing new problems. Also, planning permission may be refused if the development would increase the traffic inordinately.

10.12 European Union law

Origins

European Union law is a relatively new source of English law. The European Economic Community (EEC) was set up in 1957 by the first Treaty of Rome, which was signed by the original six member states. Two other communities were also set up, the European Coal and Steel Community (ECSC) by the Treaty of Paris 1951 and the European Atomic Energy Community (Euratom) by a second Treaty of Rome 1957. The administrative aspects of the three communities were merged in 1967 but they remain separate legal entities. The first Treaty of Rome was amended by the Single European Act 1986. As a result the EEC was from then on referred to as the European Community (EC) and the first Treaty of Rome as the EC Treaty. Also environmental issues were given proper Treaty authority which they had hitherto lacked. The EC thus is the source of environmental policy and law. The Treaty on European Union (the Maastricht Treaty) 1991 was finally adopted by all the member states in 1993. The Maastricht Treaty amended the EC Treaty and now the collective communities are referred to as the *European Union* (EU).

Objects

By Article 2 of the first Treaty of Rome (the Treaty) as amended, the main object of the Community is to 'promote throughout the Community a harmonious development of economic activities, sustainable and non-inflationary growth respecting the environment, a high degree of convergence of economic performance, a high level of employment and of social protection, the raising of the standard of living and quality of life, and of economic and social cohesion and solidarity among Member States'.

In 1985 a further step was taken by all member states signing the Single European Act 1986 in order to create the Single European Market on 1 January 1993. With this move trade barriers were abolished, border controls for Community nationals were eliminated and employees were permitted to move freely between the states. Article 130r of the Treaty inserted by the Act stated that the objectives of the Community's environmental action were to be:

(i) To preserve, protect and improve the quality of the environment.
(ii) To contribute towards protecting human health.
(iii) To ensure a prudent and rational utilization of natural resources.

The Treaty on European Union (the Maastricht Treaty) 1991 also emphasized the importance of environmental issues when implementing the main aims of the EU. Thus continued prosperity in Western Europe is being encouraged but with an environmental dimension. The Fifth Environmental Action programme 'Towards Sustainability' (1993–2000) has now been implemented, the previous four programmes all having instigated much environmental law.

Accession by the United Kingdom

In 1971 the United Kingdom, Ireland and Denmark signed the Brussels Treaty of Accession by which they agreed to join the three communities. Like all treaties, this would have remained a mere political agreement had the UK's Parliament not passed the European Communities Act 1972 which came into effect on 1 January 1973. The Act ratified the Treaty and incorporated the three treaties as part of the law of the United Kingdom (note not just England and Wales) as well as all the case law from the European Court of Justice. At one fell swoop there was a new source of English, Scottish and Northern Irish law. The Single European Act 1986 and the Maastricht Treaty were ratified in the UK by the European Communities (Amendment) Act 1986 and the European Communities (Amendment) Act 1993 respectively. Thus, the English courts must apply EU law to a case before them, if it is applicable. If there is conflict, EU law prevails.

EU institutions

The EU is run by four main institutions, the Council of Ministers of the European Union, the European Parliament, the European Commission of the European Union and the European Court of Justice. Each member state has voluntarily given up to these bodies some of its own rights to make law (sovereignty). This is because the communities are **supra-national**, having authority to make decisions for all the member states on certain community matters, e.g. the Common Agricultural Policy.

The **Council of Ministers** of the European Union which sits at Brussels is the principal executive body of the EU. The Council of Ministers contains one appropriate representative from the government of each member state. Thus, if an environmental matter is to be discussed then an environment representative is sent. The Council of Ministers has the final decision on proposals put forward by the Commission and legislation may be passed with a qualified majority (achieved by weighting the votes among the states) but complete unanimity is required in foreign and security policy, justice and home affairs, after consultation with the European Parliament. The Council also comprises a number of working groups of officials from the member states.

The European **Commission** is the executive of the EU. It sits in Brussels and consists of twenty independent members appointed by the member states, two from Germany, France, Italy, Spain and the United Kingdom and one from each of the other states. The president is appointed by common accord between Governments after consulting the European Parliament, but there is a power of veto which applies to commissioners as well as to the president. John Major the United Kingdom Prime Minster exercised his veto at Corfu in 1994. Commissioners are nominated by Governments in consultation with the new president and the resultant list is then sent for approval to the European Parliament.

The European Commission implements EU policy, initiates and draws up proposals for legislation for the Council to approve. It also polices and enforces EU law and has extensive investigative powers. It can also start proceedings before the European Court of Justice. The European Commission draws up the environment action programme and drafts proposed EU legislation. Decisions are reached by majority. Member states must inform the Commission of proposed domestic law. It is divided into a number of directorates. Directorate General XI deals with Environment, Consumer Protection and nuclear safety.

The **European Parliament** sits at Strasbourg. It has evolved from a mere advisory assembly into a directly elected representative parliament having equal rights as the European Council on budgetary matters. Its powers *vis-à-vis* the introduction of new legislation have been extended by the Single European Act and the Maastricht Treaty so that there is now a conciliation committee designed, where possible, to produce a joint statement from the Commission and Parliament. It can set up inquiry committees to investigate maladministration and any citizen of the EU can directly petition Parliament on community matters that affect him or her. He or she can do this either in a private capacity or as a member of an organisation, including companies. It has been suggested that this right is not used enough.

The member states

Since 1 January 1995 there are now fifteen member states: Spain, Portugal, United Kingdom, Denmark, Belgium, Italy, France, Finland, Ireland, Netherlands, Germany, Luxembourg, Austria, Greece and Sweden (mnemonic SPUD BIFFING LAGS).

European Union law

The three Treaties are obviously the main source of community law, which the Single European Act 1986 and the Maastricht Treaty supplement. The European Commission and Council are allowed to make three types of law called regulations, directives and decisions.

1. Regulations These are rules which once made are immediately binding on all the member states, without reference to their legal systems.

2. Directives These are orders or requirements directed

to all or some of the member states, the results of which are binding but the means to achieve them are left to each member state. This is useful because of the difficulties in translating legal ideas between different legal systems. Because of its nature, most environmental law will be created in this way. An example of this was 86/188/EEC. This was the Council Directive 'The protection of workers from noise' which in the UK was implemented under the Health and Safety at Work etc. Act 1974 in the Noise at Work Regulations 1989.

Proposals for directives may come from the Council of Ministers or the Commission. Preparation of drafts of directives is undertaken by the Commission who may set up technical working groups where necessary. Directives must be approved by the Council of Ministers. In some cases ministers may have the power of veto but on others only a majority agreement is required.

Implementation of a directive is normally within a specified time, usually two years. Once the domestic law has changed there has been formal compliance.

Decisions These are decisions on some aspect of community law which are binding on the person or body to whom they are addressed.

European Court of Justice

This court situated at Luxembourg is a court specially set up to deal with problems of community law. Each state sends at least one judge. The court rules on the meaning of the Treaties and under the Treaty of Rome; if there is any doubt as to the meaning of the Treaties a national case must be sent for clarification of the community law involved to the court (Art. 177 EC Treaty). The court also has jurisdiction over disputes involving aspects of community law between states, corporations and even individuals. It is therefore another source of judicial precedent, binding national courts by its decisions. The ECJ itself is not bound by its own previous decisions.

There have been many pieces of legislation introduced in relation to noise as a result of the EU. Examples include the Lawnmowers (Harmonisation of Noise Emission Standards) Regulations 1992, the Household Appliances (Noise Emission) Regulations 1990, the Town and Country Planning (Assessment of Environmental Effects) Regulations 1988 and the Noise at Work Regulations 1989.

10.13 Covenants in a lease

A covenant is a promise contained in a document, in this case, a lease. When someone acquires the lease of a house or flat, the agreement usually contains many covenants with which the lessee must comply, e.g. to pay the rent, to repair and not to sublet the property without permission of the lessor. Often such leases also contain particular covenants dealing with the prevention of nuisances, e.g. not to operate a washing-machine or other domestic appliance between the hours of 11:00 PM and 8:30 AM the following day so as to cause a nuisance or annoyance to the neighbours. These sorts of covenants are especially important in leases of flats within a block, as noise can be potentially contentious. Where there is a breach of the covenant then the lessor may be able to forfeit the lease, or in less extreme cases, take other forms of action such as obtaining an injunction.

10.14 Restrictive covenants

Restrictive covenants are different to covenants in leases; they must not be confused as there are complicated rules relating to whether or not they are enforceable. In essence they are covenants of a negative nature, e.g. not to commit a nuisance or carry on a trade or business on particular land, or even not to build on a particular piece of property, and they can apply to leasehold or freehold land. They are usually imposed by a person selling part only of his or her land for building purposes. Obviously he or she wishes to keep the remaining land free from nuisances or generally to maintain the original character of the neighbourhood and thus he or she imposes the restrictive covenant on the purchaser. It is clear that the original parties to the covenants must be bound by what they agreed. But what if the seller sells the land he or she retained, or the purchaser sells the land he or she bought, which was subject to the covenant? Equity allows both the benefit and the burden of the covenant to pass to subsequent purchasers of either the retained land or the part sold, and appropriate action may be taken where there is a breach. Where a restrictive covenant no longer benefits the original property, e.g. because the character of the district has gone down, then the Lands Tribunal may, on application, discharge or modify it. It may also merely fall into disuse. It is probable nowadays that action under the Environmental Protection Act is more useful to a person affected by noise. However, if the noise would not amount to a nuisance, but is in breach of a restrictive covenant, then the action taken would be in relation to that.

10.15 Byelaws

Like restrictive covenants, byelaws can be effectively used against a particular type of noise which may not perhaps amount to a nuisance. Byelaws are local laws made by the local authority with power delegated to it by Act of Parliament, i.e. in most cases by S235 of the Local Government Act 1972, and confirmed by the appropriate authority, the Secretary of State for the Environment. (But note all the byelaws made in relation to civil aviation, above.)

In order to maintain some sort of uniformity, model byelaws have been issued by the Home Office covering such matters as music near houses, churches and hospitals, the playing of organs, wireless sets and gramophones and noisy hawking, unruly behaviour in places of entertainment, noisy conduct at night and noisy animals. Breach of any byelaw will result in a summary conviction and fines of up to £200 may be imposed.

Questions

(1) Control of neighbourhood noise from construction sites may be effected by use of special or general provisions laid down in the Control of Pollution Act, 1974. State what these are, and give an example of each situation in which you would expect one procedure to be used more appropriately than any of the others.

Describe briefly those supporting regulations and codes of practice which would need to be considered during application of these procedures.

(2) An application is received by the planning authority from a company which intends to make reinforced concrete fencing posts and panels. The site in question is only a few yards from good quality housing which surrounds it, but it has been used for general industrial purposes before without causing complaint. This manufacturing process, however, will be carried out mainly in the open air and will involve the use of noisy machinery. Experience indicates that noise nuisance will be created if the development proceeds: there is a local unemployment problem which may influence the authority's decision.

What powers are available to the local planning authority to restrict or control noise from development of this type, what alternative action may be available to the local authority to control noise levels thought likely to be generated, and what advice would you give to the planning officer?

(3) The licensee of the public house known as the 'Pheasant and Owl' has, over the last two years, been regularly hiring out a room in the premises, known as the 'Partridge Room', to various pop groups. The groups perform during the evenings three or four times every week. A relevant licence covering musical and dancing entertainment has been in force for many years.

The public house is situated in Conduit Street, a narrow street in a densely populated residential area. Percy bought 8 Conduit Street in 1939 and has lived there continuously ever since. He has recently written a letter to the licensee complaining about the intolerable noise arising from the premises.

Percy has received a letter from the licensee's solicitor stating that as Percy has not complained about noise from music on any previous occasion, the 'Pheasant and Owl' has aquired a prescriptive right to commit a noise nuisance. The letter is brought to the notice of the environmental health officer. Discuss the position.

(4) Mr Smith is buying his house on a mortgage and finds that a new motorway is to be constructed quite close to his house. How will it be determined whether Mr Smith's house is eligible for improved insulation against traffic noise?

What are the main components of the standard package which will be installed if his house is eligible? What other compensation might Mr Smith be able to claim? How would you advise Mr Smith on the procedure for claiming this compensation?

(5) Distinguish between private, public and statutory nuisance.

Section 80 of the Environmental Protection Act gives a local authority powers to deal with nuisance caused by noise but does not define *nuisance*. By reference to case law indicate what principles you would need to consider before taking action under section 58.

(6) You have been asked to prepare a proposal for your district council's first noise abatement zone. What factors would you need to take into account in making your recommendations?

What information must be included in the noise level register? Is there any additional information which you consider it important to record?

(7) Describe the procedure for producing noise legislation in the United Kingdom with particular reference to the Environmental Protection Act Part III.

Briefly outline the procedure for the production of EU directives.

(8) Outline the options available to a local authority for controlling noise from construction sites. Discuss the advantages and disadvantages of different options.

(9) As an environmental health officer, you have received a complaint about high levels of amplified music from an adjoining flat. Describe how you would investigate such a complaint to determine whether the amplified music caused a statutory nuisance and what action would be appropriate.

If you decided that it would not be appropriate for the local authority to take action, what legal action could the complainant take if he/she wished to pursue the matter?

(10) Distinguish between private, public and statutory nuisance.

Discuss the concept of assessment, as opposed to measurement, in the context of noise nuisance investigation.

(11) In the context of initiatives on noise control originating from the European Union, what are (i) the Council of Ministers, (ii) a directive, (iii) a derogation, (iv) a regulation?

Using the example of the Noise at Work Regulations 1989, explain the process by which EU requirements affect UK law.

(12) A developer receives conditional planning permission from the local planning authority in respect of a development proposal. You are consulted to assess the action of the planning authority. Explain the features which should be used to test appropriateness of a planning condition.

(13) (i) Describe the main provisions of the Noise at Work Regulations 1989.
(ii) Do the Noise at Work Regulations 1989 comply fully with the requirements of Directive 86/188/EEC? Discuss.

(14) Breaches under planning law may be controlled by either a *Stop notice* or an *Enforcement notice*. Explain:
(i) Who is empowered to serve these notices.
(ii) Which types of breaches can be controlled by these notices.
(iii) The main provisions related to each notice.

(15) Case law has established a number of factors which are relevant to assessing the existence of a nuisance. One of them is the *Standard of comfort*.
(i) What are the other factors? Briefly explain each.
(ii) Discuss the concept of *Standard of comfort* in relation to noise nuisance.

(16) Outline the main provisions of the Building Regulations covering sound transmission in dwellings.

What differences exist between the way the regulations deal with conversions compared to new build? (Students in Scotland or Northern Ireland should specify which regulations they are addressing.)

(17) You are asked to investigate, on behalf of a local authority, an alleged noise nuisance occurring between two adjoining flats. Describe how you would determine your opinion on the following:
(i) Whether or not the noise constitutes a nuisance.
(ii) Whether the nuisance has arisen from inadequate sound insulation, antisocial behaviour, or both.
(iii) Who is responsible for the nuisance.
(iv) What statutory controls are available.

(18) Under what circumstances may a local authority apply for an injunction in respect of an alleged noise nuisance? Illustrate your answer with a practical example.

Outline the relevant legislation.

What is an interlocutory injunction?

(19) Discuss the meaning of *precise* in relation to (a) noise nuisance notices and (b) planning conditions.

In the case of planning conditions, outline all appeal provisions which are available.

APPENDIX 1 Glossary of acoustical terms

absorption see under sound absorption
absorption coefficient see under sound absorption coefficient
acceleration rate of change of velocity (in m/s^2)
accelerometer a transducer which measures acceleration
acoustic calibrator a device for producing an accurately known sound pressure level; used for the calibration of sound level meters
acoustic enclosure a structure built around a machine to reduce noise
acoustic impedance of a surface or acoustic source; the (complex) ratio of the sound pressure averaged over the surface to the volume velocity through it; the volume velocity is the product of the surface area and acoustic particle velocity; see also under characteristic, mechanical and specific acoustic impedance
acoustic lagging materials applied externally to the surface of pipes and ducts to reduce the radiation of noise; not to be confused with thermal lagging
acoustic particle velocity the velocity of a vibrating particle in an acoustic wave
acoustic reactance the imaginary part of the complex acoustic impedance
acoustic resistance the real part of the complex acoustic impedance
acoustic trauma sudden permanent hearing damage caused by exposure to a burst of high level noise
acoustics (1) the science of sound; (2) of a room: those factors which determine its character with respect to the quality of the received sound
action level a noise exposure level in the workplace above which certain actions are required under the Noise at Work Regulations
active filter a filter which contains transistors, integrated circuits or other components requiring a power supply
active noise control a noise control system which uses antiphase signals from loudspeakers to reduce noise by destructive interference
airborne sound sound or noise radiated directly from a source, such as a loudspeaker or machine, into the surrounding air (in contrast to structure-borne sound)
airborne sound insulation the reduction or attenuation of airborne sound by a solid partition between source and receiver; this may be a building partition, e.g. a floor, wall or ceiling, a screen or barrier or an acoustic enclosure
aliasing introduction of false spectral lines into a spectrum by having the maximum frequency of the signal greater than one-half the digital sampling frequency
ambient noise the totally encompassing noise in a given situation at a given time; it is usually composed of noise from many sources, near and far (defined in BS 4142)
amplitude the maximum value of a sinusoidally varying quantity
analogue-to-digital converter (A/D converter or ADC), a device which samples and digitizes analogue signals, preparatory for digital signal processing; the continuously varying analogue signal is converted into a finite number of discrete steps or levels then represented as a series of numbers
anechoic literally 'without echo', i.e. without any sound reflections. An anechoic room is one in which all the interior surfaces (walls, floor and ceiling) are lined with sound-absorbing materials so that there are no reflections; it provides a standard environment for acoustic tests
angular frequency the product 2π times frequency; symbol ω; measured in radians per second
anti-aliasing filter a low pass filter inserted in an instrument, before the ADC, in order to prevent aliasing
antinode a point, line or surface of an interference pattern at which the amplitude of the sound pressure or particle velocity is at a maximum
antivibration (AV) mounts springs or other resilient materials used to reduce vibration (and noise) by isolating the source from its surroundings

articulation index a measure of the intelligibility of speech; the percentage of words correctly heard and recorded in an intelligibility test

attenuation a general term used to indicate the reduction of noise or vibration, by whatever method or for whatever reason, and the amount, usually in decibels, by which it is reduced

attenuator a device introduced into air or gas flow systems in order to reduce noise; absorptive types contain sound-absorbing materials; reactive types are designed to tune out noise at particular frequencies

audibility the ability of a sound to be heard; the concept of audibility has been used as a criterion for setting limits to noise levels, particularly from amplified music; it is a subjective criterion, i.e. one which can only be determined by the ear of the listener, not by measurement of sound levels; also used as a criterion to determine the degree of privacy between rooms (e.g. offices)

audibility threshold the minimum sound pressure which can just be heard at a particular frequency by people with normal hearing; usually taken to be 20 μPa at 1000 Hz

audible range frequencies from 20 Hz to 20 kHz (approx.); sound pressures from 20 μPa to 100 Pa (approx.)

audiogram a chart or graph of hearing level against frequency

audiometer an instrument which measures hearing sensitivity

audiometry the measurement of hearing

auditory cortex the region of the brain which receives signals from the ear

aural of or relating to hearing or the hearing mechanism

A weighting a frequency weighting devised to attempt to take into account the fact that human response to sound is not equally sensitive to all frequencies; it consists of an electronic filter in a sound level meter, which attempts to build in this variability into the indicated noise level reading so that it will correlate, approximately, with human response

axial mode the room modes associated with each pair of parallel surfaces

background noise level defined in BS 4142 as the L_{A90} value of the residual noise (see under residual noise)

band pass filter a filter which provides zero attenuation to all frequencies within a certain band but which attenuates completely all other frequencies

band sound pressure level the sound pressure level of the sound signal within a certain frequency band

bandwidth the range of frequencies contained within a signal, passed by a filter, or transmitted by a structure or device

basilar membrane a membrane inside the cochlea of the inner ear which vibrates in response to sound, thus exciting the hair cells

beats periodic variations which are heard when two pure tones of slightly differing frequencies are superimposed

bel ten decibels; a unit of level on a logarithmic scale which is based on the ratio of two powers, or of power-related quantities such as sound intensity or the square of sound pressure

bending or flexural waves elastic waves in plates, panels, beams, etc., which are a combination of compression and shear waves and which are responsible for the transmission of structure-borne sound in buildings and other structures

binaural relating to hearing using both ears, e.g. binaural localization; the use of both ears to locate the direction of sounds

bit abbreviation of binary digit; the smallest possible unit of information in binary form, i.e. on or off, yes or no, 0 or 1

broadband containing a wide range of frequencies

byte a binary word or group of bits

capacitor one of the basic elements of an electrical circuit consisting of two conducting plates separated by a gap containing an insulator, or dielectric; it has the property of capacitance, measured in farads (F), microfarads (μF) or picofarads (pF); also known as a condenser

centre frequency the centre of a band of frequencies; in the cases of octave or one-third octave it is the geometric mean of the upper and lower limiting frequencies of the band

characteristic acoustic impedance (of a medium) the ratio of sound pressure to acoustic particle velocity at a point in the medium during the transmission of a plane wave; it is the product of the speed of sound in the medium and its density

charge amplifier a type of preamplifier suitable for use with piezoelectric accelerometers; it gives an output which is proportional to the electric charge present in the input signal

cochlea a coiled, snail-shaped structure in the inner ear; it is fluid-filled and contains a complex arrangement of membranes and hair cells which convert mechanical vibrations of the fluid into electrical impulses transmitted to the brain

coincidence effect an effect which leads to increased transmission of sound by panels and partitions when the speed (and wavelength) of flexural waves in the panel coincide with the speed (and wavelength) of the sound waves exciting the panel

compressional wave an elastic wave in a fluid or solid in which the elements of the medium are subjected to deformations which are purely compression, i.e. which do not contain any element of rotation or shear, and of which sound waves in air are an example

condenser see under capacitor
conductive deafness hearing loss which is caused by some defect or fault in the outer or middle ear
continuous equivalent noise level, L_{Aeq} of a time-varying noise; the steady noise level (usually in dB(A)) which, over the period of time under consideration, contains the same amount of (A-weighted) sound energy as the time-varying noise, over the same period of time
continuous spectrum a sound or vibration spectrum whose components are continuously distributed over the particular frequency range, for example, random noise; contrast with a line spectrum from a harmonic sound
coulomb damping a form of damping in which the damping force is constant, independent of either displacement or velocity (also called dry friction damping)
crest factor of a signal; the ratio of the peak to the root mean square (RMS) value
criterion the basis on which a noise or vibration is to be judged, e.g. damage to hearing, interference with speech, annoyance
critical damping the amount of viscous damping in a system which will allow the system to return to its equilibrium position, in the minimum time, without overshoot, i.e. without oscillation; the boundary between overdamping and underdamping
critical frequency the lowest frequency at which the coincidence effect takes place for a particular panel or partition, and above which the sound insulation performance starts to deteriorate
cortex see under auditory cortex
crosstalk a signal from one track, channel or circuit which is transmitted, unwanted, into another track, channel or circuit
Curie point the temperature above which a piezoelectric material becomes polarized, and loses its piezoelectric properties
C weighting one of the frequency weightings defined in BS 5969 (IEC 651); it corresponds to the 100-phon contour and is the closest to the linear or unweighted value
cycle of a periodically varying quantity; the complete sequence of variations of the quantity which occurs during one period
cycle per second unit of frequency; one cycle per second is one hertz (Hz)

damping a process whereby vibrational energy is converted into heat through some frictional mechanism, thus causing the level of vibration to decrease
damping ratio the ratio of the amount of damping in a vibrating system to the amount of damping when critical
day–night level an index of environmental noise which is a 24 h L_{eq}, but with a 10 dB weighting added to the night-time noise levels (2200 h to 0700 h) to allow for increased sensitivity to noise during the night-time
dB(A) the A-weighted sound pressure level; see under A weighting
decade a range of ten to one, e.g. from 100 Hz to 1000 Hz
decibel (dB) the decibel scale is a scale for comparing the ratios of two powers, or of quantities related to power, such as sound intensity; on the decibel scale the difference in level between two powers, W_1 and W_2 is N dB, where $N = 10 \log_{10} (W_1/W_2)$; the decibel scale may also be used to compare quantities, whose squared values may be related to powers, including sound pressure, vibration displacement, velocity or acceleration, voltage and microphone sensitivity; in these cases the difference in level between two signals, of magnitude S_1 and S_2, is given by $N = 20 \log_{10} (S_2/S_1)$; the decibel scale may be used to measure absolute levels of quantities by specifying reference values which fix one point in the scale (0 dB) in absolute terms; a decibel is one-tenth of a bel
degrees of freedom the number of degrees of freedom of a mass–spring model of a vibrating system is the minimum number of coordinates required to specify all the different possible modes of vibration of the system
deterministic a deterministic signal is one whose value can be predicted with certainty from a knowledge of its behaviour at previous times, as opposed to a random signal, where this is not possible
dielectric a material which is an electrical non-conductor or insulator; it is used between the plates of a capacitor
diffraction the process whereby an acoustic wave is disturbed and its energy redistributed in space as a result of an obstacle in its path; the relative size of the sound wavelength and the object are always important in diffraction; reflection may be considered to be a special case of diffraction when the size of the obstacle is very large compared to the wavelength; the combined effects of diffraction from an irregular array of objects in the path of the sound is also known as scattering; diffraction theory deals with all aspects of the interactions between matter (i.e. obstacles) and waves, so it also determines the directional patterns of sound radiation from vibrating objects
diffuse sound field a sound field of statistically uniform energy density in which the directions of propagation of waves are random from point to point
digital signal a signal having a discrete number of values, which can be represented as a sequence of numbers; see also analogue-to-digital converter and digital-to-analogue converter
digital-to-analogue converter (DAC) an electronic device which converts digital signals into analogue signals
digital audio tape recorder (DAT) a tape recorder which includes an ADC (and a DAC) and which records analogue signals on tape in coded digital form

directivity factor the ratio of the sound intensity at a given distance from the source, in a specified direction, to the average intensity over all directions, at the same distance
directivity index the directivity factor (DF) of a source, expressed in decibels, i.e. $10 \log_{10}$ (DF)
direct sound sound which arrives at the receiver having travelled directly from the source, without reflection
direct sound field that part of the sound field produced by the source where the effects of reflections may be neglected
distortion a lack of faithfulness in a signal, such as the introduction of harmonics into the frequency spectrum, introduced, for example, because of non-linearity or of overload of some component of the measurement system
D_{nT} standardized level difference a measurement of airborne sound insulation, corrected according to BS 2750 for receiving room characteristics; a complete set of measurements consists of 16 third-octave band values, from 100 Hz to 3150 Hz
D_{nTW} weighted standardized level difference a single-figure value of airborne sound insulation performance, derived according to procedures in BS 5821, used for rating and comparing partitions and based on the values of D at different frequencies; values of D_{nTW} are specified in the Building Regulations
Doppler effect the change in the observed frequency of a wave caused by relative motion between source and receiver
dynamic magnification factor (*Q* factor) a quantity which is a measure of the sharpness of resonance of an oscillating system (either mechanical or electrical); it is related to the amount of damping in the system
dynamic range the range of magnitudes of a signal which a measuring system, or component of a system, can faithfully record, process or measure, from highest to lowest; usually expressed in decibels
dynamic stiffness the ratio of change of force to change of displacement in a vibrating system; it may be different from the static stiffness of the system

ear defenders or ear protectors earmuffs or earplugs worn to provide attenuation of sounds reaching the ear and reduce the risk of noise-induced hearing loss
echo a sound reflection whose magnitude and time delay is such that it is perceived as a separate, distinguishable repetition of the direct sound, as opposed to reverberation which is perceived as part of the original sound
electret or prepolarized microphone a type of condenser microphone in which a prepolarized layer of electret polymer is used as a dielectric between the diaphragm and backing plate which form the condenser
electrostatic actuator a device which fits over a microphone, close to the diaphragm, and is used for remote calibration
equal loudness contours a standardized set of curves which show how the loudness of pure tone sounds varies with frequency at various sound pressure levels
equivalent continuous noise level see under continuous equivalent noise level, L_{Aeq}
eVDV estimated vibration dose value; a measure of a cumulative amount of vibration based upon weighted RMS acceleration values and durations; for signals of limited crest factor, the eVDV approximates to the vibration dose value, VDV; see also under VDV and RMQ
Eustachian tube the passage from the middle ear to the back of the throat which serves to equalize the pressure across the eardrum
Eyring's formula a modified version of Sabine's formula, for reverberation time, which takes into account the discrete nature of sound reflections; also known as the Norris–Eyring formula

F (fast) time weighting an averaging time used in sound level meters, and defined in BS 5969 (IEC 651)
far field of a sound source; that part of the sound field of the source where the sound pressure and acoustic particle velocity are substantially in phase, and the sound intensity is inversely proportional to the square of the distance from the source
fast Fourier transform (FFT) an algorithm or calculation procedure for the rapid evaluation of Fourier transforms; an FFT analyser is a device which uses FFTs to convert digitized waveform signals into frequency spectra, and vice versa
fatigue-decreased proficiency boundary a criterion, based on task performance, for evaluating human response to vibration, defined in ISO 2631
field measurements measurements carried out on-site, away from controlled laboratory conditions; the results of field tests of sound insulation may include the effects of flanking paths as well as direct sound transmission, which would not be the case for laboratory tests
filter a device which transmits signals within a certain band of frequencies but attenuates all other frequencies; filters may be electrical, mechanical or acoustical
flanking transmission the transmission of airborne sound between two adjacent rooms by paths other than via the separating partition between the rooms, e.g. via floors, ceilings and flanking walls
flutter echo a series of repeating echoes caused by parallel reflecting surfaces
forced vibration steady-state vibration of a system caused by a continuous external force
form factor the ratio of the RMS value of a signal to the mean value between two successive zero-crossings
Fourier analysis/series/spectrum Fourier's theorem

shows that any periodic function may be broken down (or analysed) into a series of discrete harmonically related frequency components which may be represented as a line spectrum

Fourier transform a mathematical process which transforms a non-periodic function of time into a continuous function of frequency, and vice versa (in the case of the inverse transform)

fractional dose a fractional component of a total noise exposure, defined in Noise Guide No. 3 of the Noise at Work Regulations

free-field conditions a situation in which the radiation from a sound source is completely unaffected by the presence of any reflecting boundaries; see also under anechoic

frequency of a sinusoidally varying quantity such as sound pressure or vibration displacement; the repetition rate of the cycle, i.e. the reciprocal of the period of the cycle, the number of cycles per second; measured in hertz (Hz)

frequency analysis the separation and measurement of a signal into frequency bands

frequency response of measurement system or component of such a system, e.g. a sound level meter or microphone; the variation in performance, e.g. sensitivity, with change of frequency

frequency spectrum a graph resulting from a frequency analysis and showing the different levels of the signal in the various frequency bands

frequency weighting an electronic filter built into a sound level meter according to BS 5969; see also under A and C weighting

fundamental frequency the lowest natural frequency of a vibrating system; the repetition rate of a harmonic waveform

Haas effect a psycho-acoustic phenomenon in which precedence is given to the direction of the first arrival of direct sound in attributing the direction from which the sound is coming

hair cells biological cells in the cochlea of the inner ear where vibration is turned into a neural signal which is transmitted to the brain

harmonic a signal having a repetitive pattern

hearing level a measured threshold of hearing, expressed in decibels relative to a specified standard threshold for normal hearing

hearing loss any increase of an individual's hearing levels above the specified standard of normal hearing

Helmholtz resonator a vibrating system having a single degree of freedom; it consists of an air-filled enclosure connected to the open air by a narrow column; the air in the enclosure acts as the spring and the air in the column acts as the mass

henry (H) the unit of electrical inductance

hertz (Hz) the unit of frequency; the number of cycles per second

high pass filter a filter which transmits frequency components of a signal that are higher than a certain cut-off frequency but which attenuates those below the cut-off

hysteresis damping a type of damping that occurs within materials as a result of phase changes which occur between stress and strain during the vibration cycle

impact noise sound resulting from the collision between colliding bodies

impact sound insulation the resistance of a floor to the transmission of impact sound; measured according to BS 2750

impedance see under acoustic impedance

impedance matching the use of a device to act as a buffer between a system, or component of a system, with a high output impedance and a system, or component of a system, with a low input impedance.

impulse a transient signal of short duration; impulsive noise is often described by words such as bang, thump, clatter

incus or anvil, the middle of the three bones in the middle ear

inductance the property of an electrical coil, or inductor, associated with the rate of change of magnetic field; measured in henrys (H)

inertia base a concrete slab used under antivibration mounts to provide additional mass, rigidity and stability

infrasound acoustic waves with frequencies below the audible range, i.e. below about 20 Hz

insertion loss a measure of the effectiveness of noise control devices such as silencers and enclosures; the insertion loss of a device is the difference, in dB, between the noise level with and without the device present

insulation see under sound insulation

integrating circuit an electrical circuit which converts an acceleration signal into a velocity or displacement signal

integrating sound level meter a sound level meter which electrically integrates sound pressure signals to measure the equivalent continuous sound level, L_{Aeq}

intelligibility of speech signals; the degree to which each individual syllable of speech can be identified and understood

intensity see under sound intensity

interference (1) the principle of interference governs how waves interact; the combined wave disturbance is the algebraic sum of the individual wave disturbances, leading to the possibility of constructive and destructive interference; (2) the disturbing effect of unwanted signals, often electrical in nature

isolation see under vibration isolation

isolation efficiency a measure of the effectiveness of a vibration isolation; isolation efficiency = $(1 - T) \times 100\%$, where T is the transmissibility of the system; see also under transmissibility

jerk the rate of change of acceleration

just noticeable difference (jnd) a concept used in psycho-acoustic measurement; the difference between two (acoustic) stimuli which is just noticeable in some defined condition

L **(level)** sound pressure level, SPL; in general, it implies the use of decibels related to the ratio of powers, or power-related quantities such as sound intensity or sound pressure

L_A see under A-weighted sound pressure level

L_{AE} see under sound exposure level, SEL

$L_{Aeq,T}$ see under continuous equivalent sound level

L_{Amax} the maximum RMS A-weighted sound pressure level occurring within a specified time period; the time weighting, fast or slow, is usually specified

$L_{AN,T}$ percentile level, i.e. the sound pressure level in dB(A) which is exceeded for $N\%$ of the time interval T, e.g. in L_{A10} and L_{A90}

$L_{EP,d}$ see under personal daily noise exposure level

L'_{nT} see under standardized impact sound pressure level

$L'_{nT,W}$ see under weighted standardized impact sound pressure level

L_{peak} see under peak sound pressure level

L_W see under sound power level

level difference BS 2750 uses the difference in level between two rooms as the basic measure of airborne sound insulation

level recorder an instrument for registering and measuring the variation of signals, such as sound pressures, with time

linear a measurement device is linear if its output is directly proportional to its input; in the case of a microphone, for example, this means that the sensitivity is constant and does not change with sound pressure level; linear SPL means unweighted

linearity the degree to which a device is linear

logarithmic decrement (δ) a measure of the amount of damping in a vibrating system, based on the rate of decay of natural vibrations of the system

longitudinal wave a wave in which the vibratory movement of the particles in the medium is parallel to the direction in which the wave is travelling; compressional waves in a fluid medium are longitudinal

long-term sound level, long-term average rating level, long-term time interval terms used in connection with the description and measurement of environmental noise, and defined in BS 7445

loss factor a term used to describe the amount of damping in a system, or material; it is twice the damping ratio

loudness the measure of the subjective impression of the magnitude or strength of a sound

loudness level the loudness level of a sound is the sound pressure level of a standard pure tone, of specified frequency, which is equally as loud, according to the assessment of a panel of normal observers

low pass filter a filter which transmits signals at frequencies below a certain cut-off frequency and attenuates all higher frequencies

lumped parameter model a model of a vibrating system in which mass, stiffness and damping are represented as discrete elements

magnetic tape recorder a device for capturing, storing and replaying analogue signals on to a tape medium containing ferromagnetic metal oxide particles

malleus or hammer; one of the three bones of the middle ear, connected to the eardrum

masking the process whereby the threshold of hearing for one sound is raised due to the presence of another, thus rendering the first sound inaudible

mass law an approximate relationship for predicting the sound reduction index of panels and partitions, based only on the surface density of the panel and the frequency of the sound

mean free path a term used in the statistical treatment of sound in rooms, relating to the average distance between reflections

measurement time interval a term used in standards on the measurement and rating of environmental noise (BS 4142 and BS 7445)

mechanical filter a resilient pad or layer which prevents the transmission of high frequency vibration acts as a low pass, mechanical filter

mechanical impedance the (complex) ratio of force to velocity at a point in a vibrating system

mel a unit of pitch; the pitch of any sound judged by listeners to be n times that of a 1 mel tone is n mels; 1000 mels is the pitch of a 1000 Hz tone at a sensation level of 40 dB

micron (μm) one-thousandth of a millimetre or one-millionth of a metre

microphone a transducer which converts acoustic signals into electrical (voltage) signals

middle ear an air-filled space which connects the eardrum of the outer ear to the oval window of the inner ear by three small bones, called ossicles

milli- a standard metric prefix meaning one-thousandth

mode of vibration a pattern of vibration of a vibrating system, characterized by a series of nodes and antinodes

mode shape the shape of a particular mode of vibration is usually represented as the maximum displacement of the system from its mean or equilibrium position

modulus of elasticity the stress divided by the strain for an elastic medium; an important factor in determining the speed of elastic waves in the medium; there are different types of elastic modulus, e.g. shear modulus, compression or bulk modulus and torsional modulus, for the different types of elastic deformation

monopole a model or idealized point source of sound which radiates spherical waves

nano- a standard metric prefix meaning one-thousand-millionth (i.e. 10^{-9})

narrowband filter a band pass filter with a small band-width, i.e. less than one-third octave

natural frequency the frequency of free or natural vibrations of a system

near field of a sound source; the region of space surrounding the source where sound pressure and acoustic particle velocity are not in phase, and the sound pressure varies with position in a complex way

newton (N) the SI unit of force; the force required to produce an acceleration of 1 m/s^2 in a mass of 1 kg

node a point, line or surface in a standing-wave pattern where some characteristic of the vibration, e.g. the displacement, is zero

noise unwanted sound or unwanted signal (usually electrical) in a measurement or instrumentation system

noise criteria (NC) curves a method devised by Beranek in the 1940s for rating or assessing internal (mainly office) noise; it consists of a set of curves relating octave-band sound pressure level to octave-band centre frequencies; each curve is given an NC number, which is numerically equal to its value at 1000 Hz; the NC value of a noise is obtained by plotting the octave-band spectrum against the family of curves; in order to meet a particular NC specification the noise level must be either below or equal to the SPL in each octave band

noise dose an amount of noise energy, usually A-weighted, received by a person, resulting from a combination of sound pressure level and exposure time; see also under personal daily noise exposure level, $L_{EP,d}$

noise exposure category a term used in Planning Policy Guidance Note 24: Planning and Noise

noise exposure forecast (NEF) a noise index used mainly in the United States for aircraft noise

noise index a method of evaluating or rating a noise, usually by assigning a single number to it, based on some combination of its physical characteristics (sound pressure level, frequency, duration) and other factors such as time of day, tonal characteristics and impulsive characteristics

noise limit a maximum or minimum value imposed on a noise index, e.g. for some legal purpose or to determine eligibility for some benefit

noise pollution level, L_{PN} an index devised in the 1960s for assessing environmental noise, based on a combination of its L_{Aeq} value and its variability, expressed in terms of its standard deviation; it is now rarely used

noise rating (NR) curves a method of rating noise which is similar to the NC system but intended to be applicable to a wider range of situations; the method was defined in ISO R1996, now withdrawn, but the NR system continues to be used, particularly for offices

noise reduction coefficient a single-figure number sometimes used to describe the performance of sound absorbing materials, based on a combination of its absorption coefficient at various frequencies

non-linear in general there is a non-linear relationship between two quantities if they are not directly proportional to each other; if in measurement and instrumentation systems the input exceeds the linear range, then non-linearity results in a distorted output

normalized corrected or standardized in some way

normal mode a natural mode of a vibrating system

normal threshold of hearing the modal value of the thresholds of hearing of a large number of otologically normal observers between 18 and 25 years of age

noy a unit of noisiness related to the perceived noise level in PNdB by the formula: PNdB = 40 + 10 $\log_2$ (noy)

Nyquist frequency the frequency which corresponds to half the sampling rate of digitized data, above which aliasing occurs

octave the range between two frequencies whose ratio is 2 : 1

organ of Corti a complex structure in the cochlea of the middle ear, supported by the basilar membrane and containing the hair cells

oscillation a to-and-fro motion; a fluctuation of a quantity or value about a mean

oscilloscope a device for displaying oscillatory signals on a cathode-ray screen

ossicles the three small bones of the middle ear which connect the eardrum with the oval window in the cochlea

outer ear the outer part of the hearing mechanism which collects and guides airborne sound down the ear canal to the eardrum

output impedance the impedance of a device measured at its output

oval window diaphragm connecting the cochlea to the middle ear

overdamping (1) an amount of damping, in excess of critical, which is sufficient to prevent oscillation in a mass–spring system; (2) producing a damping ratio greater than one

overload a situation in which a component or system is used beyond its range of linearity

overload indicator a device which indicates when an

instrument is likely to read incorrectly because it is being overloaded
overtone a higher (i.e. not the lowest) harmonic or natural frequency of a vibrating system

P-wave a longitudinal compression wave in an elastic medium
particle velocity see under acoustic particle velocity
pascal a unit of pressure equal to 1 N/m^2
pass band a band of frequencies which are transmitted by a band pass filter
passive a device which does not require a source of power for its operation, e.g. a passive filter or a passive noise control (cf. active)
peak the maximum deviation of a signal from its mean value within a specified time interval
peak-to-peak the algebraic difference between the extreme values of a signal occurring within a specified time interval
perceived noise level of a sound; the sound pressure level of a reference sound which is assessed by normal observers as being equally noisy; the reference sound consists of a band of random noise centred on 1000 Hz
percentile level, $L_{AN,T}$ the sound level, in dB(A) which is exceeded for N% of the time interval T, for example in L_{A10} and L_{A90}
period of a repetitive signal; the time for one cycle
periodic signal one which repeats itself exactly
permanent threshold shift the component of threshold shift which shows no progressive reduction with the passage of time when the apparent cause has been removed
personal daily noise exposure level, $L_{EP,d}$ that steady or constant level which, over 8 h, contains the same amount of A-weighted sound energy as is received by the subject during the working day
phase of a sinusoidal signal; an angle whose value determines the point in the cycle, i.e. the magnitude of the signal, at some reference time
phase difference the difference between the phase angles (of two sinusoidal signals of the same frequency)
phon the unit of loudness level; the loudness level of a sound, in phons, is the sound pressure level of a 1000 Hz pure tone judged by the average listener to be equally loud
piezoelectric the behaviour of certain crystalline materials whereby a deformation of the material (caused by force or stress) results in the production of electric charge on the stressed faces, and a voltage difference between them
pink noise a random broadband signal which has equal power per percentage bandwidth and therefore has a flat, i.e. horizontal, frequency spectrum when plotted on a logarithmic frequency scale (cf. white noise)
pinna the external part of the ear leading to the ear canal
pitch that attribute of auditory sensation in terms of which sounds may be ordered on a scale related primarily to frequency; the unit of pitch is the mel
plane wave a wave in which the wavefronts are plane and parallel everywhere, so the sound energy does not diverge with increasing distance from the source
plenum a chamber or space used to collect air prior to its distribution via a duct system
PNdB the unit of perceived noise level
point source an idealized concept of an acoustic source which radiates spherical waves
polarization a property of transverse waves but not longitudinal waves; it relates to the direction of the particle displacement in the plane normal to the direction of propagation
preamplifier a circuit which acts as an electrical impedance-matching device between a transducer with a high output impedance, such as a microphone or accelerometer, and the signal processing circuits of the sound level or vibration meter
precedence effect see under Haas effect
preferred speech interference level the arithmetic average of the sound pressure levels in the three octave bands 500 Hz, 1000 Hz and 2000 Hz
prepolarized microphone see under electret microphone
presbycusis hearing loss, mainly of high frequencies, that occurs with advancing age
progressive wave a wave that travels outwards, from its source, and is not being reflected
psycho-acoustics the study of the relationship between the physical parameters of a sound and its human perception
pure tone a sound for which the waveform is a sine wave, i.e. for which the sound pressure varies sinusoidally with time
pure tone audiometer an instrument for measuring hearing acuity to pure tones by determination of hearing levels

***Q* factor** a quantity which measures the sharpness of the resonance of a single degree of freedom mechanical or electrical vibrating system; in a mechanical system it is related to the damping ratio, the amplification produced at resonance and the shape of the resonance peak

random noise/vibration/signal a noise, vibration or signal which has a random waveform, with no periodicity
RASTI rapid analysis speech transmission index; an instrument for measuring the articulation index in a room
rating level a noise index defined in BS 4142 and BS 7445; the equivalent continuous A-weighted sound pressure level during a specified time period, adjusted for tonal and impulsive characteristics of the sound
ray a straight line representing the direction in which a sound is travelling, used in situations where the size of reflecting surfaces is large compared to the sound

wavelength
rayl the unit of specific acoustic impedance
Rayleigh wave a type of elastic wave which propagates close to the surface of a solid
Raynaud's disease a disorder affecting the blood vessels, nerves, connective tissues and bones of the fingers; one of its causes is prolonged exposure to high levels of vibration
reactance the complex component of impedance associated with energy being stored and converted from one form to another (e.g. from potential to kinetic, or from electrostatic to electromagnetic) rather than being converted to heat
reactive silencer a silencer which reduces sound levels by using changes in impedance instead of sound absorbing materials
real time, in quickly enough to observe changes in a situation as they happen
real-time analyser a device which is capable of analysing signals (usually in the frequency domain) in real time
recruitment an aspect of certain forms of perceptive deafness; an abnormally rapid increase in the sensation of loudness with increasing sound pressure level
reference time interval the time interval over which a noise index is measured or calculated for assessment purposes in BS 4142 and BS 7445
reference value standardized values used as the basis for decibel scales of sound pressure, sound intensity, sound power, vibration acceleration, velocity and displacement
reflection the redirection of waves which occurs at a boundary between media when the size of the boundary interface is large compared with the wavelength
refraction the change in direction of waves caused by changes in the wave velocity in the medium
repeatability the variability of measurements when repeated
residual noise the ambient noise remaining when the specific is suppressed, defined in BS 4142 and BS 7445; see also specific noise and ambient noise
resonance the situation in which the amplitude of forced vibration of a system reaches a maximum at a certain forcing frequency (called the resonance frequency)
resonance frequency the frequency at which resonance occurs, i.e. at which the forced vibration amplitude in response to a force of constant amplitude is a maximum; for an undamped system the resonance frequency is the same as the natural frequency of the system; for a damped system the resonance frequency is slightly reduced
reverberant room a standard acoustic test environment designed to produce diffuse sound conditions throughout the space
reverberant sound/reverberation the sound in an enclosed space which results from repeated reflections at the boundaries
reverberant sound field the region in an enclosed space in which the reverberant sound is the major contributor to the total sound pressure level
reverberation time the time required for the steady sound pressure level in an enclosed space to decay by 60 dB, measured from the moment the sound source is switched off
ringing transient free vibration of bodies caused by impact
room constant, R_c a constant used in the calculation of reverberant sound pressure level in a room: $R_c = S\alpha/(1 - \alpha)$, where S = total area of room surfaces and α = average absorption of room surfaces
room mode a three-dimensional standing-wave sound pressure pattern, i.e. a mode shape, associated with one of the natural frequencies of a room
root mean quad (RMQ) the RMQ value of a set of numbers is the fourth root of the average of the fourth powers of the numbers; for a vibration waveform the RMQ value over a given time period is the fourth root of the average value of the fourth power of the waveform over that time period
root mean square (RMS) the RMS value of a set of numbers is the square root of the average of their squares; for a sound or vibration waveform the RMS value over a given time period is the square root of the average of the square of the waveform over that time period
round window a diaphragm or membrane at the end of the cochlea which connects with the middle ear

S (slow) time weighting one of the standard averaging times for sound level meter displays, defined in BS 5969
Sabine's formula a formula for predicting reverberation times of rooms
sabin unit of sound absorption; one sabin is the amount of absorption equivalent to one square metre of perfect absorber
sampling frequency of a digitized signal; the number of samples per second
sampling interval the time interval between samples
scalar a quantity that may be completely defined by its magnitude alone, i.e. it has no direction
semi-anechoic a room with anechoic walls and ceiling, but with a sound-reflecting floor
semi-reverberant a room which is neither completely anechoic nor reverberant, but somewhere in between
sensation value of a specified sound; the sound pressure level when the reference sound pressure corresponds to the threshold of hearing for the sound
sensitivity of a transducer; the ratio of output to input, e.g. for a microphone the sensitivity is measured as output voltage (V)/input pressure (Pa)
shear wave a transverse wave of shear stress propagating in an elastic medium

shock a sudden transient disturbance to a vibrating system caused by a rapid change in force, displacement, velocity or acceleration
signal-to-noise ratio a measure of the strength of a signal, indicating its magnitude relative to the background electrical noise in the measurement system; usually expressed in decibels
silencer a device for reducing noise in air and gas flow systems; silencers are either absorptive or reactive; also called attenuators or mufflers
simple harmonic motion a single frequency vibration, i.e. one in which the displacement varies sinusoidally with time
simple source an idealized model of an acoustic source, which radiates spherical waves, under free-field conditions; see also under point source, monopole
sine wave the graph of a sinusoidal function, which indicates the simplest possible repeating waveform, characterized by a single frequency and constant amplitude
single degree of freedom system a vibrating system consisting of only one mass, one spring and one dashpot (damper); such a system has one natural frequency and one mode of vibration; its motion can be completely described by one variable
sinusoidal relating to a sine wave
snubber a device used to restrict the relative displacement of a vibrating system
sociocusis hearing loss arising from everyday activities
sone the unit of loudness; the tone scale is devised to give numbers which are approximately proportional to the loudness; it is related to the phon scale as follows: $P = 40 + 10 \log_2 S$, where P represents phons and S represents sones
sound (1) pressure fluctuations in a fluid medium within the (audible) range of amplitudes and frequencies which excite the sensation of hearing; (2) the sensation of hearing produced by such pressure fluctuations
sound absorbing material material designed and used to maximize the absorption of sound by promoting frictional processes; the most commonly used materials are porous, such as mineral fibre materials or certain types of open-cell foam polymer materials
sound absorption (1) the process whereby sound energy is converted into heat, leading to a reduction in sound pressure level; (2) the property of a material which allows it to absorb sound energy
sound absorption coefficient a measure of the effectiveness of materials as sound absorbers; it is the ratio of the sound energy absorbed or transmitted (i.e. not reflected) by a surface to the total sound energy incident upon that surface; the value of the coefficient varies from 0 (for very poor absorbers and good reflectors) to 1 (for very good absorbers and poor reflectors)
sound exposure level, SEL (L_{AE}) a measure of A-weighted sound energy used to describe noise events such as the passing of a train or aircraft; it is the A-weighted sound pressure level which, if occurring over a period of one second, would contain the same amount of A-weighted sound energy as the event
sound insulating material material designed and used as partitions in order to minimize the transmission of sound; the best materials are those which are dense and solid, such as wood, metal or brick, although lightweight panels can also be effective when in the form of double-skin constructions
sound insulation the reduction or attenuation of airborne sound by a solid partition between source and receiver; this may be a building partition (e.g. a floor, wall or ceiling), a screen or barrier, or an acoustic enclosure
sound intensity the sound power flowing per unit area, in a given direction, measured over an area perpendicular to the direction of flow; its units are W/m^2
sound intensity level, L_I sound intensity measured on a decibel scale: $L_I = 10 \log_{10} (I/I_0)$, where I_0 is the reference value of sound intensity, 10^{-12} W/m^2
sound level a frequency-weighted sound pressure level, such as the A-weighted value
sound level meter an instrument for measuring sound pressure levels
sound power the sound energy radiated per unit time by a sound source, measured in watts (W)
sound power level, L_W sound power measured on a decibel scale: $L_W = 10 \log_{10} (W/W_0)$, where W_0 is the reference value of sound power, 10^{-12} W
sound pressure the fluctuations in air pressure, from the steady atmospheric pressure, created by sound, measured in pascals (Pa)
sound pressure level, SPL (L_p) sound pressure measured on a decibel scale: $L_p = 20 \log_{10} (p/p_0)$, where p_0 is the reference sound pressure, 20×10^{-6} Pa
sound propagation the transmission or transfer of sound energy from one point to another
sound reduction index, R a measure of the airborne sound insulating properties, in a particular frequency band, of a material in the form of a panel or partition, or of a building element such as a wall, window or floor; it is measured in decibels: $R = 10 \log_{10} (1/t)$, where t is the sound transmission coefficient; it is measured under laboratory conditions according to BS 2750; also known as transmission loss
sound transmission the transfer of sound energy across a boundary from one medium to another
sound transmission coefficient the ratio of the sound energy transmitted by a partition, or across a boundary, to the sound energy incident upon the partition or the boundary

sound wave a pressure wave in a fluid which transmits sound energy through the medium by virtue of the inertial, elastic and damping properties of the medium
specific acoustic impedance at a point in a sound field; the complex ratio of sound pressure to the acoustic particle velocity
specific noise the particular component of the ambient noise which is under consideration or investigation, e.g. in connection with a planning application or noise complaint; defined in BS 4142
spectrum a frequency spectrum is a graph showing variation of sound pressure level (or other quantity) with frequency
speech intelligibility the ability of speech to be understood; the concept of intelligibility is used as a criterion to determine the degree of acoustic privacy between rooms
speech interference level a measure of the ambient noise level in offices which gives an indication of the degree to which speech will be intelligible; it is based on the arithmetic mean of the octave-band sound pressure levels 500 Hz, 1000 Hz and 2000 Hz, which are most significant for good speech intelligibility
spherical waves an idealized model of how sound propagates in free-field conditions, and used as the basis of certain sound level prediction methods
standard deviation a measure of the deviation or scatter of a set of values (e.g. sound pressure level measurements) from the mean value
standardized impact sound pressure level, L'_{nT} a measurement of impact sound insulation, corrected according to BS 2750 for room characteristics; a complete set of measurements consists of 16 values, one for each third-octave frequency band from 100 Hz to 3150 Hz
standardized level difference, D_{nT} a measurement of airborne sound insulation, corrected according to BS 2750 for receiving room characteristics; a complete set of measurements consists of 16 values, one for each third-octave frequency band from 100 Hz to 3150 Hz
standing waves a wave system characterized by a stationary pattern of amplitude distribution in space arising from the interference of progressive waves; also called stationary waves
stapes or stirrup; one of the three bones of the middle ear, connected to the oval window of the inner ear
static deflection the deflection produced in the spring of a mass–spring system by the weight of the mass; it is related to the natural frequency of the system and is used to specify the stiffness of springs for vibration isolation
stationary waves see under standing waves
steady noise noise for which the fluctuations in time are small enough to permit measurement of average sound pressure level to be made satisfactorily without the need to measure L_{Aeq} using an integrating sound level meter; defined in BS 4142
strain the fractional change in shape due to an elastic deformation in a material caused by an applied stress
stress force per unit area, measured in N/m^2; stress applied to elastic materials causes strain
structure-borne sound sound which reaches the receiver after travelling from the source via a building or machine structure; structure-borne sound travels very efficiently in buildings, and is more difficult to predict than airborne sound
subjective depending upon the response of the individual
superposition according to the principle of superposition, the wave disturbances in a medium caused by different sources may be combined algebraically

tangential mode a room mode which involves reflections between two pairs of parallel surfaces (e.g. walls)
temporary threshold shift the component of threshold shift which shows progressive reduction with the passage of time, when the apparent source has been removed
threshold of hearing for a given listener the lowest sound pressure level of a particular sound that can be heard under specified measurement conditions, assuming the sound reaching the ears from other sources is negligible
threshold of pain for a given listener the minimum sound pressure level of a specified sound which will produce the sensation of pain in the ear
threshold shift the deviation, in decibels, of a measured hearing level from one previously established
timbre the quality of a sound which is related to its harmonic structure
time constant of a process or quantity which decays exponentially with time; the time required for the value to reduce by a factor of 1/e, where e is the exponential number 2·7183 . . .
time weighting one of the standard averaging times (F, S, I) used for the measurement of RMS sound pressure level in sound level meters, specified in BS 5969
tinnitus a subjective sense of noises in the head or ringing in the ears for which there is no observable cause
tone a sound which produces the sensation of pitch; see also under pure tone
traffic noise index an index used for the assessment of environmental noise in the 1960s, based on a combination of the L_{A10} value and the L_{A90} value; it is now rarely used
transducer a device for converting signals from one form to another; frequently, the requirement is to convert changes in some physical variable, such as temperature or sound pressure, into analogous changes in electrical voltage or charge
transfer function of a vibrating system; the ratio of the output or response of the system to the input excitation, usually expressed as a complex function of frequency
transfer standard a calibrated noise source designed to

fit over a microphone
transient a noise or vibration signal which is not continuous but which decreases to zero then remains zero
transmissibility of a vibrating system; the non-dimensional ratio of vibration amplitude at two points in the system; frequently, the two points are on either side of springs used as antivibration mounts, and the transmissibility is used as an indicator of the effectiveness of the isolation
transmission coefficient see under sound transmission coefficient
transmission loss see under sound reduction index
transverse sensitivity of an accelerometer; the sensitivity to vibration in a direction perpendicular to the axis of the accelerometer
transverse wave a wave in which the direction of vibration of the particles of the medium is perpendicular to the direction of wave travel; an example is a shear wave in a solid medium
triboelectric effect the production of electric charge as a result of vibration in an accelerometer cable, leading to electrical noise in the accelerometer signal, unless the cable is secured to prevent movement

ultrasonics the study of ultrasound
ultrasound acoustic waves with frequencies which are too high to be heard by human ears
unweighted sound pressure level a sound pressure level which has not been frequency weighted, sometimes known as the linear sound pressure level; symbol L_p

vector a quantity which has a direction as well as a magnitude, e.g. force, displacement, velocity and acceleration
velocity the rate of change of displacement, measured in m/s or mm/s
vibration a to-and-fro motion; a motion which oscillates about a fixed equilibrium position
vibration dose value, VDV a measure of vibration exposure; the fourth root of the integral, over the measurement period, of the fourth power of the frequency-weighted time-varying acceleration; see also under eVDV and RMQ
vibration isolation the reduction of vibration and structure-borne sound by the use of resilient materials inserted in the transmission path between source and receiver
vibration white finger blanching of the fingers and other symptoms caused by exposure to hand-transmitted vibration; see also under Raynaud's disease
viscous damping damping of the sort which occurs in viscous fluid layers, in which the damping force is proportional to the velocity of the fluid element
volt (V) the unit of electrical potential; the difference in electrical potential between two points on an electric conductor which is carrying a constant electric current of one ampere (A) when the power dissipated between the points is one watt (W)

watt (W) the unit of power; the power dissipated when one joule of energy is expended in one second
wave in an elastic medium; a mechanism whereby a disturbance, and the energy associated with it, is propagated through an elastic medium; the disturbance results in vibrations of the particles of the medium, vibrations transmitted to nearby regions as a result of the elastic and inertial nature of the medium, resulting in a disturbance which is a function of both position and time
waveform a graph showing how a variable at one point in a wave (e.g. sound pressure or particle velocity) or vibration varies with time
wavefront the leading edge of a progressive wave, along which the vibration of the particles of the medium are in phase
wavelength the minimum distance between two points that are in phase within a medium transmitting a progressive wave
Weber–Fechner law a law of psychology which states that the change of subjective response to a physical stimulus is proportional to the logarithm of the stimulus
weighted sound reduction index, R_W a single-figure value of sound reduction index, derived according to procedures given in BS 5821, used for rating and comparing partitions and based on the values of sound reduction index at different frequencies
weighted standardized impact sound pressure level, $L'_{nT,W}$ a single-figure value of impact sound insulation performance, derived according to procedures in BS 5821, used for comparing and rating floors and based on the values of L'_{nT} at different frequencies; values of $L'_{nT,W}$ are specified in the Building Regulations
weighted standardized level difference, $D_{nT,W}$ a single-figure value of airborne sound insulation performance, derived according to procedures given in BS 5821, used for rating and comparing partitions and based on the values of D_{nT} at different frequencies; values of $D_{nT,W}$ are specified in the Building Regulations
weighting see under frequency weighting and time weighting
white finger see under vibration white finger
white noise a random broadband noise which contains equal power per unit bandwidth, so it has a flat, i.e. horizontal, frequency spectrum when plotted on a linear frequency scale (cf. pink noise)
whole-body vibration vibration transmitted to the body as a whole

APPENDIX 2
List of formulae

Chapter 1

Relationship between frequency, sound velocity and wavelength:

$V = f\lambda$

Simple harmonic motion formulae:

displacement $s = a \sin 2\pi ft$

velocity $= 2\pi fa \cos 2\pi ft$

acceleration $= -4\pi^2 f^2 a \sin 2\pi ft$

Weber and Fechner relationships between subjective response and stimulus:

$$\delta R \propto \frac{\delta S}{S}$$

$$R = k \log S$$

Intensity pressure and decibel relationships:

$$\text{dB} = 20 \log_{10} \frac{p_1}{p_0}$$

$$I \propto P^2$$

$$\frac{I_1}{I_0} = \left(\frac{p_1}{p_0}\right)^2$$

$$\text{dB} = 10 \log_{10} \frac{I_1}{I_0}$$

$$I = \frac{P^2}{\rho c}$$

Relationships involving sound power and sound power level:

$$L_W = 10 \log_{10} \frac{W}{W_0}$$

$$\text{SPL} = L_W - 20 \log_{10} r - 11$$

(8 in place of 11 for hard ground)

$$I = \frac{W}{4\pi r^2} \quad \text{for spherical propagation}$$

$$I = \frac{W}{2\pi r^2} \quad \text{for hemispherical propagation}$$

Combining decibels:

$$L_T = 10 \log_{10} (10^{L_1/10} + 10^{L_2/10} + \ldots + 10^{L_n/10})$$

Calculating L_{eq}:

$$L_{eq} = 10 \log_{10} (t_1 \times 10^{L_1/10} + t_2 \times 10^{L_2/10} + \ldots + t_n \times 10^{L_n/10})/T$$

Chapter 2

Calculation of total loudness in sones:

$$S_t = S_m + F(\Sigma S - S_m)$$

Calculation of noise exposure and presbycusis loss:

$$E = L + 10 \log_{10} T$$

Presbycusis loss:

$$= \frac{K}{1000} (N - 20)^2$$

Continuous equivalent noise level (L_{eq}):

$$E_i = \frac{\Delta t_i}{40} 10^{0 \cdot 1(L_i - 70)}$$

$$L_{eq} = 70 + 10 \log_{10} \Sigma E_i$$

Calculation of PNdB:

$$N = N_{max} + 0 \cdot 3(\Sigma N - N_{max})$$

$$N = 2^{(x-40)/10}$$

Noise and number index:

$$\text{NNI} = \text{average peak noise level} + 15 \log_{10} N - 80$$

$$\text{average peak noise level} = 10 \log_{10} \frac{1}{N} \sum_{1}^{N} 10^{L/10}$$

Chapter 3

Stephens and Bate formula for optimum reverberation time:

$$t = r(0{\cdot}012\sqrt[3]{V} + 0{\cdot}107)$$

Sabine's formula:

$$t = \frac{0{\cdot}16V}{A}$$

Eyring's formula:

$$t = \frac{0{\cdot}16V}{-S \log_e (1 - \bar{\alpha})}$$

Measurement of absorption coefficient by the reverberant-chamber method:

$$\delta A = 0{\cdot}16V\left(\frac{1}{t_2} - \frac{1}{t_1}\right)$$

$$\alpha = \frac{\delta A}{S}$$

Measurement of absorption coefficient by the impedance-tube method:

$$\alpha = \frac{4A_1A_2}{(A_1 + A_2)^2}$$

Resonant frequency of panel and Helmholtz resonators:

$$f = \frac{60}{\sqrt{md}}$$

$$f = \frac{cr}{2\pi}\sqrt{\left[\frac{2\pi}{(2l + \pi r)V}\right]}$$

Chapter 4

Velocity of sound in a bar:

$$V_1 = \sqrt{\frac{E}{\rho}}$$

Velocity of sound in plates:

$$V = \sqrt{\left[\frac{E}{\rho(1 - \mu^2)}\right]}$$

Resonant frequency of partitions:

$$f = 0{\cdot}45V_L h\left[\left(\frac{N_x}{l_x}\right)^2 + \left(\frac{N_y}{l_y}\right)^2\right] \text{Hz}$$

Coincidence frequency:

$$f = \frac{c^2}{1{\cdot}8hV_L \sin^2 \theta}$$

Resonant frequency of mounts:

$$f = \sqrt{\frac{250}{h}}$$

Impact sound insulation octave-band corrected sound pressure level for measurements in one-third octaves:

$$L_c = L_R = 10 \log_{10} \frac{t}{0{\cdot}5} + 5$$

Chapter 5

Airborne sound insulation measurement:

$$D_n = L_S - L_R + 10 \log_{10} \frac{10}{A}$$

$$D_n = L_S - L_R + 10 \log_{10} \frac{t}{0{\cdot}5}$$

Sound insulation of composite partitions:

$$R \text{ (dB)} = 10 \log_{10} \frac{1}{T}$$

$$T_{AV} \times A = T_1 \times A_1 + T_2 \times A_2 + T_3 \times A_3 + \ldots$$

Mass law (approx.):

$$R_{AV} = 10 + 14{\cdot}5 \log_{10} m$$

Reduction by absorbents:

$$R = 10 \log_{10}\left(\frac{A + a}{A}\right)$$

Duct attenuation:

$$R_1 = \left(\frac{P}{S}\right)\alpha^{1{\cdot}4}$$

$$R_2 = 10 \log_{10} \frac{A}{S}$$

Chapter 6

$$\text{TNI} = 4L_{10} - 3L_{90} - 30$$

$$\text{Reduction in TNI} = 45 \log_{10} \frac{d_1}{d_0}$$

$$L_{10}\ (18\ \text{h}) = L_{10}\ (3\ \text{h}) - 1\ \text{dB(A)}$$

Chapter 7

Relationships between displacement velocity and acceleration:

$V = 2\pi f X$

$A = 2\pi f V$

$A = 4\pi^2 f^2 X$

Relationship between peak and RMS values for sinusoidal signals:

$X_{RMS} = \sqrt{2} \,.\, X_{Peak} = 0{\cdot}707 X_{Peak}$

Use of decibels to describe vibration levels:

$$N = 10 \log_{10} \left(\frac{A}{A_0}\right)^2 = 20 \log_{10} \left(\frac{A}{A_0}\right)$$

Natural frequency of a mass–spring system:

$$f_0 = \frac{1}{2\pi} \sqrt{\frac{k}{m}}$$

Angular frequency:

$\omega = 2\pi f$

Frequency of damped natural vibrations:

$f = f_0 \sqrt{(1 - \xi^2)}$

Definitions and relationships between damping terms:

$$\delta = \frac{1}{n} \log_e \frac{x_1}{x_n}$$

$\delta = 2\pi\xi$

$$Q = \frac{f_0}{f_1 - f_2}$$

$$Q = \frac{1}{2\xi}$$

Definitions of force and displacement transmissibility:

$$T = \frac{F_T}{F_0} = \frac{X_T}{X_0}$$

Transmissibility formula:

$$T = \sqrt{\frac{1 + 4\xi^2 (f/f_0)^2}{[1 - (f/f_0)]^2 + 4\xi^2 (f/f_0)^2}}$$

$$T = \frac{1}{(f/f_0)^2 - 1} \quad \text{for } \xi = 0$$

Isolation efficiency:

$\eta = (1 - T) \times 100\%$

Relation between static deflection and natural frequency:

$f_0 = 15{\cdot}8\sqrt{1/X_{st}}$

Sensitivities of an accelerometer:

charge sensitivity = voltage sensitivity × capacitance

$$\text{RMQ} = \left[\int_0^T \frac{a^4(t)\,dt}{T}\right]^{0\cdot25}$$

$$\text{VDV} = \left[\int_0^T a^4(t)\,dt\right]^{0\cdot25}$$

$\text{VDV} = aT^{0\cdot25} \qquad [a(t) = a]$

Combining events:

$V_T = [(V_1)^4 + (V_2)^4 + \ldots + (V_N)^4]^{0\cdot25}$

N identical events:

$V_T = N^{0\cdot25} \cdot V$

$\text{eVDV} = 1{\cdot}4aT^4 \qquad [a(t) = a]$

$\text{eVDV} = 1{\cdot}4[a_1^4 n_1 t_1 + a_2^4 n_2 t_2 + \ldots + a_N^4 n_N t_N]$

$$A(8) = \left[\frac{1}{8} \int_0^T a_{h,w}^2(t)\,dt\right]^{0\cdot5}$$

$A(8) = (T/8)^{0\cdot5} \cdot A(T) \qquad (T \text{ in hours})$

Chapter 8

Radiation of sound from a vibrating plate:

$p = ZV$

Definition of directivity factor, directivity index:

$$Q = \frac{I}{I_{av}}$$

$D = \text{SPL} - \text{SPL}_{av}$

$D = 10 \log_{10} Q$

Free-field radiation from a source:

$\text{SPL} = L_W - 20 \log_{10} r - 11 + D$

$L_W = \text{SPL} + 20 \log_{10} r + 11 - D$

Averaging decibels:

$$L_{AV} = 10 \log_{10} \left(\frac{10^{L_1/10} + 10^{L_2/10} + \ldots + 10^{L_n/10}}{n}\right)$$

Reverberant field sound pressure level:

$$\text{SPL} = L_W + 10 \log_{10} \frac{4}{R_c}$$

$$R_c = \frac{S\bar{\alpha}}{1 - \bar{\alpha}} \simeq S\bar{\alpha} = A \quad (\text{for small } \bar{\alpha})$$

Total sound pressure level in a room:

$$SPL = L_W + 10 \log_{10}\left[\frac{Q}{4\pi r^2} + \frac{4}{R_c}\right]$$

Sound power measurement by reverberant-room method:

$$L_W = SPL + 10 \log_{10} V - 10 \log_{10} T - 14$$

Definition of sound reduction index:

$$R = 10 \log_{10}\left(\frac{1}{t}\right)$$

Sound transmission between two adjacent rooms:

$$SPL_1 - SPL_2 = R - 10 \log_{10} S + 10 \log_{10} A$$

Sound transmission from indoors to outside (close to partition):

$$SPL_1 - SPL_2 = R - 6$$

Relations between sound intensity, area and sound power:

$$P = I \times S$$

$$L_W = SPL + 10 \log_{10} S$$

Sound transmission from outside to indoors:

$$SPL_{in} = SPL_{out} - R + 10 \log_{10} S - 10 \log_{10} A + K$$

Performance of large enclosures:

Noise reduction $= SPL_{in} - SPL_{out}$

$= R - 10 \log_{10} S + 10 \log_{10} A$ (outside enclosure)

Insertion loss $= SPL_{before} - SPL_{after}$

$= R - 10 \log_{10} S + 10 \log_{10} A$ (inside enclosure)

Pure tones from fans:

$$f_N = \frac{RPM}{60} \times N \times n$$

Chapter 9

Relative microphone sensitivities, in dB:

$$N = 20 \log_{10} (S_2/S_1)$$

Time–bandwidth product theorem:

$$\epsilon = \frac{1}{2\sqrt{BT}}$$

FFT relationships:

number of samples	$= N$
interval between successive samples	$= \Delta T$
total duration of samples	$= T = N\Delta T$
sampling rate	$= N/T$
	$= 1/\Delta T$ samples per second
line spacing	$= \Delta f = 1/T$
number of frequency lines	$= N/2$
maximum frequency, f_{max}	$= N/2 \times 1/T$

Acoustic intensity:

$$I = pu$$

Equation of motion:

$$\rho\frac{\Delta u}{\Delta t} = -\frac{\Delta p}{\Delta x}$$

Acoustic pressure gradient:

$$\frac{\Delta p}{\Delta x} = \frac{p_1 - p_2}{x}$$

APPENDIX 3
Experiments

Experiment 1 Use of a sound level meter to measure sound pressure level and sound level

Apparatus

Sound level meter with linear (or 'C') and 'A' weightings, calibrator, tripod stand.

Method

The meter is set up on the tripod stand and switched on to the 'battery check' position. If the batteries are in full working order the switch is then turned to the 'lin' or 'C' response setting. The attenuator is then adjusted to the nearest '10' below the calibrator level and the calibrator (or pistonphone) placed carefully on the microphone and switched on. The meter should read the calibrator level (i.e. attenuator setting + meter reading = calibrator level). If necessary, the fine-adjustment screw is used to ensure this. The calibrator is removed and the meter on the tripod placed at the position where it is desired to measure a steady noise.

The attenuator is adjusted to obtain a reading between '0' and '10' on the meter. The sound pressure level (SPL) is then the attenuator setting + meter reading. With some meters the attenuator setting is displayed on the meter scale and with digital meters the SPL is read directly. The SPL is noted.

The response is now changed to 'A' weighting and the sound level similarly found.

This should be repeated for different sounds such as an electric drill, stationary diesel engine or motor car, band saw, etc. Any noticeable difference between the types of noise should be observed such as low pitch or high pitch. This may be compared with the differences between dB and dB(A).

Theory

The *sound pressure level* (SPL) of a sound, in decibels, is equal to 20 times the logarithm to the base 10 of the ratio of the RMS sound pressure to the reference sound pressure (2×10^{-5} Pa in air).

Sound level is the 'A' weighted value of the sound pressure level, and is written as dB(A).

The dB(A) scale is an attempt to obtain a reading with a sound level meter which correlates with loudness. The weighting is shown in Fig. A3.1.

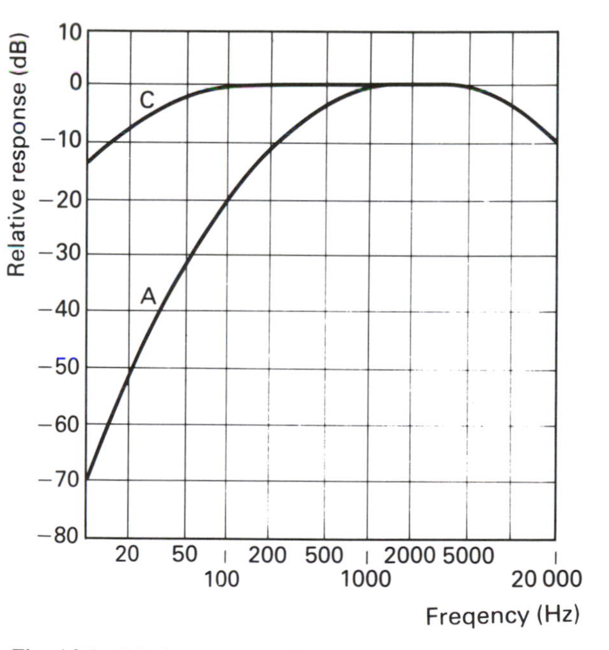

Fig. A3.1 Relative response of A and C weighting scales

Results

Noise source	Measuring distance	Sound pressure level (dB)	Sound level (dB(A))

Experiment 2 Measurement of the L_{10} levels of traffic noise using a sound level meter

Apparatus

Sound level meter (ideally this should be of precision grade but this is not important for the purpose of this experiment), wind shield, tripod stand and acoustic calibrator.

Method

The sound level meter is set up on the tripod stand at the appropriate position. For traffic noise the microphone should be 1·2 m above the ground. The meter is then switched on to the 'battery check' position to ensure that the battery is functioning correctly. The calibrator is then placed on the microphone, the meter attenuator being set at the appropriate position for the calibration. Thus if the calibrator produces 94 dB, the attenuator is set at 90 dB. The meter is now switched to the 'slow' linear position, or if the calibrator is of 1 kHz frequency, to slow 'A' position. Hence the meter response is checked and if necessary adjustment is made using the fine-adjustment screw.

The meter should now be set on 'dB(A) slow' with a suitable attenuation to ensure that the needle does not go beyond the top of the scale for the highest levels of noise to be measured. This attentuation setting is entered in the centre box of the results chart. The wind shield must be used for all external measurements.

The meter reading is taken by glancing at the meter and entering a tick on the chart at the appropriate level (the 'five-bar gate' method is convenient). This is repeated for 15 minutes, ensuring that a reading is taken at least once every 4 seconds. The total counts at each level are obtained and the grand total, as shown in Table A3.1 and thus the level in dB(A) exceeded for 10 per cent of the time (counts) is obtained.

Theory

The L_{10} level is the level of noise in dB(A) exceeded for 10 per cent of the measurement time. For traffic noise this is normally measured 1 m from the façade of dwellings at

Table A3.1

60	Counts	Total counts
1		
2		
3	1	1
4	111	3
5	HHT HHT HHT HHT HHT	25
6	HHT HHT HHT HHT HHT HHT HHT HHT 1	41
7	HHT HHT HHT	15
8	HHT HHT HHT HHT HHT HHT HHT HHT 1	41
9	HHT HHT HHT HHT HHT HHT HHT HHT HHT HHT 111	53
70	HHT HHT HHT HHT HHT HHT HHT 11	37
1	HHT HHT HHT HHT HHT HHT 1111	34
2	HHT HHT HHT HHT 1	21
3	HHT HHT HHT	15
4	HHT 111	8
5	111	3
6	11	2
7	1	1
8		
9		
80		
		Total 300

↑ 30 counts

1·2 m from the ground for ground floors, each hour for the 18 hours from 06·00 hrs to 24·00 hrs. The 18 measurements are then arithmetically averaged to determine the 18 hour L_{10} value.

The L_{10} value is easily obtained by counting 10 per cent of the total counts from the top and reading the level in dB(A) to the nearest 0·5 dB(A)

It will be realized that the levels shown are actually a range of levels, e.g. 72 is a range from 71·5 dB(A) to 72·5 dB(A). Thus the L_{10} shown is (72·5 − 1/21) dB(A) = 72·5 dB(A) to the nearest 0·5 dB(A).

The L_{50} or L_{90} values may be found in a similar manner, where L_{50} = level in dB(A) exceeded for 50 per cent of the measurement time; L_{90} = level in dB(A) exceeded for 90 per cent of the measurement time.

Results

dB(A)		Number of counts							Total counts
	0								
−9	1								
−8	2								
−7	3								
−6	4								
−5	5								
−4	6								
−3	7								
−2	8								
−1	9								
	0								
	1								
	2								
	3								
	4								
	5								
	6								
	7								
	8								
	9								
	0								

Total

L_{10} = ________ dB

Experiment 3 Determination of the sound power level of a machine in a reverberation room

Apparatus

Reverberation room, electric drill, tape measure, microphone, extension cable, audio-frequency spectrometer, level recorder, sine random generator, amplifier (75 to 100 W), 2 loudspeakers (50 W each).

Method

The drill whose sound power level is to be measured is mounted on a retort or tripod stand at least 1·5 m from any wall. The sound level is measured in dB at four different positions around the drill at distances of 0·5 m and the average value obtained. This is repeated at a distance of 1 m and any other distances desired. For all these measurements it is desirable that the microphone is set up in the reverberation room and the audio-frequency spectrometer is remote from it. It is important that the apparatus is calibrated initially.

The reverberation time of the room is measured as described in Experiment 5.

The measurements are repeated in dB(A) and in octave bands with centre frequencies at 250 Hz, 500 Hz, 1kHz and 2 kHz.

Theory

If a sound is produced in a reverberation room, the sound pressure level increases until equilibrium is reached where the rate of production of sound energy equals the rate of absorption by the room surfaces. Then:

$$\text{Sound power level } (L_W) = \text{sound pressure level (SPL)} + 10 \log V - 10 \log T - 14 \text{ dB}$$

where V = volume of room in m^3

T = reverberation time in s

It can be seen that the equation is independent of distance, and the measurement positions do not matter provided they are not too close to the machine or walls. The method described, while not conforming to the international standard, is capable of reasonably accurate results with the minimum of equipment. If a reverberation room is not available, any fairly reverberant room may be used.

The average is obtained by adding together the four sound pressure levels in the usual way and subtracting 6.

(*References:* ISO 3741, 3742, 3743, 3744, 3745, 3756)

Results

1. *dB (linear)*

Position	*SPL (dB)*	*Total*	*Total*
1	________	________	________
2	________	________	________
3	________	________	________
4	________	________	________

∴ average sound pressure level = Total − 6

= ________ − 6

= ________ dB

Volume of room, V = ________ m^3

Reverberation time, T = ________ s

Now: L_W = SPL + $10 \log_{10} V$ − − $10 \log_{10} T$ −14

= ________ + $10 \log_{10}$ − $10 \log_{14}$ −14

= ________ + ________ − ________ −14

= ________ dB

2. *A weighted sound power level*

Position	*SPL (dB)*	*Total*	*Total*
1	________	________	________
2	________	________	________
3	________	________	________
4	________	________	________

∴ average sound pressure level = Total − 6
= ________ − 6
= ________ dB
Volume of room, V = ________ m^3
Reverberation time, T = ________ s
Now: L_W = SPL + 10 $\log_{10}$ V − − 10 $\log_{10} T$ −14
= ________ + 10 $\log_{10}$ − 10 $\log_{10}$ −14
= ________ + ________ − ________ −14
= ________ dB

3. 250 *Hz*

Position	*SPL (dB)*	*Total*	*Total*
1	________	________	________
2	________	________	________
3	________	________	________
4	________	________	________

∴ average sound pressure level = Total − 6
= ________ − 6
= ________ dB
Volume of room, V = ________ m^3
Reverberation time, T = ________ s
Now: L_W = SPL + 10 $\log_{10}$ V − − 10 $\log_{10} T$ −14
= ________ + 10 $\log_{10}$ − 10 $\log_{10}$ −14
= ________ + ________ − ________ −14
= ________ dB

4. 500 *Hz*

Position	*SPL (dB)*	*Total*	*Total*
1	________	________	________
2	________	________	________
3	________	________	________
4	________	________	________

∴ average sound pressure level = Total − 6
= ________ − 6
= ________ dB
Volume of room, V = ________ m^3
Reverberation time, T = ________ s
Now: L_W = SPL + 10 $\log_{10}$ V − − 10 $\log_{10} T$ −14
= ________ + 10 $\log_{10}$ − 10 $\log_{10}$ −14
= ________ + ________ − ________ −14
= ________ dB

5. 1 *kHz*

Position	*SPL (dB)*	*Total*	*Total*
1	________	________	________
2	________	________	________
3	________	________	________
4	________	________	________

∴ average sound pressure level = Total − 6
= ________ − 6
= ________ dB
Volume of room, V = ________ m^3
Reverberation time, T = ________ s

Now: L_W = SPL + 10 $\log_{10}$ V − − 10 $\log_{10} T$ − 14
= ______ + 10 $\log_{10}$ − 10 $\log_{10}$ − 14
= ______ + ______ − ______ − 14
= ______ dB

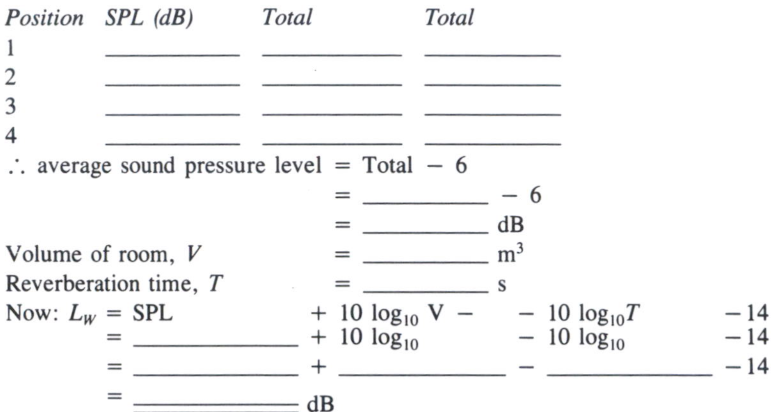

6. *2kHz*

Position	*SPL (dB)*	*Total*	*Total*
1	______	______	______
2	______	______	______
3	______	______	______
4	______	______	______

∴ average sound pressure level = Total − 6
= ______ − 6
= ______ dB

Volume of room, V = ______ m^3
Reverberation time, T = ______ s

Now: L_W = SPL + 10 $\log_{10}$ V − − 10 $\log_{10} T$ − 14
= ______ + 10 $\log_{10}$ − 10 $\log_{10}$ − 14
= ______ + ______ − ______ − 14
= ______ dB

Experiment 4 Determination of the sound power level of a machine under free field conditions outdoors

Apparatus

Sound level meter with octave filters, tape measure, quiet field (empty football pitch or car park) with no reflecting surfaces.

Method

The background level of sound should be measured first at several positions in the area being used and the mean obtained. It should be at least 15 dB quieter than the lowest level to be measured from the machine.

The sound pressure level is measured at four positions at 1 m distance from the machine. The average is obtained. Measurements are repeated at 2 m, 3 m, 4 m, etc., up to four times the original distance. A graph is plotted on logarithmic/linear paper of distance against sound pressure level. The sound power level is calculated for each measurement position plotted on the straight-line part of the graph.

This may be repeated in dB(A) and octave bands.

Theory

Sound power level, $L_W = \log_{10} \dfrac{W}{W_0}$ decibels.

where W = sound power output of the machine in watts
W_0 = reference sound power
$= 10^{-12}\ W$

$$\therefore L_W = 10 \log_{10} W - 10 \log_{10} 10^{-12}$$
$$= 10 \log_{10} W + 120$$

If the sound pressure level of the source is measured in space, then

$$\text{intensity, } I = \frac{W}{4\pi^2}$$

where r = radius of sphere or distance from source

$$\therefore L_W = 10 \log_{10} 4\pi r^2 I + 120$$
$$= 10 \log_{10} I + 10 \log_{10} 4\pi r^2 + 120$$

$$\text{But SPL} = 10 \log_{10} \frac{I}{I_0}$$

where $I_0 = 10^{-12} W/m^2$

$$\therefore L_W = 10 \log_{10} \frac{I}{I_0} + 10 \log_{10} 4\pi r^2$$

$$\therefore L_W = \text{SPL} + 10 \log_{10} 4\pi + 10 \log_{10} r^2$$
$$= \text{SPL} + 10 \log_{10} 12{\cdot}57 + 20 \log_{10} r$$
$$= \text{SPL} + 10{\cdot}97 + 20 \log_{10} r$$
$$= \text{SPL} + 20 \log_{10} r + 11$$

If measurements are taken in the practical situation on a hard reflecting ground, then the sound pressure level would be 3 dB higher and in consequence the equation becomes:

$$L_W = \text{SPL} + 20 \log_{10} r + 8$$

If a graph is plotted on log/linear graph paper of distance against SPL, a straight-line graph should result in the 'far' field of the machine where the gradient is 6 dB/doubling of distance.

Results

Linear

Background SPL = ________ dB

1. *Distance* = 1 *m*

Position	*SPL* (dB)	*Total*
1	________	________
2	________	________
3	________	________
4	________	________

Mean SPL = Total − 6
= ________ − 6
= ________ dB

2. *Distance* = 2 *m*

Position	*SPL* (dB)	*Total*
1	________	________
2	________	________
3	________	________
4	________	________

Mean SPL = Total − 6
= ________ − 6
= ________ dB

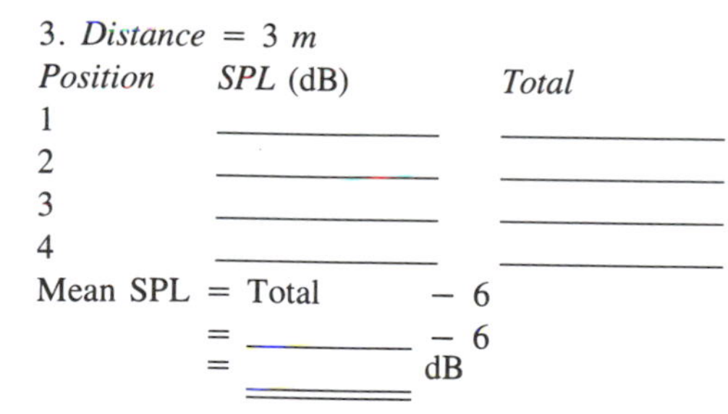

3. *Distance* = 3 *m*

Position	*SPL* (dB)	*Total*
1	________	________
2	________	________
3	________	________
4	________	________

Mean SPL = Total − 6
= ________ − 6
= ________ dB

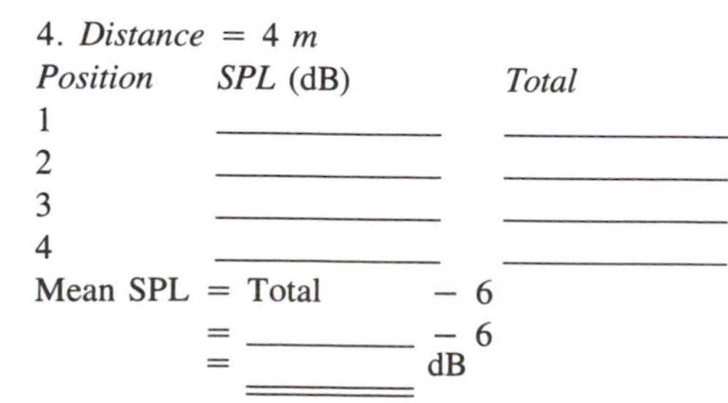

4. *Distance* = 4 *m*

Position	*SPL* (dB)	*Total*
1	________	________
2	________	________
3	________	________
4	________	________

Mean SPL = Total − 6
= ________ − 6
= ________ dB

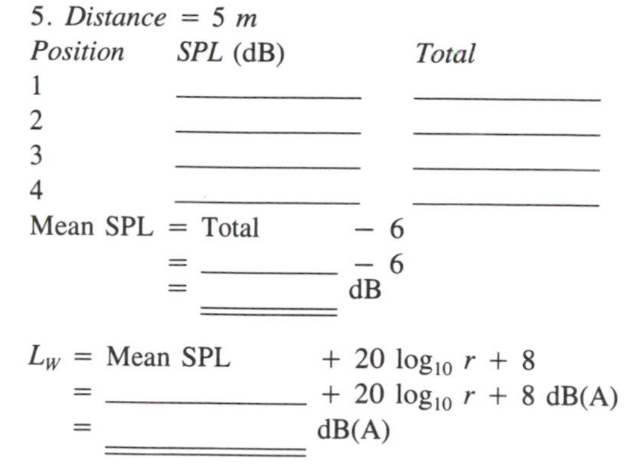

5. *Distance* = 5 *m*

Position	*SPL* (dB)	*Total*
1	________	________
2	________	________
3	________	________
4	________	________

Mean SPL = Total − 6
= ________ − 6
= ________ dB

L_W = Mean SPL $+ 20 \log_{10} r + 8$
= ________ $+ 20 \log_{10} r + 8$ dB(A)
= ________ dB(A)

Experiment 5 Measurement of the reverberation time of a room using white noise (Method 1)

Apparatus

White-noise generator or tape recorder with white noise recorded, amplifier, two 50 W loudspeakers, condenser microphone, audio-frequency spectrometer (third-octave filters) and level recorder.

Method

The apparatus is set up as shown in Fig. A3.2. The white-noise generator is set to produce a bandwidth greater than one-third of an octave centred around 100 Hz. By adjustment of the amplifier, a sound pressure level of about 100 dB is produced within the room. A suitable microphone position is selected greater than 1 m from any surface and the frequency analyser set at a one-third octave centred around 100 Hz. The level recorder using a 50 dB potentiometer is then adjusted to give full-scale deflection for the sound pressure level produced. With the recording paper running at 30 mm/s (10 mm/s in a room with a very long reverberation time), the noise is switched off instantly. The decay line is recorded on the moving paper, from which the reverberation time may be calculated or determined using protractors.

This procedure should be repeated in one-third octave intervals from 100 Hz to 3150 Hz. The microphone is then moved to other room positions and the whole procedure repeated, six times for frequencies below 500 Hz and three times for frequencies of 500 Hz and above.

Theory

Reverberation time is the time in seconds for a 60 dB decay in the sound pressure level after the source has been switched off.

Standing waves make accurate measurement difficult,

particularly for low frequencies, and measurements need to be made at several different room positions. It is for this reason that white noise is used.

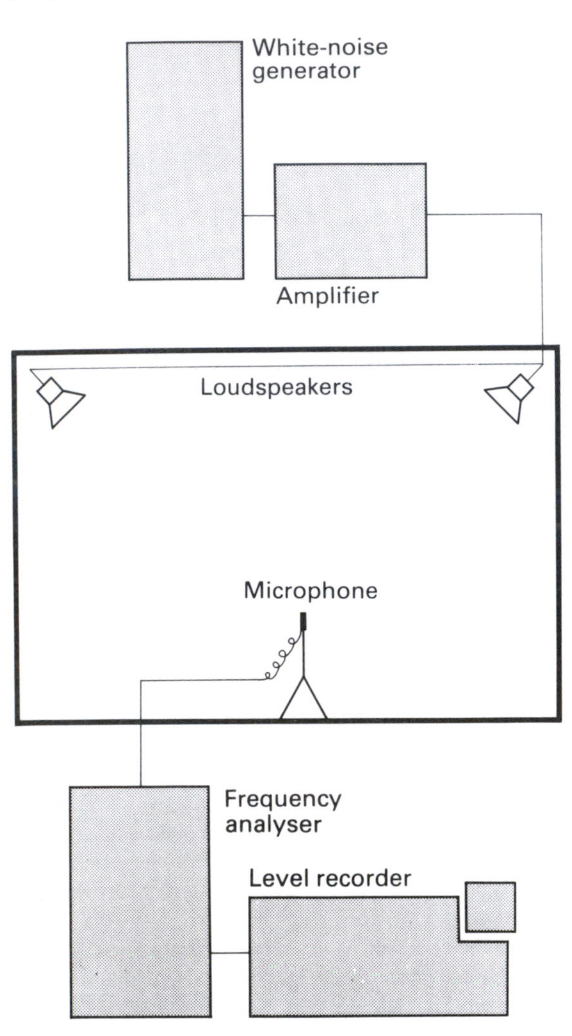

Fig. A3.2 Measurement of reverberation time using a white-noise generator as sound source

Results

Centre frequency (Hz) $\frac{1}{3}$-octave bandwidth	Reverberation time (s) 1	2	3	4	5	6	Mean value
100							
125							
160							
200							
250							
315							
400							
500							
630							
800							
1000							
1250							
1600							
2000							
2500							
3150							

Experiment 6 Measurement of the reverberation time of a room using a starting pistol (Method 2)

Apparatus

Pistol with blank cartridges (a starting pistol is suitable for small rooms), condenser microphone, audio-frequency spectrometer or third-octave filters and level recorder. (Ear defenders.)

Methods

The apparatus is set up as shown in Fig. A3.3. A suitable microphone position is selected greater than 1 m from any surface and the frequency analyser set at one-third octave centred around 100 Hz. The level recorder using a 50 dB potentiometer is then adjusted to give full-scale deflection for the sound produced by the pistol. With the recording paper running at 30 mm/s, the pistol is fired. The decay line is recorded on the moving paper from which the reverberation time may be calculated or determined using protractors.

This procedure should be repeated in one-third octave intervals from 100 Hz to 3150 Hz. The microphone is then moved to other room positions and the whole procedure repeated, six times for frequencies below 500 Hz and three times for frequencies of 500 Hz and above.

It is undoubtedly preferable for those concerned to wear ear defenders during these measurements.

Theory

Reverberation time is the time in seconds for a 60 dB decay in the sound pressure level after the source has been switched off.

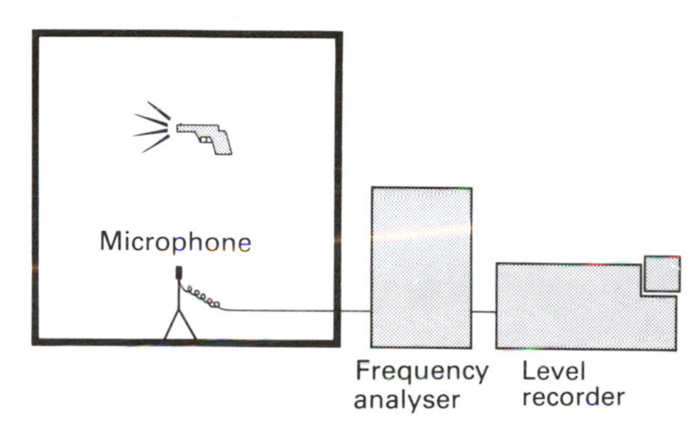

Fig. A3.3 Measurement of reverberation time using a pistol as sound source

Results

Centre frequency (Hz) $\frac{1}{3}$-octave bandwidth	Reverberation time (s)						Mean value
	1	2	3	4	5	6	
100							
125							
160							
200							
250							
315							
400							
500							
630							
800							
1000							
1250							
1600							
2000							
2500							
3150							

Experiment 7 Measurement of the airborne sound insulation of a partition

Apparatus

White-noise generator or tape recorder with white noise recorded, amplifier (about 100 W), two 50 W loud-speakers, 2 microphones, extension leads, switchbox, third-octave filters, microphone stands, level recorder.

Method

The apparatus is set up as shown in Fig. A3.4 with the two microphones at position 1 in each room. Each of the microphone positions is chosen so as to be at least 1 m from any large object. White noise is produced at a level of approximately 100 dB in the source room. The filters are adjusted to one-third octave centred around 100 Hz and the

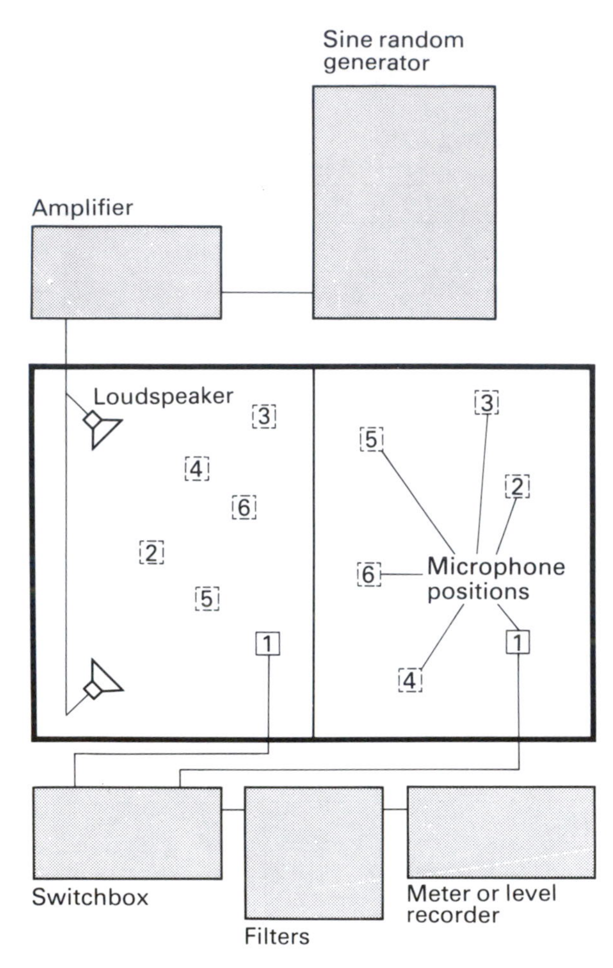

Fig. A3.4 Measurement of airborne sound insulation

levels in the source and receiving room recorded on paper. This is then repeated for each third-octave bandwidth up to 3150 Hz.

The whole is then repeated for a total of six microphone positions in each room up to and including 400 Hz, and three positions each above this frequency.

The mean sound reduction at each frequency is calculated.

Reverberation time is measured as described in Experiment 5 and the correction term applied to obtain the sound reduction index for the partition for each bandwidth.

The results may be compared with the airborne sound insulation requirements of the Building Regulations (Part E).

Theory

Normalized level difference in dB

= sound pressure level in the source room
− sound pressure level in the receiving room
+ correction for the absorption provided by the receiving room

or Normalized level difference

$$= L_S - L_R + \text{correction term}$$

For dwellings, the correction term is

$$+ 10 \log_{10} \frac{T}{0 \cdot 5}$$

where T = reverberation time of the receiving room in seconds

∴ Normalized level difference

$$= L_S - L_R + 10 \log_{10} \frac{T}{0 \cdot 5}$$

(*Reference:* Building Regulations, BS 2750:1993)

Results

$\frac{1}{3}$-octave centre frequency (Hz)	dB difference			Mean diff. (dB) (1)	Reverberation times (s)	$10 \log_{10} \frac{T}{0 \cdot 5}$ (2)	Normalized sound level difference (dB) (1) + (2)
100							
125							
160							
200							
250							
315							
400							
500							
630							
800							
1000							
1250							
1600							
2000							
2500							
3150							

Experiment 8 Measurement of the impact sound insulation of a floor

Apparatus

Footsteps machine, sound level meter, third-octave filters, calibrator, tripod stand, level recorder.

Method

The apparatus is set up as shown in Fig. A3.5 with the footsteps machine near the centre of the floor whose insulation is to be measured. If the floor is liable to be damaged by the metal feet they are removed and replaced by the rubber ones and then adjusted for height. The sound level meter with attached filters is calibrated using the acoustic calibrator, and set up on the tripod stand. The microphone should be more than 1 m from any large objects or people.

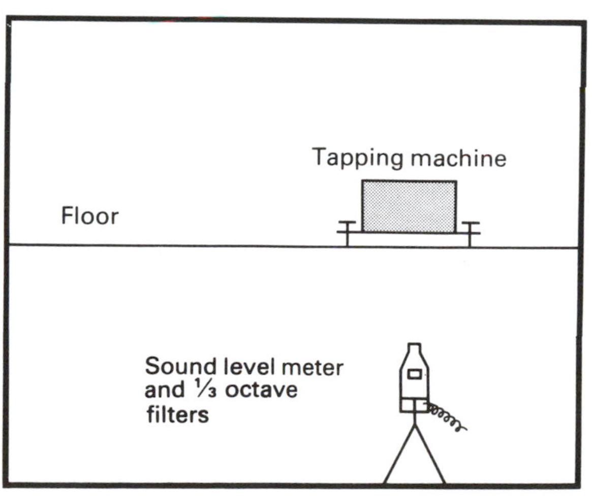

Fig. A3.5 Measurement of impact sound insulation of a floor

The level produced in the receiving room by the footsteps machine is then measured in third-octave bandwidths from 100 Hz to 3150 Hz centre frequencies. This is repeated for six different microphone positions up to 400 Hz, and three positions above this. If the range is greater than 6 dB at any frequency then further readings are taken to reduce the standard deviation.

The reverberation time of the receiving room is then measured as described in Experiment 5 to enable the correction term to be applied.

Theory

Standardized sound pressure level in the receiving room in dB = measured sound pressure level in dB − correction terms.

The correction term allows for the absorption of receiving room.

For dwellings the correction is $10 \log_{10} \dfrac{T}{0{\cdot}5}$

$$\therefore \text{ standardized SPL in dB } = L - 10 \log_{10} \frac{T}{0{\cdot}5}$$

where L = measured SPL in receiving room
T = receiving room reverberation time in seconds

(*Reference:* Building Regulations Part E, BS 2750:1980)

Results

Frequency (Hz)	Readings dB difference		Mean (dB)	Reverberation times T	$10 \log_{10} 2T$	Rec. room level Corr. $\frac{1}{3}$-oct.	Rec. room level
100							
125							
160							
200							
250							
315							
400							
500							
630							
800							
1000							
1250							
1600							
2000							
2500							
3150							

Experiment 9 To investigate the way in which loudness varies with frequency

Apparatus

Two audio-frequency generators, two-way switch, amplifier, loudspeaker, sound level meter, tripod stand, a reasonably quiet room.

Method

The apparatus is set up as shown in Fig. A3.6 with the subject seated so that he can conveniently reach the switchbox and the gain controls on the generators. The sound level meter should be set up on its tripod stand to measure as nearly as possible the sound as heard at the subject's ears.

One generator is set to produce a frequency of 1000 Hz and adjusted to produce a sound pressure level of 20 dB at the meter. (If the room is too noisy start at 40 dB.) The second generator is set to 500 Hz and using its gain control the level adjusted until judged equally loud with the 1000 Hz note. This may be compared by quickly switching back and forth between the two generators. The sound pressure level for equal loudness is noted. This is repeated for 250 Hz, 125 Hz, 63 Hz, 2000 Hz, 4000 Hz and 8000 Hz.

The first generator is then adjusted to 40 dB at 1000 Hz and the experiment repeated. The same may be repeated for 1000 Hz levels of 60 dB, 80 dB and possibly 100 dB. (At 100 dB it is advisable that students do not spend more than about half an hour because of potential hearing damage, apart from nuisance to others and temporary threshold shift which will invalidate the results.)

A graph is plotted of equal loudness which may be compared with the equal-loudness contours shown in Fig. A3.7.

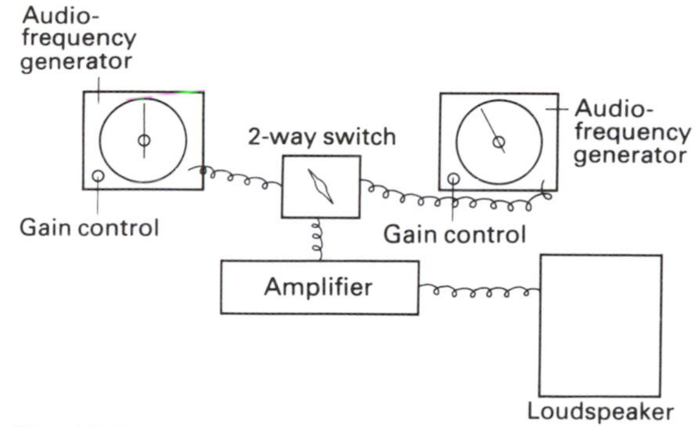

Fig. A3.6

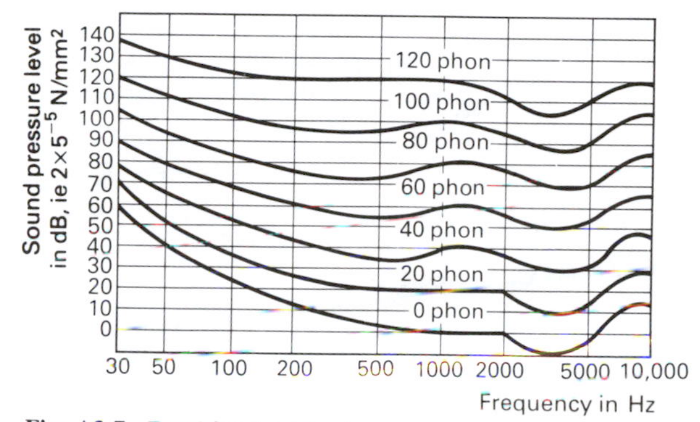

Fig. A3.7 Equal-loudness contours

Theory

Loudness is defined as an observer's auditory impression of the strength of a sound. The loudness of a sound depends upon its amplitude and frequency. Thus for equal loudness at different frequencies the amplitude will have to change. The *phon* is the unit of loudness level for a pure tone of 1000 Hz measured in decibels above 2×10^{-5} N/m^2.

Thus the number of phons for any pure tone is the sound pressure level of a 1000 Hz pure tone which is equally loud.

Results

Sound pressure levels (dB)

1000 Hz	63 Hz	125 Hz	250 Hz	500 Hz	2000 Hz	4000 Hz	8000 Hz
20 dB							
40 dB							
60 dB							
80 dB							
100 dB							

Experiment 10 Determination of the L_{10} levels of traffic noise using a calibrated tape recording

Apparatus

Sound level meter, tripod stand, calibrator, portable tape recorder (e.g. Uher 4000 or Nagra have suitably good response), 13 cm, 270 m long-play tape, connecting leads (Fig. A3.8). High-speed level recorder, statistical distribution analyser.

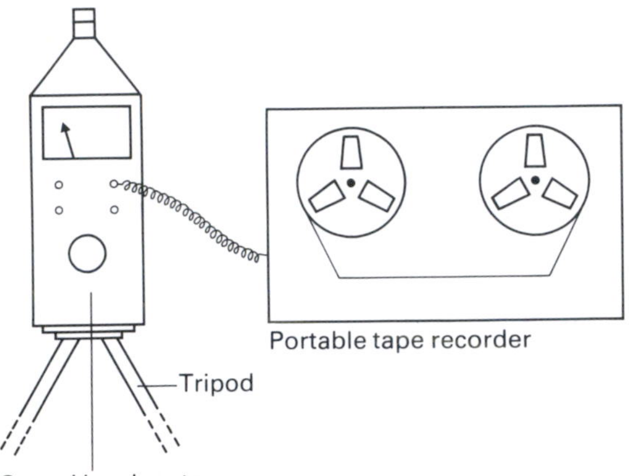

Fig. A3.8 Analysis of a sound that has been previously recorded on tape

Method

The sound level meter and tape recorder are set up as shown in Fig. A3.8 at the position where it is required to find the L_{10} value. The meter is switched on and the batteries checked. The weighting switch is then turned to the 'A' position and the attenuator adjusted so that the needle is indicating on the meter without overloading. This is not easy with some traffic flows and it is best not to rush this part. Having achieved the best attenuator setting for the traffic flow, the tape recorder is switched on to record with the pause button on. The recording level is then adjusted so that the tape recorder does not overload.

The calibrator of known sound level (in dB(A)) is placed over the microphone, the meter attenuator adjusted appropriately (e.g. 90 for a 94 dB(A) level or 110 for a 115 dB(A) level. The meter is checked to see that it is reading correctly and then a recording is made of the calibration signal for about 15 s. After this the information is recorded on tape to include:

(i) the calibrator level in dB(A);
(ii) the attenuation at the meter during calibration;
(iii) the attenuation setting for the traffic noise;
(iv) place, time, etc.

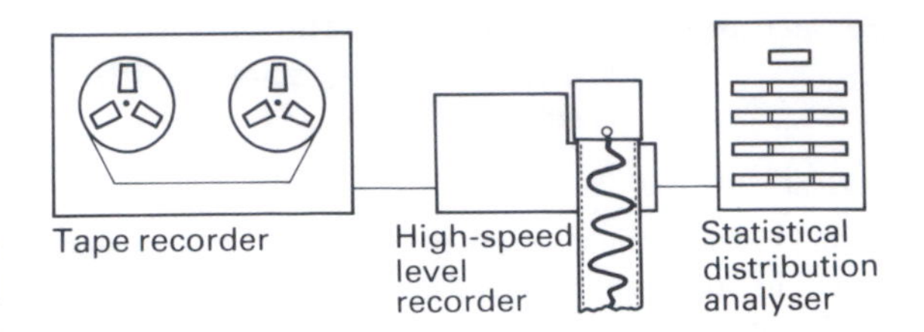

Fig. A3.9 Measurement of reverberation time using a beat-frequency oscillator or white-noise generator as sound source

It is important that no change is made to the tape recorder level control.

The meter attenuator is now readjusted to that which was found suitable for the traffic concerned and a 15-minute recording made.

In the laboratory the analysis apparatus is set up as shown in Fig. A3.9. The level recorder is set to a writing speed of 100 or 160 mm/s and paper speed to 0·3 mm/s with a 50 dB dynamic range. The statistical distribution analyser should be set to count at, say, 0·1 s intervals. The calibration signal is played to the level recorder and the level recorder attenuators adjusted so that the signal is at a convenient position near the top side of the paper. Thus the paper is calibrated. It helps to make certain that the bottom of the paper is a round number in dB(A). The recording attenuation was adjusted and the analysis levels therefore change by the same amount. Hence the actual levels at different positions on the paper are written on. The tape, level recorder and statistical analyser are started. Thus a list of counts on a normal distribution basis are obtained and the level exceeded for 10 per cent of the time determined.

The method of determination can be similar to the calculation procedure used in Experiment 2 or by plotting graphs on linear or probability paper.

The L_{50} or L_{90} or any other L_n level may be determined from the results provided the whole of the trace is recorded on the paper.

Theory

The L_{10} level is the level of noise in dB(A) exceeded for 10 per cent of the measurement time. For traffic noise, this is normally measured in 1 m from the façade to dwellings at 1·2 m from the ground for ground floors, each hour for 18 hours from 06.00 hrs to 24.00 hrs. The 18 measurements are then arithmetically averaged to determine the 18-hour L_{10} level.

Specimen result

Counter	Sound level (dB(A))	Counts
1	Up to 30	—
2	30–35	5
3	35–40	36
4	40–45	487
5	45–50	1860
6	50–55	4224
7	55–60	1981 ← L_{10}
8	60–65	294
9	65–70	95
10	70–75	18
11	75–80	—
12	Above 80	—
	Total	9000

(Arrow from rows 8–12 up to L_{10} marked 900)

Hence L_{10} = 59 dB(A)
Also L_{50} = 52·5 dB(A)
L_{90} = 46 dB(A)

Results

Counter	Sound level (dB(A))	Counts
1	Up to . . .	
2	. . . to . . .	
3	. . . to . . .	
4	. . . to . . .	
5	. . . to . . .	
6	. . . to . . .	
7	. . . to . . .	
8	. . . to . . .	
9	. . . to . . .	
10	. . . to . . .	
11	. . . to . . .	
12	Above . . .	
	Total	______

∴ L_{10} = ________ dB(A)
L_{50} = ________ dB(A)
L_{90} = ________ dB(A)

Experiment 11 Experiment to study the effect of masking sound on speech intelligibility

Apparatus

Two tape recorders, tape of white noise or other suitable constant sound, tape of monosyllabic words, sound level meters.

Method

Lists of suitable words are recorded on tape at the rate of about one every four seconds using a normal voice. Sets of suitable words are listed in Table A3.3. It is important to use a series of unrelated monosyllabic words because if sentences are used, words not heard can be worked out.

The tape recorder with the words recorded is set up with its loudspeaker at a height of about 1·5 m, and the one with the white noise nearby. With the sound level meter(s) the sound level in dB(A) is measured with no white noise at each listening position. The first list of recorded words is then played at the level of a normal teaching voice and written down by the subjects, after which they are checked against the originals. The percentage correct is noted.

The white noise is run so that the sound level is approximately 45 dB(A). It will vary throughout the room and the level at each subject position noted. The second list is now played at the same level as before. Again the percentage correct is noted.

The experiment is repeated for levels of masking noise of approximately 50 dB(A), 55 dB(A), 60 dB(A), 65 dB(A) and the percentage intelligibility determined.

Table A3.2 Speech interference levels

Distance from speaker to hearer (m)	Normal voice (dB)
0·1	73
0·2	69
0·3	65
0·4	63
0·5	61
0·6	59
0·8	56
1·0	54
1·5	51
2·0	48
3·0	45
4·0	42

Raised voice: add 6 dB to each
Very loud voice: add 12dB to each
Shouting: add 18 dB to each
In the case of a female voice, all levels should be reduced by 5 dB

Theory

Background sound can interfere with the hearing of speech. Table A3.2 gives a guide to the maximum background levels before speech interference starts.

Results

List 1. Background level = _______ dB(A)
% intelligibility = _______
Distance from words = _______ m

List	2	3	4	5	6
White-noise sound level (dB(A))					
% intelligibility					

Distance from speaker = m
(masking sound).

Table A3.3 Monosyllabic word lists

List 1

1. in	18. pack	35. judge
2. roost	19. wash	36. fast
3. theme	20. gob	37. walk
4. sigh	21. gang	38. mow
5. web	22. hump	39. mouse
6. ace	23. fair	40. wise
7. duke	24. soak	41. eye
8. salve	25. get	42. cart
9. slice	26. skid	43. beard
10. rout	27. rouge	44. brass
11. quip	28. slush	45. cork
12. did	29. ramp	46. joke
13. retch	30. through	47. crate
14. tilt	31. lid	48. puss
15. pew	32. flash	49. clog
16. base	33. seed	50. click
17. pad	34. robe	

List 2

1. shave	18. fade	35. thaw
2. chill	19. ouch	36. chip
3. note	20. chaff	37. sack
4. gush	21. lip	38. got
5. chain	22. loud	39. waste
6. flare	23. grab	40. crab
7. trod	24. rob	41. lunge
8. fat	25. art	42. his
9. grew	26. dot	43. lynch
10. claws	27. thine	44. weed
11. grew	28. camp	45. axe
12. debt	29. freeze	46. rose
13. hush	30. thorne	47. bale
14. hide	31. dice	48. cat
15. sieve	32. fool	49. bless
16. sash	33. cub	50. claw
17. aims	34. lime	

List 3

1. fright	18. lay	35. tire
2. turn	19. din	36. thou
3. aid	20. sheik	37. sheep
4. wield	21. part	38. stab
5. gab	22. had	39. ink
6. rogue	23. sang	40. soar
7. droop	24. knee	41. three
8. map	25. hash	42. dub
9. hose	26. house	43. rye
10. stress	27. pump	44. cheese
11. rug	28. pitch	45. kind
12. book	29. crews	46. next
13. leash	30. tuck	47. closed
14. cliff	31. ton	48. gas
15. fifth	32. rock	49. drape
16. thresh	33. suit	50. nap
17. barge	34. dame	

List 4

1. dead	18. dash	35. tent
2. wrist	19. isle	36. vague
3. waif	20. car	37. purge
4. grate	21. shook	38. slab
5. thy	22. heat	39. tray
6. shoot	23. darn	40. bust
7. hunk	24. clip	41. stuff
8. cute	25. life	42. rid
9. tell	26. muss	43. quack
10. vote	27. news	44. foam
11. wag	28. prude	45. at
12. curve	29. kick	46. coax
13. oft	30. nod	47. nick
14. shrug	31. fife	48. me
15. group	32. douse	49. lathe
16. barn	33. soil	50. howl
17. dung	34. sing	

List 5

1. ball	18. ass	35. romp
2. gnash	19. foot	36. sledge
3. chink	20. flood	37. knife
4. chafe	21. laugh	38. jolt
5. dime	22. jaw	39. chap
6. vine	23. set	40. done
7. cling	24. fought	41. reek
8. wove	25. lash	42. out
9. tile	26. ripe	43. depth
10. sky	27. clutch	44. bluff
11. jazz	28. hunch	45. shut
12. and	29. chair	46. hear
13. rove	30. loose	47. sod
14. greet	31. cave	48. cad
15. frill	32. fed	49. park
16. wage	33. hug	50. throb
17. priest	34. flog	

List 6

1. back	18. value	35. champ
2. tree	19. hat	36. mope
3. thug	20. rip	37. void
4. bug	21. page	38. scrub
5. jay	22. nudge	39. cord
6. bash	23. rape	40. wake
7. earth	24. slug	41. clothe
8. gull	25. goose	42. tag
9. ears	26. hurt	43. plus
10. snipe	27. wade	44. force
11. rush	28. chance	45. line
12. staff	29. bob	46. cue
13. those	30. real	47. put
14. maze	31. daub	48. etch
15. flight	32. pink	49. youth
16. cow	33. flaunt	50. ail
17. fir	34. lap	

Experiment 12 Measurement of sound absorption coefficients in a reverberation room

Apparatus

Beat-frequency oscillator, 75–100 W amplifier, two 50 W loudspeakers, omnidirectional microphone, audio-frequency spectrometer, high-speed level recorder, 10 m^2 absorbent material, reverberation room (ideally this should be specially constructed with a volume of about 200 m^3, but for the purpose of this experiment a fairly reverberant room of about this volume should be satisfactory).

Method

The reverberation time, t_1, is first measured for the room (Fig. A3.10) without the absorbent material as described in Experiment 5. Six measuring positions are chosen for each octave interval with mean frequencies at 125, 250, 500, 1000, 2000 and 4000 Hz. The absorbent material is then placed in the room, preferably covering a single area as near as possible to 10 m^2. The reverberation time measurements are repeated at each frequency for six different room positions. The amount of absorption added is then calculated for each bandwidth and hence from the superficial area the absorption coefficients are calculated.

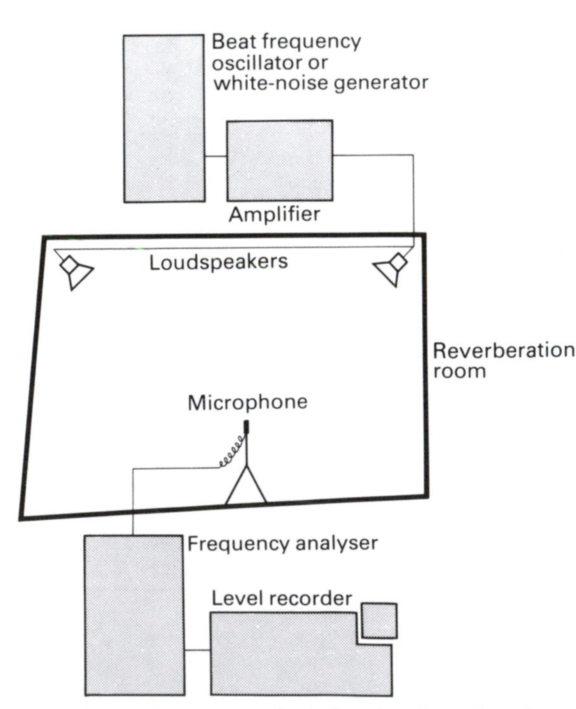

Fig. A3.10 Measurement of reverberation time using a beat-frequency oscillator or white-noise generator as sound source

Theory

Sabine's formula states that for a room,

$$T = \frac{0 \cdot 16V}{A}$$

where

T = reverberation time (s)
V = volume of the room (m^3)
A = area of absorption (m^2)

Thus for the empty room,

$$A = \frac{0 \cdot 16V}{T_1}$$

and for the room containing the extra absorption, δA,

$$A + \delta A = \frac{0 \cdot 16V}{T_2}$$

where

T_1 = original reverberation time
T_2 = reverberation time with the absorbent material added

$$\therefore\ \delta A = 0 \cdot 16\, V \left\{ \frac{1}{T_2} - \frac{1}{T_1} \right\}$$

Hence the absorption coefficient,

$$\alpha = \frac{\delta A}{s} = \frac{0 \cdot 16\, V}{s} \left\{ \frac{1}{T_2} - \frac{1}{T_1} \right\}$$

where s = superficial area of the material.

(*Reference:* BS 3638:1987 *Method for the Measurement of Sound Absorption Coefficients (ISO) in a Reverberation Room.*)

Results

Area of absorbent = __________ m^2

Centre frequency of octave band (Hz)	Reverberation time (s)							Extra absorption (m^2)	Absorption coefficient α
	1	2	3	4	5	6	Mean		
125									
250									
500									
1000									
2000									
4000									

Experiment 13 Determination of absorption coefficients using a standing-wave tube

Apparatus

Loudspeaker, 100 mm and 30 mm diameter measuring tubes, microphone and probe, pure-tone oscillator, measuring amplifier and third-octave filters.

Method

The apparatus is set up as shown in Fig. A3.11 but with the larger standing-wave tube. A 100 mm diameter sample of the absorbent material is inserted in the end of this tube by means of the holder. The pure-tone oscillator is adjusted to 100 Hz at a suitable amplitude and the microphone and probe moved until a maximum is measured by the measuring amplifier. This value $(A + B)$ is noted (see Fig. A3.12). The microphone and probe are then moved to obtain the minimum $(A - B)$, which is noted. Hence $n = (A + B)/(A - B)$ is calculated, and by use of the graphs shown in Figs. A3.13 or A3.14 the absorption coefficient is determined. This is repeated for frequencies up to 1600 Hz.

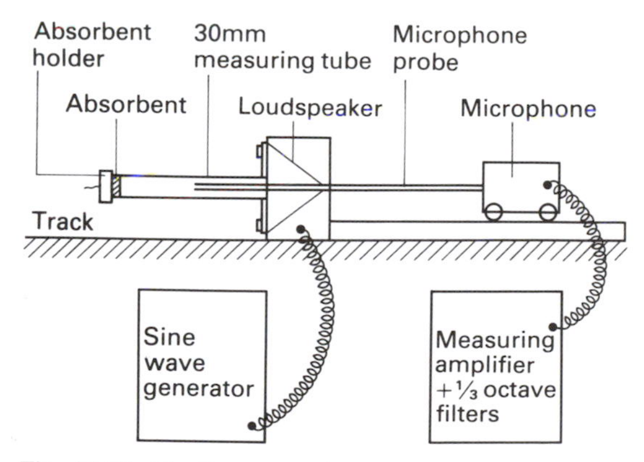

Fig. A3.11 Standing-wave tube

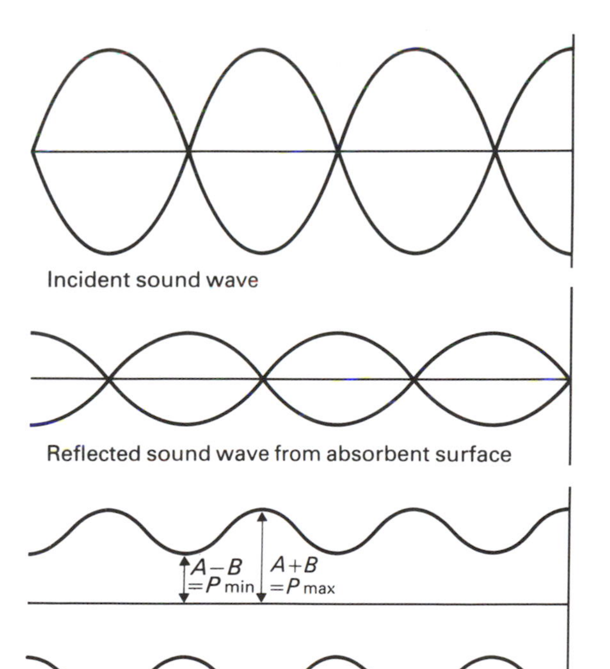

Fig. A3.12 Formation of the standing wave pattern in the impedance tube

The large tube is then changed for the smaller tube with a 30 mm diameter sample of the same absorbent material. The process is repeated for frequencies from 800 to 6300 Hz. It will be noticed that the overlap enables the results from the two different tubes to be compared.

Theory

This method is convenient for the measurement of absorption coefficients at normal incidence for fibrous materials. It is not suitable for absorbers which rely on resonance. It works on the basis of comparing the incident and reflected pure tones on a small sample of the absorbent. As there is a quarter wavelength phase shift on reflection from an absorbent material, this is relatively easy. It means that the maximum amplitude for the reflected wave coincides with the position of minimum for the incident wave. Similarly, the maximum for the incident wave coincides with the minimum for the reflected wave. The result is illustrated in Fig. A3.12.

The definition of absorption coefficient is the proportion of the incident energy which is not reflected from the surface.

Thus absorption coefficient $\alpha = 1 - \left(\frac{B}{A}\right)^2$

where A = amplitude of reflected wave
B = amplitude of incident wave

Note the fact that the energy is proportional to the square of amplitude or pressure.

The maximum measured is $A + B$ while the minimum is $A - B$.

$$\therefore \frac{\text{maximum pressure}}{\text{minimum pressure}} = \frac{P_{max}}{P_{min}} = \frac{A + B}{A - B}$$

Now let $n = \frac{A + B}{A - B}$

i.e. the ratio of maximum to minimum pressure.

It can be shown that $\alpha = \frac{4n}{n^2 + 2n + 1}$

With some equipment, the voltage scales on the measuring amplifier are calibrated to enable the coefficient to be read directly. If not, the graphs shown in Figs A3.13 and A3.14 may be used.

The maximum diameter of standingwave tube that can be used is approximately half the wavelength under investigation ($0{\cdot}586\lambda$). The shortest length possible is a quarter wavelength. Thus to make measurements down to 63 Hz, a tube of about 1·25 m minimum is needed. At the same time, a 100 mm diameter tube can only be used up to 1800 Hz. A 30 mm diameter tube of length at least 100 mm can be used from about 800 Hz to 6·5 kHz.

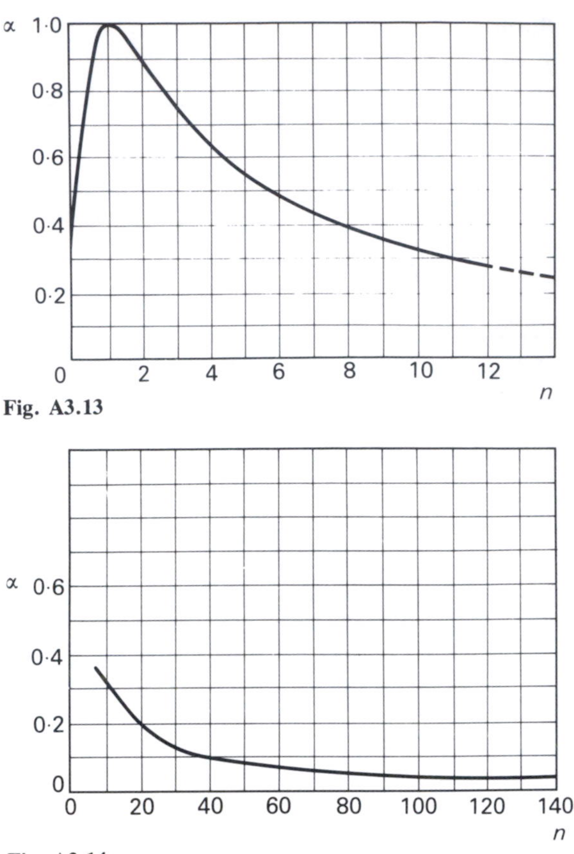

Fig. A3.13

Fig. A3.14

Results

Frequency (*Hz*)	*Max* ($A + B$)	Min ($A - B$)	$\therefore n = \dfrac{A + B}{A - B}$	Absorption coefficient α
100				
125				
160				
200				
250				
315				
400				
500				
630				
800				
1000				
1250				
1600				
2000				
2500				
3150				
4000				
5000				
6300				

Experiment 14 Determination of the NR and NC ratings of a noise from its octave-band spectrum

Apparatus

Sound level meter with octave-band filters, calibrator, noise source, tripod stand.

Method

A suitable noise source is chosen. This may be an electric drill, extractor fan fixed in a room or any other convenient source. A suitable position in the room is chosen for the measurements. The sound level meter batteries are checked and the meter calibrated. Readings of sound pressure level are taken in octave bands with centre frequencies from 63 Hz to 8 Hz.

The values obtained are then superimposed upon the noise rating (NR) and noise criteria (NC) curves. The lowest NR or NC curve just above the values obtained determines the NR number or NC number. This is best found by plotting the results graphically together with the appropriate NC and NR curves, (Figs. A3.15 and A3.16).

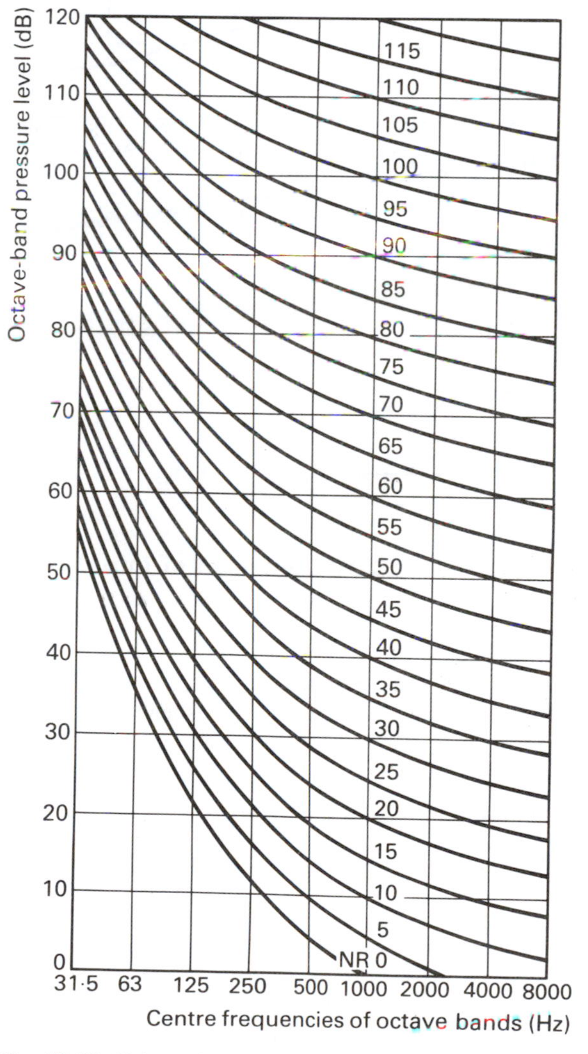

Fig. A3.15 Noise rating curve (NR)

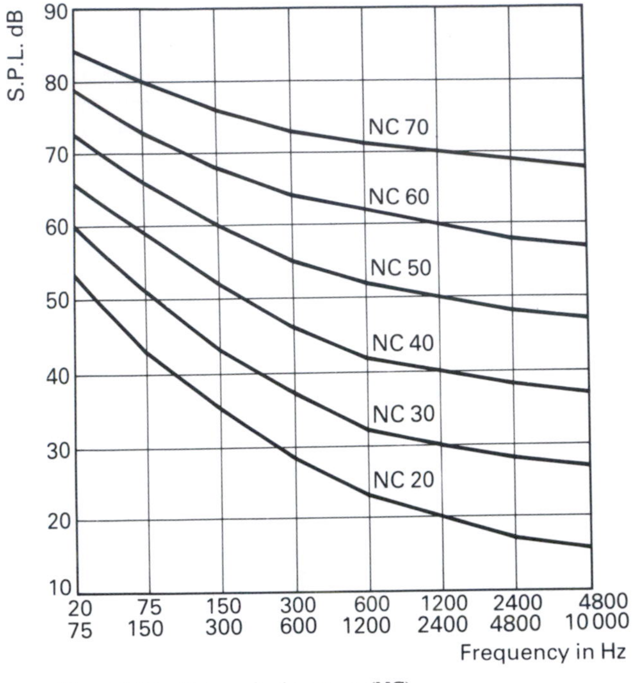

Fig. A3.16 Noise criteria curves (NC)

Theory

Results

Noise source:
NR = __________
N C = __________

Experiment 15 Measurement of impulsive noise

Apparatus

Sound level meter fitted with S, F, and I time weightings, and with the facility to measure peak sound pressure level (P); various sources of transient and impulsive noise.

Method

The purpose of the experiment is simply to investigate what differences, if any, are produced by using the S, F, I and P time weightings. A variety of different transient sounds should be used, with different degrees of impulsiveness, for example, tape recording of fast car or motor cycle passing, handclaps, clinking together of two bottles, sharp crack produced by slapping a metre ruler (or similar) on a wooden bench, children's toy gun or starting pistol, tape recordings of industrial impact noise, such as staple guns, riveting or a punch press, and tape recordings of gunfire.

Each type of noise should be repeated several times, and the average of the maximum values of sound pressure level noted for each of the four settings S, F, I and P. The frequency weighting of the meter should be set to linear, or unweighted because this is the only correct setting for peak (P) measurement. The sound level meter may have a hold facility for the I and P settings, in which case it will be necesssary to reset in between each reading. Several measurements may be required for each time weighting in order to establish a good average value, depending upon the repeatability of the source.

The results should be tabulated and inspected, and conclusions drawn about the effect of using different time weightings for the various noises. Attention should be focused, in particular, on the difference between the peak values and the other settings (S, F and I).

The important conclusion which should be demonstrated by this experiment is that only the peak setting can give a true measurement of peak sound pressure level for sharply impulsive sounds, and that values taken using either F or I will be very inaccurate.

Results

Type of noise	S	F	I	P
Handclap				
Ruler on bench				
Toy pistol				

Experiment 16 Experiments with a digital sound level meter

Apparatus

A sound level meter capable of measuring L_{10}, L_{50}, L_{90} and L_{eq} (and if possible SEL too), various noise sources, as available, sound level meter handbook, sound level meter calibrator.

Method

Modern digital sound level meters can at first seem complicated. Time should be spent carefully reading the operating instructions in the handbook provided by the manufacturer, and practising using the meter before taking readings in earnest.

The purpose of this experiment is to provide the user with some additional exercises which can help to build confidence in the use of the instrument, and also to increase understanding of some of the various measurement indices and parameters.

Simple checks using a calibrator and stopwatch

Exercise A3.1 (L_{10}, L_{50}, L_{90})

Switch on the sound level meter and calibrate. Leave the calibrator fitted over the microphone, then switch it off (if it has an off switch) or wait until the display indicates that it has switched off (if it operates on a time switch).

Reset the meter and start the measurement of L_{10}; simultaneously start the stopwatch. The meter should indicate only a very low sound pressure level because the microphone is being shielded from the ambient level by the calibrator body.

After 60 s from the start of the measurement period, switch on the calibrator and observe the L_{10} value on the display. The value should suddenly rise from its low background level to the calibrator level (e.g. 94 dB(A)) after the calibrator has been on for just under 7 s (6·66 s).

When this happens, switch the meter to L_{50}, with the calibrator remaining on, and note the time on the watch. Observe the display; the display value should have dropped back to the background level when the meter was switched to L_{50}.

The L_{50} value should suddenly rise to the calibration value after just under a further 23 s (23·33 s).

When this happens, switch the meter to L_{90}; with the calibrator remaining on, and note the time. Observe the display; the indicated value should once again have dropped back to the background value.

How long will it take before the L_{90} value rises to 94 dB(A)? Check your answer by observation.

This exercise should give confidence that the meter is correctly calculating the percentile levels. The L_{90} value should rise to 94 dB(A) after a further 510 s, provided that the calibrator remains switched on. The test could be repeated using a different initial off time, e.g. for 100 s instead of 60 s.

Exercise A3.2 (L_{eq})

Switch on the sound level meter and calibrate. Leave the calibrator fitted over the microphone and wait for it to switch off (if on a time switch) or switch it off (if it has an off switch).

Reset the meter and start measurement of L_{eq}.

At the same time, switch on the calibrator and note the time on the stopwatch.

Observe the display, which should indicate the calibration value (e.g. 94 dB(A)), and measure for how long the calibrator remains on, if on a time switch (e.g. 60 s); otherwise switch off after 60 s (if it has an off switch).

Note the value of L_{eq} at the moment the calibrator switches off (e.g. 94 dB).

Observe the falling L_{eq} value and measure how long it takes for the L_{eq} to fall by 3 dB. This should be 60 s.

How long should it take for the L_{eq} value to drop by a further 3 dB.

This exercise should give confidence that the meter is calculating L_{eq} correctly. The L_{eq} value should drop by 3 dB after a further 2 min, i.e. after a total measurement time of 4 min, and by a further 3 dB after a total measurement period of 8 min. A graph of L_{eq} versus the logarithm of measurement time should be a straight line.

Simple checks for L_{eq} and SEL (L_{AE}) A source which can provide a repeatable burst of noise is required. An electric drill turned on for exactly 10 s each time would be a suitable source.

The experiment consists of setting the meter to measure L_{eq} and observing how the measured value varies as the number of bursts of noise increases, from 1 to up to 20, with constant time interval (of background noise, which should be relatively low) between each burst. This procedure should be repeated but with the meter set to measure SEL.

You should think about, and try to predict, what will happen in each case, then confirm by carrying out the experiments.

In both cases the indicated value should increase, at first, as the number of bursts increases, but eventually the rate of increase will reduce, i.e. the indicated value will level off or flatten out. However, this happens for different reasons, and at different levels, in the two cases.

For the SEL measurement the indicated value continues to increase, but at a diminishing rate, as the number of noise bursts increases. The successive values may be compared with the results of decibel addition. Thus the SEL value should increase by 3 dB after the second burst, and by almost another 2 dB after the third burst. After 4 bursts the value will be 6 dB higher than after the first burst; it will be 10 dB higher after 10 bursts and 13 dB higher after 20 bursts. A graph of the accumulating SEL value against log N (N is the number of bursts) should be a straight line.

For the L_{eq} measurement the indicated value is an average which will at first fluctuate, but these fluctuations will reduce as the L_{eq} value stabilizes with increasing numbers of bursts. After the first burst, the L_{eq} value will fall until the second burst, when it will rise. After the second burst, it will fall until the third burst, and so on. If the duration of each burst of noise is known, together with its level and the time between bursts, it is possible to calculate the L_{eq} after any number of bursts and to compare it with the measurements.

L_{eq} and SEL for train and aircraft noise This experiment requires a tape recording of a few train or aircraft noise events over a total period of a few minutes, although bursts of some other sort of noise would also be acceptable. During this experiment, which is a practice for real measurements in the field, the tape recording is played through an amplifier and loudspeaker in the laboratory, and the measurements are carried out on the sounds produced in the laboratory by the loudspeaker. It is possible to make accurate measurements by taking an a.c. output signal directly from the (calibrated) tape recorder into the sound level meter, removing the microphone and dispensing with the loudspeaker. But that is another experiment.

First of all, the SEL of each event should be measured. To do this the meter should be reset, and the SEL measurement started when the noise of the event, i.e. the train or aircraft is first heard above the background noise, and continued until after the event has finished, when the noise has returned to its background level. Provided that the highest levels of the noise event are well in excess of 10 dB above the background level, it is acceptable to judge the start and finish of the event by ear in this way.

If the sound level meter can measure L_{eq}, but not SEL, then the above procedure should be carried out, measuring instead the L_{eq} for each event. The measurement duration should also be determined in each case, using a stopwatch if the meter has no facility for measuring the time. The SEL value is obtained simply by adding 10 log T to the L_{eq} value, where T is the measurement duration in seconds.

When the SEL values of each event have been measured, the L_{eq} value due to the events over the total measurement period may be calculated. The tape should then be played continuously from beginning to end, and the period L_{eq} value measured directly. If the levels of background noise between the events are low compared to the levels during the events, the L_{eq} values determined by the two methods should be very close. If there are some high levels of background nose, the directly measured period L_{eq} will be higher than the value calculated from the individual SEL values.

The usefulness of the SEL method is that it enables predictions of L_{eq} to be made for different combinations of events.

L_{10}, L_{50} and L_{90} for traffic noise The digital sound level meter is set up alongside some simpler equipment for measuring traffic noise, either by the basic tick-test method (Experiment 2) or by making a calibrated tape recording (Experiment 10). A comparison should show similar results from the three methods.

The effect of F and S on various noise indices The F and S time weightings are built into many noise measurements. The purpose of this part of the experiment is to investigate what effect, if any, the selection of F or S has on the measurement of L_{10}, L_{90}, L_{eq} and L_{Amax} (L_{Amax} is the maximum A-weighted level occurring during the measurement period). A variety of continuous noises may be investigated, such as machinery noise (e.g. electric drill) or various types of music, but interesting and thought-provoking results may be obtained using rapidly repetitive noises such as buzzers, bleepers and rattles. Suitable noises of this type could include an electric bell, a smoke alarm, footsteps or tapping machine (used in impact sound insulation measurements), fishing-rod reel or rattle (baby's toy or football supporter's). An ideal source, if available, is an electronic signal generator which can deliver short duration pulses to a loudspeaker, so that each pulse produces a repeatable, audible click. The click repetition rate should be variable from one click every few seconds to several clicks per second.

The experiment simply consists of measuring the same noise in different ways, using first the F weighting then the S weighting, to see whether any significant differences occur for the particular index being measured, and perhaps explaining the results.

Some measurement codes and standards will specify whether F or S must be used. Examples are the use of S for L_{90} measurement in BS 4142, and F for L_{10} in traffic noise measurement according to the document *Calculation of Road Traffic Noise*. This experiment should help to illustrate why it is necessary to distinguish between F and S, and to emphasize how important it is to record their details in the noise measurement report.

Experiment 17 The performance of an acoustic enclosure

Apparatus

A small noise source such as a high-pitched electronic buzzer, a box to enclose it, with a sound-absorbing foam lining, and a resilient pad to fit under the buzzer, as shown in Fig. A3.17. A suitable box could have internal dimensions of 30 cm × 30 cm × 30 cm and be made from 18 mm chipboard with a separate sixth side to act as base. The foam lining (5 cm thick) should be joined together to act as a smaller, foam box.

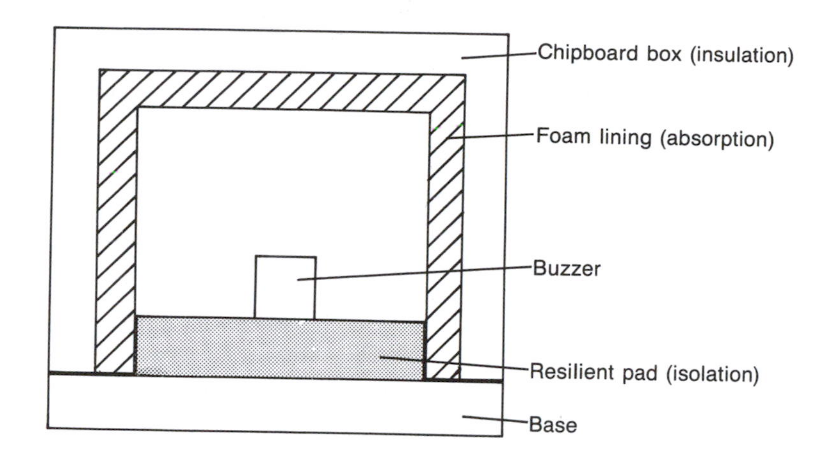

Fig. A3.17

Method

The aim of the experiment is to investigate the performance of the box as an acoustic enclosure for the buzzer, and to demonstrate the different roles of sound absorption, sound insulation and vibration isolation.

The box can be used as a simple aural demonstration, or as a student experiment, with dB(A) sound level measurements being taken at the various stages described below.

1. Set the buzzer on the base and measure or listen to the noise.
2. Cover the buzzer with the (inverted) foam box. A small reduction in noise level will be observed. Although the foam may be a good sound absorber, it is not a good sound insulator because it has a low density.
3. Replace the foam box by the chipboard box. A much greater noise reduction is observed, demonstrating that the dense chipboard is a good sound insulator.
4. If one edge of the box is lifted very slightly, about a millimetre, even this small crack will reduce the effect of the insulation almost to zero.
5. Fitting the foam lining inside the chipboard box produces a further reduction in the noise. This is because the foam is a good sound absorber, so it reduces the build-up of reverberant sound inside the box. Therefore insulation and absorption combined give improved sound reduction. Lifting the corner of the box again demonstrates that the absorbing lining has made the enclosure performance less sensitive to the effects of gaps.
6. The final stage is to fit resilient pads under the buzzer,

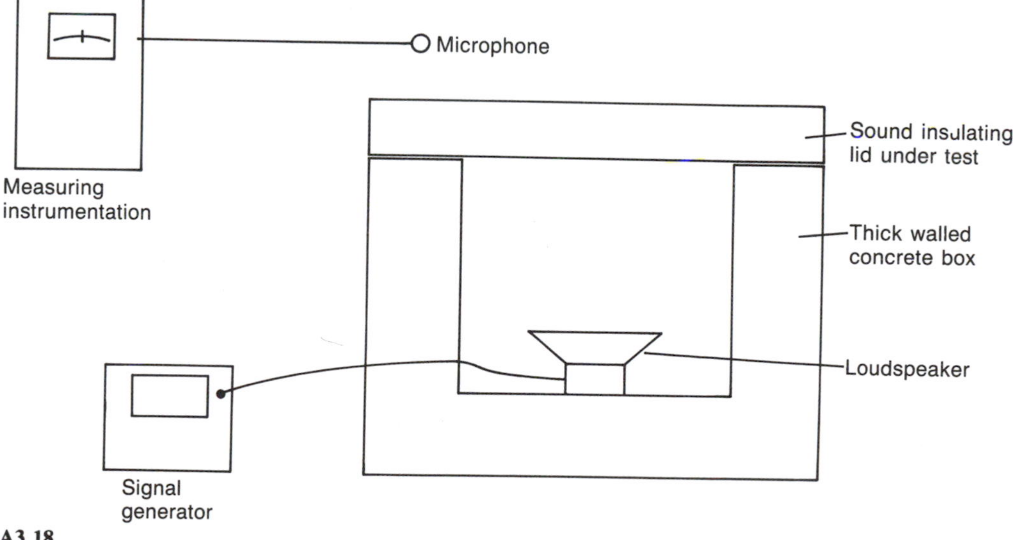

Fig. A3.18

isolating the buzzer from the base. This reduces structure-borne sound transmission, and leads to a further decrease in the noise level.

Thus maximum effectiveness requires all three: absorption, insulation and isolation.

This experiment may be modified and refined in a number of ways:

(a) By using a microphone close to the buzzer, in stage 1, and actually inside the box, in stages 2 and 3.
(b) By comparing results for boxes made of different materials and different wall thicknesses.
(c) By using a double-skin box, i.e. using a second, larger box, fitted over the original, with sound-absorbing material filling the cavity in between.
(d) By using a duct lined with sound-absorbing material to let air in and out of the box, e.g. for cooling, without substantially reducing the enclosure performance.
(e) By using different noise sources inside the box. A loudspeaker connected to a signal generator which produces bands of noise allows the effect of frequency to be determined. A machine such as a motor will produce more structure-borne noise than the buzzer, so it will make the final stage of the demonstration particularly effective.
(f) A variation which compares the sound insulation performance of many different materials, in sheet or pane form, uses a moulded, thick walled (25 mm) concrete box containing a loudspeaker as the noise source (see Fig. A3.18). The sound insulating performance of lids made of various materials may be investigated.

APPENDIX 4 Measurement of sound power according to ISO 3740–46

N.B. These notes are intended only as a background to and explanation of some of the points covered in the ISO standards. The relevant standard, in full, should be consulted before making measurements.

There are seven standards. The first one serves as a guide to the other six. The choice of measurement procedure (and of relevant standard) will depend on the following factors:

(a) *The aim or purpose of the measurements* (which in turn will determine the accuracy required and the data required, i.e. A-weighted power levels only or octave bands as well — overall sound power or directivity information also).
(b) *The size of the noise source* (in relation to the room) *and its nature* (i.e. steady broadband, tonal, impulsive, intermittent).
(c) *The test environment*, i.e. whether or not the machine can be moved to a specialist test room (reverberant, anechoic or semi-anechoic) and if not then the acoustic characteristics of the *in situ* test environment.

ISO 3740 deals with these points.

The aim or purpose of the measurements might include one of the following:

(a) *Verification* that emitted noise lay between certain limits (e.g. a manufacturer's specification or a code of practice).
(b) *Comparison* testing — between different machines of the same type.
— between machines of a different type.
(c) *Prediction* of noise levels at a distance from the machine.
(d) *Assessment* of occupational and environmental noise problems produced by the machine.
(e) *Testing* the effectiveness of noise control measures which have been applied to the machine.

The methods are classified into three grades: **Precision** (most accurate), **engineering** and **survey** (least accurate). The accuracy or uncertainty in determining sound power levels is expressed as a standard deviation in decibels. The main factor affecting the classification is the quality of the test environment available (either anechoic, semi-anechoic or reverberant). The one survey method is for testing machines *in situ*, i.e. no special test environment. It is considered that this method is only suitable for overall, A-weighted sound power levels with an accuracy of 5 dB(A), whereas the other methods allow for octave or maybe even third-octave band measurements, with much greater accuracy.

The basis of the anechoic and semi-anechoic room methods is that free-field measurements (i.e. direct sound unaffected by wall, ceiling and other reflections) are taken far enough away from the noise source for them to be in the far field. Sound measurements are taken on a measurement surface surrounding the machine, and averaged ('pressure' or logarithmic averaging). The sound power level L_W is calculated from the average sound pressure level (SPL) and the area (S) of the measurement surface in m^2:

$$L_W = \text{SPL} + 10 \log_{10} S$$

The measurement surface may be a hemisphere, a parallelepiped or a conformal surface, i.e. one which follows the general shape of the noise source. The standards give methods for testing (or 'qualifying') the room, by measuring how closely it agrees with the 6 dB per doubling of distance inverse square law.

In a reverberant room there are two ways of measuring sound power level, either directly or using a calibrated reference source of known power level. It is not possible to obtain information about the directionality of the source by either method. In both methods it is necessary to measure the 'space-time' averaged sound pressure in the room. This may be done in one of three ways: by moving a microphone from place to place inside the room, by using an array of microphones in different places in sequence or by time averaging as a microphone is moved through the room. In the direct method it is necessary to measure the reverberation time of the room, and also its volume. The sound power level may then be calculated from formulae given in the standards relating sound power level to reverberation time, room volume and average sound pressure level. In the comparison method the reverberation time data is not needed and the sound power level is found by comparing the sound pressure levels in the room for the reference source and the source under test. The standards give methods for assessing the suitability (or 'qualifying') the test room. The requirements of the test room include conditions as to its volume, acoustic absorption, reverberation time and the variability of levels within the room when a test source is operating. A test source is used also to check that the sound power of the source is unaffected by coupling between the room and the source. This coupling between room and source is likely to cause a particular problem if the source emits pure tones which can cause standing waves to be set up in the room. There is also the additional problem that the sound distribution in the room will be less uniform because of the standing waves. Because of these difficulties a separate standard is devoted to the measurement of sources radiating discrete frequencies or narrow bands of noise. The difficulties are overcome by using more microphone positions, and also more source positions than would be necessary for broadband sources. Alternatively, rotating diffusers may be used to break up the standing waves.

The standards contain many specifications concerning the noise source and the positioning of the microphones during tests. Some of the items specified are given below.

A4.1 Source

(a) *Its size*. This should be less than 0·5 or 1 per cent of room volume. If the source is too big for the room it is difficult to get into the far field (for anechoic and semi-anechoic rooms) and into the reverberant field (reverberant rooms).
(b) *Its acoustic characteristics*, i.e. steady broadband, impulsive, tonal, time-varying. It is suggested that impulsive noises can be 'recognized' by noting the difference between readings taken with slow and impulse meter functions.
(c) *Its location*. It should not be too close to a wall (1·0 m) because the proximity of the wall may affect the sound power output. Different source locations should be a minimum distance apart.
(d) *Its operating conditions*, e.g. normal load, full load, no load, etc. Another possibility is the loading condition giving rise to maximum sound generation.
(e) *Its mounting conditions*. The way in which the source is mounted can have a great effect on the sound power emitted.

A4.2 Microphone positions

(a) *Number of positions* required depends on accuracy required and upon acoustic conditions.
(b) *Minimum distance of microphone from noise source*. This is specified to ensure that measurements are made in the far field (in the case of anechoic and semi-anechoic rooms) and in the reverberant field (in reverberant rooms).
(c) Minimum distance from room and other reflecting surfaces.
(d) Minimum distance between different microphone positions.

A4.3 Near-field measurements

It sometimes happens that with large immovable machines, possibly with other noise sources nearby, only near-field measurements are practical. In some standards, measurements are taken around a surface which follows the general outline shape of the machine a short distance away from it (often 1·0 m). Measurements are also taken at operator positions and where the sound levels are maximum. BS 4813: 1972 on the 'Method of measuring noise form machine tools (excluding testing in anechoic chambers)' is an example of such a standard.

Such measurements are very useful for many purposes, but estimations of sound power made from them, using the equation given earlier (based on the area of the measurement surface and the average sound pressure level) only give an approximate value, even when background levels from other machines are low. There are two reasons for this: firstly, the calculation of sound power assumes that the sound energy from the source is flowing outwards through the measurement surface, and perpendicular to it; this is true for far-field conditions, but may not be so in the near field. The second reason relates to the assumption that the measured sound pressure levels are equal to the intensity levels at the measurement points. This is true for plane waves and approximately true for sound radiated from real sources in the far field, but it is not true for near-field measurements.

In the survey method (ISO 3746) an imaginary reference surface in the form of a parallelepiped is defined, which just encloses the source. The measurement surface over which the microphones are positioned can be either a hemisphere or a rectangular parallelepiped. The radius of the hemisphere should be twice the largest dimension of the reference surface. When the reference surface is a rectangular parallelepiped the distance between this surface and the reference surface should normally be 1 m. The hemisphere is preferred for small and cubical sources, whereas the rectangular parallelepiped is more suitable when measurements have to be performed close to the test source. The minimum numbers of microphone positions for the hemisphere and the rectangular parallelepiped are 5 and 6 respectively. If the range of measured levels in dB exceeds 1·5 times the number of microphone positions then additional positions as prescribed in the standard should be used.

Even though a specialist room is not required there is a requirement for the room, which is that $A/S \geq 1$, where A is the sound absorption area of the room and S is the area of the measurement surface.

A4.4 Substitution method in a semi-reverberant room

Earlier standards (ISO R495 and BS 4196) contained this method whereby machines could be tested *in situ* in rooms having limited absorption, such as factories, shops, offices, etc. Levels produced by the machine were compared with levels produced when the machine was substituted by a reference source of known acoustic power.

A4.5 Measurement of noise from machines (sound power)

Information to be recorded and reported

(Taken from ISO 3740–45)

Information to be recorded

Source

- Description of sound source including dimensions. (Give details if there are multiple sources or ancillary equipment which produces noise.)
- Operating conditions (speed, load, feed rate, work piece, etc.).
- Mounting conditions (floor type, use of AV mounts, etc.).
- Location in test room.

Acoustic environment

- Description of test room.
- Dimensions. Surface treatments.
- Sketch showing locations and room contents.
- Acoustical qualification of room (see the ISO standards) either as a reverberant or anechoic/semi-anechoic room.
- Relative humidity, pressure, temperature.

Instrumentation

- Equipment list (name, type, serial number).
- Bandwidth of frequency analyser (octave or one-third octave).
- Calibration methods. Date and place.
- Calibration of microphone, room, frequency response.

Acoustic data

- Location and orientation of microphone.
- Corrections applied, e.g. for background noise, for frequency response of microphone, etc.
- Tabulated sound power levels (and directivity index).
- The difference between slow and impulsive readings (where applicable for impulsive sounds); see the ISO standards.

Remarks on the subjective impression of the noise

Audible discrete tones (whistle, whine, hum).
Impulsive character of noise (bang, thump, clatter).
Spectral characteristics (i.e. mainly high/low frequency).
Temporal characteristics (on/off time, cycle times, etc.).

Information to be reported

Date and time of measurement.
Sound power levels at all frequency bands and at all operating conditions.
Location of sound source.
Reference to the ISO standard used.

APPENDIX 5
Some electrical principles

These notes are intended for the reader with little or no knowledge of electrical terminology and principles; they aim to introduce some of the technical terms and ideas met in Chapter 9. They offer only the very briefest introduction to the subject, and readers requiring a more comprehensive treatment should consult an appropriate textbook on electrical or physical science.

A5.1 Ohm's law and resistors: volts, amps and ohms

Figure A5.1 shows the simplest possible electrical circuit; it consists of a cell or battery, as found in a car or a transistor radio, connected by copper wire to a **resistor**. It is possible to draw analogies with a simple water circuit in which water is pumped via copper pipes through a radiator. The pump provides a difference in water pressure, which drives the water through the circuit. The radiator provides resistance to the flow of water, and if more radiators are added, the flow rate decreases. The cell is the equivalent of the pump and provides a difference in electrical pressure or **electric potential**; it is measured in **volts** (V). The electric current is related to the rate at which electrons flow through the circuit; it is measured in **amperes** (A). The resistor provides a **resistance** to the current flow; it is measured in **ohms** (Ω). Resistors commonly found in electrical circuits (e.g. in a radio or television) can vary from a few ohms to millions of ohms. The resistance of the copper wire in the circuit is probably a few hundredths of an ohm and is taken as negligible by comparison with the resistor. **Ohm's law** relates the current (I in amps) flowing through a resistor (R in ohms) to the potential difference (V in volts) across it

$$V = IR$$

In the case of the simple circuit in Fig. A5.1, the voltage drop across the resistor is the voltage provided by the cell. Suppose it is 12 V, as in a car battery, and that the resistance is 3 Ω, then $12 = I \times 3$, from which the current is 4 A.

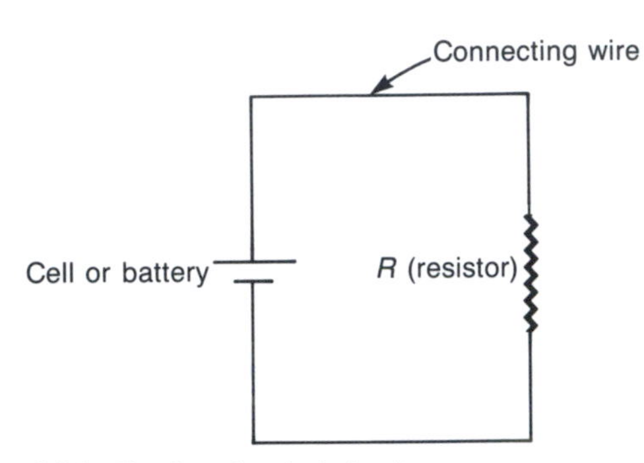

Fig. A5.1 Simplest electrical circuit

A5.2 Resistors in series and parallel: attenuator chains

A second resistor has been added in Fig. A5.2. The same current flows through each resistor in turn. The resistors are said to be **in series**. An alternative arrangement (Fig. A5.3) shows the two resistors **in parallel**. In this arrangement the current divides and a different portion flows through each resistor. Returning to the series arrangement, it can be shown that the total effective resistance is simply $R_1 + R_2$, the sum of the individual resistances. If, for example, both resistances were equal to 3 Ω, the total resistance would be 6 Ω and the current, according to Ohm's

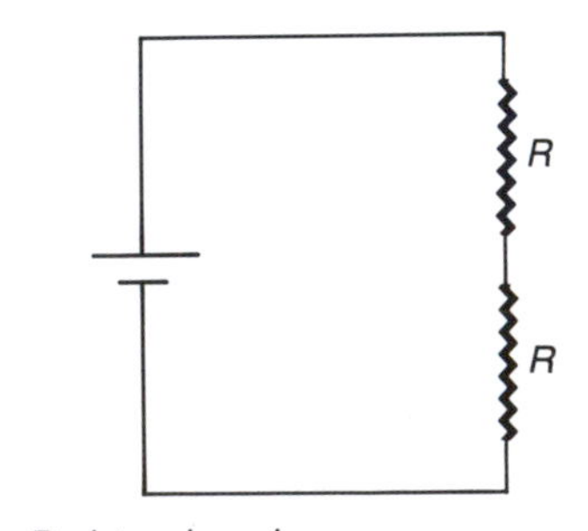

Fig. A5.2 Resistors in series

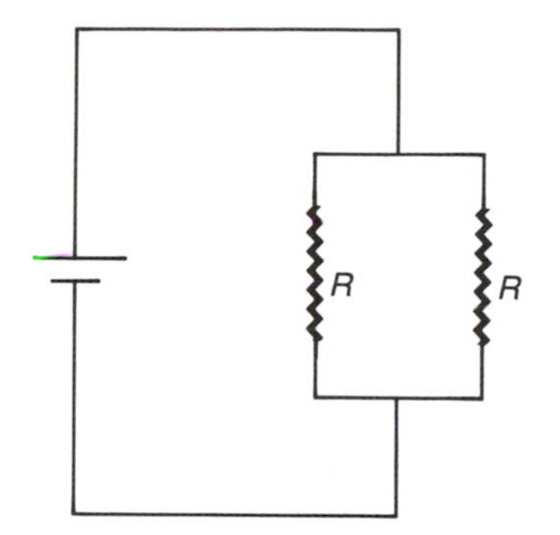

Fig. A5.3 Resistors in parallel

law, would be 2 A. Another point to notice about this two-resistor circuit is that the potential or voltage drop across each resistor is 6 V, making a total drop around the circuit of 12 V as before.

Figure A5.4 shows the same idea extended to a chain or ladder of 10 identical resistors in series. The voltage drop across each individual resistor is exactly one-tenth of the voltage provided by the cell or battery. Chains of resistors can therefore be used to divide an electrical voltage signal. This is the principle of the **attenuator**, used in sound level meters, to provide changes in range of exactly 10 dB, 20 dB, 30 dB, etc. Attenuators can be made with great accuracy and stability because they do not rely on components such as transistors whose performance can vary with age and with variations in supply (i.e. battery) voltage.

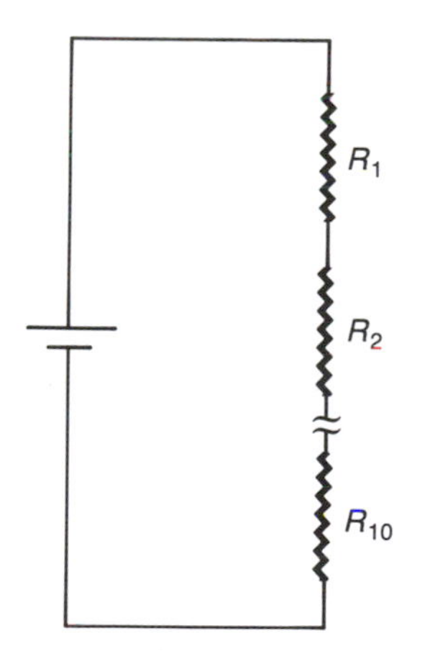

Fig. A5.4 Attenuator: a chain of resistors in series

A5.3 Electrical loads: input and output resistance

Figure A5.5 shows a slightly more complex but more realistic version of Fig. A5.2. The cell is connected to an external device which has an effective resistance R_L. This device, which draws current from the cell, is called the **load**. However, this is not the only resistance in the circuit

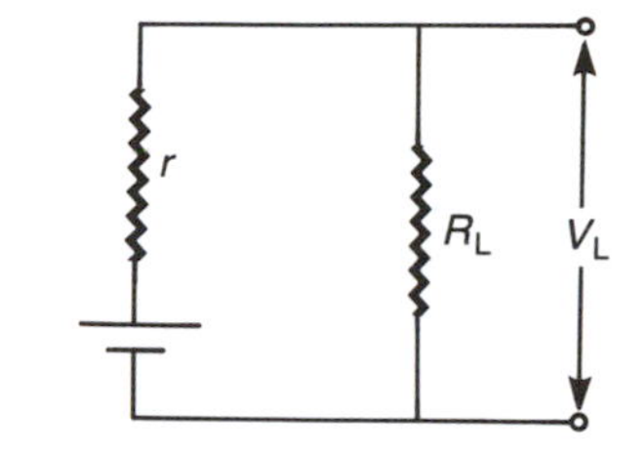

Fig. A5.5 Series circuit including internal resistance of the cell

because the cell itself provides a certain amount of resistance to current flow. This is called the internal resistance of the cell and is shown as r in Fig. A5.5. The load R_L behaves in the circuit as though it were a simple single resistor, but it could be a complicated device such as a transistor radio circuit or the amplifiers in a sound level meter. R_L is the resistance as seen by the cell looking into the load; it is called the **input resistance** of the load. The cell itself presents a resistance r to the load; it is called the **output resistance** of the cell. In this particularly simple case the output resistance of the cell is the same as its internal resistance.

The maximum potential difference provided by the battery is V_B and the total resistance in the circuit is $r + R_L$. The current in the circuit is therefore given by $I = V_B/(R + r)$. The voltage drop across the load, V_L, is given by Ohms law:

$$V_L = IR_L = V_B \cdot \frac{R_L}{R + r}$$

Let us consider two extreme cases. First, suppose the internal resistance r is negligible compared to the load. The potential drop across the load, and across the terminals of the cell, will be the maximum possible, i.e. V_B, and the current through the load (V_B/R_L) will depend on the value of the load itself. **The smaller the value of the load resistance, the higher the current drawn from the cell.** Now suppose the cell has a high internal resistance compared to the load. The potential drop across the load, and across the terminals of the cell, will be given, approximately, by $V_L = V_B R_L/r$ and will be much less than V_B. The current through the load will be given by $I = V_B/r$ and will be limited by the internal resistance of the cell. **The larger the output resistance of the voltage source, the lower the current it can deliver.**

To illustrate the two cases, consider two 12 V cells: a lead-acid accumulator (i.e. a car battery) and a dry cell (or bank of cells) such as used in a transistor radio. Both cells identically show 12 V when a voltmeter is connected across their terminals. Both cells are capable of providing the current to operate the transistor radio, and in both cases the voltmeter across the terminals remains unchanged at 12 V when the radio is switched on. But when the cells are connected to a car headlamp, there is a different story. The accumulator lights up the bulb and there is very little drop in the voltmeter reading when the lamp is switched on. The dry cells are not capable of providing enough current to light up the headlamp, so the voltmeter reading drops rapidly from the 12 V mark when the lamp is switched on. The difference in the performance of the two cells arises from the difference in their output resistances. The accumulator has a very low internal resistance, about 1 Ω, so it is capable of delivering a current of several amps, sufficient to operate the lamp. The dry cells have a much higher internal resistance, so they can only deliver a maximum current of a few thousandths of an amp, sufficient for the transistor radio but not for the lamp.

Although we must not push the analogy too far, the condenser microphone is rather like the dry cell whose output voltage drops drastically under the heavy load of the headlamp bulb. The microphone acts as a cell (a source of alternating voltage) and, without a suitable preamplifier, its output signal may be drastically reduced by the loading effect of the sound level meter circuits.

A5.4 Alternating current

So far we have considered circuits in which the electric current flows in one direction only, i.e. direct current (d.c.). Circuits involving only resistors will operate in exactly the same way if the cell is replaced by a source which generates an alternating potential difference. The mains supply of electricity provides a voltage which varies sinusoidally with a frequency of 50 Hz and a voltage amplitude of 240 V. This produces an alternating current (a.c.) in devices connected to it. In a circuit containing only resistors, the voltage and current will always be in phase. There are, however, two other circuit components, inductors and capacitors, which have to be considered in a.c. circuits; their voltage–current relationships are more complicated.

A5.5 Inductors

The simplest *inductor* is a piece of wire wound into the form of a coil. The first law of electromagnetism says that a piece of wire carrying an electric current generates a magnetic field around itself. In a long, straight wire this is not very significant. But when the wire is wound into a coil, the magnetic field is concentrated in and around the coil. This is the principle of the electromagnet, and the coil behaves like a bar magnet while the current is switched on.

Changes in current through the coil cause changes in the magnetic field it produces, so when an alternating voltage is applied to the coil, the magnetic field has to go through a cycle of building up, decreasing, then building up again in the opposite direction, and so on. A magnetic field takes its energy from the electrical energy of the voltage source; therefore, an inductor or coil in a circuit will hamper or impede the flow of current. This is rather similar to the effect of a resistance in a d.c. circuit, except that the electrical energy needed to overcome resistance in a circuit is converted to heat instead of magnetic field energy. The electrical *impedance* (Z) provided by the inductor is given by the expression:

$$Z = 2\pi fL$$

where f is the frequency of the alternating voltage and L is the *inductance* of the coil, which depends on its physical dimensions and the number of turns. Inductance is measured in henrys (H). Electrical impedance is measured in ohms, like electrical resistance. In fact resistance is just one special form of impedance. Unlike inductance, resistance does not depend on the frequency of the current. Another difference is that the current through an inductor and the voltage across it are out of phase by 90°. Ohm's law also applies to inductors (between current and voltage amplitudes); impedance (Z) replaces resistance (R) in $V = IR$.

A5.6 Input and output impedance

Input and output resistance have already been introduced. In general they may be replaced by input impedance and output impedance. The *output impedance* is an important property of any device which *delivers* a signal, e.g. a battery, a microphone or an amplifier. Devices with a low output impedance can deliver higher currents than those with high output impedance. The *input impedance* is an important property of any device which *receives* a signal (i.e. of any load), such as a loudspeaker, a level recorder or an amplifier. Devices with a low input impedance draw a higher current from the source than those with a high input impedance.

A5.7 The electrodynamic loudspeaker and electrodynamic microphone

The operation of electrodynamic loudspeakers and microphones is based upon the behaviour of a coil in a magnetic field. The loudspeaker operates on the same principle as the electric motor. The loudspeaker cone, which radiates the sound, is attached to a coil that is free to move in the magnetic field produced by a permanent magnet. When an alternating current signal is supplied to the coil, it becomes a small electromagnet. The coil is alternately attracted and repelled by the permanent magnet as the polarity of the coil's magnetic field changes with the alternating current of the signal. As the coil vibrates, it vibrates the attached loudspeaker cone, radiating a sound pressure waveform similar to the alternating current.

The microphone operates on the reverse principle, which is also the basis of the generator or dynamo. Sound waves strike the microphone diaphragm, causing it to vibrate to and fro. The vibratory motion is transferred to the coil attached to the diaphragm and situated in the field of a permanent magnet. According to the principle of electromagnetic induction (Faraday's law), a current is induced to flow in a coil moving in a magnetic field. This induced current is the microphone signal, and if the operation is faithful, it has the same waveform as the incoming sound pressure wave.

A number of vibration transducers are coil-operated and work in a similar way to the electrodynamic microphone. They produce an electrical signal (a voltage) which is proportional to the vibration velocity.

A5.8 Capacitors

A **capacitor**, also called a condenser, consists of two conducting surfaces, often parallel plates, separated by a non-conductor such as air. A continuous direct current cannot flow through the non-conductor, called a **dielectric**, but if the plates are connected to a d.c. battery, a momentary or transient current flows from the cell to the plates until they reach the same electrical potential as the cell terminals. Once the potentials are equal, the flow of current ceases and the capacitor is fully charged. The capacitor can be thought of as a device which stores electric charge. The amount of charge it can store depends on the voltage difference across its plates and upon its **capacitance**. The capacitance depends on the physical dimensions of the capacitor, i.e. the area of the plates, the separation of the plates and the dielectric material between them.

Capacitance is measured in farads (F), but one farad is a very very large unit, so capacitors used in electronic circuits usually have values measured in microfarads ($1\ \mu F = 10^{-6}$ F) or in picofarads ($1\ pF = 10^{-12}$ F). A value of $1{\cdot}0\ \mu F$ represents a fairly large capacitance, whereas 1 pF is a very small capacitance indeed. A typical value for the capacitance of a one-inch condenser microphone is about 60 pF.

If an alternating voltage is applied to a circuit containing a condenser, an alternating electric current flows in the circuit even though electric charge cannot pass across the gap between the plates. Electric charge is continually moving to and from the plates in response to the cyclic variations in the voltage. The capacitance presents an impedance to the flow of current; it is measured in ohms and is given by:

$$Z = \frac{1}{2\pi fC}$$

where C is the capacitance in farads. For a condenser microphone of 60 pF operating at 1000 Hz:

$$Z = \frac{1}{2 \times \pi \times 1000 \times 60 \times 10^{-12}} = 2{\cdot}65 \times 10^{6}\ \Omega$$

$$= 2{\cdot}65\ \text{M}\Omega$$

A5.9 Resistance–capacitance (*RC*) circuits

Figure A5.6 shows a d.c. battery, a resistor and a capacitor connected in series; a switch is also in the circuit. When the switch is closed, the voltage at point A rises instantaneously to the battery voltage. The voltage also rises at point B, i.e. the voltage rises on the plate of the capacitor as the capacitor becomes charged, but it rises more gradually than at A. The rate of increase of the voltage at B is in fact exponential; it depends on the **time constant** of the circuit, which is given in seconds by RC, i.e. by the production of R in ohms and C in farads. As the voltage at B reaches the battery voltage, it rises more and more slowly and the flow of current becomes less and less. If the voltage at A suddenly drops to zero, e.g. if the battery is suddenly shorted out, the voltage at B also falls, but more

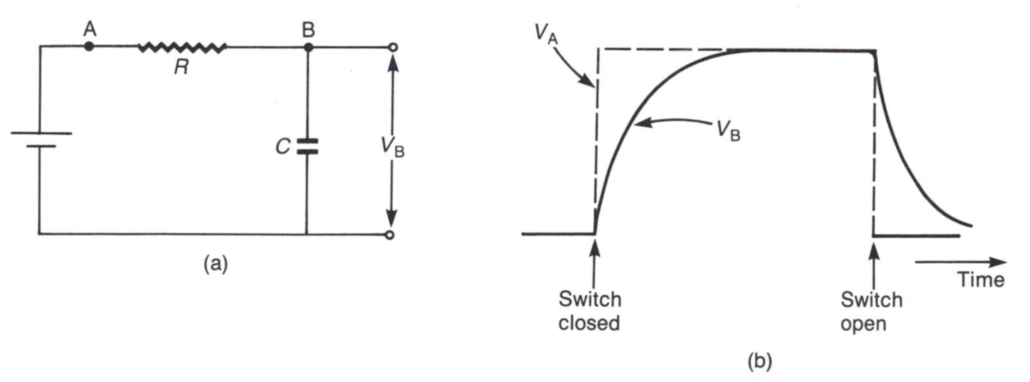

Fig. A5.6 (a) RC circuit with d.c. supply and (b) graph of voltages at A and B against time

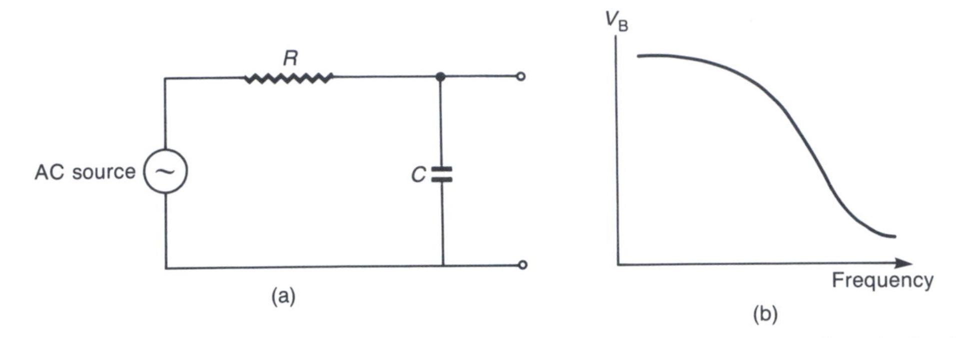

Fig. A5.7 (a) RC circuit with a.c. supply and (b) graph of V_B against supply frequency. This configuration is a low pass filter

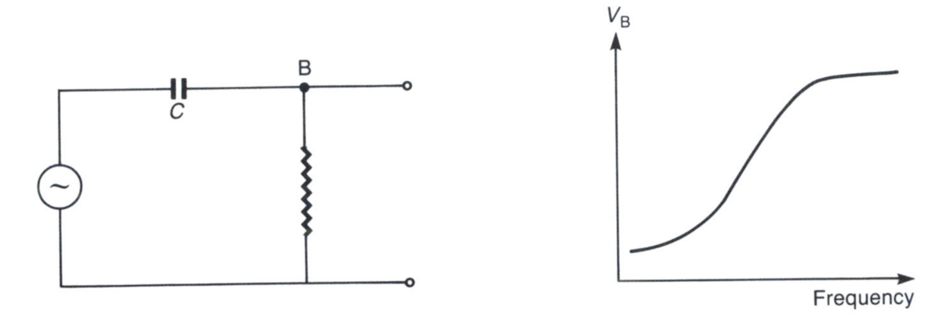

Fig. A5.8 (a) RC circuit where V_B measures voltage across the resistor and (b) graph of V_B against supply frequency. This configuration is a high pass filter

slowly (exponentially in fact) as the capacitor discharges. **A capacitor in a circuit can have the effect of smoothing out or slowing down rapid changes in voltage signals.** This is similar to the way a large reservoir smooths out variations of flow in air or water systems. The same property of capacitors can be used to smooth out or average time-varying signals; this is the basis for the averaging circuits in sound level meters. The averaging time (which is 1·0 s for slow mode and 0·125 s for fast mode) refers to the time constant of the circuit.

If the battery is replaced by an a.c. voltage source (Fig. A5.7), a steady current flows through the circuit. The voltage amplitude at B, at the capacitor plate, depends on the relative magnitudes of the impedances of C and R. The situation is similar to that of Fig. A5.5, but with two impedances instead of the two resistances R and r. At low frequencies the impedance of the capacitor will be far greater than the impedance of the resistor, so the voltage across the capacitor will be almost the source voltage. The voltage across the resistor will be small. As the frequency increases, the voltage across the capacitor decreases and the voltage across the resistor increases. If we consider the voltage at A as an input to the *RC* circuit and the voltage at B as the output, then the circuit is acting as a very crude **low pass filter**. If R and C are interchanged (Fig. A5.8), similar reasoning shows that the circuit now behaves as a crude **high pass filter**. The filters used in sound measuring equipment are much more complex and have a much sharper cut-off rate, but some of them are based on multi-stage *RC* networks.

A5.10 Oscillating electrical circuits

Inductors store energy as magnetic field energy. Capacitors store energy in the electric field between their two plates. In a circuit containing an inductor and a capacitor, the energy continually flows from one to the other and back again, resulting in an oscillating or alternating current flow. This electrical oscillation or vibration is analogous to the vibration of a mechanical oscillator, i.e. a mass on a spring. In both cases there wil be a loss of energy. In the electrical system this arises because of resistance in the circuit and in the mechanical system it is due to damping.

Bibliography

Chapter 1

British, European and International Standards

All British Standard and ISO Recommendations may be obtained from the British Standards Institution, British Standards House, 2 Park Street, London W1A 2BS.

Standard	ISO	Title
BS 2497		
Part 5: 1988	ISO 389	Specification for a standard reference zero using an acoustic coupler complying with BS 4668
Part 7: 1988	ISO 389/Add 2	Specification for a standard reference zero at frequencies intermediate between those given in Parts 5 and 6
BS 3425: 1966	ISO 362	Method of measurement of noise emitted by accelerating motor vehicles — engineering method
BS 3593: 1986	ISO R266	Recommendation on preferred frequencies for acoustical measurements
BS 4196		
Part 0: 1981	ISO 3740	Guide for the use of basic standards and for the preparation of noise test codes
Part 1: 1991	ISO 3741	Precision methods for determination of sound power levels for broadband sources in reverberation rooms
Part 2: 1991	ISO 3742	Precision methods for determination of sound power levels for discrete-frequency and narrowband sources in reverberation rooms
Part 3: 1991	ISO 3743	Engineering methods for determination of sound power levels for sources in special reverberation test rooms
Part 4: 1981	ISO 3744	Engineering methods for the determination of sound power levels for sources in free-field conditions over a reflecting plane
Part 5: 1981	ISO 3745	Precision methods for the determination of sound power levels in anechoic and semi-anechoic rooms
Part 6: 1981	ISO 3746	Survey methods for determination of sound power levels of noise sources
Part 7: 1988	ISO 3747	Survey methods for determination of sound power levels of noise sources using a reference sound source
Part 8: 1991	ISO 6926	Specification for the performance and calibration of reference sound sources
BS 4727		
Part 3 Group 8: 1985 IEC 50(60), IEC 50(801)		Acoustics and electro-acoustics terminology

BS 4813: 1972		Method of measuring noise from machine tools excluding testing in anechoic chambers
BS 5331: 1976	ISO 2249	Method of test measurement of the physical properties of sonic bangs
BS 5721: 1979	ISO 537	Specification for frequency weighting for the measurement of aircraft noise (D weighting)
BS 5793		
Part 8: 1991	(IEC 543-8-2)	Method for laboratory measurement of noise generated by hydrodynamic flow through control values
BS 5944		Measurement of airborne noise from hydraulic fluid power systems and components
Part 1: 1980	ISO 4412/1	Method of test for pumps
Part 2: 1980	ISO 4412/2	Method of test for motors
Part 3: 1980		Guide to the application of Part 1 and Part 2
Part 4: 1984		Method of determining sound power levels from valves controlling flow and pressure
Part 5: 1985		Simplified method of determining sound power levels from pumps using an anechoic chamber
BS 5969: 1981	IEC 651	Specification for sound level meters
BS 6056: 1991	IEC 551	Determination of transformer and reactor sound levels
BS 6086: 1981	ISO 5128	Method of measurement for noise inside motor vehicles
BS 6686		Methods for determination of airborne acoustical noise emitted by household and similar electrical appliances
Part 1: 1986	IEC 704-1	General requirements for testing
Part 2		Particular requirements
Part 2.3: 1991		Dishwashers
Part 2.4: 1991		Washing machines and spin-driers
BS 6812		Airborne noise emitted by earth-moving machinery
Part 1: 1987	ISO 6393	Method of measurement of exterior noise in a stationary test condition
Part 2: 1987	ISO 6394	Method of measurement at the operator's position in a stationary test condition
Part 3: 1991	ISO 6395	Method of measurement of exterior noise in dynamic test conditions
BS 6916		Chain-saws
Part 6: 1988	ISO 7182: 1984 EN 27182: 1991	Method of measurement of airborne noise at the operator's position
BS 7135		
Part 1: 1989	ISO 7779: 1988 EN 27779: 1991	Acoustics — measurement of airborne noise emitted by computer and business equipment
Part 2: 1989	ISO 9295: 1988 EN 29295: 1991	Acoustics — measurement of high-frequency noise emitted by computer and business equipment
BS 7136		
Part 1: 1989	ISO 7917 EN 27917	Method of measurement at the operator's position of emitted airborne noise emitted by brush-saws
BS 7189: 1989	IEC 942	Specification for sound calibrators
ISO 362		Measurement of noise emitted by passenger cars under conditions representative of urban driving
ISO 8960: 1991		Refrigerators, frozen-food storage cabinets and food freezers for household and similar use — measurement of emission of airborne acoustical noise
EN 21680		Acoustics — test code for the measurement of airborne noise emitted by rotary electrical machinery
ISO 11094: 1991		Acoustics — test code for the measurement of airborne noise emitted by power lawnmowers, lawn tractors, lawn and garden tractors, professional mowers and lawn and garden tractors with moving attachments. (No BS but see Statutory Instrument 1986, No. 1795.)

ISO 140-10: 1991	Acoustics — measurement of sound insulation in buildings and of building elements, Part 10: Laboratory measurement of airborne sound insulation of small building elements

US Standards

ANSI S1.4	Specification for sound level meters
S1.13	Methods for the measurement of sound pressure levels
S1.23	Method for the designation of sound power 5 emitted by machinery and equipment

Pneurop publications

Obtainable from British Compressed Air Society, 8 Leicester Street, London WC2H 7BN. These EN standards are preliminary at the time of writing and bear the designation prEN.

5604 (1969)	Measurement of sound from pneumatic equipment

General

Noise measurement techniques. *Notes on Applied Science No. 10*, National Physical Laboratory, Department of Scientific and Industrial Research

Porges G, *Applied Acoustics*, Edward Arnold, 1977

Chapter 2

British, European and International Standards

BS 2497: 1992	ISO 389: 1991	Standard reference zero for the calibration of pure tone, air conduction audiometers
Part 6: 1988	ISO 389/Add 1	Specification for a standard reference zero using an artificial ear complying with BS 4669
BS 3045: 1991	ISO 131	Method of expression of physical and subjective magnitudes of sound or noise in air
BS 4668: 1971	IEC 303	Specification for an acoustic coupler (IEC reference type) for calibration of earphones used in audiometry
BS 4669: 1971	IEC 318	Specification for an artificial ear of the wideband type for the calibration of earphones used in audiometry
BS 5108		Sound attenuation of hearing protectors
Part 1: 1991	ISO 4869-1	Subjective method of measurement
BS 5228		Noise control on construction and open sites
Part 1: 1984		Code of practice for basic information and procedures for noise control
BS 5330: 1976	ISO 1999	Method of test for estimating the risk of hearing handicap due to noise exposure
BS 5727: 1979	ISO 3891	Method for describing aircraft noise heard on the ground
BS 5966: 1980	IEC 645	Specification for audiometers
BS 6083		Hearing-aids
Part 10: 1988	IEC 11810	Guide to hearing-aid standards
BS 6344		Industrial hearing protectors
Part 1: 1989		Specification for earmuffs
Part 2: 1988		Specification for earplugs
BS 6402: 1983		Specification for personal sound exposure meters
BS 6655: 1986	ISO 6189: 1983 EN 26189: 1991	Acoustics — pure tone, air conduction threshold audiometry for hearing conservation purposes

BS 7750: 1992		Specification for environmental management systems
BS 6950: 1988	ISO 7566: 1987 EN 27566: 1991	Acoustics — standard reference zero for the calibration of pure tone, bone conduction audiometers
BS 6951: 1988	ISO 7029: 1984 EN 27029	Acoustics — threshold of hearing by air conduction as a function of age and sex for otologically normal persons
BS 7113: 1989	ISO 8798: 1987 EN 28798: 1991	Acoustics — reference levels for narrowband masking noise
BS 7527		Classification of environmental conditions
Part 1: 1991	IEC 721-1	Environmental parameters and their severities
ISO 2204		Guide to International Standards on the measurement of airborne acoustical noise and evaluation of its effects on human beings
EN 352		Safety requirements and testing
EN 3521		Earmuffs
EN 3522		Earplugs
EN 3523		Helmet-mounted muffs
EN 3524		Level-dependent muffs
EN 3525		Hearing protectors — safety requirements and testing, alternative performance
EN 457: 1992		Safety of machinery — auditory danger signals General requirements, design and testing
EN 458		Recommendations for selection, use, care and maintenance of hearing protectors

US Standards

ANSI S3.19	Methods for the measurement of real ear protection of hearing protectors
S3.28	Evaluation of the potential effect on human hearing of sounds with peak A-weighted sound pressure levels above 120 dB and peak C-weighted sound pressure levels below 140 dB
S12.6	Method for the measurement of real ear attenuation of hearing protectors

General

Burns W, *Noise and Man*, John Murray

Noise and the worker. *Safety, Health and Welfare (New Series No. 25)*, HMSO

Noise. Final report for the Committee on the Problem of Noise, HMSO, Cmnd 2056:1966

Rodda M, *Noise and Society*, Oliver & Boyd

Van Bergeijk W A, Pierce J R and David E E, *Waves and the Ear*, Heinemann

Molitor, D L, *Noise Abatement*. A Public Health Problem Report for the Council of Europe, Strasbourg, 1964

Burns W and Robinson D W, *Hearing and Noise in Industry*, HMSO, 1970

Jergen, J, *Modern Developments in Audiology*, Academic Press, 1973

Ventry I M, Chaiklin J B and Dixon R F, *Hearing Measurement*, Meredith, 1971

Health and Safety Executive, *Audiometry in Industry*, HMSO

Chapter 3

British, European and International Standards

BS 3638: 1987	ISO 354	Method for the measurement of sound absorption in a reverberation room
ISO 9053: 1991		Acoustics — materials for acoustical applications Determination of airflow resistance
ISO 10053: 1991		Acoustics — measurement of office screen sound attenuation under specific laboratory conditions

US Standards

ASTM C384	Test method for impedance and absorption of acoustical materials by the impedance tube method

General

Doelle L L, *Acoustics in Architectural Design*, National Research Council Canada, Division of Building Research Bibliography, No. 29, 1965
Evans E J and Bazely E N, *Sound Absorbing Materials*, HMSO
Parkin P H and Humphreys H R, *Acoustics: Noise and Buildings*, Faber
Purkis H J, *Building Physics: Acoustics*, Pergamon
Knudsen V O and Harris C M, *Acoustical Designing in Architecture*, Wiley
Moore J E, *Design for Good Acoustics*, Architectural Press
Beranek L L, *Music, Acoustics and Architecture*, McGraw-Hill, 1960
Egan M D, *Concepts in Architectural Acoustics*, McGraw-Hill, 1973
Ford R D, *Introduction to Acoustics*, Applied Science Publishers, 1970
Gilford C, *Acoustics for Radio and Television Studios*, IEE Monograph, 1972
Kuttmuff M, *Room Acoustics*, Applied Science Publishers, 1975
Lawrence, *Architectural Acoustics*, 1970
Mankorsky V S, *Acoustics for Studios and Auditoria*, Focal Press, 1971
Mackenzie R, *Auditorium Acoustics*, Applied Science Publishers, 1975
Sound Research Laboratories, *Practical Building Acoustics*, 1976

Chapters 4 and 5

British, European and International Standards

BS 2750		Measurement of sound insulation in buildings of building elements
Part 1: 1980	ISO 140/1	Recommendations for laboratories
Part 2: 1980	ISO 140/2	Statement of precision requirements
Part 3: 1980	ISO 140/3	Laboratory measurements of airborne sound insulation of building elements
Part 4: 1980	ISO 140/4	Field measurements of airborne sound insulation between rooms
Part 5: 1980	ISO 140/5	Field measurements of airborne sound insulation of façade elements and façades
Part 6: 1980	ISO 140/6	Laboratory measurements of impact sound insulation of floors
Part 7: 1980	ISO 140/7	Field measurements of impact sound insulation of floors
Part 8: 1980	ISO 140/8	Laboratory measurements of the reduction of transmitted impact noise by floor coverings on a standard floor
Part 9: 1980	ISO 140/9	Method for laboratory measurement of room-to-room airborne sound insulation of a suspended ceiling with a plenum above it
BS 4773		
Part 2: 1989	ISO 5135: 1984 EN 25135: 1991	Acoustics — determination of sound power levels of noise from air terminal devices, high/low velocity/pressure assemblies, dampers and valves by measurement in reverberation rooms
BS 5821		Methods for rating the sound insulation in buildings and of building elements
Part 1: 1984	ISO 717/1	Method for rating airborne sound insulation in buildings and of interior building elements
Part 2: 1984	ISO 717/2	Method for rating the impact sound insulation
Part 3: 1984	ISO 717/3	Method for rating the airborne sound insulation of façade elements and façades

BS 7458		
Part 1: 1991	ISO 1680-1 EN 21680-1: 1991	Engineering methods for free-field conditions over a reflecting surface
Part 2: 1991	ISO 1680-2: 1986 EN 21680-2: 1991	Survey method
BS 8233: 1987		Code of practice for sound insulation and noise reduction for buildings
BS AU 193a		Specification for replacement motor cycle and moped exhaust systems
EN 23741: 1991	ISO 3741: 1988	Precision method for discrete frequency and narrowband sources in reverberation rooms

Codes of Practice

CP 153	Windows and roof-lights
Part 3: 1972	Sound insulation

The Control of Noise (Measurement and Registers) Regulations 1976
Building Regulations 1985 SI 1065
Noise Insulation Regulations 1975 SI 1763

US Standards

ANSI S12.8	Methods for the determination of insertion loss of outdoor noise barriers
ASTM E90	Standard test method for laboratory measurement of airborne sound transmission loss of building materials
E413	Classification for rating sound insulation
E1124	Standard test method for free-field measurement of sound power level by the two-surface method
E336	Test method for the measurement of airborne sound insulation in buildings

General

BRE (Building Research Establishment) documents
Current Paper 6/73. Designing offices against road traffic noise
Current Paper CP 20/78. Sound insulation performance between buildings built in the early 70s
Digest No. 266. Thermal visual and acoustic requirements in buildings
Digest No. 293. Improving the sound insulation of party-walls and floors
Berarek L L, *Noise reduction*, McGraw-Hill
Bazley E N, *The Airborne Sound Insulation of Partitions*, HMSO
Hines W A, *Noise Control in Industry*, Business Publications
King A J, *The Measurement and Suppression of Noise*, Chapman and Hall
Moore J E, *Design for Noise Reduction*, Architectural Press
London Noise Survey, Building Research Station, Ministry of Building and Works, HMSO
Noise in Factories, Factory Building Studies 6, HMSO, 1962
Anthrop D, *Noise Pollution*, Lexington Books, 1973
Cremer L M and Under E E, *Structure-borne Sound*, Springer-Verlag, 1973
Croome D, *Noise, Buildings and People*, Pergamon, 1977
Sharland I, *Woods Practical Guide to Noise Control*, Woods of Colchester
Sound Research Laboratories, *Noise Control in Industry*, 1976
Sound Research Laboratories, *Noise Control in Mechanical Services*, 1976
Warring R H (ed), *Handbook of Noise and Vibration Control*, Trade & Technical Press (Morden)

Chapter 6

British, European and International Standards

BS 4142: 1990	ISO 1996-1 ISO 1996-2 ISO 1996-3	Method of rating industrial noise affecting mixed residential and industrial areas
BS 5228		
Part 2: 1984		Guide to noise control legislation for construction and demolition, including road construction and maintenance
Part 3: 1984		Code of practice for noise control applicable to surface coal extraction by open-cast methods
Part 4: 1986		Code of practice for noise control applicable to piling operations
BS 6840		
Part 16: 1989		Guide to the RASTI method for the objective rating of speech intelligibility in auditoria
ISO/TR 4870: 1991		Acoustics — the construction and calibration of speech intelligibility tests

Codes of practice

Code of Practice on Noise from Ice-Cream Van Chimes 1982
Code of Practice on Noise from Audible Intruder Alarms 1982
Code of Practice on Noise from Model Aircraft 1982
Code of Practice on Noise from Construction and Open Sites 1984
DOE 10/73 Planning and Noise. This has been superseded by Planning Policy Guidance: Planning and Noise

US Standards

ANSI S3.5	Methods for the calculation of the articulation index
S3.14	Rating noise with respect to speech interference

General

Alexandre A et al. *Road Traffic Noise*, Applied Science Publishers, 1975
Duerden C, *Noise Abatement*, Butterworths, 1970
Doelle L L, *Environmental Acoustics*, McGraw-Hill, 1972
Timber Research & Development Association, *Noise and Woodworking Machinery*, Publication TYOM/1B3, 1971
Department of the Environment, Welsh Office, *Calculation of Road Traffic Noise*, HMSO

Chapter 7

British, European and International Standards

BS 3015: 1991	ISO 2041	Glossary of terms relating to mechanical vibration and shock
BS 3852		Balancing machines
Part 1: 1979	ISO 2953	Method of description and evaluation
BS 4675		Mechanical vibration in rotating machinery
Part 1: 1976	ISO 2372	Basis for specifying evaluation standards for rotating machines with operating speeds from 10 to 200 revolutions per second
Part 2: 1978	ISO 2954	Requirements for instruments for measuring vibration severity
BS 5265		Mechanical balancing of rotating bodies
Part 2: 1981	ISO 5406	Methods for the mechanical balancing of flexible rotors
Part 3: 1984	ISO 5343	Recommendations for criteria for evaluating flexible rotor balance
BS 6055: 1981	ISO 5008	Methods of measurement of whole-body vibration of operators of agricultural wheeled tractors and machinery

BS 6140		Test equipment for generating vibration
Part 1: 1981	ISO 5344	Methods of describing characteristics of electrodynamic test equipment for generating vibration
Part 2: 1982	ISO 6070	Methods for describing characteristics of auxiliary tables for test equipment for generating vibration
BS 6177: 1982		Guide to selection and use of elastomeric bearings for vibration isolation of buildings
BS 6294: 1982	ISO 7096	Method for measurement of body vibration transmitted to the operator of earth-moving machinery
BS 6414: 1983	ISO 2017	Method for specifying characteristics of vibration and shock isolators
BS 6472: 1984		Guide to the evaluation of human exposure to vibration in buildings (1 Hz to 80 Hz)
BS 6611: 1985	ISO 6897	Guide to the evaluation of the response of occupants of fixed structures, especially buildings and offshore structures to low frequency horizontal motion (0·63 Hz to 1 Hz)
BS 6749		Measurements and evaluation of vibration on rotating shafts
Part 1: 1986	ISO 7919/1	Guide to general principles
BS 6841: 1987	ISO 2631-1	Guide to measurement and evaluation of human exposure to whole-body mechanical vibration and repeated shock
BS 6842: 1987	ISO 5349	Guide to measurement and evaluation of human exposure to vibration transmitted to the hand
BS 6916		
Part 8	ISO 7505	Method of measurement of hand-transmitted vibration
BS 6955		Calibration of vibration and shock pick-ups
Part 0: 1988	ISO 5347-0	Guide to basic principles
BS 7085: 1989		Guide to safety aspects of experiments in which people are exposed to mechanical vibration and shock
BS 7129	ISO 5348	Recommendations for mechanical mounting of accelerometers for measuring mechanical vibration and shock
BS 7285: 1990	ISO 8626	Method for describing characteristics of servohydraulic test equipment for generating vibration
BS 7347: 1990	ISO 8568	Guide for characteristics and performance of mechanical shock-testing machines
BS 7385		
Part 1: 1990	ISO 4866	Guide for measurement of vibration and evaluation of their effects in buildings
Part 2: 1993		Guide to damage levels from ground borne vibrations
BS 7482: 1991 Parts 1, 2, 3		Instrumentation for the measurement of vibration exposure of human beings
BS 7527		
Section 2.6 1991	IEC 721-2-6	Earthquake vibration and shock
ISO 6258: 1985	ISO 8662	Nuclear power plants — design against seismic hazards
		Measurement of vibrations in hand-held power tools
Part 1		General (only part published at present)
Part 2		Chipping hammers, riveting hammers
Part 3		Rotary hammers and rock drills
Part 4		Grinding machines
Part 5		Breakers and hammers for construction
Part 6		Impact drills
Part 7		Impact wrenches
Part 8		Orbital sanders
Part 9		Rotary drilling tools

ISO 10816	Evaluation of mechanical vibration by measurement on non-rotating parts
Part 4	Industrial gas turbine sets
Part 5	Hydraulic turbine sets
Part 6	Reciprocating machines
ISO 10819	Method of measurement of vibration attenuation
Part 2	Resilient materials used for the protection of workers against risk from vibration exposure to the hands

US Standards

ANSI S2.8	Guide for describing the characteristics of resilient mountings
S2.61	Guide to the mechanical mounting of accelerometers
S3.40	Guide for the measurement and evaluation of gloves which are used to reduce exposure to vibration transmitted to the hand
S9.1	Guide for the selection of mechanical devices used in monitoring acceleration induced by shock

Pneurop publications

Obtainable from the British Compressed Air Society, 8 Leicester Street, London WC2H 7BN. These EN standards are preliminary at the time of writing and bear the designation prEN.

6610 (1978)	Test procedure for the measurement of vibration from hand-held power-driven grinding machines
66160 (1985)	Test procedure for the measurement of vibration from chipping hammers

General

Griffin M J, *Handbook of Human Vibration*, Academic Press, 1990

Brock J T, *Mechanical Vibration and Shock Measurements*, Bruel and Kjaer, 1984

Beranek L L (ed), *Noise and Vibration Control*, McGraw-Hill, 1971

Petrusewicz S A and Longmore D K (eds), *Noise and Vibration Control for Industrialists*, Elek Science, 1974, Chs 6–13

Baker J K, Vibration isolation. *Engineering Design Guide 13*, Oxford University Press, 1975

Middleton A H, Machinery noise. *Engineering Design Guide 22*, Oxford University Press, 1977

Steffens R J, *Structural Vibration and Damage*, Building Research Establishment Report, HMSO, 1974 (republished 1992)

Building Research Establishment, *Digest 353: Damage to Structures from Ground-borne Vibration*, 1990

Wasserman D E, *Human Aspects of Occupational Vibration*, Vol. 8 of *Advances in Human Factors/Ergonomics*, Elsevier, 1987

Nelson P M (ed), *Transportation Noise Reference Book*, Butterworth, 1987

Cempel C, *Vibroacoustic Condition Monitoring*, Ellis Horwood, 1991

Hunt J B, *Dynamic Vibration Absorbers*, Mechanical Engineering Publications, 1979

Health and Safety Executive, *Hand-Arm Vibration*, HMSO, 1994

Smith J D, *Vibration Measurement and Analysis*, Butterworth, 1989

Piezoelectric Accelerometers and Vibration Preamplifiers, Bruel and Kjaer, 1976

Broch J T, *Mechanical Vibration and Shock Measurements*, Bruel and Kjaer, 1973

Noise Control and Vibration Isolation, monthly journal

Handbook of Noise Measurement, General Radio Company

Michael Neale & Associates, *A Guide to the Condition Monitoring of Machinery*, HMSO

Chapter 8

British, European and International Standards

Standard	ISO/EN	Title
BS 4196		See p 307 for Parts 0 to 8
BS 4718: 1971		Methods of test for silencers for air distribution systems
BS 5512: 1991	ISO 281	Method of calculating dynamic load ratings and rating life of rolling bearings
BS 6107		Rolling bearings: tolerances
Part 1: 1981	ISO 1132	Glossary of terms
Part 2: 1987	ISO 492	Specification for tolerances of radial bearings
Part 3: 1992	ISO 5753	Specification for radial internal clearance
BS 6085		Statistical methods for determining and verifying stated noise emission values of machinery and equipment
Part 1: 1987	ISO 7574/1	Glossary of terms
Part 2: 1987	ISO 7574/2 EN 27 574-2	Method for determining and verifying stated values for individual machines
Part 3: 1987	ISO 7574/3 EN 27 574-3	Method for determining and verifying stated values for batches of machines using a simple (transition) method
Part 4: 1987	ISO 7574/4 EN 27 574-4	Method for determining and verifying stated values for batches of machines
BS 6861		Balance quality requirements of rigid rotors
Part 1: 1987	ISO 1940/1	Method for determination of residual unbalance
BS 7025: 1988	ISO 6081	Method for preparation of test codes of engineering grade for measurements at the operator's or bystander's position of noise emitted by machinery

EC directives

Directive	Title
70/157/EEC 77/212/EEC 81/334/EEC 84/372/EEC 84/424/EEC	Permissible sound level and the exhaust system of motor vehicles. Implemented by the Road Vehicles (Construction & Use) Regulations 1986 SI 1986:1078
74/151/EEC	Tractors — implemented by the Agricultural Forestry Tractors and Tractor Components (Type Approval) Regulations 1979 SI 1979:221
77/311/EEC	Driver-perceived noise level of wheeled agricultural or forestry tractors — implemented by SI 1979:221
78/1015/EEC 89/235/EEC	Permissible sound level and exhaust systems of motor cycles
79/113/EEC 81/1051/EEC 84/532/EEC	Determination of the noise emission of construction plant and equipment — implemented by SI 1985:1968 Construction Plant and Equipment (Harmonisation of Noise Emission Standards) Regulations 1985
84/533/EEC 85/406/EEC	Acoustic power level admissible for motor compressors — implemented by SI 1985:1968
84/534/EEC 87/405/EEC	Acoustic power level admissible for tower cranes — implemented by SI 1985:1968
84/535/EEC 885/407/EEC	Acoustic power level admissible for welding generating sets — implemented by SI 1985:1968
84/536/EEC 85/408/EEC	Acoustic power level admissible for generating sets — implemented by SI 1985:1968
84/537/EEC 85/409/EEC	Acoustic power level admissible for hand-held concrete breakers and pick hammers — implemented by SI 1985:1968
86/594/EEC	Airborne noise emitted by household appliances — implemented by The Household Appliances (Noise Emission) Regulations 1990 SI 1960:161

86/662/EEC	Limitation of sound emission by hydraulic and rope-operated excavators, tractors with dozer equipment, loaders and backhoe loaders — implemented by SI 1988:361 Noise: Construction Plant and Equipment (Harmonisation of Noise Emission Standards) 1988
86/663/EEC	Self-powered industrial trucks
84/538/EEC 87/252/EEC 88/180/EEC 86/188/EEC	Acoustic power level admissible for lawnmowers — implemented by the Lawnmowers (Harmonisation of Noise Emission Standards) Regulations 1986 and 1987, SI 1986:1795 and SI 1987:876
86/188/EEC	The protection of workers from risks related to exposure to noise at work — implemented by SI 1989:1790 Noise at Work Regulations 1989
89/391/EEC	Improvements of health and safety at work
89/392/EEC	Machinery
89/629/EEC	The limitation of noise emission from civil subsonic jet aeroplanes — implemented by the Air Navigation (Noise Certification) Order 1990 and SI 1990:1514
85/337/EEC	The assessment of the effects of certain public and private works on the environment — implemented by the Town and Country Planning (Assessment of Environmental Effects) Regulations 1988 European Standards (includes EC and EFTA)

Code of practice

Code of Practice for Noise Control on Construction and Demolition Sites: BS 5228:1975

General

Sharland I J, *Woods Practical Guide to Noise Control*, Woods of Colchester, 1972
Webb J D (ed), *Noise Control in Industry*, Faber, 1958
Parkin P H and Humphreys H R, *Acoustics Noise and Buildings*, Faber, 1958
Petrusewicz S A and Longmore D K, *Noise and Vibration Control for Industrialists*, Elek Science, 1974
Porges G, *Applied Acoustics*, Edward Arnold, 1977
Middleton A H, Machinery noise. *Engineering Design Guide 22*, Oxford University Press, 1977
Beranek L L (ed), *Noise and Vibration Control*, McGraw-Hill, 1971
Warring R H (ed), *Handbook of Noise and Vibration Control*, Trade and Technical Press (Morden)
Bazely E N, *The Airborne Sound Insulation of Partitions*, HMSO, 1966
Institute of Heating and Ventilating Engineers (now Chartered Institute of Building Services), *Sound Control*, IHVE Guide, Section B12
King A J, *The Measurement and Suppression of Noise*, Chapman and Hall, 1965
Hassall J R and Zaveri K, *Acoustic Noise Measurements*, Bruel and Kjaer, 1979, Ch. 5
Timber Research & Development Association, *Noise and Woodworking Machinery*, Information Bulletin Three, 1975
Building Research Establishment, *A Noise-Reducing Enclosure for a Panel-Planer*, Information Sheet IS 27/76

Chapter 9

British, European and International Standards

BS 3539: 1986		Specification for sound level meters for the measurement of noise emitted by motor vehicles
BS 5969: 1981	IEC 651: 1979	Specification for sound level meters
BS 5647: 1979	IEC 561	Specification for electro-acoustical measuring equipment for aircraft noise certification
BS 6402: 1983		Specification for personal sound exposure meters

BS 6698: 1986	IEC 804: 1984	Specification for integrating–averaging sound level meters
BS 7189: 1989	IEC 942: 1988	Specification for sound calibrators
BS 7580: 1992		The verification of sound level meters
BS 7703	ISO 9614	Acoustics — determination of sound power levels of noise sources using sound intensity
Part 1: 1993		Measurement at discrete points
	IEC 119-5	Digital audio tape cassette system (DAT)
Part 5		DAT for professional use
BS 2475: 1964	IEC 225: 1966	Octave, half-octave and third-octave band filters intended for the analysis of sounds and vibrations
	ISO 266: 1975	Preferred frequencies for acoustical measurements

General

Brook D and Wynne R J, *Signal Processing: Principles and Applications*, Edward Arnold, 1988

Alton Everest F A, *The Master Handbook of Acoustics*, 3rd edn, TAB Books, 1994, Chs 6, 24

Anderson J S and Bratos-Anderson M, *Noise: Its Measurement, Analysis, Rating and Control*, Avebury Technical, 1993, Ch. 2

Foreman J E K, *Sound Analysis and Noise Control*, Van Nostrand Reinhold, 1990, Ch. 3

Beranek L L (ed), *Noise and Vibration Control*, McGraw-Hill, 1971 (contains chapters on sound and vibration transducers, field measurements: equipment and techniques, data analysis)

Condenser Microphones: Data Handbook, Bruel and Kjaer, 1982

Beauchamp K G, *Signal Processing: Using Analogue and Digital Techniques*, Allen and Unwin, 1973

Marple D L, *Digital Spectral Analysis*, Prentice Hall, 1987

Chapter 10

Table of cases

AG v. Gastonia Coaches (1976) *The Times* Nov. 12 213

AG (Gambia) v. N'Jie (1961) AC 617, (1961) 2 WLR 845 225

AG v. PYA Quarries (1957) 2 QB 169 211, 212

Aitken v. South Hams District Council (1994) *The Independent* July 13 220

Allen v. Gulf Oil Refinery Ltd (1981) 2 WLR 188 215

Andreae v. Selfridge (1938) Ch. 1 212, 216

Anns v. London Merton Corporation (1977) 2 WLR 1024 232

Caparo Industries plc v. Dickman (1990) 2 WLR 358 (HL) 232

Christie v. Davey (1893) 1 Ch. 316 213

City of London Corporation v. Bovis Construction Limited (1988) 86 LGR 660 227

Cooke v. Adatia (1988) 153 JP 129 221, 223

Coventry City Council v. Cartwright (1975) 1 WLR 845 218

Donoghue v. Stevenson (1932) AC 562 232

Dunton v. Dover District Council (1977) 76 LGR 87 213

Fraser v. Booth (1949) 50 SR (NSW) 113 213

Galer v. Morrissey (1955) 1 All ER 380 218

Gillingham Borough Council v. Medway (Chatham) Dock Co. Ltd (1992) 1 PLR 123 246

Haley v. London Electricity Board (1965) AC 778, (1964) 3 WLR 479 233

Halsey v. Esso Petroleum Co. Ltd (1961) All ER 145 212, 213

Hammersmith London Borough Council v. Magnum Automated Forecourts (1978) 1 WLR 50 217, 222, 224

Hollywood Silver Fox Farm Ltd v. Emmett (1936) 2 KB 468 213

Table of statutes

Table of statutory instruments

Noise at Work Regulations 1989 210, 237, 240–241, 258
Noise Insulation Regulations 1975/1763 242, 250
Noise Insulation (Amendment) Regulations 1988/2000 242
Safety Signs Regulations 1980 241
Social Security (Industrial Injuries) (Prescribed Diseases) Regulations 1985 237
Statutory Nuisance (Appeals) Regulations 1990 221–222
Town and Country Planning (Applications) Regulations 1988 245
Town and Country Planning (Assessment of Environmental Effects) Regulations 1988 246, 252

Air Navigation Order 1989/2004 248
  Article 83 248
Air Navigation (Noise Certification) Order 1990/1514 250
Control of Noise (Code of Practice for Construction Sites) Order 1975 226, 232
Control of Noise (Code of Practice for Construction and Open Sites) Order 1984/92 and 1987/1730 226, 232
Control of Noise (Code of Practice on Noise from Ice-Cream Van Chimes, Etc.) Order 1981 232
Control of Noise (Code of Practice on Noise from Audible Intruder Alarms) Order 1981 232
Control of Noise (Code of Practice on Noise from Model Aircraft) Order 1981 232
Town and Country General Development Order 1988 245, 246
  Para. 3 245
  Schedule 2 245
Town and Country Planning (Use Classes) Order 1987 245
Town and Country Planning (Use Classes) (Amendment) Order 1991 245

Codes and circulars

BSI Code of Practice 5228
BSS 4197/1967 230

DOE Circular 10/73 Planning and Noise 244, 245
DOE Circular 2/76 230
DOE Circular 1/85 The use of conditions in planning permissions 245
DOE Circular 15/88 246

European Community legislation

Treaty of Paris 251
First Treaty of Rome 252
Second Treaty of Rome 251
Treaty on European Union (Maastricht Treaty) 251, 252
Single European Act 1986 251, 252

EC Directives

Construction Products Directives 89/100/EEC 243
The Protection of Workers from Noise 86/188/EEC 240, 252
EEC Council Directive 85/337/EEC Planning 246
EC Commission Directive 92/14/EEC 250
EC Council Directive 89/629 249

Reports

Report of the Noise Review Working Party 1990 (DOE) (Batho Report) 248

Books

General

Kerse C S, *The Law Relating to Noise*, Oyez Publishing, 1975
Penn C, *Noise Control*, Shaw & Sons, 1995
Ball S and Bell S, *Environmental law: the law and policy relating to the protection of the environment*, Blackstone, 1994
Hughes D, *Environmental Law*, Butterworth, 1992
Current Law Year Book, Sweet & Maxwell

Tort

Heuston R F V and Buckley R A, *Salmon and Heuston on the Law of Torts*, Sweet & Maxwell, 1992
Baker C D, *Tort*, Sweet & Maxwell, 1991
Jolowicz J A and Rogers W V H, *Winfield and Jolowicz on Tort*, Sweet & Maxwell, 1994

Town and country planning

Encyclopaedia of Planning Law and Practice

Local government

Wilson, D and Game C, *Local Government in the United Kingdom*, Macmillan, 1994

Health and safety at work

Health and Safety at Work (Croner)
Encyclopaedia of Health and Safety at Work (Sweet & Maxwell)
Encyclopaedia of Environmental Health Law

Land law

Megarry, Sir Robert and Thompson M P, *A Manual of the Law of Real Property*, Sweet & Maxwell

Journals

Environmental Health
Environmental Health News
Occupational Safety and Health
Health and Safety at Work
Noise and Vibration Bulletin
Noise and Vibration Worldwide
Journal of Planning and Environmental Law
Environmental Law and Management
Environmental Policy and Practice
Acoustics Bulletin

Answers

Chapter 1

1. 111 dB
2. 101·5 dB
3. 4
4. 73·4 dB, $2{\cdot}2 \times 10^{-5}$ W/m^2
5. 95·5 dB, 74·3 dB A
6. $4{\cdot}9 \times 10^{-8}$ W/m^2
7. 20·25 N/m^2
8. 70 to 77 dB re 10^{-12} W
9. $2{\cdot}5 \times 10^{-4}$ W/m^2, 84 dB
10. 96 dB
11. $6{\cdot}3 \times 10^{-2}$ W
12. 0·18 W
13. 66 dB

14(c) 64·5 dB

15. Approx 21 dB(A)
16. 89·5 dB(A)
17. 80·6 dB(A)
18. 70·3 dB(A)
19. 3 bursts

20(a) 1·6 hours
20(b) 1·0 hours

21. 102 take offs
22. 71·5 dB(A)

Chapter 2

1. 88 phons
2. 35 phons
3. 90 PNdB
4. Below 40 PNdB
5. 106 phons
6. 109 PNdB
9. 0·1 hr (6 minutes)

15(a) 34 sones
15(b) 91 phons
15(c) 77·5 dB(A)

16. A basilar membrane
    B tectorial membrane
    C arch or rods of Corti
    D hair cells
    E branch of cochlear or auditory nerve.

Chapter 3

1. 4400
2. 600 m^2, 0·8 s, 70 m^2 (approx)
3. 51
4. 0·27 (125 Hz), 0·32 (250 Hz), 0·36 (500 Hz), 0·39 (1000 Hz), 0·37 (2000 Hz)
5. Actual 1·87 s, Optimum 1·63 s, 34 extra
6. Optimum 1·2 s, Actual 1·8 s
7. 125 Hz — none; 250 Hz — 6·4 m^2; 500 Hz — 9 m^2; 1000 Hz — 10·3 m^2; 2000 Hz — 7·6 m^2; 4000 Hz — 4·3 m^2
8. 0·87 s, 54·3 m^2
9. 3·27 s, 4 dB
10. Approx 1800 m^3; RT about 1 s; absorption approx 50 m^2
11. Optimum RT 1·8 s; Actual RT 2 s
12. 0·21
13. 0·67, 0·59, 0·56
14. Volume approx 2000 m^3
    Optimum RT 1·3 s

16(b) 960 m^2

19. 0·02, 0·19, 0·65, 0·45

20(a) 0·031
(b) 0·49

Chapter 4

1. 30 Hz
2. 16 mm
3. 2 m square
4. 5810 N/mm^2
5. 56 dB, meets Building Regulations requirements
6. 400 m/s

Chapter 5

1. 106·9 dB

2. 58 m
3. $8 \cdot 1 \times 10^{-6}$ W
4. $1 \cdot 9$ m^2
5. 5 dB
6. 36·5 dB
7. 17, 17, 14, 13, 9, 8, 6 dB
8. 1·6, 10, 14, 25, 29, 32, 29
9. 27·5 dB
10. 99 dB
11. 13 dB
12. 5 kg/m^2
13. 42 dB
14. 93 dB(A)
15(b) 43 dB(A)
16(b) 51 dB (airborne)
66·2 dB (impact)
17(i) 28·5 dB
(ii) 26 dB
24. Sound level difference 17, 23, 24 dB
Insertion loss 6, 21, 36 dB

Chapter 6

1. 39 dB
2. 24, 37, 45, 50, 47, 45, 44, 37 dB
3. 25 dB; 8, 13, 18, 23, 28, 33, 38, 43 dB
4. 30, 38, 36, 31, 23, 20, 20
5. 260 k/m^2; 1268 kg/m^2 (impossibly high); 562 m^2
6. 36 m

7(a)(i) Dumper only working
Site 1 54 dB(A)
2 62 dB(A)
3 53 dB(A)
4 61 dB(A)
(ii) Both working
Site 1 70 dB(A)
2 79 dB(A)
3 68 dB(A)
4 64 dB(A)
(b) 67 dB(A); 25·5 dB(A)

8(a) 28, 33, 41, 45, 42, 40, 38, 31 dB; 15%
(b) Volume approx 2000 m^3
Low frequencies approx 1·4 s
Medium frequencies approx 1 s
High frequencies approx 1 s
240 m^2

12. (7am–4pm) L_{eq} = 92·8 dB(A)
(4pm–10pm) L_{eq} = 89·5 dB(A)
(10pm–7am) L_{eq} = 89·3 dB(A)
24 hr L_{eq} = 91·0 dB(A)
14. 0·73 hours = 43·6 minutes

15(i) 11·2 dB(A)
(ii) 5 hr 3 minutes
16. 68 dB (31·5 Hz), 52 dB (63 Hz), 48 dB (125 Hz), 40 dB (250 Hz), 33 dB (500 Hz), 24 dB (1000 Hz); problem at 30 Hz caused by vibration from pump suitable isolators would produce T = 0·125, fo = 10 Hz
19. 88 dB(A)
20. NR25 just achieved; noise at 2 kHz probably caused by turbulence from obstructions to smooth flow, eg from grille and/or damper

Chapter 7

1. 1·6 mm/s, 8·4 microns
2. 0·63 mm/s, 0·2 m/s^2
3. 159 microns, 0·628 m/s^2
4. 22·0 m/s^2, 5·52 m/s^2, 147 dB, 135 dB
5. 1·58 m/s^2
6. 0·14 mm/s
7. 930 Hz
8. (a) 8·0 dB, 0·4
(b) 25 Hz, 0·4 mm
19. 0·263 m/$s^{1\cdot75}$ (low probability of adverse comment)
20. 0·057 m/s^2
21. 0·113 m/$s^{1\cdot75}$ (low probability of adverse comment)
22. 130 bursts
23(a) 1 Hz
(b) 250 mm
(c) 7·11 kN/m
24. 24·8 mm/s
25. above 36 Hz
26. 93·5 dB
27. 1·4 m/s, $3 \cdot 9 \times 10^{-5}$ m
28(i) 3·97 m/s^2
(ii) 12·6 mm/s, 40·1 μm
29(a) 0·0138 m/s
(b) $3 \cdot 88 \times 10^{-5}$ m
(c) 123 dB

Chapter 8

1. 0·24 microns
2. 65·9 dB
3. 96·5 dB, 63 dB
4. 40·7 dB, 33·5 dB
5. 38 dB
6. 43·5 dB
10(c) 95·7 dB
11(c) 22 dB
13. 0·0762 W, 129 dB re 1 μPa
14(a) 112 dB
(b) 112 dB

15. 27 dB (less a nominal allowance for sound absorbing ground)
26. 61·6 dB (125 Hz), 61·5 dB (250 Hz), 54·0 dB (500 Hz), 57·5 dB (1 kHz), 55·0 dB (2 kHz), 54·0 dB (4 kHz), 49·0 dB (8 kHz)

Chapter 9

11. not real time
12. 24·8 mm/s
15. 0·447 microvolts (assuming 38 dB is 5 dB above noise floor of instrument), 60 dB to 162 dB
17. 123 dB, $3{\cdot}9 \times 10^{-5}$ m
20(ii) (a) 89·2 dB(A), (b) 90·3 dB(A)
(iii) 99 dB(A), 87·3 dB(A)
21. 40 dB to 150 dB (assumed to be 10 dB above noise level to 10 dB below distortion level)
22(a) (i) 4·0 m/s^2, (ii) 12·6 mm/s, 40·1 μm
(b) 1122 Hz, 891 Hz
27. 9 m/s^2

Index